T0260378

Neurochemistry
of the Vestibular System

Neurochemistry of the Vestibular System

Edited by

Alvin J. Beitz and John H. Anderson

CRC Press
Taylor & Francis Group
Boca Raton London New York

CRC Press is an imprint of the
Taylor & Francis Group, an **informa** business

CRC Press
Taylor & Francis Group
6000 Broken Sound Parkway NW, Suite 300
Boca Raton, FL 33487-2742

© 2000 by Taylor & Francis Group, LLC
CRC Press is an imprint of Taylor & Francis Group, an Informa business

No claim to original U.S. Government works

ISBN 13: 9780849376795 (hbk)

Visit the Taylor & Francis Web site at
http://www.taylorandfrancis.com

and the CRC Press Web site at
http://www.crcpress.com

The Editors

Alvin J. Beitz, Ph.D., is professor of anatomy and neuroscience in the Department of Veterinary Pathobiology at the University of Minnesota, St. Paul. He received his BS in biology at Gannon University, Erie, PA, and his Ph.D. at the University of Minnesota.

In 1978, following postdoctoral training with Drs. Sanford Palay and Victoria Chan-Palay at Harvard Medical School, Dr. Beitz joined the faculty of the University of South Carolina College of Medicine as an assistant professor. In 1982 he joined the faculty of the University of Minnesota College of Veterinary Medicine as an assistant professor. In 1984 he was promoted to associate professor, and attained full professorship in 1988.

Dr. Beitz has written more than 100 publications in referred journals in a variety of neuro-science-related fields, including vestibular neurochemistry and neuroanatomy, cerebellar anatomy and connectivity, and brainstem and spinal cord neurotransmitter systems. His laboratory was one of the first to describe the effects of vestibular compensation and hypergravity stimulation on immediate early gene expression in the brainstem. One of his major research interests is the effect of labyrinthectomy on neuropeptide and neurotransmitter receptor expression in the vestibular nuclei

Dr. Beitz has been a frequent speaker at national and international conferences and workshops. He was also a member of the Expert Panel on Balance and Balance Disorders for the National Institute on Deafness and Other Communication Disorders, and was involved in developing the institute's National Strategic Research Plan.

Dr. Beitz is currently on the editorial boards of Synapse, Journal of Chemical Neuroanatomy, and Journal of Histochemistry and Cytochemistry. He is a member of the Society for Neuroscience, International Brain Research Organization, American Association of Veterinary Anatomists, American Pain Society, International Association for the Study of Pain, and the American Association for the Advancement of Science.

John H. Anderson, M.D., Ph.D., is associate professor of Otolaryngology and a member of the graduate faculty in the Neuroscience Program at the University of Minnesota. He received his BS in chemistry and mathematics at St. Olaf College, his MD at the University of Minnesota, and his Ph.D. at the University of Minnesota.

Dr. Anderson did postdoctoral training with Wolfgang Precht at the Max Planck Institute for Brain Research in Frankfurt/Main, Germany, and with Carlo Terzuolo in the Neurophysiology Laboratory in the Physiology Department at the University of Minnesota. Dr. Anderson then joined the faculty of the Otolaryngology Department at the University of Minnesota Medical School. He is currently director of the Equilibrium and Balance Laboratory, and director of Advanced Vestibular Testing and Research.

Dr. Anderson has investigated the vestibular system in a multidisciplinary manner, including studies of single neuron electrophysiology and motor behavior of vestibular reflexes, modeling central nervous control of the vestibulo-ocular reflex, and clinical studies of eye movement and postural abnormalities in cerebellar ataxia patients. He has collaborated with Dr. Beitz on characterizing neurochemical changes in the brainstem during vestibular compensation and hypergravity stimulation

Dr. Anderson has been a member of the Communicative Disorders Review Committee in the National Institute on Deafness and Other Communication Disorders, and was involved in developing the institute's first National Strategic Research Plan. He was a member of the Equilibrium Subcommittee of the American Academy of Otolaryngology, and is currently a member of the Society for Neuroscience, International Brain Research Organization, American Academy of Otolaryngology, Association for Research in Otolaryngology, American Academy of Neurology, and the Neural Network Society.

Contributors

John H. Anderson
Department of Otolaryngology
University of Minnesota Hospital
Minneapolis

Alexander Babalian
Laboratoire de Physiologie de la Perception et
 de l'Action, CNRS–Collège de France, Paris

S. Bankoul
Institute of Anatomy and Special Embryology,
 University of Fribourg, Switzerland

Neil H. Barmack
Neurological Sciences Institute, Oregon Health
 Sciences University, Portland

Alvin J. Beitz
Department of Veterinary Pathobiology
University of Minnesota
St. Paul

Robert H. I. Blanks
Departments of Anatomy and Neurobiology
 and Otolaryngology—Head and Neck
 Surgery, College of Medicine, University of
 California, Irvine

Alan M. Brichta
Department of Otolaryngology—Head and
 Neck Surgery, University of Chicago

J. A. Büttner-Ennever
Institute of Neuropathology, Ludwig-
 Maximilians University, Munich

Stephen L. Cochran
Departments of Otolaryngology and Anatomy
 and Neuroscience, University of Texas
 Medical Branch, Galveston

Cynthia L. Darlington
Department of Psychology
Neuroscience Research Centre
University of Otago
Dunedin, New Zealand

Norbert Dieringer
Department of Physiology, Ludwig-
 Maximilians University, Munich

Maria R. Diño
Northwestern University Institute for
 Neuroscience, Northwestern University
 Medical School, Chicago

L. N. Eisenman
Department of Physiology, Northwestern
 University Medical School, Chicago

Roland A. Giolli
Departments of Anatomy and Neurobiology,
 College of Medicine, University of California,
 Irvine

Donald A. Godfrey
Department of Otolaryngology—Head and
 Neck Surgery, Medical College of Ohio,
 Toledo

Jay Goldberg
Departments of Pharmacological and
 Physiological Sciences and
 Otolaryngology—Head and Neck Surgery,
 University of Chicago

Volker Henn
Department of Neurology
University of Zurich
Switzerland

G. R. Holstein
Departments of Neurology and Cell
 Biology/Anatomy, Mount Sinai School of
 Medicine, New York

J. C. Houk
Department of Physiology, Northwestern
  University Medical School, Chicago

Dick Jaarsma
Northwestern University Institute for
  Neuroscience, Northwestern University
  Medical School, Chicago

Galen D. Kaufman
Department of Otolaryngology, University of
  Texas Medical Branch, Galveston

Golda Anne Kevetter
Department of Otolaryngology, University of
  Texas Medical Branch, Galveston

Hongyan Li
Department of Otolaryngology—Head and
  Neck Surgery, Medical College of Ohio,
  Toledo

G. A. Kinney
Department of Physiology, Northwestern
  University Medical School, Chicago

James G. McElligott
Department of Pharmacology, Temple
  University School of Medicine, Philadelphia

Enrico Mugnaini
Northwestern University Institute for
  Neuroscience, Northwestern University
  Medical School, Chicago

Michel Mühlethaler
Département de Physiologie, Centre Médical
  Universitaire, Geneva

Winifried L. Neuhuber
Anatomy Institute of Erlangen-Nürnberg,
  Germany

Adrian Perachio
Department of Otolaryngology, University of
  Texas Medical Branch, Galveston

B. W. Peterson
Department of Physiology, Northwestern
  University Medical School, Chicago

O. Pompeiano
Dipartimento di Fisiologia e Biochimica,
  Università di Pisa, Italy

Ingrid Reichenberger
Department of Physiology, Ludwig-
  Maximilians University, Munich

David J. Rossi
Department of Physiology, Northwestern
  University Medical School, Chicago

Allan M. Rubin
Department of Otolaryngology—Head and
  Neck Surgery, Medical College of Ohio,
  Toledo

Dale Saxon
Department of Veterinary Pathobiology,
  University of Minnesota, St. Paul

Mauro Serafin
Département de Physiologie, Centre Médical
  Universitaire, Geneva

N. T. Slater
Department of Physiology, Northwestern
  University Medical School, Chicago

Paul F. Smith
Department of Pharmacology, School of
  Medical Sciences, University of Otago
  Medical School, Dunedin, New Zealand

Robert F. Spencer
Departments of Anatomy and
  Otolaryngology—Head and Neck surgery,
  Medical College of Virginia, Virginia
  Commonwealth University, Richmond

Hans Straka
Department of Physiology, Ludwig-
  Maximilians University, Munich

Nicolas Vibert
Laboratoire de Physiologie de la Perception et
  de l'Action, CNRS–Collège de France, Paris

Pierre-Paul Vidal
Laboratoire de Physiologie de la Perception et
  de l'Action, CNRS–Collège de France, Paris

Philip A. Wackym
Department of Otolaryngology and
  Communication Sciences, Medical College of
  Wisconsin, Milwaukee

Catherine de Waele
Laboratoire de Physiologie de la Perception et
  de l'Action, CNRS–Collège de France, Paris

Johannes J. L. van der Want
Department of Cell Biology and Electronmicroscopy,
  Medical Faculty,Rijksuniversiteit Groningen,
  Groningen, The Netherlands

Preface

Recent advances in the neurochemistry, neuropharmacology and molecular biology of the vestibular hair cells of the inner ear, the vestibular afferent fibers, the efferent vestibular system, the vestibular nuclei and other central vestibular components have had a major impact on our understanding of the vestibular system. This volume represents a concerted effort by a number of investigators to summarize the most recent information related to the neurochemical organization of the vestibular system and the behavioral implications. The ability to quantify the sensory stimuli and resulting behavioral responses has enabled researchers over the past thirty years to gain an in depth understanding of the physiological processes and neural pathways involved in the control of eye and head movements and posture. However, the molecular mechanisms and neurochemical processes that underlie neurotransmission of the vestibular and oculomotor pathways have begun to be defined only during the past decade. With technical advances in the areas of molecular biology and neurochemistry that have occurred within the past five years, there has been a proliferation of new information. That is the focus of this book.

There are a few reviews of the neurochemistry and pharmacology of central vestibular pathways (deWaele et al., *Brain Res. Rev.* 20:24-46, 1995; Smith and Darlington, *Trends Pharmacol. Sci.* 17:421-427, 1996; and Vibert et al., *Prog. Neurobiology* 50:243-286, 1997) and a recent book that summarizes the effects of vestibular stimulation on the autonomic nervous system (Yates and Miller, *Vestibular Autonomic Regulation*, CRC Press, 1996). However, there is no comprehensive review of the neurochemical organization of the vestibular system including both peripheral and central vestibular structures. It is our hope that this volume will provide an up-to-date review of this topic and identify future directions for research in the areas of neurotransmitters, second messengers, transcription factors and molecular mechanisms involved with both normal vestibular function and compensation.

This volume is organized as follows: the anatomy and physiology of the vestibular system is reviewed in Chapter 1; the neurochemistry of peripheral vestibular transmission is reviewed in Chapters 2-4; and the neurochemistry of central vestibular transmission at the level of the vestibular nuclei is reviewed in Chapters 5-9. Chapters 10-13 review the neurochemical mechanisms underlying the vestibular control of movement and posture, while Chapters 14-16 examine the neurochemical organization of sensory inputs to the vestibular system. The final two chapters review the neurochemical basis of plasticity and adaptation. This book should be especially useful to researchers, clinicians and students working on the vestibular system, but should also help to bring new investigators into the field and provide examples and useful discussions for other related areas of research.

Contents

Section I

Perspectives

1

Overview of the Vestibular System: Anatomy

J.A.Büttner-Ennever

CONTENTS

1.1 INTRODUCTION

The vestibular system is the sensorimotor complex in the brain which senses the movement and position of the head and uses that information to control posture, stabilize vision, and register the orientation and perception of the body in space.[34] The vestibular signals are generated in the labyrinth and transmitted to the central nervous system via the vestibular portion of the VIIIth cranial nerve. At the first major relay in the vestibular nuclei, they are integrated with sensory information from postural musculature and visual (optokinetic) pathways and modulated by inputs from the cerebellar and cerebral cortex. Through a complex set of intrinsic connections and parallel pathways, the vestibular nuclei feed back the appropriate information to the extraocular and postural muscles, as well as the cerebellum, thalamus, and cerebral cortex. There have been several recent reviews on the vestibular system[14,29,35,36,70] and its influence on the sympathetic nervous system.[143]

In this chapter we will outline the general anatomical features of the vestibular system, concentrating on the connectivity between the component cell groups.

1.2 VESTIBULAR NERVE

1.2.1 THE LABYRINTH

The membranous labyrinth contains three semicircular canals, each lying in a different plane and providing a three-dimensional representation of any *angular* acceleration of the head in space (see Chapter 16, Godfrey et al.). The otolith organs — the utricle and saccule — also lie in the labyrinth and detect *linear* accelerations of the head in the horizontal and vertical planes respectively. This includes changes in head position with respect to gravity. The sensory signals are generated in modified hair cells in the cristae ampullaris of the canals and maculae of the otoliths (see Section II). Each hair cell has a bundle of microvilli (stereocilia and one kinocilium) protruding from the apical surface. Any movement between the hair bundles, embedded in a gelatinous mass called the "cupula," and the overlying endolymph in the semicircular canals, which causes the kinocilium to bend "backward," opens local Na^+, K^+ and especially Ca^{++} channels and causes an activation of the primary afferent fibers synapsing on the hair cell.[42] Head motion leading to the kinocilium bending "forward" causes an inhibition in the primary afferents' resting discharge. This direction sensitivity of the kinocilia, along with the identical orientation of all the hair bundles in one crista, provide the basis for the specific direction sensitivity of the semicircular canals. In the maculae, the polarized hair cells are arranged around a striola, or central ridge, in a highly organized fashion and provide responses to shear forces in all directions within the plane of the otolith (see Section II).

1.2.2 TERMINATION OF FIRST-ORDER (1°) AFFERENTS IN THE VESTIBULAR NUCLEI

Studies of the afferent innervation of the hair cells in the cristae and maculae recognize three different classes: calyx units, bouton units, and dimorphic units (see Chapters 2, 3, and 4). However, the response dynamics of the first-order (1°) afferents are more closely correlated to the location in the sensory epithelium than to the terminal morphology.[58] They have different resting firing patterns and can also be divided into three groups (regular, irregular, and intermediate), which relay the vestibular signals to central vestibular neurons.[7,58,59,71]

The 1° vestibular afferents are bipolar neurons with their cell bodies in the ganglion of Scarpa.[114] Morphological studies show that 1° afferent fibers enter the medulla at the level of the lateral vestibular nucleus (LVN), but there is some controversy about the extent of their termination within the vestibular nuclei.[14,15,22,28-30,50,55,64,70,85-87,116] Some afferents terminate directly in the abducens nucleus (nVI).[88] There is consensus that the 1° vestibular afferents divide into an ascending branch, which mainly innervates the central part of the superior vestibular nucleus (SVN) and frequently sends axon collaterals to the cerebellum, and a descending branch, which terminates in the medial and inferior vestibular nuclei (MVN and IVN). The LVN and nucleus prepositus hypoglossi (NPH) receive very few 1° afferent terminals. Canal afferents, according to Gacek,[50] also supply a small group of floccular projecting[90] neurons called the "interstitial nucleus of the vestibular nerve" (INT), which lie within the VIIIth nerve lateral to the LVN. These features are elegantly illustrated in the reconstructions of single horizontal canal afferents by Sato et al.[116] (See Figure 1.1). They demonstrate that regular and irregular afferents possess similar patterns of connectivity within the vestibular complex, but they find clear differences in the mode of terminal arborization. These authors also describe horizontal canal afferents which terminate selectively within the small-celled, ventral y-group and the ipsilateral fastigial nucleus. Previous studies, which include vertical canal and otolith afferents, have reported substantial projections to the anterior and posterior vermis, mainly in the nodulus (lobule X) and the uvula (lobule IXd).[10,14,28,30,87] There are conflicting reports concerning afferents to the flocculus; if they are present, they are very small.[55] Other studies describe afferents to the reticular formation, external cuneate nucleus,[86] and saccule afferents to the ventral y-group.[28,50]

A summary diagram of the termination sites of mammalian 1° vestibular afferents on a generalized horizontal outline of the vestibular nuclei is illustrated in Figure 1.1.

The MVN, SVN, and IVN are the main sites of termination, but there are also significant projections to the interstitial nucleus of the VIIIth nerve, the y-group, the nodulus, and the uvula. There is some differentiation in the pattern of termination of the individual canals and otoliths, especially in the region of the SVN, but the central region of the vestibular complex receives all types of vestibular inputs.

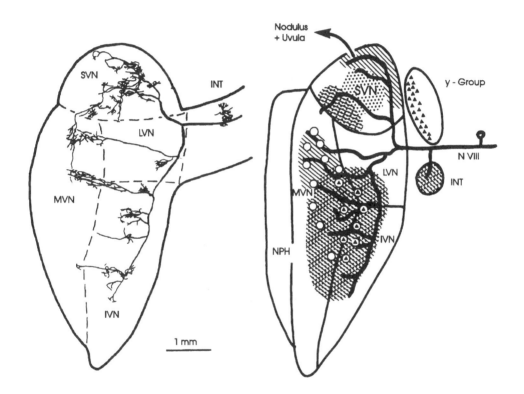

FIGURE 1.1 The termination pattern of vestibular nerve fibers within the vestibular nuclei. On the left: an identified irregular-type horizontal canal fiber is reconstructed on a horizontal view of the vestibular complex to show the location of terminals (adapted from Sato et al., 1989[116]; up is rostral, right is lateral). On the right: a summary of several studies[14,15,22,28-30,50,55,64,70,85-88,116] showing a generalized scheme of termination sites from vestibular nerve afferents: horizontal and anterior canal afferents (stripes), from the posterior canal , from the utricle (open circles), from the saccule (filled triangles). The central region around the MVN_m receives all types of inputs; whereas, in SVN, there is a differential pattern of termination. Some vestibular nerve afferents extend into the cerebellum (for Abbreviations see Section IV).

1.2.3 VESTIBULAR EFFERENT SYSTEM

All vertebrates have a vestibular efferent, or centrifugal, system (see Chapter 4, Goldberg et al.). It originates from a compact group of somata located in the rostral medulla at the same coronal levels as the vestibular complex.[69] In the monkey it is situated on the lateral border of the abducens nucleus (nVI)[57] (See Figure 1.3).

In the rat the vestibular efferent cell group is referred to as the "paragenual nucleus" by Paxinos and Watson.[110] In terms of cytoarchitecture, the neurons resemble facial motorneurons, and they project centrifugally to innervate the end organ via the VIIIth nerve.[57,69,137] An increased activity of the efferent fibers, for example during elevated attentional states, modifies the resting rate and the dynamics of the primary vestibular afferents.[57] The efferent neurons are multisensory and, at present, the function of the vestibular efferent system is thought to be the enhancement of the processing of the peripheral endorgans.

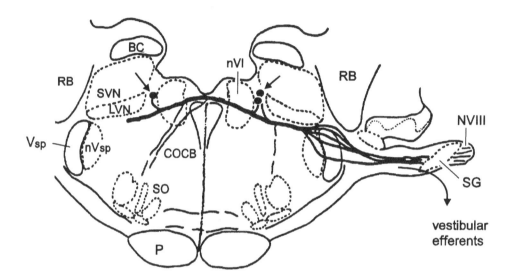

FIGURE 1.2 Drawing of a tranverse section through the caudal brainstem of the monkey to show the location, lateral to the abducens nucleus, of the cell soma (arrows) which provide the vestibular efferent fibers (thick lines) (adapted from Goldberg and Fernandez[57]).

1.3 VESTIBULAR NUCLEI

1.3.1 CYTOARCHITECTURE AND NOMENCLATURE

The MVN, SVN, LVN, and IVN are well established in all mammals, and along with several smaller cell groups, form the vestibular complex (Figure 1.3), the subject of several recent neuroanatomical reviews.[15,22,29,54,70] From a historical perspective, Clarke's (1861) early descriptions of the vestibular complex divided it into an "inner" and an "outer nucleus." But Deiters' descriptions of the giant cells in the rostro-dorsal part of the vestibular complex resulted in the current name, the "lateral vestibular nucleus of Deiters" (LVN).

The more caudal part of Clarke's outer nucleus has retained the name subsequently given to it by Cajal (1896), "the spinal vestibular nucleus," but it is also referred to as the "descending" or "inferior" vestibular nucleus (IVN). Von Bechterew originally described a separate cell group located between the superior cerebellar peduncle and LVN, which he referred to as the "angular nucleus." This nucleus is currently referred to as the superior vestibular nucleus (SVN) following the terminology of Gray (1926). The remaining medial part of Clarke's inner nucleus was given

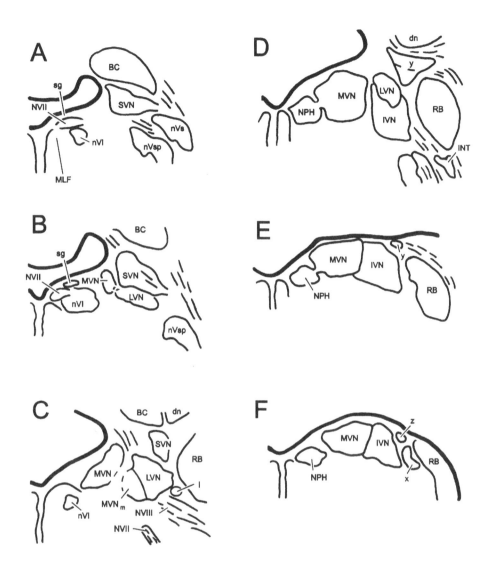

FIGURE 1.3. Drawing of the classical cytoarchitectural subdivisions of the vestibular complex of the monkey on transverse sections; A-F are arranged from rostral to caudal. Group f on the caudal tip of IVN is not shown (adapted from Brodal[15] to show the magnocellular region of MVN, MVN$_m$).

the name "nucleus triangularis" by Schwalbe (1881), and was subsequently designated as the dorsal or principal vestibular nucleus. This nucleus is now universally called the "medial vestibular nucleus" (MVN). For historical references, see Gerrits.[54]

The MVN contains predominantly small neurons (parvicellular) except for a ventrolateral group, which was originally considered part of the LVN. This magnocellular region (MVN$_m$) contains many secondary (2°) vestibular neurons[54,70] (Figure 1.3(c)). The SVN has a small-celled perimeter and a magnocellular core, but the giant spinal-projecting cells of the LVN make it the most conspicuous feature of the vestibular complex. The IVN is characterized by a multitude of longitudinally running fibers, and is bordered laterally by the restiform body and medially by the MVN. In addition to these four main nuclei are three subsidiary cell groups: (1) the *y-group*, which consists of a small-celled ventral division receiving sacculus afferents[50] and a dorsal premotor division which projects to the oculomotor nucleus (called the "infracerebellar nucleus" by Gacek); (2) *the interstitial nucleus of the vestibular nerve* (Figures 1.1-1.3); and (3) several minor nuclei called groups f, x, l, and z,.[15] Groups f (lying at the caudal tip of the IVN and not shown in Figure 1.3) and x are reciprocally connected to the cerebellum (in particular with the flocculus, nodulus, and uvula) and receive afferents from the red nucleus; group l is associated with the LVN and projects to the spinal cord; and group z is known to be a relay for hindlimb afferents in the spinothalamic pathway.[15] Unfortunately, the functional properties of the vestibular nuclei seldom conform to the cytoarchitectural borders on which the nomenclature is based.

1.3.2 VESTIBULO-OCULOMOTOR PATHWAYS

A convenient way to consider the neuroanatomy of eye movements is to recognize five different types of eye movements, each associated with relatively distinct central pathways: vestibulo-ocular responses, optokinetic responses, smooth pursuit eye movements, saccades, and vergence eye movements. Figure 1.4 is a highly simplified diagram illustrating this: it shows that the premotor circuits of the first three types utilize the vestibular nuclei. There are several parallel, direct pathways that run from the vestibular nuclei to the oculomotor nuclei.[22,93] The best documented is a three-neuron arc also illustrated.

This pathway begins with a 1° canal afferent, which projects onto a 2° afferent located in the vestibular nuclei. In the simplest vestibulo-ocular reflex, this 2° neuron terminates on a motorneuron supplying an extraocular eye muscle. Three groups of vestibular neurons will be considered here: (1) 2° vestibular neurons projecting to the oculomotor nuclei (VO neurons); (2) non-2° vestibular neurons projecting to the oculomotor nuclei; and (3) 2° vestibular neurons bifurcating to the oculomotor nuclei and spinal cord (VOC neurons). In addition, in Section 1.3.3 below, we will consider a fourth group of vestibular neurons, 2° vestibular neurons projecting to the spinal cord (VC neurons).

1.3.2.1 2° Vestibulo-Ocular Neurons (VO)

Refer to Figures 1.5 and 1.6 for this section.

Each 2° vestibular neuron has a dominant input from one canal, which is clear under anaesthesia, but in the awake state several additional inputs from other canals or otoliths converge on these cells.[49,113] There is mounting evidence that there is no bilateral convergence from vertical canals onto 2° vestibular neurons, and that these cells control the vestibular responses of only one eye.[38,62,129]

Figures 1.5 and 1.6 show the location of 2° VO canal neurons which project to the abducens nucleus (nVI), the trochlear nucleus (nIV), and the oculomotor nucleus (nIII), but not to the spinal cord. The figures represent a compilation of data from a large number of recent studies.[21,60,61,68,72,77,78,80,99-101,104,105,108,132,133] According to these studies, the pattern of distribution of VO cell bodies varies very little among species. In addition, these 2° neurons are localized to the following two main areas (see Figure 1.5):

> **1.** Centered in and around the magnocellular region of the MVN (MVN$_m$): vertical 2° neurons tend to lie slightly caudal to the horizontal ones, and their activity encodes eye **P**osition and head **V**elocity and they **P**ause with saccades — so-called "**PVP** neurons." The efferent projections of these 2° vestibular neurons conform to a clear pattern illustrated in Figure 1.6. Excitatory efferent projections cross the midline and ascend in the medial longitudinal fasciculus (MLF) to terminate on the appropriate pair of synergistic motoneuron-subgroups (e.g., superior oblique (SO) of one side and inferior rectus (IR) of the other side; superior rectus (SR) of one side and the inferior oblique (IO) of the other; or the lateral rectus motorneurons (LR) and the internuclear neurons in nVI, which, in turn, activate the medial rectus (MR) subgroup in nIII).[21,23] For a simple diagram of

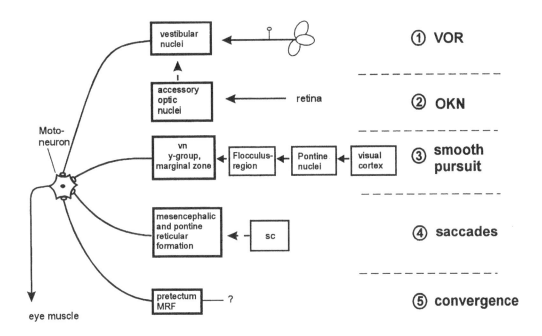

FIGURE 1.4 Simplified diagram of the inputs to extraocular motorneurons, to emphasize that the premotor networks for the five main types of eye movements are relatively independent. VOR, OKN and smooth pursuit all access the motorneuron through the vestibular nuclei.

the planes of action of the canals with respect to the muscle-pulling planes, see References[21,24] (Figure 3). In the same area of the MVN lie cell bodies giving rise to efferents entering the ipsilateral ascending tract of Deiters (ATD)(the squares in Figure 1.5), located in the lateral wing of the MLF. The axons arising from these cells terminate directly on MR motorneurons. It has been suggested that the ATD also carries utricular information, but this has not yet been tested.[67, 70] The central pathways involved in inducing otolith responses are poorly understood in comparison to the canal pathways. However, it is important to point out that many neurons in the IVN respond predominantly to otolith inputs and may be involved in the central relay of otolith information.[22,38,67,70]

2. The magnocellular central region of the SVN: the 2° projection neurons in the SVN give rise to a variety of different efferents (see Figure 1.6). There are inhibitory neurons whose axons ascend in the MLF ipsilaterally carrying eye position-related signals derived from the vertical canals. In addition, there are anterior canal-related efferents ascending via the brachium conjunctivum (BC) in the rabbit (or slightly ventral to the BC in the decussation of the ventral tegmental tract, CVTT), which provide an additional excitatory input to inferior oblique motorneurons.[68,118,134]

1.3.2.2 Non-Second Order Vestibulo-Ocular Neurons

Refer to the shaded areas in Figure 1.5 for this section. There is far less information available regarding cells in the vestibular nuclei which project to the extraoculomotor nuclei but lack a direct input from a 1° vestibular afferent. These neurons are referred to as "non-second-order VO neurons" and they appear to be quite numerous.[131] Non-second-order VO neurons have been located in the rostral MVN, in the marginal zone between the NPH and the MVN, and in the central region of the SVN (their locations are designated by the hatched area). It is possible that the rostral MVN

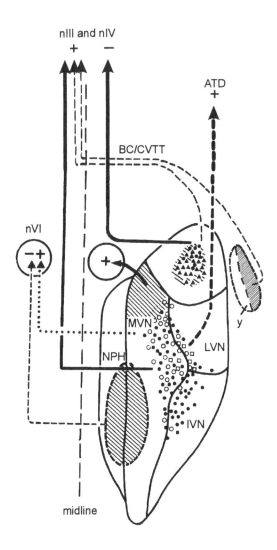

FIGURE 1.5 The location of 2^0 vestibular neurons (i.e., with direct input from nVIII) which project monosyn-aptically onto extraocular motorneurons (VO neurons): dots and triangles indicate soma with vertical direc-tional sensitivity; circles, horizontal sensitivity; squares, neurons of the ascending tract of Deiters (ATD). The hatched regions indicate the location of *non*-second-order eye position-related neurons which project directly onto motorneurons. The y-group projection to nIII arises from *non*-second order neurons.

group of Keller and colleagues[82,83] is identical to the large group of MVN neurons that are retrogradely filled by injections into the ipsilateral nVI.[70,92,94] There is a greater number of non-second-order VO neurons located in the marginal zone of primates as compared to cats. This population of non-second-order VO neurons has been shown to project contralaterally, to provide an inhibitory input to the nIV,[92,102] and to carry eye position signals. This type of activity is typically associated with neurons that receive inputs from a common neural integrator (i.e., whose velocity signals have been integrated to encode position).

The dorsal y-group neurons, but not the ventral y-group, are non-2^0 floccular target-neurons which are inhibited by their input from the flocculus.[117] Their efferent axons provide a monosynaptic, predominantly contralateral excitatory input to eye muscle motoneurons in the trochlear and oculomotor nuclei. These efferent fibers decussate in the ventral part of the brachium conjunctivum (BC), or even further ventrally in the "crossing ventral tegmental tract" (CVTT).[31,52,118] In the monkey, the y-group provides one of the strongest inputs to the nIII.[22] These neurons were found to fire in relation to upward eye velocity during smooth pursuit, and upward head velocity during supression of the VOR, but showed little activity during the VOR itself.[33,130] These and other results support a role for the y-group in the adaptive changes of the VOR.[109]

1.3.2.3 Vestibulo-Oculomotor Collic Neurons (VOC)

The axons of another group of vestibulo-oculomotor projecting neurons branch and also send a projection to the spinal cord. These neurons are called "vestibulo-ocular collic neurons" (VOC). Their location occupies a large region of MVN and IVN (see circles in Figure 1.7) and they mainly project caudally via the contralateral $MVST_c$[103], although some inhibitory VOC cells send their axons down the ipsilateral $MVST_i$.[79] Unlike the VO and vestibulo-colic neurons (see below), they have no floccular inputs.[73]

1.3.3 VESTIBULOSPINAL (VESTIBULO-COLLIC, VC) Pathways

With the exception of SVN, y-group, and NPH, the vestibular nuclei project to the spinal cord through the lateral vestibulospinal tract (LVST), the medial vestibulospinal tract (MVST), and the caudal vestibulospinal tract (CVST). Many of the neurons are 2° vestibular neurons receiving inputs from both regular and irregular 1° vestibular afferents.[13] The LVST has its origin mainly in the LVN and its fibers descend in the ipsilateral ventrolateral funiculus and excite neck, axial, and extensor muscle motoneurons, but inhibits muscles which act as flexors (Figure 1.7). There is some somatotopic organization in that the neurons projecting as far caudally as the lumbar spinal cord lie dorsally, and those to the cervical region lie ventrally in the vestibular complex.[1] The MVST arises primarily from the contralateral MVN and, to a lesser extent, from the LVN and IVN. It also regulates the activity of neck and axial motoneurons.[74,120-122,127] In addition to the LVST and MVST, there is a third descending projection called the "caudal vestibulospinal tract" (CVST),[111] which travels in the dorsal columns and dorsolateral funiculi to terminate in the upper cervical dorsal horn. This termination zone in the dorsal horn contrasts with the ventral horn termination sites of the LVST and MVST.[140] The cell bodies of vestibulo-colic neurons (VC), which provide a selective projection to the spinal cord, lie laterally around the border between the LVN and rostral IVN. Their location is shown diagramatically by the triangles in Figure 1.7. Note that their location contrasts with that of the VOC neurons, described above, which lie in the MVN and throughout the rostrocaudal extent of the IVN.

Finally, experimental studies of pontomedullary reticulospinal neurons demonstrate that their activity is also related to canal and/or otolith signals. Unlike vestibulospinal neurons, they appeared to be predominantly sensitive to otolith signals and could play a significant role in the gravity-dependent postural reflexes of the neck and limbs.[12]

1.3.4 COMMISSURAL INTRINSIC AND VESTIBULAR CONNECTIONS

The vestibular nuclei possess an extensive set of intrinsic and commissural pathways (Figure 1.8).

These pathways serve to integrate the activity of the individual nuclei of both sides and enable them to operate synchronously. This extensive commissural network is different for the canal and otolith pathways and provides the basis for changes in sensitivity of the canal response.[112,113,119] It participates in the common neural integrator network,[32,47] and the velocity storage mechanism.[81] The MVN is interconnected with the contralateral MVN throughout the length of the nucleus, as is the NPH, but to a lesser extent.[97,98] The small-celled peripheral region of SVN also has prominent commissural connections to its contralateral counterpart and to the small-celled parts of MVN. There

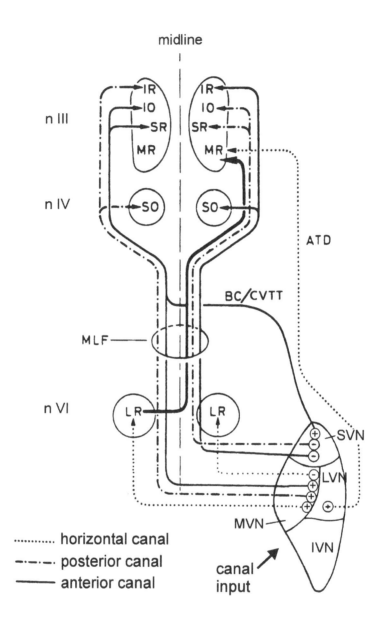

FIGURE 1.6 Connectivity between canal inputs of one side and individual muscle motorneurons. In general, pathways ascending in the ipsilateral MLF are inhibitory, in the contralateral MLF are excitatory; the ATD is also excitatory. In cat and monkey, there are additional collaterals to other motorneurons.[101]

are some discrepancies in the studies of these pathways,[27,31,39,51,97,112,115] but Epema and colleagues[39] have managed to interpret the results into a consistent pattern: the peripheral portions of the vestibular complex are linked by intrinsic pathways and are the origin of the major commissural connections. They feed into the central regions of the vestibular complex which provide the major output pathways from the complex (Figures 1.5 and 1.8). Interestingly, the y-group projects primarily to the contralateral SVN but does not appear to have a commissural projection to the y-group on the collateral side.[31]

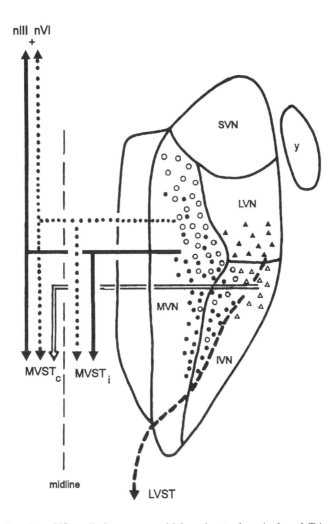

FIGURE 1.7 The location of 2^0 vestibular neurons which project to the spinal cord. Triangles indicate neurons which project only to the cord (VC neurons), open symbols for the crossed MVST component and filled for LVST. Cells supplying both the oculomotor nuclei and the spinal cord (VOC neurons) have dominant inputs from horizontal (open circles) or vertical canals (filled circles), and project mainly contralaterally in MVST$_c$.

1.3.5 Vestibulo-Cerebellar Interconnections

1.3.5.1 Afferents

Several studies have reviewed the mossy fiber input from the vestibular nuclei to the cerebellar cortex and the fastigial nucleus.[14,29,54,135] Other studies have investigated the projections in inspecific1species: rat,[11] rabbit,[5,9,40,128,142] cat,[87,96] and monkey.[16,27,31,91]. Figure 1.9 provides a summary of the common pattern of vestibulo-cerebellar projections.

The majority of terminals are found in the nodulus, uvula (lobule !Xd), flocculus and adjacent ventral paraflocculus region, and the proximal parts of lobules I-V. There is no clear topographic pattern in the origin of these connections. The terminals exhibit a certain degree of clustering but they do not seem to form longitudinal zones over successive folia as do the climbing fiber projections from the inferior olive[40,96,128,136] (see Chapter 12). The afferents to the nodulus arise predominantly from the caudal MVN and IVN, from the SVN, and possibly from group x. Projections to the floccular and ventral para floccular region arise bilaterally from MVN and NPH, central SVN, IVN, and the ventral y-group. In the monkey, the caudal NPH makes the most significant contribution to the flocculus, and reciprocal projections are provided by the little-known basal interstitial nucleus of the cerebellum.[90,91] Very little cerebellar input comes from the magnocellular MVN, and none arises from LVN.

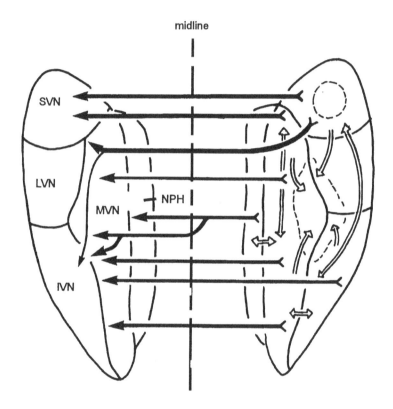

FIGURE 1.8 The intrinsic and commissural pathways of the vestibular complex are well developed and strongly interconnect SVN and MVN, NPH, the peripheral vestibular areas with each other and to the magnocellular core of the vestibular complex (dotted areas). In contrast, the "core" is the main source of vestibular output pathways.[39,40,128]

1.3.5.2 Efferents

All the major subdivisions of the vestibular complex receive projections from the Purkinje cells of the cerebellum, providing an important inhibitory "control" of vestibular function (see Chapter 8).[66,135,138] The LVN is contacted by efferents of the anterior lobe (zone B).[135] The flocculus and ventral paraflocculus project strongly to the dorsal y-group, central SVN, and rostral MVN, whereas axons from the nodulus terminate in the peripheral region of SVN, in the caudal MVN and IVN, and in group x. The uvula efferents differ slightly from those of the nodulus. There is mounting evidence for a zonal arrangement of the cerebellar output from Purkinje cells of the flocculus and nodulus,[37,141] which is similar in many respects to zonal organization of the olivocerebellar projection (see Chapter 12).[136] These output zones appear to encode activity related to rotation in the individual canal planes. This association between specific zones and certain semicircular canalis further

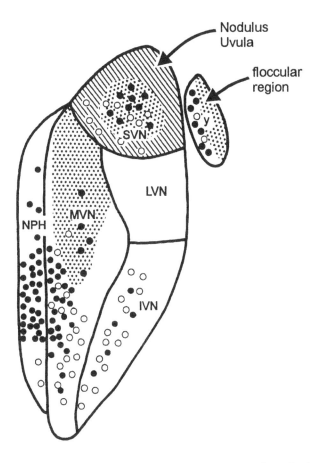

FIGURE 1.9 Connections between the cerebellum and vestibular nuclei. Cells projecting to the floccular region are represented as filled circles, cells projecting to the nodulus and uvula as open circles. Areas where cerebellar efferents from the floccular region terminate are indicated by fine dots, and inputs from the nodulus and uvula are depicted by cross-hatching. The location of the floccular-projecting neurons in the marginal zone between MVN and NPH does not overlap with the MVN area receiving floccular efferents.

supported by anatomical studies showing that the Purkinje axons arising from a particular.zone project to the appropriate vestibular regions. Thus, "horizontal" zones project to the magno- and parvicellular MVN, while "vertical" zones project to the dorsal y-group, SVN, and the ventral dentate nucleus.[136]

1.3.6 PROJECTIONS TO AND FROM THE INTERSTITIAL NUCLEUS OF CAJAL

The interstitial nucleus of Cajal (iC) lies as a compact group of cells dorsolateral to the rostral half of the oculomotor nucleus (nIII), embedded in the medial longitudinal fasciculus (MLF), and bordered laterally by the mesencephalic reticular formation. The rostral and caudal iC have different cytoarchitectures and their connections, for example with the lower brainstem, differ, but the significance of these differences is not yet understood.[125] The rostral border of the iC is the rostral interstitial nucleus of the MLF (rostral iMLF) containing the premotor neurons for vertical saccades.[23,126] The iC is functionally associated with the control and coordination of vertical gaze (eye and head position).[45-47]

The major input to iC is from axon collaterals of the vestibular neurons with monosynaptic projections to vertical oculomotor (or trochlear) motoneurons. It is not clear if the y-group also projects to iC,[117] but several studies confirm inputs from the contralateral iC, from all vertical premotor neurons in the rostral iMLF, and from the nucleus of the posterior commisure (nPC) and the frontal eye-fields of the cerebral cortex.[23,45]

The efferent fibers from iC cross in the posterior commissure (PC) and innervate the rostral iMLF and nPC, the contralateral iC, and vertical motoneuron subgroups in the contralateral nIII and troclear nucleus (nIV). The ipsilateral iC efferents innervate the vertical motoneuron subgroups of the same side in nIII and nIV others descend within the MLF and supply terminals to the vestibular nuclei, the cervical spinal cord (C1 and C2), and the midline cell groups of the paramedian tracts (PMT), including those in the abducens rostral cap.[23,25,74,76]

There is an ipsilateral input from the iC, but not rostral iMLF, to the vestibular nuclei.[48] It is not well delineated, but existing evidence suggests that it supplies the central portion of the SVN and the MVN/IVN border, a region that is interconnected with central SVN.[31,76] These inputs have been shown to be excitatory,[48] with major effects on both vertical and horizontal *type II neurons*. The type II cells are important for the commissural interaction between the vestibular nuclei of each side and for sustaining the "gain of the response" to head rotation in vestibular neurons (see above). These connections probably underlie the function of iC as a neural integrator, which serves to transform velocity -coded signals from vertical premotor neurons to vertical eye position signals, and to maintain a stable eye position.[47] In this way, the vestibular nuclei act as an essential part of the integrator feedback loop. The descending pathways from the iC also modulate vestibulospinal activity,[45] and underlie its control of postural muscles and head position.

1.3.7 PROJECTIONS TO AND FROM THE CEREBRAL CORTEX

Earlier attempts to trace vestibular nuclear pathways to the cerebral cortex in the monkey showed that they ascended bilaterally to the thalamus and terminated in diffuse, scattered patches predominantly in nucleus ventroposterior lateralis, pars oralis (VPL_o), and, to a lesser extent, in ventroposterior inferior nucleus (VPI) and ventroposterior centralis lateralis (VL_c).[89] Efferents of these thalamic nuclei project to vestibular cortical areas.[2,3]

Several cortical regions have been reported to process vestibular signals (Figure 1.10).

The parietal area, 2v, is small and contains neurons which respond to vestibular and optokinetic stimuli.[19] The parietal areas 7a and 7b contain a few scattered "vestibular" neurons comprising only 2% of the population. They are independent of eye movements evoked by the head in space rotations. Vestibular, optokinetic, and sensorimotor responses have been identified in single neurons of the neck region of area 3a.[107] The most extensive vestibular field with neurons responding to complex vestibular, optokinetic, and somatosensory stimuli lies buried in the lateral sulcus in the parieto-insular and retroinsular cortex (PIVC, parieto-insular vestibular cortex).[63,43] Afferents to PIVC include the "vestibular" thalamic nuclei frontal eye fields, cortical areas 3a (neck region), 7a and 7b, the contralateral PIVC, the cingulate gyrus, and the supplementary motor area for the eye and neck.[2,65]

Projections from cortical neurons predominantly in layer 5 back to the vestibular complex are found in PIVC, 2a, and the vestibular field of area 3a.[3,4,41] Collectively these three projection areas are called the "inner cortical vestibular circuit."[65] In addition, the temporal area (T3), the postarcuate premotor area (6pa), and the anterior cingulate gyrus (6c and 23c), which are known to project to the inner cortical vestibular circuit, provide efferents to the vestibular nuclei. These three cortical areas are collectively termed the "outer cortical vestibular circuit." It is speculated that the inputs from this closely interconnected network of cortical areas to the vestibular complex could regulate the vestibulo-ocular, vestibular-spinal and optokinetic responses so that they do not counteract voluntary movements.

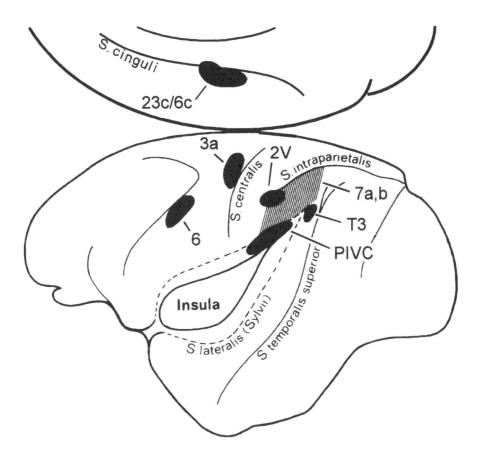

FIGURE 1.10 Lateral and medial view of the monkey cerebral cortex. Cortical areas containing vestibular sensitive neruons are indicated in black. (Adapted from Grüsser et al.[63])

1.3.8. OTHER INTERCONNECTIONS OF THE VESTIBULAR NUCLEI

The vestibular nuclei also send efferent projections to the inferior olivary nucleus of the medulla these projections are discussed in further detail in Chapter 12. Additional efferent projections from the vestibular nuclei are known to innervate the midline cell groups of the paramedian tracts (PMT), which in turn project to the flocculus and might play a role in gaze-holding.[25]

In terms of afferents, there are small cell groups near iC, around the tractus retroflexus,[8] and others on the midline between the cell groups of the oculomotor nuclei, that project to the vestibular nuclei,[94,125] but their significance is not known. Anatomical and physiological studies have demonstrated a relationship between the vestibular nuclei and the primary afferent receptors in the extraocular eye muscles.[6,17,18] In this regard, injections of retrograde tracers into the vestibular nuclei have labeled ganglionic neurons in the ipsilateral mesencephalic trigeminal nucleus.

1.4 OPTOKINETIC INPUTS TO THE VESTIBULAR SYSTEM

It is now well established that 2° vestibular neurons respond to both vestibular and optokinetic stimuli, and that these vestibular neurons act as premotor sources for both vestibulo-ocular and optokinetic responses.[20,44,84] However, it is not clear which afferents carry the optokinetic signals to the vestibular complex.[95,123] Appropriate visual information (from large, slowly moving visual fields) is known to enter the accessory optic nuclei (the medial, lateral, dorsal, and interstitial terminal nuclei).[44,123,124] Recently, sensitive tracing experiments have been able to show that these nuclei (and the closely associated visual tegmental relay zone, VTRZ) all project to the vestibular complex,[26,56,124] and could provide at least part of the direct optokinetic signals to the vestibular nuclei.[26] This scheme is shown in a very simplified manner in Figure 1.3. Nevertheless, the role of the nucleus of the optic tract (NOT) is not entirely clear. The NOT is considered a pretectal nucleus rather than an accessory optic nucleus, but it projects to the vestibular complex. Our anatomical studies of NOT confirm previous reports and emphasize that it is more strongly connected with upper brainstem structures such as the lateral geniculate nucleus, the superior colliculus, the accessory optic nucleus, and pontine nuclei, than to the NPH and MVN of the vestibular complex.[26,53,75,106,139] It also projects strongly to the inferior olive, but this nucleus is not essential for the generation of optokinetic responses. The nucleus of the optic tract does, however, provide afferents to all five eye movement circuits illustrated in Figure 1.3. Based on this connectivity, we propose that it coordinates the activity of these visuomotor circuits during optokinetic stimulation under the control of the cerebral cortex.[26] Current knowledge supports this hypothesis, but many other interpretations are possible (see Chapter 14).

ACKNOWLEDGMENTS

Grateful thanks are due to Professor P. Mehraein for his enthusiastic and continuous support, and to Mrs. U. Schneider for her excellent technical assistance. This work was supported by the Deutsche Forschungsgemeinschaft SFB 220/D8.

REFERENCES

1. Akaike, T, (1983) Neuronal organization of the vestibulospinal system in the cat. *Brain Res.,* 259:217,1983.
2. Akbarian, S, Grüsser, OJ, Guldin, WO, Thalamic connections of the vestibular cortical fields in the squirrel monkey (*saimiri sciureus*). *J. Comp. Neurol.,* 326:423,1992.
3. Akbarian, S, Grüsser, OJ, Guldin, WO, Corticofugal projections to the vestibular nuclei in squirrel monkeys — further evidence of multiple cortical vestibular fields. *J. Comp. Neurol.,* 332:89,1993.
4. Akbarian, S, Grüsser, OJ, and Guldin, WO, Corticofugal connections between the cerebral cortex and brainstem vestibular nuclei in the macaque monkey. *J. Comp. Neurol.,* 339:421,1994.
5. Alley, K, Baker R, and Simpson, JI, Afferents to the vestibulo- cerebellum and the origin of the visual climbing fibres in the rabbit. *Brain Res.,* 98:582,1994.
6. Ashton, JA, Boddy, A, Dean, SR, Milleret, C, and Donaldson, IM, Afferent signals from cat extraocular muscles in the medial vestibular nucleus, the nucleus prepositus hypoglossi and adjacent brainstem structures. *Neurosci.,* 26:131,1988.
7. Baird, RA, and Schuff, NR, Peripheral innervation patterns of vestibular nerve afferents in the bullfrog utriculus. *J. Comp. Neurol.,* 342:279,1994.
8. Barmack, NH, Henkel, CK, and Pettorossi, VE, A subparafascicular projection to the medial vestibular nucleus of the rabbit. *Brain Res.,* 172:339,1979.
9. Barmack, NH, Baughman, RW, and Eckenstein, FP, Shojaku H Secondary vestibular cholinergic projection to the cerebellum of rabbit and rat as revealed by choline acetyltransferase immuno-histochemistry, retrograde and orthograde tracers. *J. Comp. Neurol.,* 317:250,1992.
10. Barmack, NH, Baughman, RW, Errico, P, and Shojaku, H, Vestibular primary afferent projection to the cerebellum of the rabbit. *J. Comp. Neurol.,* 327:521,1993.
11. Blanks, RHI, Precht, W, and Torigoe, Y, Afferent projections to the cerebellar flocculus in the pigmented rat demonstrated by retrograde transport of horseradish peroxidase. *Exp. Brain Res.,* 52:293,1983.

12. Bolton, PS, Goto, T, Schor, RH, Wilson, VJ, Yamagata, Y, and Yates, BJ, Response of pontomedullary reticulospinal neurons to vestibular stimuli in vertical planes — role in vertical vesti-bulospinal reflexes of the decerebrate cat. *J Neurophysiol.,* 67:639,1992.

13. Boyle, R, Goldberg, JM, Highstein, SM, Inputs from regularly and irregularly discharging vestibular nerve afferents to secondary neurons in squirrel monkey vestibular nuclei. Correlation with vestibulospinal and vestibuloocular output pathways. *J. Neurophysiol.,* 68:471,1992.

14. Brodal, A, Anatomy of the Vestibular Nuclei and their Connections. *Handbook of Sensory Physiology,* Kornhuber, HH, Ed., Springer, New York, 1974.

15. Brodal, A, The vestibular nuclei in the macaque monkey. *J. Comp. Neurol.,* 227:252,1984.

16. Brodal, A, Brodal, P, Oberservations on the secondary vestibulo-cerebellar projections in the macaque monkey. *Exp. Brain Res.,* 58:62,1985.

17. Buisseret-Delmas, C, and Buisseret, P, Central projections of extraocular muscle afferents in cat. *Neurosci., Lett.,* 109:48,1990.

18. Buisseret-Delmas, C, Epelbaum M, Buisseret, P, The vestibular nuclei of the cat receive a primary afferent projection from receptors in extraocular muscles. *Exp. Brain Res.,* 81:654,1990.

19. Büttner, U, Buettner, UW, Parietal cortex (2v) neuronal activity in the alert monkey during natural vestibular and opto-kinetic stimulation. *Brain Res.,* 153:392,1978.

20. Büttner, U, Büttner-Ennever, JA, Present Concepts of Oculomotor Organization, *Neuroanatomy of the Oculomotor System,* Büttner-Ennever, JA, Ed, Elsevier, Amsterdam, 1981.

21. Büttner-Ennever, JA, Vestibular-Oculomotor Organization. *Progress in Oculomotor Research,* Fuchs, AF, and Becker, W, Eds,), Elsevier, Amsterdam, 1981.

22. Büttner-Ennever, JA, Patterns of Connectivity in the Vestibular Nuclei, *Sensing and Controlling Motion. Vestibular and Sensorimotor Function,* Cohen, B, Tomko, DL, and Guedry, F, Eds., The New York Academy of Sciences, New York, 1992.

23. Büttner-Ennever, JA, Büttner, U, The Reticular Formation, *Neuroanatomy of the Oculomotor System,* Büttner-Ennever, JA, Ed., Elsevier, Amsterdam, 1988.

24. Büttner-Ennever, JA, Büttner, U, Neuroanatomy, of the Ocular Motor Pathways. *Ocular Motor Disorders of the Brain Stem,* Büttner U, and Brandt T, Eds., Bailliere Tindall, London, 1992.

25. Büttner-Ennever, JA, Horn, AKE, and Schmidtke, K, Cell groups of the medial longitudinal fasciculus and paramedian tracts. *Rev. Neurol,* (Paris), 145:533, 1989.

26. Büttner-Ennever, JA, Cohen, B, Horn, AKE, and Reisine, H, Efferent pathways of the nucleus of the optic tract (NOT) in monkey and their role in eye movements. *J. Comp. Neurol. 373:90, 1996.*

27. Carleton, SC, and Carpenter, MB, Afferent and efferent connections of the medial, inferior and lateral vestibular nuclei in the cat and monkey. *Brain Res.,* 278:29, 1983.

28. Carleton, SC, and Carpenter, MB, Distribution of primary vesibular fibers in the brainstem and cerebellum of the monkey. *Brain Res.,* 294:281,1984.

29. Carpenter, MB, Vestibular nulei: afferent and efferent projections. *Prog. Brain Res.,* 76:5,1988.

30. Carpenter, MB, Stein, BM, and Peter, P, Primary vestibulocerebellar fibers in the monkey: distribution of fibers arising from distinctive cell groups in the vestibular ganglia. *Am. J. Anat.,* 135:221,1972.

31. Carpenter, MB, Cowie, RJ, Connections and oculomotor projections of the superior vestibular nucleus and cell group "y". *Brain Res.,* 336:265,1985.

32. Cheron, G, and Godaux E, Disabling of the oculomotor neural integrator by kainic acid injections in the prepositus-vestibular complex of the cat. *J Physiol, Lond.,* 394:267,1987.

33. Chubb, MC, and Fuchs, AF, Contribution of y group of vestibular nuclei and dentate nucleus of cerebellum to generation of vertical smooth eye movements. *J. Neurophysiol.,* 48:75,1982.

34. Cohen, B, The Vestibulo-Ocular Reflex Arc, *Handbook of Sensory Physiology,* Kornhuber, HH, Ed., pp. 477-540, Springer, New York, 1974.

35. Cohen, B, Tomko, DL, and Guedry, F, Sensing and Controlling Motion. *Vestibular and Sensorimotor Function.* The New York Academy of Sciences, New York, 1992.

36. Cohen, B, Highstein, SM, Büttner-Ennever, JA, *New Directions in Vestibular Research.* The New York Academy of Sciences, New York, in press, 1996.

37. De Zeeuw, CI, Wylie, DR, Digiorgi, PL, and Simpson, JI, Projections of individual Purkinje cells of identified zones in the flocculus to the vestibular and cerebellar nuclei in the rabbit. *J. Comp. Neurol.,* 349:428,1994.

38. Endo, K, Thomson DB, Wilson, VJ, Yamaguchi, T, and Yates, BJ, Vertical vestibular input to and projections from the caudal parts of the vestibular nuclei of the decerebrate cat. J. Neurophysiol., 74:428,1995.

39. Epema, AH, Gerrits, NM, and Voogd, J, Commissural and intrinsic connections of the vestibular nuclei in the rabbit: a retrograde labeling study. *Exp. Brain Res.,* 71:129,1988.

40. Epema, AH, Gerrits, NM, and Voogd, J, Secondary vestibulocerebellar projections to the flocculus and uvulonodular lobule of the rabbit: a study using HRP and double fluorescent tracer techniques. *Exp. Brain Res.,* 80:72,1990.

41. Faugier-Grimaud, S, and Ventre, J, Anatomic connections of inferior parietal cortex (area 7) with subcortical structures related to vestibulo-ocular function in a monkey (*Macaca fascicularis*). *J. Comp. Neurol.,* 280:1,1989.

42. Fettiplace, R, Crawford, AC, Evans, MG, The Hair Cell`s Mechanoelectrical Transducer Channel. Sensing and Controlling Motion. *Vestibular and Sensorimotor Function,* Cohen, B, Tomko, DL, and Guedry, F Eds., The New York Academy of Sciences, New York, 1992.

43. Friberg, L, Olsen, TS, Roland, RE, Paulson, OB, and Lassen, NA, Focal increase of blood flow in the cerebral cortex of man during vestibular stimulation. *Brain,* 108:609,1985.

44. Fuchs, AF, and Mustari, MJ, The optokinetic response in primates and its possible neuronal substrate. *Visual Motion and Its Role in the Stabilization of Gaze,* Miles, FA, Wallmann, J, Eds., Elsevier, Amsterdam, 1993.

45. Fukushima, K, The interstitial nucleus of Cajal and its role in the control of movements of head and eyes. *Prog. Neurobiol.,* 29:107,1987.

46. Fukushima, K, Fukushima, J, The interstitial nucleus of Cajal is involved in generating downward fast eye movement in alert cats. *Neurosci. Res.,* 15:299,1992.

47. Fukushima, K, and Kaneko, CRS, Vestibular integrators in the oculomotor system. Neurosci. Res., 22:249,1995.

48. Fukushima, K, and Takahashi, K, Kato, M, Responses of vestibular neurons to stimulation of the interstitial nucleus of Cajal in the cat. *Exp. Brain Res.,* 51:1,1983.

49. Fukushima, K, Perlmutter, SI, Baker, JF, and Peterson, BW, Spatial properties of second order vestibulo-ocular relay neurons in the alert cat. *Exp. Brain Res.,* 81:462,1990.

50. Gacek, RR, The course and central termination of first order neurons supplying vestibular end organs in the cat. *Acta Otolaryngol.,* 254:1,1969.

51. Gacek, RR, Location of commissural neurons in the vestibular nuclei of the cat. *Exp. Neurol* . 59:479,1978.

52. Gacek, RR, Location of trochlear vestibulo-ocular neurons in the cat. *Exp. Neurol.,* 66:692,1979.

53. Gamlin, PDR, Cohen, DH, Projections of the retinorecipient pretectal nuclei in the pigeon (*Columba livia*). *J. Comp. Neurol.,* 269:18,1988.

54. Gerrits, NM, Vestibular nuclear complex. *The Human Nervous System,* Paxinos, G, Ed., Academic Press, San Diego, 1990.

55. Gerrits, NM, Epema, AH, van Linge, A, Dalm, E, The primary vestibulocerebellar projection in the rabbit: absence of primary afferents in the flocculus. *Neurosci. Lett.,* 105:27,1989.

56. Giolli, RA, Blanks, RHI, and Torigoe, Y, Pretectal and brainstem projections of the medial terminal nucleus of the accessory optic system of the rabbit and rat as studied by anterograde and retrograde neuronal tracing methods. *J. Comp. Neurol.,* 227:228,1984.

57. Goldberg, JM, and Fernandez, C, Efferent vestibular system in the squirrel monkey: anatomical location and influence on afferent activity. *J. Neurophysiol.,* 43:986,1980.

58. Goldberg, JM, Desmadryl, G, Baird, RA, and Fernandez, C, The vestibular nerve in the chinchilla. v. Relation between afferent response properties and peripheral innervation patterns in the utricular macula. *J. Neurophysiol.,* 63:781,1990.

59. Goldberg, JM, Lysakowski, A, and Fernandez, C, Structure and function of vestibular nerve fibers in the chinchilla and squirrel monkey. Sensing and controlling motion. Vestibular and sensorimotor function, *Ann. NY. Acad. Sci.,* 92, 1992.

60. Graf, W, and Ezure, K, Morphology of vertical canal related second order vestibular neurons in the cat. *Exp. Brain Res.,* 63:35,1986.

61. Graf, W, McCrea, RA, and Baker, R, Morphology of posterior canal-related secondary vestibular neurons in rabbit and cat. *Exp. Brain Res.,* 52:125,1983.

62. Graf, W, Baker, J, and Peterson, BW, Sensorimotor transformation in the cat's vestibulooocular reflex system. I. Neuronal signals coding spatial coordination of compensatory eye movements. *J. Neurophysiol.*, 70:2427,1993.

63. Grüsser, OJ, Guldin, W, Harris, L, Lefebre, J, and Pause, M, Cortical Representation of Head-in-Space Movement and Same Psychophysical Experiments on Head Movement. *The Head-Neck Sensory Motor System,* Berthoz A, Graf, W, and Vidal PP, Eds, Oxford University Press, New York, 1992

64. Gstoettner, W, Burian, M, and Cartellieri, M, Central projections from singular parts of the vestibular labyrinth in the guinea pig. *Acta Oto-Laryngol.*, 112:486,1992.

65. Guldin, WO, Akbarian, S, and Grüsser, OJ, Cortico-cortical connections and cytoarchitectonics of the primate vestibular cortex: a study in squirrel monkeys (*Saiiri sciureus*). *J. Comp. Neurol.,* 326:375,1992.

66. Haines, DE, Cerebellar corticovestibular fibers of the posterior lobe in a prosimian primate, the lesser bushbaby (*Galago senegalensis*). *J. Comp. Neurol.,* 160:363,1975.

67. Hess, BJM, and Dieringer, N, Spatial organization of linear vestibuloocular reflexes of the rat — responses during horizontal and vertical linear acceleration. *J. Neurophysiol.*, 66:1805,1991.

68. Highstein, SM, Organization of the vestibulo-oculomotor and trochlear reflex pathways in the rabbit. *Exp. Brain Res.,* 17:285,1973.

69. Highstein, SM, The efferent control of the organs of balance and equilibrium in the toadfish, *Opsanus tau.* Sensing and controlling montion. Vestibular and sensorimotor function, *Ann. NY Acad. Sci.*,108,1992.

70. Highstein, SM, and McCrea, RA, The Anatomy of the Vestibular Nuclei. *Neuroanatomy of the Oculomotor System,* Büttner-Ennever, JA, Ed., Elsevier, Amsterdam,1988.

71. Highstein, SM, Goldberg, JM, Moschovakis, AK, and Fernandez, C, Inputs from regularly and irregularly discharging vestibular nerve afferents to secondary neurons in the vestibular nuclei of the squirrel monkey. II. Correlation with output pathways of secondary neurons. *J. Neurophysiol.*, 58:719,1987.

72. Hirai. N, Uchino. Y. Superior vestibular nucleus neurones related to the excitatory vestibulo-ocular reflex of anterior canal origin and their ascending course in the cat. *Neurosci. Res.*, 1:73,1984a.

73. Hirai, N, Uchino, Y, Floccular influence on excitatory relay neurons of vestibular reflexes of anterior semicircular canal origin in the cat. *Neurosci. Res.*, 1:327,1984a.

74. Holstege, G, Brainstem-spinal cord projections in the cat, related to control of head and axial movements. *Rev. Oculomot. Res.*, 2:431,1988.

75. Holstege, G, and Collewijn, H, The efferent connections of the nucleus of the optic tract and the superior colliculus in the rabbit. *J. Comp. Neurol.,* 209:139,1982.

76. Holstege, G, and Cowie, RJ, Projections from the rostral mesencephalic reticular formation to the spinal cord. *Exp. Brain Res.,* 75:265,1989.

77. Isu, N, and Yokota, J, Morphophysiological study on the divergent projection of axon collaterals of medial vestibular nucleus neurons in the cat. *Exp. Brain Res.,* 53:151,1983.

78. Isu, N, Uchino, Y, Nakashima, H, Satoh, S, Ichikawa, T, and Watanabe, S, Axonal trajectories of posterior canal-activated secondary vestibular neurons and their coactivation of extraocular and neck flexor motoneurons in the cat. *Exp. Brain Res.,* 70:181,1988.

79. Isu, N, Sakuma, A, Hiranuma, K, Uchino, H, Sasaki, S, Imagawa, M, and Uchino, Y, The neuronal organization of horizontal semi-circular canal activated inhibitory vestibulocollic neurons in the cat. *Exp. Brain Res.,* 86:9,1991.

80. Ito, M, Nisimaru, N, and Yamamoto, M, The neuronal pathways relaying reflex inhibition from semicircular canals to extra-ocular muscles of rabbits. *Brain Res.,* 55:189,1973.

81. Katz, E, DeJong, JMBV, Büttner-Ennever, JA, and Cohen, B Effects of midline medullary lesions on velocity storage and the vestibulo-ocular reflex. *Exp. Brain Res.,* 87:505,1991.

82. Keller, EL, and Daniels, PD, Oculomotor related interaction of vestibular and visual stimulation in vestibular nucleus cells in the alert monkey. *Exp. Neurol.,* 46:187,1975.

83. Keller, EL, and Kamath, BY, Characteristics of head rotation and eye movement related neurons in alert monkey vestibular nucleus. *Brain Res.,* 100:182,1975.

84. Keller, EL, and Precht, W, Visual vestibular responses in vestibular nuclear neurons in the intact and cerebellectomized, alert cat. *Neurosci.,* 4:1599,1979.

85. Kevetter, GA, and Perachio, AA, Distribution of vestibular afferents that innervate the sacculus and posterior canal in the gerbil. *J. Comp. Neurol.,* 254:410,1986.

86. Korte, GE, The brainstem projection of the vestibular nerve in the cat. *J. Comp. Neurol.,* 184:279,1979.

87. Kotchabhakdi, N, and Walberg, F, Primary vestibular afferent projections to the cerebellum as demonstrated by retrograde axonal transport of horseradish peroxidase. *Brain Res.,* 142:142,1978.

88. Lang, W, and Kubik, S, Primary vestibular afferent projections to the ipsilateral abducens nucleus in cats. An autoradiographic study. *Exp. Brain Res.,* 37:177,1979.

89. Lang, W, Büttner-Ennever, JA, and Büttner, U, Vestibular projections to the monkey thalamus: an autoradiographic study. *Brain Res.,* 177:3,1979.

90. Langer, T, Fuchs, AF, Chub,b MC, Scudder, CA, and Lisberger, SG, Floccular efferents in the rhesus macaque as revealed by auto-radiography and horseradish peroxidase.*J. Comp. Neurol.,* 235:26,1985a.

91. Langer, T, Fuchs, AF, Scudder, CA, and Chubb, MC, Afferents to the flocculus of the cerebellum in the rhesus macaque as revealed by retrograde transport of horseradish peroxidase. *J. Comp. Neurol.,* 235:1,1985b.

92. Langer T, Kaneko CRS, Scudder CA, and Fuchs AF Afferents to the abducens nucleus in the monkey and cat. *J. Comp. Neurol.,* 245:379,1986.

93. Lisberger, SG, The neuronal basis for learning of simple motor skills. *Science.,* 242:728,1988.

94. Maciewicz, RZ, Eagen, K, Kaneko, CRS, and Highstein, SM, Vestibular and medullary brainstem afferents to the abducens nucleus in the cat. *Brain Res.,* 123:229,1977.

95. Magnin, M, Courjon, JH, and Flandrin, JM, Possible visual pathways in the cat vestibular nuclei involving the nucleus prepositus hypoglossi. *Exp. Brain Res.,* 51:298,1983.

96. Margras, IN, and Voogd, J, Distribution of secondary vestibular fibers in the cerebellar cortex. An autoradiographioc study in the cat. *Acta Anat.* 123:51,1985.

97. McCrea, RA, The nucleus prepositus. *Neuroanatomy of the Oculomotor System,* Büttner-Ennever, JA, Ed., Elsevier, Amsterdam,1988.

98. McCrea, RA, and Baker, R, Anatomical connections of the nucleus prepositus of the cat.*J. Comp. Neurol.,* 237:377,1985.

99. McCrea, RA, Yoshida, K, Berthoz, A, and Baker, R, Eye movement, related activity and morphology of second order vestibular neurons terminating in the cat abducens nucleus. *Exp. Brain Res.,* 40:468,1980.

100. McCrea, RA, Strassman, A, May, E, and Highstein, SM, Anatomical and physiological characteristics of vestibular neurons mediating the horizontal vestibulo-ocular reflex of the squirrel monkey.*J. Comp. Neurol.,* 264:547,1987.

101. McCrea, RA, Strassman, A, and Highstein, SM, Anatomical and physiological characteristics of vestibular neurons mediating the vertical vestibulo-ocular reflexes of the squirrel monkey. *J. Comp. Neurol.,* 264:571,1987.

102. McFarland, JL, Fuchs, AF, and Kaneko, CRS, The Nucleus Prepositus and Nearby Medial Vestibular Nucleus and the Control of Simian Eye Movements. *Vestibular and Brainstem Control of Eye, Head and Body Movements* Shinoda, Y, and Shimazu H, Eds., Japan Scientific Societies Press, Tokyo.

103. Minor, LB, McCrea, RA, and Goldberg, JM, Dual projections of secondary vestibular axons in the medial longitudinal fasciculus to extraocular motor nuclei and the spinal cord of the squirrel monkey. *Exp. Brain Res.,* 83:9,1990.

104. Mitsacos, A, Reisine, H, Highstein, SM, The superior vestibular nucleus: an intracellular HRP study in the cat. I. Vesti-bulo-ocular neurons. *J. Comp. Neurol.,* 215:78,1983a.

105. Mitsacos, A, Reisine, H, and Highstein, SM, The superior vestibular nucleus: an intracellular HRP study in the cat. II. Non- vestibulo-ocular neurons. *J. Comp. Neurol.,* 215:92,1983b.

106. Mustari, MJ, Fuchs, AF, Kaneko, CRS, and Robinson, FR, Anatomical connections of the primate pretectal nucleus of the optic tract. *J. Comp. Neurol.,* 349:111,1994.

107. Ödkvist, L, Schwarz, DWF, Fredrickson, JM, and Hassler, R, Projections of the vestibular nerve to the area 3a arm field in the squirrel monkey (*Saimiri sciureus*). *Exp. Brain Res.,* 21:97,1974.

108. Ohgaki, T, Curthoys, IS, and Markham, CH, Morphology of physiologically identified second-order vestibuar neurons in cat, with intracellulary injected HRP. *J. Comp. Neurol.,* 276:387,1988.

109. Partsalis, AM, Zhang, Y, and Highstein, SM, Dorsal Y group in the squirrel monkey. I. Neuronal responses during rapid and long-term modifications of the vertical VOR. *J. Neurophysiol.,* 73:615,1995.

110. Paxinos, G, and Watson, C, *The Rat Brain in Stereotaxic Coordinates,* Academic Press, Orlando, 1986.

111. Peterson, BW, Maunz, RA, and Fukushima, K, Properties of a new vestibulospinal projection, the caudal vestibulospinal tract. *Exp. Brain Res.,* 32:287,1978.

112. Pompeiano, O, Mergner, T, and Corvaja, N, Commisural, perihypoglossal, and reticular afferent projections to the vestibular nuclei in the cat: an experimental anatomical study with horseradish peroxidase. *Arch. Ital. Biol.,* 116:130,1978.
113. Precht, W, Vestibular mechanisms. *Ann. Rev. Neurosci.,* 2:265,1979.
114. Richter, E, and Spoendlin, H, Scarpia`s ganglion in the cat. *Acta. Otolaryngol. ,* 92:423,1981.
115. Rubertone, JA, Mehler, WR, and Cox, GE, The intrinsic organization of the vestibular complex: evidence for internuclear connectivity. *Brain Res.,* 263:137,1983.
116. Sato, F, Sasaki, H, Ishizuka, N, Sasaki, S, and Mannen, H, Morphology of single primary vestibular afferents originating from the horizontal semicircular canal in the cat. *J. Comp. Neurol.,* 290:423,1989.
117. Sato, Y, and Kawasaki, T, Target neurons of floccular caudal zone inhibition in Y-group nucleus of vestibular nuclear complex. *J. Neurophysiol.,* 57:460,1987.
118. Sato, Y, Yamamoto, F, Shojaku, H, and Kawasaki, T, Neuronal pathway from floccular caudal zone contributing to vertical eye movements in cats — role of group y nucleus of vestibular nuclei. *Brain Res.,* 294:375,1984.
119. Shimazu, H, and Precht, W, Inhibition of central vestibular neurons from the contralateral labyrinth and its mediating pathway. *J. Neurophysiol.,* 29:467,1966.
120. Shinoda, Y, Ohgaki, T, Futami, T, and Sugiuchi, Y, Vestibular projections to the spinal cord: the morphlgy of single vestibulo-spinal axons. *Prog. Brain Res.,* 76:17,1988.
121. Shinoda, Y, Ohgaki, T, Sugiuchu, Y, Futami, T, Spatial Innervation Patterns of Single Vestibulospinal Axons in Neck Motor Nuclei. *The Head-Neck Sensory Motor System,* Berthoz, A, Graf, W, and Vidal, PP, Eds., Oxford University Press, New York, 1992.
122. Shinoda, Y, Sugiuchi, Y, Futami, T, Ando, N, and Kawasaki, T, Input patterns and pathways from the six semicircular canals to motoneurons of neck muscles. 1. The multifidus muscle group. *J. Neurophysiol.,* 72:2691,1994.
123. Simpson, JI, The accessory optic system. *Annu. Rev. Neurosci.,* 7:13,1984.
124. Simpson, JI, Giolli, RA, Blanks, RHI, The Pretectal Nuclear Complex and the Accessory Optic System. *The Neuroanatomy of the Oculomotor System,* Büttner-Ennever JA, Ed., pp 335-364. Elsevier, Amsterdam,1988.
125. Spence, SJ, and Saint-Cyr, JA, Mesodiencephalic projections to the vestibular complex in the cat. *J. Comp. Neurol.,* 268:375,1988.
126. Steiger, HJ, and Büttner-Ennever, JA, Oculomotor nucleus afferents in the monkey demonstrated with horseradish peroxidase. *Brain Res.,* 160,1979.
127. Sugiuchi, Y, Izawa, Y, and Shinoda, Y, Trisynaptic inhibition from the contralateral vertical semicircular canal nerves to neck motoneurons mediated by spinal commissural neurons. *J. Neurophysiol.,* 73:1973,1995.
128. Thunnissen, IE, Epema, AH, and Gerrits, NM, Secondary vestibulo-cerebellar mossy fiber projection to the caudal vermis in the rabbit. *J. Comp. Neurol.,* 290:262,1989.
129. Tomlinson, RD, McConville, K, King, WM, Paige, G, and Na, EQ, Eye Position Signals in the Vestibular Nuclei. *Contemporary Ocular Motor and Vestibular Research: A Tribute to David A. Robinson,* Fuchs, AF, Brandt, T, Büttner, U, and Zee, D, Eds., Thieme, 1994.
130. Tomlinson, RD, and Robinson, DA, Signals in vestibular nucleus mediating vertical eye movements in the monkey. *J. Neurophysiol.,* 51:1121,1984.
131. Uchino Y, and Hirai, N, Axon collateral of anterior semicircular canal-activated vestibular neurons and their coactivation of extraocular and neck motoneurons in the cat. *Neurosci. Res.,* 1:309,1984.
132. Uchino, Y, Hirai, N, Suzuki, S, and Watanabe, S, Properties of secondary vestibular neurons fired by stimulation of ampullary nerve of the vertical, anterior or posterior semicircular canals in the cat. *Brain Res.,* 223:273,1981.
133. Uchino, Y, Hirai, N, and Suzuki, S, Branching pattern and properties of vertical- and horizontal-related excitatory vestibulo-ocular neurons in the cat. *J. Neurophysiol.,* 48:891,1982.
134. Uchino, Y, Sasaki, M, Isu, N, Hirai, N, Imagawa, M, Endo, K, and Graf, W, Second-order vestibular neuron morphology of the extra-Mlf anterior canal pathway in the cat. *Exp. Brain Res.,* 97:387,1994.
135. Voogd, J, Feirabend, HKP, AND Schoen, JHR, Cerebellum and Precerebellar Nuclei, *The Human Nervous System,* Paxinos G, Ed., Academic Press, San Diego, 1990.

136. Voogd, J, Gerrits, NM, and Ruigrok, A, Organization of the Vestibulocerebellum. *New Directions in Vestibular Research*, Cohen B, Highstein SM, and Büttner-Ennever JA, Eds., The New York Academy of Sciences, 1996.
137. Warr, WB, Olivocochlear and vestibular efferent neurons of the feline brain stem: their location, morphology and number determined by retrograde axonal transport and acetylcholinesterase histochemistry. *J. Comp. Neurol.*, 161:159, 1975.
138. Waespe, W, Cohen, B, and Raphan, T, Dynamic modification of the vestibulo-ocular reflex by the nodulus and uvula. *Science*, 228:199,1985.
139. Weber, JT, and Harting, JK, The efferent projections of the pretectal complex: an autoradiographic and horseradish peroxidase analysis. *Brain Res.*, 194:1,1980.
140. Wilson, VJ, and Melvill-Jones, G, *Mammalian Vestibular Physiology*, Plenum, New York, 1976.
141. Wylie, DR, De, Zeeuw, CI, Digiorgi, PL, and Simpson, JI, Projections of individual Purkinje cells of identified zones in the ventral nodulus to the vestibular and cerebellar nuclei in the rabbit. *J. Comp. Neurol.*, 349:448,1994.
142. Yamamoto, M, Vestibulo-ocular reflex pathways of rabbits and their representation in the cerebellar flocculus. *Prog. Brain Res.*, 50:451,1979.
143. Yates, BJ, Vestibular influences on the sympathetic nervous system. *Brain Res.*, Rev 17:51,1992.
144. Yokota, J, Reisine, H, and Cohen, B, *Velocity Storage and Effects of Injection of GABAergic Substances in the Prepositus Hypoglossi Nuclei. Contemporary Ocular Motor and Vestibular Research: A Tribute to David A. Robinson*, Fuchs, AF, Brandt, T, Büttner U, and Zee, D, Eds., Thieme, New York, 1993.

Section II

Peripheral Vestibular Transmission

2 Vestibular Hair Cell-Afferent Transmission

Stephen L. Cochran

CONTENTS

2.1. INTRODUCTION

Encased within the temporal bone lies the membranous inner ear and its associated sensory endorgans, the cochlea, sacculus, utriculus, lagena (in some species), and the horizontal, anterior, and posterior semicircular canals. The sensory epithelium of each endorgan is geometrically organized and differentiated to provide the specific functional sensitivity attributable to that endorgan. While the cochlea, sacculus, and perhaps lagena transduce and transmit sensory signals relating to audition and vibration, the other inner ear vestibular endorgans are more involved with sensing head movement and position, which are then relayed to the brain.

Common to each epithelium are the sensory cells, the hair cells which straddle the K^+-rich endolymphatic compartment and the Na^--rich, plasma-like, perilymphatic compartment. The hair cells are embedded in a matrix of supporting cells and contact the distal processes of Scarpa's ganglion cell afferents that project to various second-order vestibular neurons within the central nervous system. Contemporary research has revealed that not only is each endorgan highly specialized functionally, but that each hair cell is highly differentiated dependent upon the specific

0-8493-7679-3/00/$0.00+$.50

endorgan within which it resides, and also upon where it is located within a given endorgan. This specialization is mirrored by the associated differentiation of the afferent processes in that specific types of afferents contact specific hair cell types, resulting in functional specialization of the afferent fibers (cf. Ref. 46).

Hudspeth and colleagues have discovered some of the basic mechanisms of sensory transduction by vertebrate hair cells (for reviews, see refs. 58, 60, and 90). Mechanical deflection of stereocilia modulates the opening of nonspecific cationic channels[21] at the tips of the stereocilia[28,59]. This modulation is direction selective in that deflection of the stereocilia toward the kinocilium opens these channels, while deflection away from the kinocilium closes these channels.[102] The opening of these channels results in an influx of cations from the endolymph into the hair cell. This influx generates a receptor potential that is transmitted rapidly and with little decrement to the basal portion of the hair cell.[90]

Between the stereocilia and the synapse are a variety of voltage-dependent channels that act to dictate the resting membrane potential of the hair cells and to modify and filter the receptor potential. These channels vary from hair cell to hair cell, from species to species, and from endorgan to endorgan, and involve a number of K^+ channels, including Ca^{2+}-dependent K^+ channels, A-type K^+ channels, delayed rectifying K^+ channels (see Ref. 22 for review), and other inwardly rectifying K^+ channels.[57] When the receptor potential reaches the basal region of the hair cell, it triggers voltage-gated Ca^{2+} channels localized to the presynaptic active zones[91] and transmitter release ensues. The entry of Ca^{2+} at these sites also triggers Ca^{2+}-dependent K^+ conductances, which are colocalized with the Ca^{2+} channels, at least with respect to frog saccular hair cells[91]. The Ca^{2+} channels have properties similar to L-type channels in other neurons (cf. Ref. 112) by virtue of their functional[91] and pharmacological properties[84,108]. Similar channels have been found in chick cochlear hair cells.[41,131] It is presumed that the Ca^{2+} entering the cell through these channels triggers transmitter release.

Hair cells have traditionally been classified into two types.[126] The Type I hair cell is characterized by being almost totally enveloped by the afferent fiber so that it forms a calyx-like structure around the hair cell. Type II hair cells are contacted by afferent boutons. In addition, Type II hair cells are contacted by centrally arising efferent axons which do not contact the Type I hair cells, but rather the calyceal afferents (cf. Refs. 126-129). Type II hair cells are found in all vertebrates, but Type I hair cells appear to be lacking in anamniotes (e.g., frogs and fish,[44,127] but see Ref. 12), although efferent contacts with afferents are found in toadfish,[56] but apparently not in frogs.[44] While the basic dichotomy exists between these two hair cell types, more recent studies indicate that hair cell types can be further subdivided according to morphological (e.g., Refs. 9, 52, 84, and 88) and physiological (cf. Ref. 47) criteria.

Vestibular afferents have been classified into three types according to their innervation patterns.[8,37,38,47,70] Some afferents contact, via boutons, Type II hair cells exclusively. Other afferents form only calyces with Type I hair cells. A third type of afferent, termed the dimorph, is a composite of the two in that it forms a calyx around the Type I hair cell, but branches as well to establish bouton contacts with Type II hair cells. There is some evidence to suggest that these three types of afferents can be further subdivided on morphological grounds (e.g., Ref. 11).

The model hair cell then has a dual role. It acts a transducer to convert the extent of stereocilial deflection into a receptor potential by virtue of the mechanico-electric transduction channels at the tips of the stereocilia, and also as a nerve terminal in that this receptor potential acts to modulate the membrane potential of the hair cell and thereby affect the number of voltage-gated Ca^{2+} channels that are open and which couple depolarization to transmitter secretion. This chapter will review this process of transmission, the transmitters involved in relaying hair cell activity to the vestibular afferent, and how endogenous chemical substances can affect this communication.

2.2 FUNCTIONAL ASPECTS OF HAIR CELL-AFFERENT FIBER TRANSMISSION

Furukawa and Ishii[42] were the first to record intracellularly from inner ear afferents. They found that in goldfish sacculus, afferents there were EPSPs that gave rise to action potentials in the afferents. This suggested that synaptic transmission between hair cells and afferents was chemically mediated and exhibited similar mechanisms of operation as the neuromuscular junction in that the sound-evoked synaptic potentials were comprised of summated, small, spontaneously occurring EPSPs that represented the quantal components of the evoked response[61]. Rossi and co-workers,[95] with intracellular recordings from canal afferents in the frog, first showed the presence of small depolarizing potentials that gave rise to action potentials. These depolarizing potentials were resistant to TTX, their frequency of occurrence was modulated by canal rotation, and they were blocked in high concentrations of Mg^{2+} and low concentrations of Ca^{2+}, suggesting that these were EPSPs generated by transmitter release from the hair cells. Subsequent studies[5] indicated that the amplitude of the potentials was relatively unchanged in high Mg^{2+} (prior to complete blockade), although their frequency of occurrence was reduced to zero, confirming that the Mg^{2+} acted to block transmitter release rather than acting postsynaptically. Similar studies with intracellular recordings from lizard[98] and chick[130] canal afferents indicated that both bouton afferents innervated by Type II and calyceal afferents innervated by Type I hair cells exhibited depolarizing subthreshold potentials, the frequencies of occurrence of which were modulated by caloric[98] and by mechanical[130] stimulation. Such potentials were reversibly abolished in the presence of Co^{2+}[98] suggesting that they were consequent from Ca^{2+}-dependent release of transmitter from the hair cells at the sites of synaptic contact with the afferents.

As at the neuromuscular junction (cf. Ref. 62), analysis of the frequency of occurrence and the amplitudes of the spontaneously occurring EPSPs can distinguish between pre-and postsynaptic mechanisms. Perturbations, such as movement of stereocilia to depolarize or hyperpolarize hair cells by modulation of the transduction channel (e.g., Ref. 130), or exogenously applied agents that act to depolarize the hair cell (e.g., increased K^+ [5,17]) alter the frequency of occurrence of the EPSPs, but not their amplitudes, indicating a presynaptic action. Alternatively, agents that appear to block the subsynaptic receptor (e.g., Ref. 5) or the transmitter-gated subsynaptic channel[17] reduce the amplitudes of the EPSPs without affecting their frequency of occurrence, indicating a postsynaptic effect.

Taking advantage of this ability to discriminate pre- versus postsynaptic mechanisms from intracellular afferent recordings, the effects of altered concentrations of the elemental cations K^+, Na^+, and Ca^{2+} have been investigated in order to assess the role these cations have in hair cell transmission.[17]

Intracellular recordings from frog canal and lagena afferents have shown that small increases (e.g., 3 mM) in the extracellular K^+ concentration bathing the isolated labyrinth result in a roughly sixfold increase in the frequency of occurrence of the EPSPs, while leaving EPSP amplitude relatively unaffected. These data are similar to previous findings[5] and imply that the hair cell resting membrane potential sits on the steep slope relating membrane depolarization to transmitter secretion. This is quite different from the neuromuscular junction, where much larger changes in the extracellular K^+ concentration are needed to generate such an increase in miniature endplate potential (MEPP) frequency of occurrence.[20,43,111] These findings support those of Roberts and co-workers,[91] who found that there were a number of open, voltage-dependent Ca^{2+} channels (thought to couple hair cell depolarization to transmitter release) at the basal portions of the hair cells at rest. These open channels are most likely responsible for the presence of spontaneously occurring EPSPs at rest. Consequently, small stereocilial deflections from the resting state can significantly modulate the amount of transmitter released. It is difficult as yet to relate the influence of these small extracellular changes in K^+ concentration to the various K^+ channels that have been reported to be on different types of hair cell. In the above experiments, the altered K^+ concentration was applied only transiently to achieve a maximal response. More prolonged exposures to such concentration changes could result in adaptations of the response and reveal the presence of underlying voltage-gated channels on the hair cell's membrane. Adaptive responses to prolonged K^+ application have been reported.[121] These experiments involved, for the most part, quantitation of spike discharge patterns from extracellular recordings from whole nerve branches. Pre- versus postsynaptic contributions to the responses could not be determined; in addition, the contributions of ionic currents attributable to hair cells possessing different voltage-gated channels (e.g., Ref. 72) could not be determined because the recordings were from a whole nerve branch and included multiple types of hair cells innervating all of the afferents.

Unlike K^+, increases in the extracellular Na^+ concentration did not consistently affect the frequency of occurrence of the EPSPs; rather, these increases resulted in increases in EPSP amplitude. This increase in amplitude can be explained if the subsynaptic iontophore on the afferent, gated by the hair cell transmitter, is permeable to Na^+. The increased driving force for Na^+ to enter the afferent, with increased extracellular Na^+ concentrations, would then be responsible for the increase in the synaptic potential size. That the subsynaptic iontophore is permeable to Na^+ is supported by the report that a reduction in the extracellular Na^+ concentration reduces the amplitudes of glutamate (thought to be the hair cell transmitter, see below) depolarizations.[86]

In contrast to the increase in EPSP amplitude caused by increased Na^+ concentration, increases in the extracellular Ca^{2+} concentration result in a decrease in EPSP amplitude, while decreases result in increases in EPSP amplitude. Although the mechanisms for these actions are not known, it would appear that, as has been reported at the neuromuscular junction,[1,65,72] increased Ca^{2+} can impede cation entry through the subsynaptic iontophore as well as reduce this current through charge screening. Increases in extracellular Ca^{2+} concentration also reduce the frequency of occurrence of the EPSPs. This presynaptic action is similar to that reported at the neuromuscular junction,[19,20] and may involve both charge screening and other possible mechanisms (see discussion in Ref. 17).

Griguer and colleagues[48] have conducted elegant patch clamp experiments on isolated rat Type I hair cells in which whole cell patches were examined with fura-2 in the pipette. They reported that repolarizations of the hair cell, following step depolarizations of approximately 90 mV, resulted in increases in the Ca^{2+} concentrations within the cell and that no increases were seen in response to the depolarizations themselves. They concluded from this that Type I hair cells are complementary to Type II hair cells in that Type II hair cells may release transmitter and signal afferents during mechanical stimulation, while Type I hair cells only signal when such a signal is over (i.e., during the repolarization). This hypothesis would imply that the Type I hair cell has a different synaptic release process than the Type II cell and other nerve terminals where voltage-dependent Ca^{2+} channels are coupled to the vesicular release of transmitter. It is more likely that any Ca^{2+} entry related to transmitter release occurred during the depolarizing pulses (the authors admit that such local, rapid currents are "too fast to be recorded in our experimental paradigm", Ref. 48, p. 333) and were undetected. In addition, the increased intracellular Ca^{2+} they reported may not have involved the Ca^{2+} channels associated with transmitter release (they should have closed in response to the repolarizations), but may be associated with other phenomena, such as cuticular plate movements (e.g., Ref. 49). These findings do not in themselves warrant the conclusion that the mechanisms of transmitter release from the Type II hair cell differ from that of the Type I hair cell.

Not only, then, is synaptic transmission quantal in nature at this synapse as well as at the neuromuscular junction, but both the pre- and postsynaptic ionic mechanisms involved in transmission seem similar as well. The major differences between the two junctions is the higher level of transmitter release from the hair cells over the motor nerve terminals and the different transmitter substances employed by each.

2.3 THE HAIR CELL TRANSMITTER(S)

Glutamate is currently the principal candidate as the hair cell transmitter in all endorgans of the vertebrate (and possibly invertebrate[113]) inner ear. Biochemical, immunocytochemical, and electrophysiological evidence continues to accumulate implicating glutamate as the chemical released from hair cells, which in turn excites first-order vestibular afferents. However, because glutamate plays a central role in the biochemical operations of all cells, and because there are a number of endogenous structural analogs of glutamate, which also have functional (i.e., excitatory) activity, it is difficult to definitively pinpoint the actual transmitter compound. In fact, the actual compound could be one of a number of other excitatory acidic amino acids or a small peptide composite of such. While the evidence certainly favors glutamate as the major hair cell transmitter, it is not inconceivable that a number of other excitatory amino acids or acidic dipeptides could serve this neurotransmitter role, and thus caution must be exercised in drawing final conclusions regarding the major hair cell transmitter.

In somewhat apparent conflict with the rather overwhelming candidacy of glutamate (and its family of endogenous agonists) are a number of studies suggesting that GABA could also play a role as a hair cell neurotransmitter, at least in the endorgans which comprise the vestibular labyrinth. While these studies are somewhat less convincing *in toto* than those implicating glutamate, they do implicate some role for GABA in the inner ear, which is as yet elusive.

In this section, the various investigations into the nature of the vestibular hair cell transmitter will be reviewed and discussed as they relate to glutamate and to GABA. It is worth noting that evidence regarding the cochlear hair cells' neurotransmitters has recently been reviewed by Eybalin.[34]

2.3.1 EVIDENCE FOR GLUTAMATE

2.3.1.1 Demonstrations of the Presence of Glutamate in Hair Cells.

Drescher and Drescher[31] were able to demonstrate high levels of glutamate in an enriched epithelium layer (containing supporting cells, hair cells, and afferent dendrites) of trout sacculus. Utilizing immunocytochemical techniques with antibodies to glutamate conjugated to bovine serum albumin, Demêmes and collaborators[27] found intense staining of both types of hair cells and also of afferents and ganglion cells, but reported that supporting cells lacked such glutamate-like immunoreactivity in mice, rats, and cats. Utilizing a similar, but different, antibody, Panzanelli and co-workers[81] found immunoreactivity of hair cells, afferents, and possibly also supporting cells. Utilizing reportedly specific antibodies to conjugated glutamate and aspartate, Harper and coworkers[55] reported that hair cells (including cilia), calyces, nerve fibers, and ganglion cells were all labeled with both antibodies, while supporting cells were devoid of label. In an attempt to demonstrate cycling of glutamine and glutamate between supporting cells and hair cells, Usami and Ottersen[119] investigated the immunoreactivities of other antibodies to conjugated glutamine and glutamate in the rat inner ear. They found that there was an increased number of immunogold particles in both types of hair cells when utilizing the antibody to conjugated glutamate, while there was an increased number of particles in the supporting cells when the conjugated glutamine antibody was employed. These data provide strong support for their hypothesis of transmitter (glutamate) recycling between the hair cells and the supporting cells. Because of its diverse participation in the biochemical metabolism of most cells, the demonstration of the presence of glutamate in hair cells represents relatively weak evidence that glutamate is released as the transmitter. For example, the mammalian studies utilizing conjugated glutamate antibodies[27,55,119] clearly show intense labeling of the hair cell cilia, which are not thought to be involved in transmitter release. Meanwhile, Usami and Ottersen[119] were unable to demonstrate at the ultrastructural level accumulations of label at the presynaptic sites where vesicles presumably filled with transmitter accumulate. Similar findings have been reported by these authors in studies of the cochlea.[117]

2.3.1.2. Hair Cell Release of Glutamate

Release studies have been limited, largely due to the relatively small volume the hair cells occupy in the inner ear. Drescher and co-workers[33] demonstrated glutamate release from a trout saccular/macular preparation in response to a challenge of saline containing a high K^+ concentration (53.5 mM). This release, however, was not dependent on the presence of Ca^{2+} in the medium, indicating that it occurred even when synaptic release was prevented in high Mg^{2+} / low Ca^{2+}. Later,[32] this preparation was refined to a single sheet of epithelium, a lower concentration of K^+ was utilized, and a different experimental protocol was employed. They then found that out of 22 primary a-amino acids screened by HPLC, glutamate was specifically released in a Ca^{2+}-dependent fashion, suggesting that it could be the hair cell transmitter. Zucca and colleagues[133] also reported that glutamate was released from isolated frog semicircular canals by K^+ depolarization in a Ca^{2+}-dependent fashion (utilizing a bioluminescence-enzymatic method of glutamate detection), again supporting its role as a hair cell transmitter.

2.3.1.3 Immunohistochemistry of Glutamate Receptors

Perhaps one of the most promising recent developments to investigate transmitter pharmacology has been the generation of specific antibodies to neurotransmitter receptors and their subunits (see Chapter 5). Wenthold and colleagues[125] were able to isolate, clone, and develop antibodies to a kainic acid binding protein from frog brain. The antibodies were shown to bind in the inner ear epithelium exclusively to the subsynaptic thickening at the site of synaptic contact between the hair cells and the afferents in the frog[24] (in keeping with the known synaptic pharmacology of this receptor determined by electrophysiological means; see Ref. 14 and below). Subsequently, glutamate receptor subunits have been further characterized (cf. Refs. 77,and 103). Niedzielski and Wenthold,[78] utilizing *in situ* hybridization of mRNAs for various glutamate receptors, found a number of receptor subunits expressed in ganglion cells, leading them to suggest that quisqualate (or AMPA), kainate, and NMDA receptors were all present at the hair cell-afferent fiber synapse. Usami and colleagues[118] also reported that antibodies to the NMDAR1 and to the GluR2/3 receptors stained the ganglion cells in rats, guinea pigs, and both squirrel and macaque monkeys. However, they also reported that within the vestibular (and cochlear) neuroepithelia, the staining was diffuse and not localized. Demêmes and co-workers,[26] in similar studies utilizing the same antibodies, found in rats and guinea pigs that the vestibular ganglion cells were labeled with GluR2/3 and GluR4, but not GluR1 antibodies, similar to the previous studies. Within the sensory epithelia at the light level, they found staining for GluR2/3 antibodies in the cytoplasm of both Type I and II cells. At the ultrastructural level, they found GluR1 antibody staining purportedly at the subsynaptic density at the site of contact of the Type I cell and the calyx. However, it also appears that the mitochondria were also heavily stained (cf. Ref. 26; Fig. 4H), so it is somewhat unclear exactly how specific the antibody was for the receptor. With increasing development of these techniques and the specific antibodies to these receptors, this type of immunohistochemical experimentation should be extremely useful in understanding the pharmacology of the synaptic transmitter(s) and their subsynaptic receptors within the vestibular neuroepithelia.

2.3.1.4 Electrophysiological Support for Glutamate

The bulk of the electrophysiological evidence to date supports glutamate or a related compound as a hair cell transmitter. Annoni and colleagues[5] found that glutamate, aspartate, and a number of structurally related acidic amino acids were able to excite vestibular afferents when bath-applied to the isolated membranous labyrinth. With intracellular recordings, they showed that these

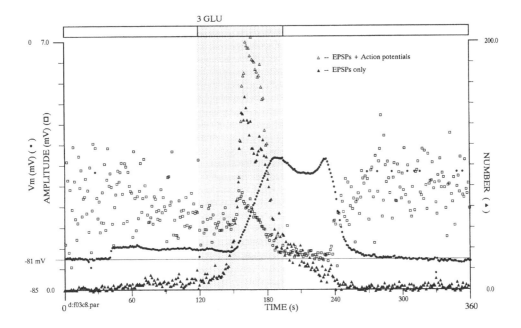

FIGURE 2.1 The influence of glutamate on activity at the hair cell afferent synapse of the frog. A canal afferent axon was recorded from intracellularly in the isolated labyrinth of the frog. The parameters number of EPSPs (closed triangles), number of EPSPs plus action potentials (open triangles), mean EPSP amplitude (open squares), and afferent membrane potential (filled dots) are plotted over the time course of the recording. Values were determined by computer algorithm.[15] Each point represents the mean value (or absolute number for frequency) in a 1-second interval. Introduction of 3 mM glutamate (3 GLU - shaded region) resulted in a reversible increase in EPSP and action potential frequency of occurrence, a decrease in EPSP amplitude, and a depolarization of the afferent. Note that the increase in EPSP frequency of occurrence in response to glutamate occurs concomitantly with the afferent depolarization.

compounds induced a depolarization of the afferent that was accompanied by an increase in the frequency of action potentials. The response of such an afferent to glutamate superfusion is illustrated in Figure 2.1

Glutamate (3 mM) was bath-applied and the result was a reversible depolarization of the afferent. This depolarization was accompanied by an increase in EPSP and action potential frequency, and by a reduction in the amplitude of the EPSPs. They then showed that, as reported previously,[95] in high Mg^{2+}/low Ca^{2+} (10 mM/1 mM), synaptic potential frequency was reduced to zero, indicating that hair cell depolarization-secretion coupling was blocked. In such a circumstance, bath-application of glutamate or aspartate still resulted in a depolarization of the afferent of a magnitude comparable to that when hair cell transmitter release was not blocked. This finding suggested that glutamate was not acting on the hair cell to increase transmitter release, but was more likely acting on the afferents directly to induce the depolarization. Glutamate and aspartate then were able to mimic the excitatory action of the endogenous transmitter. The glutamate antagonists, D-2-amino-5-phosponovaleric acid (D-APV) and kynurenic acid, which block the action of glutamate at other nervous system sites (cf. refs. 83, and 124), were able to reduce reversibly the amplitudes of the vestibular afferent EPSPs without consistently affecting their frequency of occurrence. These data suggest that the glutamate antagonists acted to block the subsynaptic receptor (thereby reducing EPSP amplitude) without influencing the release of transmitter from the hair cell (as reflected by the lack of change of the frequency of occurrence of the EPSPs). Annoni and colleagues[5] further showed that the depolarizing actions of glutamate and aspartate on the afferent (in the presence of high Mg^{2+} to block endogenous transmitter release from the hair cells) were also reversibly reduced in these same antagonists. These findings indicated that glutamate and aspartate mimic the action of the endogenous transmitter in provoking afferent depolarizations, and that both the endogenous EPSPs and the exogenously applied agonist-induced depolarizations are antagonized by the same substance. These results suggest that the endogenous compound is either glutamate or aspartate, or a similarly acting compound. Similar findings have been reported in extracellular recordings from vestibular afferents in the axolotl[107] and from afferents from the ampullae of Lorenzini in the skate.[3,4]

Other investigators have questioned whether glutamate (or related compounds) is the hair cell transmitter. Nagai and coworkers,[76] recording extracellularly from afferents to the ampullary electroreceptors in the catfish, reported that Joro spider toxin (JSTX) blocked the excitatory action of iontophoresed glutamate without blocking evoked afferent activity. However, both the evoked activity and the glutamate response were reduced by the JSTX, but the glutamate effect was reduced more. This "differential" effect of the two stimulus conditions could be explained by a difference in efficacy of the two stimuli (i.e., glutamate versus evoked) rather than two different substances or sites of action involved. In other studies[50,53,122] it has been argued that, since exogenously applied glutamate depolarizes the afferent and in doing so reduces EPSP amplitude, and with prolonged application "inhibits" action potential generation by the afferent, then these are reasons to disqualify glutamate as the endogenous transmitter. It is clear, however, that with bath-application of glutamate, the afferent depolarizes to such an extent as to provoke depolarization-inactivation of the afferent (resulting in a cessation of spiking). Combined with this depolarization there is an increase in conductance of the afferent axon, resulting in a decreased driving force for the EPSP as well as shunting of the synaptic current, so the EPSPs are smaller. These phenomena are consistent with glutamate acting as the hair cell transmitter.

The nature of the subsynaptic receptor to which the transmitter binds is less clear. A number of investigators have demonstrated that in frogs[5,14,86] and axolotls,[104,106] NMDA has little effect on afferent activity, while the agonists quisqualate and kainate are potent agonists. In these vertebrates, some[104,132] have suggested that there is some neuromodulatory action of NMDA. The antagonists D-APV, kynurenic acid, CNQX (6-cyano-7-nitro quinoxaline-2,3-dione), and streptomycin have all been reported to reduce EPSP amplitude or resting spike activity,[13,14,82,86,105,132,133] but the concentrations of these antagonists required to reduce such activity were above the level at which the antagonists are specific for a given glutamate receptor. By default, since low concentrations of specific NMDA receptor antagonists (e.g., D-APV) or quisqualate (AMPA) antagonists (e.g., CNQX) are ineffective in blocking the synapse, these electrophysiological findings would suggest that the subsynaptic receptor is of the kainate type. This suggestion concurs with the immunocytochemical evidence of Dechesne and colleagues[24] that the subsynaptic receptor is selectively labeled by an antibody to a kainate binding protein.

Most of the electrophysiological evidence derives from experiments on frogs and other anamniotes which possess only Type II hair cells. In amniotes (reptiles, birds, and mammals), which possess both types of hair cells, there is very little electrophysiological evidence largely due to the inaccessibility of the inner ear and to the fact that it is difficult to maintain viably *in vitro*, particularly in homeotherms. Dechesne and colleagues[25] demonstrated that second order vestibular neurons increased their activity following glutamate infusion into the inner ear, leading these authors to suggest that glutamate was the hair cell transmitter. More recently, in extracellular recordings from canal afferents in the turtle labyrinth maintained *in vitro*,[18] bath-applied glutamate and aspartate were reported to increase reversibly the firing of these axons, while GABA and carbachol had much weaker effects. Kynurenic acid reduced both the spontaneous activity of the afferents as well as the excitation due to exogenously applied glutamate and aspartate. These findings suggested that glutamate or aspartate (or a similar acidic amino acid) could be the hair cell transmitter, since the kynurenic acid reduced the spontaneous activity due to hair cell release of transmitter and the exogenously applied agonists. These authors also reported that NMDA had a potent excitatory action, unlike what has been reported for the frog. While these findings in the turtle concur with those from the frog, it is not clear whether the afferents innervated by Type I hair cells were also sampled in this study (only spontaneously active axons were sampled). Thus, it is unclear whether glutamate or a related compound is released by this population. Physiological studies of hair cell/vestibular afferent interactions in glutamate receptor knockout mice may provide important clues to the potential role of glutamate in hair cell transmission.

2.3.2 Evidence for GABA

Data accumulated to date would suggest that GABA plays some role in the inner ear. Whether or not GABA is released as a neurotransmitter from the hair cells is still not clear. The evidence supporting a functional role for GABA in the inner ear is reviewed below.

2.3.2.1 Demonstration of the Presence of Glutamic Acid Decarboxylase (GAD) Activity

Flock and Lam[39] demonstrated that GABA could be synthesized from extracts of the skate crista ampullaris, which led them to suggest that GABA was the hair cell transmitter. Meza and colleagues were able to replicate these findings for the crista ampullaris of the chick,[74] frog,[67] and in rats and guinea pigs.[68] They also showed that following transection of the VIIIth nerve in the frog that GAD activity remained constant, while choline acetyltransferase (ChAT, the enzyme responsible for synthesizing acetylcholine) activity was reduced (presumably due to degeneration of the efferent terminals). In addition, following streptomycin treatment to produce hair cell degeneration, GAD activity was reduced.[75] These findings were consistent with GAD being present in the hair cells and led these authors to conclude that GABA was the hair cell transmitter.

2.3.2.2 Immunocytochemical Evidence

Antibodies to GAD label hair cell cytoplasm in both chicks[114] and guinea pigs.[69] Antibodies to GABA transaminase (GABA-T, a GABA degradative enzyme) label some hair cells, calyceal endings, nerve fibers, and ganglion cells in the guinea pig,[69] and label nerve fibers and ganglion cells in the chick.[114] Antibodies to GABA conjugated to bovine serum albumin have been reported to label hair cells (including the cilia) and nerve fibers in the guinea pig.[66] Others, utilizing a different GABA-conjugated antibody in the chick, reported both Type I and II hair cells labeled, but not the afferent fibers.[116] However, when this latter antibody was utilized in the squirrel monkey, it was found to label only nerve fibers and boutons.[115] In a subsequent comparative study it was reported that, in chicks and pigeons, hair cells and efferent boutons were labeled; in squirrel monkeys, only efferents were labeled; and in rats and guinea pigs, no label was found (although labeling in the organ of Corti was detected).[73] A different group of investigators utilizing a different such antibody reported exclusive labeling of the afferent calyx.[30] This finding led them to suggest that GABA could be a compound released from the calyx to feed back upon the hair cells. A recent study investigating antibodies to the GABA-A receptor in hamsters, rats, and mice found that the calyx and some ganglion cells were labeled exclusively.[40]

2.3.2.3 Electrophysiological Studies

Felix and Ehrenberger[36] reported that GABA, when iontophoresed in the cat sacculus, excited afferents, and that this excitation was blocked by bicuculline. Annoni and colleagues[5] found in the frog that GABA and its potent agonist muscimol were able to weakly excite vestibular afferents and that picrotoxin had no effect on the amplitudes of spontaneous EPSPs. The excitation was accompanied by an increase in EPSP frequency in the afferents. They found that this excitation was absent if the labyrinth was exposed to high Mg^{2+}/low Ca^{2+} (to block depolarization-secretion coupling, and thereby hair cell release of transmitter), suggesting that GABA was acting presynaptically. Similar studies in frogs[54] and axolotls,[120] utilizing single unit recordings, showed GABA to be weakly excitatory and that picrotoxin and bicuculline had little, if any, effect on afferent activity. Recent studies in the turtle[18] also found GABA to be only weakly excitatory.

The role of GABA in the vestibular labyrinth, then, is as yet unclear. The demonstrations that GABA can be synthesized by homogenates of inner ear tissues suggests that GABA has a role, but the immunocytochemical findings indicate that some species differences may exist, and the suggestion that GABA is released from hair cells to excite afferents conflicts with many of the electrophysiological findings.

2.3.3 EVIDENCE FOR OTHER SUBSTANCES

In an attempt to isolate a possible hair cell transmitter, Sewell and Mroz[99] extracted substances from the inner ears of goldfish, trout, and skates and then fractionated these compounds according to molecular weight. These fractions were then applied to isolated frog (*Xenopus*) skin containing lateral line "stitches." Afferent axons were recorded extracellularly with a suction electrode and influences of these fractions on activity were assessed. They reported finding a low molecular weight (ca. 200) and a high molecular weight (ca. 4000) compound, each of which was found to excite the lateral line axons. There was also a high molecular weight compound that decreased the activity of the lateral line axons. The high molecular weight compound could represent calcitonin gene-related peptide (CGRP),[2] while the low molecular weight compound is as yet unidentified,[100] but is apparently not glutamate or aspartate.

Immunocytochemical evidence has suggested that CGRP may play a role in the inner ear in some associative context with the efferent system, since it is localized to efferent neurons[79,123] and is apparently colocalized with acetylcholine[80] and choline acetyltransferase (the acetylcholine synthesizing enzyme)[89] in the efferents. Sewell and Starr,[101] recording intracellularly from afferent fibers innervating the lateral line organ in *Xenopus,* found that exogenously applied CGRP excited the afferents by increasing EPSP frequency of occurrence without affecting EPSP amplitude, suggesting that CGRP was acting presynaptically to increase transmitter release from the hair cells.

2.4 HAIR CELL LIGAND-GATED CHANNELS

As stated above, various voltage-gated channels are able to set the membrane potential of the hair cells as well as filter the receptor potential, thereby interposing themselves between the transduced stimulus and transmitter exocytosis. Similarly, there is evidence suggesting that there are a number of ligand-gated channels on the hair cells which, when occupied by their appropriate agonists, can also act to modify the hair cell's conductance and, consequently, the transfer function between the stimulus and the release of transmitter.

2.4.1 ACETYLCHOLINE

Acetylcholine is thought to be the principal transmitter released by efferent fibers to the vestibular endorgans (cf. Ref. 46, this volume) and to the cochlea (cf. Ref. 45, this volume). While this hypothesis is well-established, the actions of acetylcholine in the inner ear are not as well understood. Within the cochlea, stimulation of the efferents or direct application of acetylcholine uniformly result in an increase in Ca^{2+} influx into the hair cells followed by activation of a Ca^{2+}-dependent K^+ conductance that increases the resting conductance of the hair cell and hyperpolarizes it (e.g., refs. 6, 7, 41, and 46). In the vestibular endorgans most experiments have involved the isolated frog labyrinth, and the effects of efferent stimulation and acetylcholine agonists are variable.

Efferent stimulation results in either increases or decreases in the activity of afferents in the semicircular canals,[10,92,94,109] but only in decreases in afferent activity in the sacculus.[109] Similarly, application of either nicotinic or muscarinic agonists can result in either increases or decreases in canal afferent activity,[10] but only decreases in activity in the sacculus.[51] These findings imply that the receptor-iontophore complex activated by acetylcholine in the cochlea and sacculus involves an inward Ca^{2+} current followed by an outward K^+ current. Since the same agonist can produce two different actions on the hair cell within the canals, then it is likely that the same cholinergic receptors on the hair cells are coupled to different iontophores (cf. Ref. 16).

2.4.2 GLUTAMATE

While glutamate or a related compound is thought to be the hair cell transmitter, a number of studies also suggest that hair cells possess glutamate receptors. As reported previously, application of glutamate to the isolated labyrinth of the frog results not only in a depolarization of afferents, but also an increase in EPSP frequency of occurrence.[5,14,87] Such data indicate that not only is the afferent being depolarized, but that the hair cell is too, since there is a increase in transmitter release as monitored by the quantitation of the EPSP frequency of occurrence. This effect can be seen in Figures 2.1 and 2.2, where the increase in EPSP frequency precedes, or occurs concomitantly, with the depolarization of the afferent.

It could be argued that with bath-application of glutamate, the ensuing depolarizations of the afferents and their increased spiking could lead to a significant elevation of the extracellular K^+

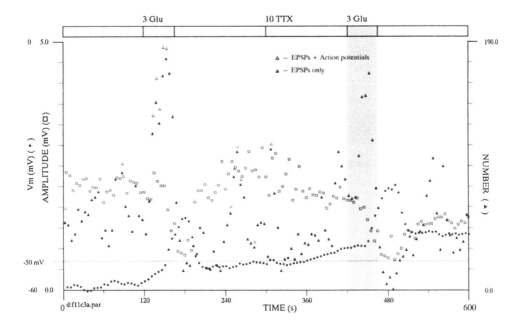

FIGURE 2.2 A comparison of the influence of glutamate on afferent activity in normal saline and in the presence of TTX. Values are plotted as in Figure 2.1 except that each data point represents a 5-second period of measurement. Bath application of 3 mM glutamate resulted in an increase in EPSP and action potential frequency of occurrence, a decrease in EPSP amplitude, and a depolarization of the afferent as in Figure 2.1. Following the introduction of 10 nM TTX (10 TTX) and after action potential blockade, 3 mM glutamate was added to the superfusing solution in the presence of TTX and still resulted in a comparable increase in EPSP frequency of occurrence as it did in normal saline.

concentration. Since a small increase in the extracellular K^+ concentration can greatly augment transmitter release (see Ref. 17 and above), then the increase in EPSP frequency seen with glutamate may be indirectly due to this K^+ release. To test if this condition is the case, glutamate was applied, as in Figure 2.2, in the presence of TTX (to block spiking and ensuing K^+ release from the afferent axons) and compared to glutamate applied in normal saline. As can be seen from Figure 2.2, in both instances the time course and extent of increase in the EPSP frequency of occurrence were relatively the same. This finding, in combination with the fact that the increase in EPSP frequency occurs simultaneously with or slightly precedes the depolarization of the afferent, suggests that glutamate is acting directly on the hair cell to depolarize it. Somewhat oddly, considering the number of investigators patch-clamping isolated hair cells, there have been no reports of the actions of glutamate on the electrical properties of the Type II hair cells. However, the study by Starr and Sewell[107] in which they utilized intracellular recording to analyze goldfish saccular afferents provides reasonable support for the hypothesis that hair cells possess glutamate receptors. These investigators showed that not only was quantal size reduced by bath application of glutamate antagonists, but quantal content was also depressed, indicating a presynaptic action of the antagonists in addition to the predicted postsynaptic action of these drugs.

There is some suggestion that vestibular afferents establish reciprocal synapses with the hair cells. Dunn[34] described an apparent reciprocal contact between Type II hair cells and bouton afferents in the bullfrog. Such feedback has not been detected electrophysiologically to date (cf. Ref. 14). Scarfone and colleagues[96,97] described vesiculated membranes located in the apex of the calyces around Type I hair cells in cats and mice. They also reported that antibodies to synapsin I and synaptophysin (proteins that can be associated with presynaptic nerve terminals; cf. Ref. 23) labeled the calyces and nerve fibers, thereby implying that the calyx could act as a nerve terminal. A clearly defined synaptic contact from the calyx to the Type I hair cell was not described, however. In support of their hypothesis of calyceal feedback, it was found that glutamate, when iontophoresed onto isolated Type I hair cells, produced an increase in intracellular Ca^{2+} as detected by the fluorescent dye (fura-2/AM) with which the cells were loaded.[29] Their evidence suggested that Type I hair cells possess AMPA, NMDA, and metabotropic receptors. They indicate that these experiments support the hypothesis that afferent activity results in glutamate release from the calyx and acts on the Type I hair cell to increase intracellular $Ca.^{2+}$

It should be noted that, because the Type I hair cell is geometrically constrained by the calyx, the transmitter released from the Type I hair cell has to somehow be inactivated, either by uptake or degradation. It is feasible that one or more of the recently described glutamate transporters could be associated with either the hair cell or the calyx, but this remains to be determined. Nonetheless, if the Type I hair cell transmitter is glutamate, then the presence of glutamate receptors on the hair cell would imply a strong positive feedback loop between the hair cell and itself, assuming that the released glutamate diffuses to the site of these receptors.

2.4.3 GABA

Lapeyre and coworkers[64] reported that GABA was able to modify Ca^{2+} and K^+ conductances of isolated, patch-clamped canal Type I hair cells from guinea pigs. The actions of GABA were mimicked by baclofen and were not antagonized by picrotoxin, suggesting that the effects were mediated through $GABA_B$ receptors. Since it was reported that the calyces were labeled with an antibody to conjugated GABA,[30] the calyx could also feed back onto the Type I hair cell by releasing GABA. However, Foster and colleagues[40] have reported that antibodies to fragments of the $GABA_A$ receptor label regions of the calyx, so the role of GABA is rather unclear as yet.

2.5. CONCLUSION

A lot happens between the deflection of hair cell stereocilia and the generation of action potentials by vestibular afferents. A general perspective of this transmission process is difficult to gain because of the extraordinary complexity of the rather small piece of tissue that comprises the endorgan epithelium, as is demonstrated in the electron micrograph of a multicalyx from the turtle horizontal crista ampullaris illustrated in Figure 2.3.

Concerted efforts by a large number of investigators from a wide variety of disciplines have focused on this tissue and an understanding of its operations is beginning to ensue. A number of mysteries remain.

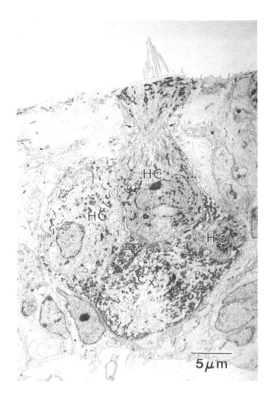

FIGURE 2.3 A multicalyx in the horizontal crista ampullaris of the turtle, demonstrating the complexity of the microorganization of the inner ear neuropil. This electronmicrograph is of an 80-nm section. Embedded in the calyx (Cx) are three Type I hair cells (HC). The fine tips of the calyx (arrows) can be seen to extend to the neck of the central hair cell.

Stereocilial deflection opens transduction channels in one direction and closes them in another. Cations flowing into the hair cell through these channels act to depolarize it. This depolarization further opens additional voltage-dependent Ca^{2+} channels, resulting in increased transmitter efflux. This transmitter, thought to be a glutamate-like compound, probably binds with a kainate-type subsynaptic receptor, resulting in the opening of a channel permeable to Na^+ and K^+ that acts to short-circuit the postsynaptic membrane and depolarize the afferent. This depolarization then acts to trigger voltage-dependent Na^+ channels in the afferent and action potentials ensue.

Many of the details of this process are obscure. It is not clear how the various voltage-dependent conductances on the hair cells' membranes influence the resting potentials of the hair cells, the receptor potentials, and the transmission process itself. Part of this lack of understanding is due to the variability of the hair cells between endorgans, between animals, and within a given endorgan itself, and by the inability to be able to investigate these phenomena *in situ*. Although the transmitter is different, much of the release process resembles that of the motor nerve terminal and other nerve terminals. The absolute identity of the hair cell transmitter is unknown because of the ubiquity in which cells utilize glutamate in their operations, and by the fact that a number of endogenous glutamate analogs mimic its transmitter actions. The specific identity of the subsynaptic receptor-iontophore complex is also unclear. This stems in part from the fact that recently developed specific kainate receptor antagonists have yet to be tested at the hair cell synapse. In addition, the currently available glutamate receptor antibodies are not directed at the binding site of the transmitter, but rather at portions of subunits of the whole receptor-iontophore complex, some of which may cross over to other such complexes. There are further areas of ignorance when it comes to the role of other compounds in transmission such as GABA. It is quite likely that GABA is in some way associated with hair cell transmission, but how is not known. Immunocytochemical studies are conflicting between investigators and between species. Electrophysiological studies indicate that GABA (at least in some species) does not act to depolarize the afferent directly, and its actual role in the operations of the neuropil remains unclear. Whether there is feedback from the afferent (regardless of type) to the hair cell or feedback from hair cells back to hair cells is still to be determined. Finally, the functional significance, if any, of the peculiar geometric arrangement of the calyx surrounding the Type I hair cell remains unclear.

Examination of the ultrastructure of the tissue (as is demonstrated in Figure 2.3) underscores the inadequate simplicities of the current approaches to understanding the chemical aspects of cellular interactions within this tissue. Hopefully, in the future, the complexities of this small, highly organized neuropil will be met with further complex and refined experimentation that will shed light on a number of these rather simple questions.

REFERENCES

1. Adams, DJ, Dwye,r TM, and Hille, B, The permeability of endplate channels to monovalent and divalent metal cations. *J. Gen. Physiol,.* 75:493,1980.
2. Adams, JC, Mroz, EA, and Sewell, WF, A possible neurotransmitter role for CGRP in a hair-cell sensory organ. *Brain Res.,* 419:347,1987.
3. Akoev, G, Andrianov, GN, Bromm, B, and Szabo, T, Effects of excitatory amino acid antagonists on synaptic transmission in the ampullae of Lorenzini of the skate *Raja clavata. J. Comp. Physiol.,* 168:647,1991.
4. Akoev, G, Andrianov, GN, Szabo, T, and Bromm, B, Neuropharmacological analysis of synaptic transmission in the Lorenzinian ampulla of the skate *Raja clavata. J. Comp. Physiol. A,* 168:639,1991.
5. Annoni, J-M, Cochran, SL, and Precht, W, Pharmacology of the vestibular hair cell-afferent fiber synapse in the frog. *J. Neurosci.* 4:2106,1984.
6. Art, JJ, and Fettiplace, R, Efferent desensitization of auditory nerve fibre responses in the cochlea of the turtle *Pseudemys scripta elegans. J. Physiol.,* 356: 507,1984.
7. Art, JJ, Fettiplace, R, and Fuchs, PA, Synaptic hyperpolarization and inhibition of turtle cochlear hair cells. *J. Physiol.,* 356: 525,1984.
8. Baird, RA, Desmadryl, G, Fernández, C, and Goldberg, JM, The vestibular nerve of the chinchilla.II. Relation between afferent response properties and peripheral innervation patterns in the semicircular canals. *J. Neurophysiol.,* 60:182,1988.
9. Baird, RA, Schuff, and NR, Peripheral innervation patterns of vestibular nerve afferents in the bullfrog utriculus. J. Comp. Neurol.,342:279,1994.
10. Bernard, C, Cochran, SL, and Precht, W, Presynaptic actions of cholinergic agents upon the hair cell-afferent fiber synapse in the vestibular labyrinth of the frog. *Brain Res.* 338:225,1985.

11. Brichta, AM, and Peterson, EH, Functional architecture of vestibular primary afferents from the posterior semicircular canal of a turtle, *Pseudemys (Trachemys) scripta elegans. J. Comp. Neurol.,* 344:481,1994.

12. Chang, JSY, Popper, AN, and Saidel, WM, Heterogeneity of sensory hair cells in a fish ear. *J. Comp. Neurol.,*324:621,1992.

13. Cochran, SL, CNQX (6-cyano-7-nitroquinoxaline-2,3-dione) blocks excitatory synaptic transmission in the vestibular periphery and vestibular nucleus of the frog. *Soc. Neurosci. Abstr.* 16:502,1989.

14. Cochran, SL, Peripheral Vestibular Transmission. *Sensing and controlling motion* (Cohen B, Tomko DI, and Guedry, F, Eds.) Ann. N. Y. Acad. Sci. 656:580,1992.

15. Cochran, SL, Algorithms for detection and measurement of spontaneous events. *J. Neurosci. Meth.,* 50:105,1993.

16. Cochran, SL, Evidence against a hypothesis of vestibular efferent function. *Brain Res.* 642:344,1994.

17. Cochran, SL., Cationic influences upon synaptic transmission at the hair cell-afferent fiber synapse of the frog. *Neurosci.,*1995.

18. Cochran, SL, and Correia, MJ, Functional support of glutamate as a vestibular hair cell transmitter in an amniote. *Brain Res.,* 670:321,1995.

19. Cooke, JD, Okamoto, K, and Quastel, DMJ, The role of calcium in depolarization-secretion coupling at the motor nerve terminal. *J. Physiol.* 228:459,1974.

20. Cooke, JD, and Quastel, DMJ, The specific effects of potassium on transmitter release by motor nerve terminals and its inhibition by calcium. *J. Physiol.* 228:435,1973.

21. Corey, DP, and Hudspeth, AJ, Ionic basis of the receptor potential in a vertebrate hair cell. Nature, 281:675,1979.

22. Correia, MJ, Filtering properties of hair cells. *Sensing and controlling motion* (Cohen, B, Tomko, DI, Guedry, F, Eds.) *Ann. N. Y. Acad. Sci.,* 656:49,1992.

23. DeCamilli, P, Benfenat, F, Valtorta, F, and Greengard, P, The synapsins. *Annu. Rev. Cell Biol.,* 6:433,1990.

24. Dechesne, CJ, Hampson, DR, Goping, G, and Wheaton, KD, Wenthold, RJ, Identification and localization of a kainate binding protein in the frog inner ear by electron microscopy immunocytochemistry. *Brain Res.,* 545:223,1991.

25. Dechesne, C, Raymond, J, and Sans, A, Action of glutamate in the cat labyrinth. *Ann. Oto. Rhino. Laryngol.,* 93:163,1984.

26. Demêmes, D, Lleixa, A, and Dechesne, CJ, Cellular and subcellular localization of AMPA-selective glutamate receptors in the mammalian peripheral vestibular system. *Brain Res.,* 671:83,1995.

27. Demêmes, D, Wenthold, RJ, Moniot, B, and Sans, A, Glutamate-like immunoreactivity in the peripheral vestibular system of mammals. *Hearing Res.* 46:261,1990.

28. Denk, W, Holt, JR, Shepherd, GMG, and Corey, DP, Calcium Imaging of single stereocilia in hair cells: localization of transduction channels at both ends of tip links. *Neuron,* 15:1311,1995.

29. Devau G, Lehouelleur J, and Sans A Glutamate receptors on type I vestibular hair cells of guinea-pig. Europ. *J. Neurosci.,* 5:1210,1993.

30. Didier, A, Dupont, J, and Cazals, Y, GABA immunoreactivity of calyceal nerve endings in the vestibular system of the guinea pig. *Cell and Tiss. Ress.,* 260:415,1990.

31. Drescher, MJ, and Drescher, DG, N-Acetylhistidine, glutamate, and b-alanine are concentrated in a receptor cell layer of the trout inner ear. *J. Neurochem.,* 56:658,1991.

32. Drescher, MJ, and Drescher, DG, Glutamate, of the endogenous primary a-amino acids, is specifically released from hair cells by elevated extracellular potassium. *J. Neurochem.,* 59:93,1992.

33. Drescher, MJ, Drescher, DG, and Hatfield, JS, Potassium-evoked release of endogenous primary amine-containing compounds from the trout saccular macula and saccular nerve in vitro. *BrainRes.,* 417:39,1987.

34. Dunn, RF, Reciprocal synapses between hair cells and first order afferent dendrites in the crista ampullaris of the bullfrog. *J. Comp. Neurol.,*193:255,1980.

35. Eybalin, M, Neurotransmitters and neuromodulators in the mammalian cochlea. *Physiol.Rev.,* 73:309,1993.

36. Felix, D, and Ehrenberger, K, The action of putative neurotransmitter substances in the cat labyrinth. *Acta Otolaryngol.* 93:101,1982.

37. Fernández, C, Baird, RA, and Goldberg, JM, The vestibular nerve of the chinchilla. I. Peripheral inner- vation patterns in the horizontal and superior semicircular canals. *J. Neurophysiol.*,60:167,1988.

38. Fernández C, Lysakowski A, and Goldberg JM Hair-cell counts and afferent innervationpatterns in the cristae ampullares of the squirrel monkey with a comparison to the chinchilla. *J.Neurophysiol.* 73:1253,1995.

39. Flock, Å, and Lam, DMK, Neurotransmitter synthesis in inner ear and lateral line sense organs. *Nature,* 249:142,1974.

40. Foster, JD, Drescher, MJ, and Drescher, DG, Immunohistochemical localization of $GABA_A$ receptors in the mammalian crista ampullaris. *Hearing Res.,* 83:203,1995.

41. Fuchs, PA, Evans, MG, and Murrow, BW, Calcium currents in hair cells isolated from the cochleaof the chick. *J Physiol.* 394: 429:553,1990.

42. Furukawa, T, and Ishii, Y, Neurophysiological studies on hearing in goldfish.. *J. Neurophysiol.,* 30:1377,1967.

43. Glavinovic, MI, Changes in miniature end-plate currents due to high potassium and calcium at the frog neuromuscular junction. *Synapse,* 2:636,1988.

44. Gleisner, L, Flock, Å, and Wersäll, J, The ultrastructure of the afferent synapse on hair cells in the frog labyrinth. *Acta Otolaryngol.* 76:199,1973.

45. Godfrey, DA, Hongyan, Ross, CD, and Rubin, AM, Auditory influences on the vestibular system. (Chapter 16, this volume),1999.

46. Goldberg, JM, Brichta, A, and Wackym, P, Neurochemistry of Vestibular Efferents (Chapter 5, this volume),1999.

47. Goldberg, JM, Lysakowski, A, and Fernández, C, Morphophysiological and ultrastructural studiesin the mammalian cristae ampullares. *Hearing Res.,* 49:89,1990.

48. Griguer, C, Chabbert, C, and Sans, A, Lehouelleur, Transient increase in cytosolic free calciumevoked by repolarization in type I vestibular hair cells of rats. *Cell Calcium,* 17:327,1995.

49. Griguer, C, Lehouelleur, J, Valat, J, Sahuquet, A, and Sans, A, Voltage dependent reversible movements of the apex in isolated guinea pig vestibular hair cells. *Hearing Res.,* 67:110,1993.

50. Guth, PS, Aubert, A, Ricci, AJ, and Norris, CH, Differential modulation of spontaneous and evokedneu- rotransmitter release from hair cells: some novel hypotheses. *Hearing Res.,* 56:69,1991.

51. Guth, PS, Dunn, A, Kronomer, K, and Norris, CH, The cholinergic pharmacology of the frog saccule. *Hearing Res.,* 75:225.1994.

52. Guth, PS, Fermin, CD, Pantoja, M, Edwards, R, and Norris, C, Hair cells of different shapes and their placement along the frog crista ampullaris. *Hearing Res.,* 73:109,1994.

53. Guth, PS, Norris, CH, and Barron, SE, Three tests of the hypothesis that glutamate is the sensoryhair- cell transmitter in the frog semicircular canal. *Hearing Res.,* 33:223,1988.

54. Guth, SL, and Norris, CH, Pharmacology of the isolated semicircular canal: effect of GABA and picro- toxin., *Exp. Brain Res.,* 56:72,1984.

55. Harper, A, Blythe, WR, Grossman, G, Petrusz, P, Prazma, J, and Pillsbury, HC, Immunocytochemical localization of aspartate and glutamate in the peripheral vestibular system. *Hearing Res.,* 86:171,1995.

56. Highstein, SM, Baker, R, Action of the efferent vestibular system on primary afferents in thetoadfish, *Opsanus tau. J. Neurophysiol.,* 54:370,1985.

57. Holt, JR, and Eatock, RA, Inwardly rectifying currents of saccular hair cells from the leopard frog.*J. Neurophysiol.,* 73:1484,1995.

58. Howard, J, Roberts, WM, Hudspeth, AJ, Mechanoelectric transduction by hair cells. *Ann. Rev. Biophys. Biophys. Chem.* 17:99,1988.

59. Hudspeth, AJ, Extracellular current flow and the site of transduction by vertebrate hair cells.*J. Neuro- sci.,* 2:1,1982.

60. Hudspeth, AJ, How the ear's works works. *Nature* 341:397,1989.

61. Ishii, Y, Matsuura, S, Furukawa, T, Quantal nature of transmission at the synapse betweenhair cells and eighth nerve fibers. *Jap. J. Physiol.,* 21:79,1971.

62. Jaramillo, F, Signal transduction in hair cells and its regulation by calcium. *Neuron,* 15:1227,1995.

63. Katz, B, Nerve, Muscle, and Synapse. McGraw-Hill. New York,1966.

64. Lapeyre, PNM, Kolston, PJ, and Ashmore, JF, $GABA_B$-mediated modulation of ionic conductances in type I hair cells isolated from guinea-pig semicircular canals. *Brain Res.,* 609:269,1993.

65. Lewis, CA, Ion-concentration dependence of the reversal potential and the single channel conductance of ion channels at the frog neuromuscular junction. *J. Physiol.*, 286:417,1979.

66. López, I, Juiz, JM, Altschuler, RA, and Meza, G, Distribution of GABA-like immunoreactivity inguinea pig vestibular cristae ampullaris. *Brain Res.*, 530:170,1990.

67. López, I, and Meza, G, Neurochemical evidence for afferent gabaergic and efferent cholinergicneurotransmission in the frog vestibule. *Neurosci.*, 25:13,1988.

68. López, I, and Meza, G, Comparative studies on glutamate decarboxylase and cholineacetyltransferase activities in the vertebrate vestibule. *Comp. Biochem. Physiol.*, 95B:375,1990.

69. López, I, Wu, J-Y, and Meza, G, Immunocytochemical evidence for an afferent GABAergicneurotransmission in the guinea pig vestibular system. *Brain Res.*, 589:341,1992.

70. Lysakowski, A, Minor, LB, Fernández, C, and Goldberg, JM, Physiological Identification of morphologically distinct afferent classes innervating the cristae ampullares of the squirrel monkey.*J. Neurophysiol.*, 73:1270,1995.

71. Magelby, KL, and Weinstock, MM, Nickel and calcium ions modify the characteristics of theacetylcholine receptor-channel complex at the frog neuromuscular junction. *J. Physiol.* 299:203,1980.

72. Masetto, S, Russo, G, and Prigioni, I, Differential expression of potassium currents by hair cells inthin slices of frog crista ampullaris. *J. Neurophysiol.*, 72:443,1994.

73. Matsubara, A, Usami, S-I, Fujita, S, and Shinkawa, H, Expression of Substance P, CGRP, and GABA in the vestibular periphery, with special reference to species differences. Acta Otolaryngol., (Suppl.) 519:248,1995.

74. Meza, G, Carabez, A, and Ruiz, M, GABA synthesis in isolated vestibulary tissue of chick inner ear. *Brain Res.*, 241:157,1982.

75. Meza, G, López, I, Paredes, MA, Peñaloza, Y, and Poblano, A, Cellular target of streptomycin inthe internal ear. *Acta Otolaryngol.*, 107:406,1989.

76. Nagai, T, Obara, S, and Kawai, N, Differential blocking effects of a spider toxin on synaptic and glutamate responses in the afferent synapse of the acoustic-lateralis receptors of *Plotosus.Brain Res.*, 300:183,1984.

77. Nakanishi, S, Molecular diversity of glutamate receptors and implications for brain function. Science., 258:597,1992.

78. Niedzielski, AS, and Wenthold, RJ, Expression of AMPA, kainate, and NMDA receptor subunits in cochlear and vestibular ganglia. *J. Neurosci.*, 15:2338,1995.

79. Ohno, K, Takeda, N, Tanaka-Tsuji, M, and Matsunaga, T, Calcitonin gene-related peptide in theefferent system of the inner ear. A review. *Acta Otolaryngol.*, 501:16,1993.

80. Ohno, K, Takeda, N, Yamano, M, Matsunaga, T, and Tohyama, M, Coexistence of acetylcholineand calcitonin gene-related peptide in the vestibular efferent neurons in the rat. *Brain Res.*, 566:103,1991.

81. Panzanelli, P, Valli, P, Cantino, D, and Fasolo, A, Glutamate and carnosine in the vestibularsystem of the frog. *Brain Res.*, 662: 293,1994.

82. Pérez, M E, Soto, E, and Vega, R, Streptomycin blocks the postsynaptic effects of excitatoryaminoacids on the vestibular system primary afferents. *Brain Res.*, 563:221,1991.

83. Perkins, MN, and Stone, TW, An iontophoresis investigation of the actions of convulsant kynurenines and the interaction with the endogenous excitant quinolinic acid. *Brain Res.*, 247:184.

84. Popper, AN, Saidel, WM, and Chang, JSY, Two types of sensory hair cell in the saccule of a teleostfish. *Hearing Res.*, 64:211,1993.

85. Prigioni, I, Masetto, S, Russo, G, and Taglietti, V, Calcium currents in solitary hair cells isolatedfrom frog crista ampullaris. *J. Vestib.*, Res. 2:31,1992.

86. Prigioni, I, Russo, G, and Masetto, S, Non-NMDA receptors mediate glutamate-induced depolarization in frog crista ampullaris. *Neuroreport,* 5:516,1994.

87. Prigioni, I, Russo, G, Valli, P, and Masetto, S, Pre- and postsynaptic excitatory action of glutamate agonists on frog vestibular receptors. *Hearing Res.*, 46:253,1990.

88. Ricci, AJ, Cochran, SL, Rennie, KJ, and Correia, MJ, (Vestibular type I and type II hair cells. Morphometric comparisons of dissociated pigeon hair cells. *J. Vestib. Res.*, 7:407, 1997.

89. Roberts BL, Maslam S, Los i, Van Der Jagt B Coexistence of calcitonin gene-related peptide and choline acetyltransferase in EEL efferent neurons. *Hearing Res.*, 74:231,1994.

90. Roberts, WM, Howard, J, and Hudspeth, AJ, Hair cells: transduction, tuning, and transmission in the inner ear. *Annu. Rev. Cell Biol.*, 4:63-92, 1988.

91. Roberts, WM, Jacobs, RA, and Hudspeth, AJ, Colocalization of ion channels involved in frequency selectivity and synaptic transmission at presynaptic active zones of hair cells. *J. Neurosci.*, 10:3664,1990.

92. Rossi, ML, Martini, M, Efferent control of posterior canal afferent receptor discharge in the frog labyrinth. *Brain Res.*, 555:123,1991.

93. Rossi, ML, Martini, M, Peluchi, B, and Fesce, R, Quantal nature of synaptic transmission at the cytoneural junction in the frog labyrinth. *J. Physiol.*, 478:17,1994.

94. Rossi, ML, Prigioni, I, Valli, P, and Casella, C, Activation of the efferent system in the isolated frog labyrinth: effects on the afferent EPSPs and spike discharge recorded from single fibers of the posterior nerve. *Brain Res.*, 180: 125,1980.

95. Rossi, ML, Valli, P, and Casella, C, Post-synaptic potentials recorded from afferent nerve fibres of the posterior semicircular canal in the frog. *Brain Res.*, 135:67,1977.

96. Scarfone, E, Demêmes, D, Jahn R, De Camilli P, and Sans, A , Secretory function of the vestibular nerve calyx suggested by presence of vesicles, synapsin I, and synaptophysin. *J. Neurosci.*, 10:3664,1988.

97. Scarfone, E, Demêmes, D, and Sans, A, Synapsin I and synaptophysin expression during ontogenesis of the mouse peripheral vestibular system. *J. Neurosci.*, 11:1173,1991.

98. Schessel, DA, Ginzberg, R, and Highstein, SM, Morphophysiology of synaptic transmission between type I hair cells and vestibular primary afferents. An intracellular study employing horseradish peroxidase in the lizard, *Calotes versicolor. Brain Res.*, 544:1,1991.

99. Sewell, WF, and Mroz, EA, Neuroactive substances in inner ear extracts. *J. Neurosci.*, 7:246,1987.

100. Sewell, WF, and Mroz, E, Purification of a low-molecular-weight excitatory substance from the inner ears of goldfish. *Hearing Res.*, 50:127, 1990.

101. Sewell, WF, Starr, and PA, Effects of calcitonin gene-related peptide and efferent nerve stimulation on afferent transmission in the lateral line organ. *J. Neurophysiol.*, 65:1158,1991.

102. Shotwell, SL, Jacobs, R, and Hudspeth, AJ, Directional sensitivity of individual vertebrate hair cells to a controlled deflection of hair bundles. *Ann. N.Y. Acad.*, Sci. 374:1,1981.

103. Sommer, B, and Seeburg, PH, Glutamate receptor channels: novel properties and new clones. *Trends Pharmacol Sci.*, 13:291,1992.

104. Soto, E, Flores, A, and Vega, R, NMDA-mediate potentiation of the afferent synapse in the inner ear. *Neurorep.*, 5:1963,1994.

105. Soto, E, Pérez, ME, and Vega, R Streptomycin blocks quisqualate and kainate effects on the vestibular system primary afferents. *Excitatory Amino Acids* (Meldrum BS, Ed.), pp 293-296. New York: Raven Press,1991.

106. Soto, E, and Vega, R, Actions of excitatory amino acid agonists and antagonists on the primary afferents of the vestibular system of the axolotl (*Ambystoma mexicanum*). *Brain Res.*, 462:104,1988.

107. Starr, PA, and Sewell, WF, Neurotransmitter release from hair cells and its blockade by glutamate-receptor antagonists. *Hearing Res.*, 52:23,1991.

108. Su, Z-L, Jiang, S-C, Gu, R, and Yang, W-P, Two types of calcium channels in bullfrog saccular hair cells. *Hearing Res.*, 87:62,1995.

109. Sugai, T, Sugitani, M, and Ooyama, H, Effects of activation of the divergent efferent fibers on the spontaneous activity of vestibular afferent fibers in the toad. *Jap. J. Physiol.*, 41:217,1991.

110. Takeuchi, A, and Takeuchi, N, Changes i,n the potassium concentration around motor nerve terminals, produced by current flow, and their effects upon neuromuscular transmission. *J. Physiol.*, 155:46,1961.

111. Teeter, JH, and Bennett, MVL, Synaptic transmission in the ampullary electroreceptor of the transparent cat-fish, *Kryptoperus. J. Comp. Physiol. A*, 142:371,1981.

112. Tsien, RW, and Tsien, RY Calcium channels stores and oscillations. *Ann. Rev. Cell Biol.*, 6:715,1990.

113. Tu, Y, and Budelmann, BU, The effect of l-glutamate on the afferent resting activity in the cephalopod statocyst. *Brain Res.*, 642:47,1994.

114. Usami, S, Hozawa, J, Tazawa, M, Igarashi, M, Thompson, GC, Wu, J-Y, and Wenthold, RJ, Immunocytochemical study of the GABA system in chicken vestibular endorgans and the vestibular ganglion. *Brain Res.*, 503:214,1989.

115. Usami, S, Igarashi, M, and Thompson, GC, GABA-like immunoreactivity in the squirrel monkey vestibular endorgans. *Brain Res.*, 417:367,1987.

116. Usami, S, Igarashi, M, and Thompson, GC, GABA-like immunoreactivity in the chick vestibular end organs. *Brain Res.,* 418:383,1987.

117. Usami, S, Lsen, KK, Nhang, N, and Ottersen, OP, Distribution of glutamate-like and glutamine-like immunoreactivities in the rat organ of Corti: a light microscopic and semiquantitative electron microscopic analysis with a note on the localization of aspartate. *Exp. Brain Res.,1992.* 91:1-11.

118. Usami, S, Matsubara, A, Fujita, S, Shinkawa, H, and Hayashi, M, NMDA (NMDAR!) and AMPA-type (GluR2/3) receptor subunits are expressed in the inner ear. *Neurorep.,* 6:1161,1995.

119. Usami, S, and Ottersen, OP, Differential cellular distribution of glutamate and glutamine in the rat vestibular endorgans: an immunocytochemical study. *Brain Res.,* 676:285,1995.

120. Vega, R, Soto, E, Budelli, R, González-Estrada, MT, Is GABA an afferent transmitter in the vestibular system? *Hearing Res.,* 29:163,1987.

121. Valli, P, Zucca, G, Botta, L, and Casella, C, Sensitivity and adaptation of ampullary receptors torapid perilymphatic potassium changes. *J. Comp. Physiol.,* A:162:173,1988.

122. Valli, P, Zucca, G, Prigioni, I, Botta, L, Casella, C, and Guth, PS, The effect of glutamate on the frog semicircular canal. *Brain Res.,* 330:1,1985.

123. Wackym, PA, Poppe,r P, Micevych, PE, Distribution of calcitonin gene-related peptide mRNA and immunoreactivity in the rat central and peripheral vestibular system. *Acta Otolaryngol.,* 113:601,1993.

124. Watkins, JC, and Evans, RH, Excitatory amino acid transmitters. *Annu. Rev. Pharmacol. Toxicol.,* 21:165,1981.

125. Wenthold, RJ, Hampson, DR, Wada, K, Hunter, C, Oberdorfer, MD, and Dechesne, CJ, Isolation,localization, and cloning of a kainic acid binding protein from frog brain. *J. Histochem. Cytochem.,* 12:1717,1990.

126. Wersäll, J, Studies on the structure and innervation of the sensory epithelium of the cristae ampullares in the guinea pig. *Acta Otolayrngol.,* Suppl. 126:1,1956.

127. Wersäll, J, Flock, Å, and Lundquist, P-G, Structural basis for directional sensitivity in cochlearand vestibular sensory receptors. *Cold Spring Harbor Symp.,* 30:115,1965.

128. Wersäll, J, Gleisner, L, and Lundquist, P-G, Ultrastructure of the vestibular end organs. *CIBA Foundation Symposium on Myotatic, Kinesthetic, and Vestibular Mechanisms.* (DeReuck AVS, Knight J, Eds.) pp. 105-120. Boston: Little, Brown, and Co.,1967.

129. Wersäll, J, and Lundquist, P-G, Morphological polarization of the mechanoreceptors of the vestibular and acoustic systems. In Second symposium on the role of the vestibular organs in space exploration. NASA,1966.

130. Yamashita, M, and Ohmori, H, Synaptic responses to mechanical stimulation in calyceal andbouton type vestibular afferents studied in an isolated preparation of semicircular canal ampullaeofchicken. *Exp. Brain Res.,* 80:475,1990.

131. Zidanic, M, and Fuchs, PA, Kinetic analysis of barium currents in chick cochlear hair cells. *Biophys J.,* 68:1323,1995.

132. Zucca, G, Akoev, GN, Maracci, A, and Valli, P, NMDA receptors in frog semicircular canals. *Neuroreport,* 4:403,1993.

133. Zucca, G, Botta, L, Milesi, V, Dagani, F, and Valli, P, Evidence for L-glutamate release in frog vestibular organs. *Hearing Res.,* 63:52, 1992.

134. Zucca, G, Vega, R, Botta, L, Pérez ME, Valli P, and Soto E, Streptomycin blocks the afferent synapse of the isolated semicircular canals of the frog. *Hearing Res.,* 59:70,1992.

3

Synaptic Transmission by Vestibular Nerve Afferent Fibers

Hans Straka, Ingrid Reichenberger, and Norbert Dieringer

CONTENTS

3.1 SYNOPSIS

Vestibular and auditory nerve fibers differ despite their common phylogenetic origin in terms of developmental requirements and anatomical, physiological, and biochemical properties. Part of the vestibular nerve afferent signals are mediated electrically via gap junctions. Mixed synapses exhibit an electrical component of the transmission that appears later in ontogeny than the chemical component. The putative transmitter of the chemical transmission is glutamate or a related substance. Compelling evidence for this statement originates from results obtained in a number of species with several methods. The possibility for co-release of other transmitter-like substances, in particular of glycine by thicker vestibular nerve afferent terminals, is discussed. Second-order vestibular neurons express a number of glutamate receptor subtypes in their membranes. Thinner vestibular nerve afferent fibers appear to activate predominantly AMPA/kainate receptors, whereas thicker vestibular afferents activate AMPA/kainate as well as NMDA receptors. The possible functional implications of the activation of NMDA receptors are discussed.

0-8493-7679-3/00/$0.00+$.50
© 2000 by CRC Press LLC

3.2 INTRODUCTION

First order vestibular afferents, often erroneously called "primary," are secondary sensory neurons that differ from auditory afferents in several aspects, in spite of their common phylogenetic origin from lateral line neuromasts. During development a brain-derived neurotrophic factor appears to be necessary for establishing or maintaining the innervation of vestibular, but not cochlear (inner and outer) hair cells by afferent fibers.[21] In mutant mice lacking this factor, only 18% of the vestibular ganglion cells are still alive at postnatal day 15. The vestibular afferents that remain in these knockout mice fail to contact hair cells in otherwise normal sensory epithelia. As a consequence, mutant mice show defective coordination of movements and balance, but respond to auditory stimuli.

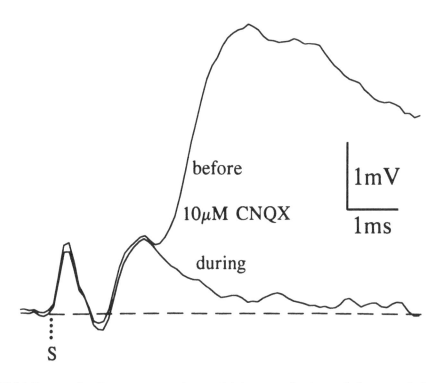

FIGURE 3.1 Compound excitatory postsynaptic potentials in a second-order vestibular neuron in frog evoked by a single electrical stimulus of the ipsilateral N. VIII. The stimulus (a) is followed by an electrically- (b) and a chemically mediated (c) excitatory potential. The chemically-mediated EPSP (control) is reduced in amplitude in the presence of the NMDA receptor antagonist 7-chloro kynurenic acid (7-Cl KYNA, 10mM). The remaining EPSP component is sensitive to the AMPA receptor antagonist 6-cyano-7-nitroquinoxaline-2,3-dione (CNQX, 10 mM).

Afferents in the vestibular nerve and their cell bodies in the ganglion of Scarpa are more heterogeneous in size than cochlear nerve afferent fibers or their cell bodies in the spiral ganglion.[1,7] Apart from size, vestibular nerve afferent fibers differ among each other in a number of interrelated morphological, histochemical, and physiological properties. A corresponding diversity of properties is observed in each of the nerve branches innervating a particular vestibular endorgan in species which possess only type II hair cells (e.g., amphibians) or both type I and type II hair cells (e.g., mammals). Consistent, fiber size-related differences in immunoreactivity,[19,55,69] in the regularity of resting discharge[4,5,25,29,45] and in the dynamic response pattern[5,26] were reported for species as diverse as guitarfish and squirrel monkey. Results from chinchilla[26] and frog[6] suggest that these matched morphological, biochemical, and physiological properties of vestibular nerve afferent fibers are related to the regional innervation pattern in the sensory epithelium. Contrasts like thin/thick, regular/irregular, or tonic/phasic are considered to mark extreme end points of a continuum rather than two distinct classes of vestibular nerve afferent fibers.

The central convergence of vestibular nerve afferent fibers with different morphophysiological properties onto second-order vestibular neurons[57] argues against a possible contrast enhancement in central vestibular neurons. Instead, range fractionation of tonic/kinetic[60] or type A/B/C[59] second-order vestibular neurons overlaps to a large extent as in ocular motoneurons and in extraocular muscle fibers.[22] However, thicker vestibular nerve afferents tend to innervate more densely vestibulo-spinal than vestibulo-ocular regions.[28,30]

3.3 ELECTRICAL TRANSMISSION

Gap junctions between vestibular nerve afferents and second-order vestibular neurons were reported for a number of non-mammalian species and for the rat. In particular, "spoon" endings of large vestibular nerve afferent fibers exhibit gap junctions in fish and bird (cf. Ref. 50). Gap junctions were found to contact the cell body and the dendrites of lateral vestibular neurons in toadfish, frog, and rat (cf. Ref. 61).

Consistent with these anatomical results is the presence of short latency electrically transmitted synaptic potentials in a number of non-mammalian species (see Figure 3.1).

Interestingly, in toadfish pure electrical EPSPs were evoked by low, but not by high, electrical stimulus intensities of vestibular nerve afferents.[38] Similarly, in the frog electrical EPSPs had very low thresholds and saturated at low stimulus intensities.[63,66] Since low electrical stimulus intensities preferentially activate thick afferent fibers, this observation suggests that electrical transmission is more a property of thicker than of thinner vestibular nerve afferent fibers.

Electrical transmission from vestibular nerve afferents to second-order vestibular neurons appears to be a common phenomenon among non-mammalian species. As far as mammals are concerned electrical transmission has been detected so far only in rat,[37,73] but not in cat.[31] This is compatible with transmission electron microscopic studies reporting the presence of gap junctions in rat (cf. Ref. 61) and their absence in cat.[43]

3.3.1 MIXED SYNAPSES AND THEIR DEVELOPMENT

Vestibular nerve afferent inputs onto Mauthner cells have been particularly well studied in terms of electrical transmission in fish and amphibian tadpoles. Vestibular nerve afferent fibers terminate as "club endings" on the lateral dendrite of the Mauthner cell.[14,44] These "club endings" have the anatomical structure of mixed synapses, that is, they exhibit characteristics of gap junctions[9,14] and of chemical junctions.[44] Upon electrical stimulation short latency electrical potentials followed by chemically-mediated potentials were recorded in Mauthner cells.[23,24,27]

Contrary to expectation, electrical synapses do not appear before chemical synapses during ontogeny. The development of vestibular nerve afferent synapses was most intensely studied in the vestibular tangential nucleus of the chicken embryo (cf. Ref. 50). The tangential nucleus, an interstitial nucleus of the vestibular nerve, receives a group of distinct axons from the vestibular nerve, the colossal fibers, which form spoon-like endings on the cell bodies of principal cells in a one-to-one relationship. Colossal fibers develop the first identified synapses on principal cells at embryonic stage E7.5. This synaptic contact is formed *en passant* on soma or lateral process of principal cells and consists of morphologically immature chemical synapses. It is not until E15 that the first gap junctions are observed on the spoon-like synaptic surface. These gap junctions are colocalized with chemical synapses and therefore form mixed synapses (cf. Ref. 50). Electrophysiological and pharmacological investigations on the nature of the synaptic transmission at the onset of the formation of mixed synapses suggested that morphologically identified gap junctions are initially non-functioning.[51] The monophasic excitatory postsynaptic potential of principal cells, generated by vestibular nerve stimulation, was chemically mediated. At later stages of development, gap junctions are abundant, but chemical synapses appear infrequently at the spoon endings.[50] As a result, the type and the number of synaptic contacts at spoon endings appear to change during development - from purely chemical synapses at early stages to mixed synapses at later stages.

3.3.2 Functional Aspects of Electrical Transmission

A prevalence of electrical transmission among heterothermic non-mammalian species may be interpreted in the context of temperature sensitivity of electrical versus chemical transmission.[38] At lower body temperatures, the synaptic delay of chemical - but not electrical - transmission is increased and the rise time of chemically mediated EPSPs is prolonged. As a consequence, spike initiation in second-order vestibular neurons is considerably delayed. Therefore, electrically mediated excitation is more reliable at variable body temperatures. Obviously, heterothermic species would benefit more from electrical transmission than homothermic species. Additionally, electrical transmission appears to facilitate a synchronous discharge in vestibular neurons,[8] which in turn tends to reduce the latency for escape and balancing responses.

Apart from electrical coupling, gap junctions could also provide a means of metabolic coupling between vestibular nerve afferent fibers and central vestibular neurons. Thereby, molecules necessary for the survival of synapses could cross from the post- to the pre-synaptic side. This possibility rests on the observation that large axon terminals of frog vestibular nerve afferent fibers with gap junctions survived a nerve transection, while most other terminals exhibited a marked degeneration.[61]

3.4 CHEMICAL TRANSMISSION

A substance has to fulfill a number of criteria before it can be called the transmitter at a specific synapse.[46,70] Primary criteria include the identity of action, the induced release, and the vesicular location of the putative transmitter. The pharmacological antagonism of specific blocking agents, the presence of a transmitter synthesizing enzyme, and the differential distribution of the specific substance form a group of secondary criteria.

The presence of glutamate, aspartate, glycine, and substance P has been shown in vestibular nerve afferent fibers by immunocytochemistry (see below). Each of the substances appears to play a role in the transmission between afferent and central vestibular neurons. However, based on the criteria discussed above, there is insufficient evidence to date to identify these substances as a primary transmitter of vestibular nerve afferent fibers. In particular, there are no reports in the literature regarding the vesicular content of identified vestibular nerve afferent terminals. According to available evidence (see below) glutamate/aspartate can be assumed to be the putative transmitter and glycine is considered to be a possible cotransmitter of vestibular nerve afferent fibers. In the following sections we summarize recent results obtained at the pre- and post-synaptic site that support this statement.

3.4.1 Glutamate is the Putative Transmitter and Glycine a Possible Cotransmitter

3.4.1.1 Evidence from Immunocytochemistry

The presence of glutamate in vestibular nerve afferent fibers and ganglion cells was demonstrated in frog, mouse, rat, and cat (for frog see Figure 3.2).[18,55]

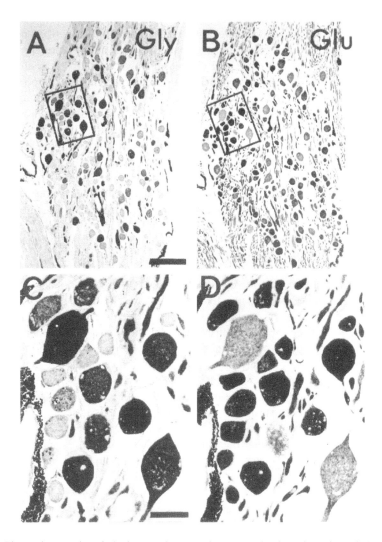

FIGURE 3.2 Photomicrographs of glycine or glutamate immunostained sections through Scarpa's ganglion of the frog. A, B: consecutive sections treated with antibodies against glycine (A) or glutamate (B). C, D: higher magnification of the areas outlined in A and B. Calibration bars in A and C represent 100mm and 20 mm, respectively, and apply also for B and D.

The intensity of the glutamate-immunoreactivity, however, was not uniform and decreased with an increase in diameter of the cell body in frog.[55] On the other hand, even in the largest ganglion cells glutamate was still present at levels above background. In addition, antibodies against mitochondrial or cytosolic aspartate aminotransferase, an enzyme involved in the synthesis of glutamate and aspartate,[54] have been reported to label vestibular ganglion cells in the mouse.

Immunocytochemical investigation of vestibular nerve afferent fibers with anti-glycine antibodies in frog and rat revealed that about 10 to 20% of these vestibular ganglion cells exhibit glycine immunoreactivity (see Figure 3.2).[55] More specifically, the intensity of the glycine-immunoreactivity increased in the frog with the diameter of vestibular ganglion cells. In comparison with the intensity of the glutamate-immunoreaction an inverse, size-related distribution pattern was observed for the glycine-immunoreactivity. A colocalization of glutamate and glycine was found on consecutive semi-thin sections of the largest ganglion cells and vestibular nerve afferent fibers (Figure 3.2). A quantitative analysis of the data suggested that vestibular nerve afferent fibers form a continuum in terms of amino acid immunocytochemistry with thin, glutamate-immunoreactive fibers marking one end and thick, glutamate- and glycine-immunoreactive fibers marking the other end of this continuum. Terminal-like structures, part of which might represent vestibular nerve afferent terminals colocalized glutamate and glycine in frogs, as well[56] as results from ongoing electron microscopic studies of labeled vestibular nerve afferent terminals, support this conclusion (Reichenberger, unpublished data).

Antibodies against substance P labeled about 85% of the vestibular ganglion cells in guinea pig.[69] The population of substance P-labeled neurons was significantly smaller in diameter than the population of substance P-immunonegative ganglion cells. A minority of small ganglion cells have been reported to exhibit substance P-immunoreactivity in rabbit,[75] cat, and monkey.[13] Antibodies against somatostatin, on the other hand, were shown to label large vestibular ganglion cells in rabbit.[72] These data suggest that small vestibular nerve afferent fibers colocalize glutamate and substance P, whereas large vestibular nerve afferent fibers colocalize glutamate, glycine, and somatostatin.

3.4.1.2 Evidence from Uptake Studies

The release of transmitter candidates from vestibular nerve afferent terminals is difficult to study directly. More indirect, suggestive evidence for the release of a substance can be obtained by uptake studies. This argument is based on the observation that terminals which release a given substance often also exhibit a high-affinity uptake mechanism for this substance.[16] Raymond et al.[53] measured the uptake of ³H-glutamate by presynaptic terminals in the vestibular nuclei of normal and eighth nerve transected cats. Eight to ten days after a section of the vestibular nerve was injected with tritiated glutamate, a significant decrease in the ³H-glutamate uptake by presynaptic terminals resulted, presumably due to the degeneration of central vestibular nerve afferent terminals. This result suggests an uptake mechanism for glutamate or a related substance by vestibular nerve afferent terminals, and thereby indirectly supports the notion of a release of this amino acid. Tritiated D-aspartate injected into the vestibular nuclear complex was taken up by, and retrogradely transported to, the cell bodies of ipsilateral vestibular nerve afferent fibers in cat[17] and frog.[64] A quantitative analysis of the data from the latter study showed that D-aspartate was preferentially taken up by smaller vestibular nerve afferents.

Glycine is also taken up by vestibular nerve afferent fibers and transported retrogradely, as evidenced by injections of this substance into the vestibular nuclei of the frog.[64] The intensity of labeling of the cell bodies in Scarpa's ganglion was heterogeneous and increased with the diameter of vestibular ganglion cells. The largest ganglion cells were more intensely labeled than smaller ganglion cells. On consecutive semi-thin sections, the amount of ³H-glycine uptake by individual vestibular nerve afferent fibers was compared with the intensity of the immunoreactivity of these fibers for glutamate. And conversely, the amount of ³H-aspartate uptake by a given fiber was compared with the intensity of its immunoreactivity for glycine. The results from both sets of experiments support each other and corroborate the view of an inverse, size-related content and uptake of both amino acids in vestibular nerve afferents. Thus thin, glutamate-immunoreactive vestibular nerve afferent fibers take up more ³H-aspartate and less ³H-glycine than thick, glutamate- and glycine-immunoreactive vestibular nerve afferent fibers.

The presence of a high-affinity uptake mechanism for a given amino acid is compatible with a possible synaptic release of this substance. However, the predictive value of uptake as an indicator for the possible release of a given substance is weakened by findings such as the absence of an uptake mechanism for GABA by GABAergic Purkinje cell terminals and the presence of an uptake mechanism for GABA by cholinergic spinal motoneurons.[16] On the other hand, electrophysiological and pharmacological studies, summarized in the following sections, strongly support the hypothesis that glutamate or a related substance is the putative transmitter of vestibular nerve afferent fibers. The potential role of glycine in vestibular nerve afferents is less clear, particularly in the absence of data concerning its presence in synaptic vesicles or its release by these terminals. Electrophysiological data preclude a possible role of released glycine as an inhibitory transmitter, since no monosynaptic, strychnine-sensitive inhibitory postsynaptic potentials were detected. A possible role of glycine as a cotransmitter and its potential involvement in the activation of NMDA receptors on second-order vestibular neurons will be discussed in Section 3.4.4. (see also Chapter 6).

3.4.2 AMPA/Kainate and NMDA Receptors in Central Vestibular Neurons

The development of specific antibodies against various AMPA/kainate receptor subunits and the development of probes against their messenger RNAs (GLUR1-GLUR7 and KA1 & 2) allowed the determination of the regional distribution of these subunits within the brainstem (cf. Chapter 5 for additional details). Antibodies against the kainate receptor subunits (KA2 and GLUR6-7) labeled neurons in almost all parts of the vestibular nuclei in rat.[48] Antibodies against the AMPA subunits of the ionotropic glutamate receptors (GLUR1, GLUR2/3, and GLUR4) labeled neurons in all vestibular nuclei in a differential manner for the four subunits in gerbil,[34] chinchilla,[52] and rat.[47] A similar distribution of AMPA receptors in the vestibular nuclei of chinchilla was described by in situ hybridization studies employing cDNA probes against different mRNAs of AMPA receptor subunits.[52] Compatible with these results is the observed high binding density of the AMPA receptor antagonist ^{3}H-CNQX in different vestibular nuclei of rat suggesting the presence of a high density of AMPA receptors.[40]

The NMDA receptor subunit NMDAR1, the major subunit of this receptor, is found in all NMDA receptor complexes. The presence of NMDA receptors in the vestibular nuclei of the rat was demonstrated with antibodies against the NMDAR1 subunit.[49] Neurons in all four major vestibular nuclei were moderately or intensely labeled by the antibodies. A similar labelling pattern was found using radioactive in-situ hybridization histochemistry for the NMDAR1 and NMDA glutamate-binding protein in the rat[58]. Neurons in all vestibular nuclei exhibit a moderate to high level of staining for both types of mRNAs, indicating the presence of a substantial amount of NMDA receptors in the vestibular nuclear complex (cf. Chapter 5). A similar distribution and density of the NMDAR1 subunit were reported in the vestibular nuclei of gerbil. However, in this study, neurons located predominantly in the medial and superior vestibular nuclei were stained.[34] Compatible with these data are the results of an autoradiographic study demonstrating a moderate level of NMDA-sensitive L-[^{3}H]glutamate-binding sites in the medial vestibular nucleus of rat.[42] So far, a specific demonstration of NMDA and/or AMPA/kainate receptor subunits at identified vestibular synapses is missing.

3.4.3 Glutamate-Activated AMPA/Kainate Receptors

Bath application of glutamate, aspartate, kainate or quisqualate evokes pronounced depolarizations in frog and rat central vestibular neurons recorded in vitro.[15,36,39] Part of the agonist-evoked responses in identified frog second-order vestibular neurons could be reversibly blocked by the glutamate-receptor antagonists kynurenic acid or gamma-D-glutamylaminomethylsulfonic acid.[15] Vestibular nerve afferent-evoked monosynaptic excitatory postsynaptic potentials in central vestibular neurons were blocked by the application of various glutamate receptor antagonists, among which kynurenic acid was most effective. These results imply that the transmission is mediated by postsynaptic receptors of the kainate/quisqualate type.[15] This result was further corroborated by data obtained from lamprey whole brain[10] and rat brainstem slices,[20,39] which revealed a kynurenic acid-sensitive component in the monosynaptic response of central vestibular neurons after electrical stimulation of the VIIIth nerve or its central root.

A further subdivision of ionotropic glutamate receptors into NMDA and AMPA/kainate receptors was rendered possible with the availability of new selective antagonists. Bath-application of the specific AMPA/kainate antagonists CNQX or DNQX reduced the intracellularly recorded monosynaptic responses in medial vestibular nucleus neurons of rat[12,20,35,67] and frog[62] reversibly. Although the antagonist concentrations were similar in these studies, the monosynaptic response was either blocked completely[12] or only partially blocked.[20,35,62,67] Chemically mediated, afferent-evoked response components of vestibular neurons in the chicken were completely blocked by the simultaneous application of the AMPA/kainate antagonist CNQX and the NMDA antagonist D-APV,[51] indicating that at least part of the responses were mediated by the AMPA/kainate receptors. For a similar situation in frog, see Figure 3.1. Taken together, these results strongly suggest that vestibular nerve afferent fibers use glutamate or a related substance as a transmitter, and that at least part of the vestibular afferent-evoked monosynaptic response component is mediated by the activation of the AMPA/kainate subtype of ionotropic glutamate receptors.

3.4.4 Glutamate-Activated NMDA Receptors

Bath application of the glutamate agonists L-homocysteate or NMDA evoked pronounced depolarizations in frog and rat central vestibular neurons recorded in vitro.[15,36,39] In frog, the specific NMDA antagonist D-APV could reversibly block NMDA-evoked responses of second-order vestibular neurons.[15] However, conflicting reports exist concerning a monosynaptic activation of NMDA receptors by vestibular nerve afferent fibers. An APV or 7-chloro kynurenic acid-sensitive component of the monosynaptic, afferent-evoked EPSP was reported for central vestibular neurons in frog[15,65] (see also Figure 3.1) and rat.[20,35,67] In these studies, the NMDA receptor-mediated component was either observed in only a few neurons[20] and/or was very small in amplitude.[15] However, in the same species the same antagonists were reported to have no effect on the afferent-evoked monosynaptic EPSP component (frog;[36] rat[12,39]). The conflict between these reports appears to dissolve if requirements for the activation of NMDA receptors in general and for the activation of NMDA receptors by vestibular nerve afferent fibers in particular, are taken into account. Since NMDA receptor activation depends, among other things, on the release from a voltage-dependent Mg^{2+} block, the concentration of Mg^{2+} in the bathing solution of slices or whole brains is of importance. In addition, the relative size of the afferent-evoked NMDA receptor-mediated EPSP component in central vestibular neurons decreases at increasingly higher electrical stimulus intensities to become negligible at high stimulus intensities.[65]

Since thick vestibular nerve afferent fibers are recruited at lower electrical currents than thin vestibular nerve afferent fibers, Straka et al.65 concluded that NMDA receptors are activated predominantly by thick vestibular nerve afferent fibers. However, at the lowest stimulus intensity, only a fraction (up to 60%) of the afferent-evoked chemically mediated EPSP was D-APV sensitive. Therefore, it is likely that thicker vestibular nerve afferent fibers coactivate NMDA as well as non-NMDA receptors, whereas thinner vestibular nerve afferent fibers activate predominantly non-NMDA receptors. The relative contribution of NMDA receptors for the afferent-evoked excitatory response in a given neuron will therefore depend on the convergence pattern of vestibular nerve afferent fibers on this cell. The afferent-evoked NMDA receptor-mediated component in frog had a particularly short delay and a fast rise time when compared with other known NMDA receptor-mediated synaptic potentials. Therefore, it is conceivable that vestibular nerve afferent fibers activate a NMDA receptor subtype with particular fast kinetics.63 Keeping in mind that this component is evoked predominantly by thicker vestibular nerve afferent fibers, which are distinguished by a more phasic response pattern, it seems that the coadapted properties of thicker vestibular afferents (see introduction) also include the activation of particular NMDA receptors.

The activation of NMDA receptors by thick vestibular nerve afferent fibers is facilitated if non-NMDA receptors and/or electrical synapses are coactivated in the vicinity. This additional excitation could originate from collaterals of the same or from converging vestibular nerve afferents and would reduce the voltage-dependent Mg^{2+} block of the NMDA receptor. Another facilitatory role could be played by glycine if it were released from thick vestibular nerve afferents, since the occupation of the glycine binding site of NMDA receptors is a prerequisite for the activation of this receptor.[3,32] On the other hand, the question arises as to whether activation of NMDA receptors is controlled by disynaptic inhibition.[63] As a rule, second -order vestibular neurons receive their monosynaptic excitation from only one of the three semicircular canal nerve branches.[33,66,71] This canal-specific organization principle includes excitatory and inhibitory side loops via vestibular interneurons.[66] Inhibitory inputs are mediated in part by glycinergic, in part by GABAergic, and in part by a combination of glycinergic and GABAergic interneurons.[63,66]

3.4.5 Possible Functional Implications

AMPA/kainate and NMDA receptors differ in terms of the conditions for their activation and in terms of the consequences, once they are activated. In addition to the binding of glutamate, activation of NMDA receptors requires the binding of glycine and a depolarization of the membrane to release the receptor ionophore from a voltage-sensitive Mg^{2+} block. In essence, the voltage-dependent activation of NMDA receptors results in a nonlinear, exponentially rising excitation that is supported by the increased permeability for calcium ions through the NMDA receptor channel. An APV-sensitive increase in the intracellular Ca^{2+} level of medial vestibular neurons of rat was in fact demonstrated after electrical stimulation of the eighth nerve root.[68]

The immediate amplification of postsynaptic potentials is of short duration but can be considerably prolonged in time by a calcium-mediated activation of second messenger systems. Typically, the latter phenomenon - long-term potentiation - requires high levels of calcium influx that can be triggered experimentally by high-frequency stimulation. Thus, NMDA receptors can modify the responses of a neuron over long periods of time and can constitute an initial step in a process associated with behavioral plasticity and adaptation.

Long-term potentiation of eighth nerve-evoked monosynaptic response components was demonstrated in the goldfish Mauthner cell.[74] This long-term potentiation affected the electrotonic coupling potential as well as the chemically mediated EPSP component. The increase in the coupling potential resulted from an increase in the conductance of the gap junctions. The potentiation of both the electrical and the chemical EPSP component required the activation of NMDA receptors, since application of the specific NMDA receptor antagonist (2-carboxypiperazine-4-yl)propyl-1-phosphonic acid or of ketamine (an NMDA channel-blocker) prevented long-term potentiation. This potentiation of electrical and chemical EPSP components was also prevented by intracellular injections of the calcium chelator BAPTA, indicating the necessity of an increased intracellular calcium level for long-term potentiation. Monosynaptically evoked field potentials were also potentiated in the rat medial vestibular nucleus following high-frequency stimulation of the eighth nerve root.[11] This long-term potentiation was prevented in the presence of the NMDA receptor antagonist D-APV. Monosynaptic responses evoked by single stimuli at a low rate were not affected by D-APV.

Given the presence of NMDA receptors on most neurons, including sensory and motoneurons, and given that a number of NMDA subtypes exist as well, different receptor subtypes may subserve different rather than a singular cellular function. Some functional specificity is suggested by the fact that not all vestibular nerve afferent fibers activate NMDA receptors. However, a specific, dynamic contribution by thick vestibular nerve afferents to responses of the vestibulo-ocular reflex has so far not been demonstrated. Minor and Goldberg[41] selectively blocked the responses of thick, irregularly discharging vestibular nerve afferent fibers from both labyrinths with galvanic currents, compared the vestibulo-ocular responses before and during the elimination of these input components, and observed virtually no difference in gain or phase of the horizontal canal-ocular reflex. This negative result was explained by vestibular interneurons that form inhibitory side loops, are activated mainly by thick vestibular nerve afferent fibers, and converge together with thick vestibular nerve afferent fibers from the same labyrinthine sensory organ on common second-order vestibular projection neurons. Thereby, the excitatory input from thick vestibular nerve afferent fibers is counterbalanced by the disynaptic inhibition via vestibular interneurons. The existence of side loops with this spatial specificity was recently shown in frogs.[66] Depending on their additional inputs, these inhibitory side loops could enable or disenable the activation of NMDA receptors on second-order vestibular projection neurons and thereby modify the weighing factors of converging inputs from afferent nerve fibers with regular and irregular discharge patterns.[41,66] Angelaki et al.[2] observed during galvanic polarization a reduction of the sustained ocular nystagmus during constant velocity off-vertical axis rotation (OVAR) in darkness. Since the same currents did not affect optokinetic after-nystagmus, the velocity storage mechanism appears to function normally during

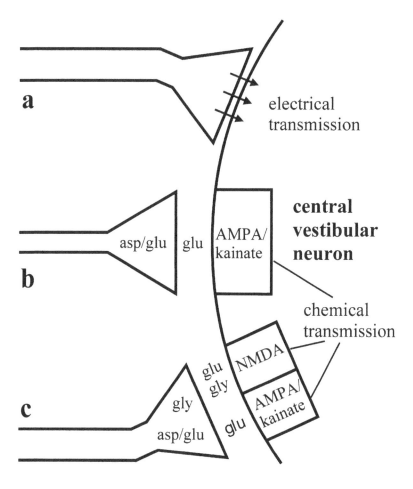

FIGURE 3.3 Summary diagram for electrical and chemical transmission of signals from vestibular nerve afferent fibers to second-order vestibular neurons, including putative transmitters and postsynaptic receptors: (a) afferent fiber electrically coupled with a central vestibular neuron; (b,c) afferent fibers mediating signals via chemical synapses. Thinner vestibular afferent fibers (b) release aspartate/glutamate and activate AMPA/kainate receptors. Thicker vestibular afferent fibers (c) release aspartate/glutamate and possibly glycine and activate NMDA and AMPA/kainate receptors.

galvanic stimulation. Because OVAR responses are mediated by otolith afferent fibers, Angelaki et al.[2] proposed that bilateral anodal polarization prevents a normal spatio-temporal convergence between regular (thinner) and irregular (thicker) otolith afferents onto central vestibular neurons. In more general terms, irregular (thicker) afferent nerve inputs could play a role in the spatio-temporal integration of inputs with different polarization vectors.

3.5 CONCLUSION

Vestibular nerve afferent input is highly organized in terms of anatomy, physiology and biochemistry. Based on their relationship to cell size, afferent fibers from each individual endorgan tend to: (1) innervate particular areas in the sensory epithelium; (2) differ in their regularity of discharge rate at rest and in their dynamics in response to head movements; and (3) contain different transmitter-like substances in addition to glutamate, their putative transmitter. Transmission of signals onto second-order vestibular neurons appears to be matched to these properties. Thinner vestibular nerve afferent fibers activate predominantly AMPA/kainate subtypes of glutamate receptors, whereas thicker vestibular nerve afferent fibers excite second-order vestibular neurons via AMPA/kainate and NMDA receptors (for a summary, see Fig. 3.3).

It is possible that NMDA receptors are activated by a corelease of glutamate and glycine from thicker vestibular nerve afferent fibers. In addition, electrical transmission of signals from thick vestibular nerve afferent fibers is frequently observed in non-mammalian species. The functional significance of this cell size-related organization in general and of the activation of NMDA receptors in particular is unclear at present.

ACKNOWLEDGMENTS

This work was supported by a grant from the Deutsche Forschungsgemeinschaft (SFB 220).

REFERENCES

1. Alexander, G, Zur Anatomie des Ganglion vestibulare der Säugetiere, *Arch Ohrenheilk,* 51: 109,1901.
2. Angelaki, DE, Perachio, AA, Mustari, MJ, and Strunk, CL, Role of irregular otolith afferents in the steady-state nystagmus during off-vertical axis rotation, *J. Neurophysiol.,* 68:1895, 1992.
3. Ascher, P, and Johnson, JW, The NMDA receptor, its channel, and its modulation by glycine, in: *The NMDA Receptor,* (Collingridge,GL and Watkins, JC, Eds.), Oxford University Press, Oxford, 1994.
4. Baird, RA, and Lewis, ER, Correspondences between afferent innervation patterns and response dynamics in the bullfrog utricle and lagena, *Brain Res.,* 369:48, 1986.
5. Baird, RA, Desmadryl, G, Fernández, C, and Goldberg, JM, The vestibular nerve of the chinchilla. II. Relation between afferent response properties and peripheral innervation patterns in the semicircular canals, *J. Neurophysiol.,* 60:182, 1988.
6. Baird, RA, and Schuff, NR, Peripheral innervation patterns of vestibular nerve afferents in the bullfrog utriculus, *J. Comp. Neurol.,* 342:279, 1994.
7. Ballantyne, J, and Engström, H, Morphology of the vestibular ganglion cells, *J. Laryngol. Otol.,* 83:19, 1969.
8. Bennett, MVL, A comparison of electrically and chemically mediated transmission, in: *Structure and Function of Synapses,* (Pappas, GD, and Purpura, DP, Eds.), Raven Press, New York, 1972.
9. Brightman, MW, and Reese, TS, Junctions between intimately apposed cell membranes in the vertebrate brain, *J. Cell. Biol.,* 40:648, 1969.
10. Bussières, N, and Dubuc, R, Phasic modulation of vestibulospinal neuron activity during fictive locomotion in lampreys, *Brain Res.,* 575:174, 1992.
11. Capocchi, G, Della Torre, G, Grassi, S, Pettorossi, VE, and Zampolini, M, NMDA receptor-mediated long term modulation of electrically evoked field potentials in the rat medial vestibular nuclei, *Exp. Brain Res.,* 90:546, 1992.

12. Carpenter, DO, and Hori, N, Neurotransmitter and peptide receptors on medial vestibular nucleus neurons, *Ann. NY Acad. Sci.*, 656:668, 1992.

13. Carpenter, MB, Huang, Y, Pereira, AB, and Hersh, LB, Immunocytochemical features of the vestibular nuclei in the monkey and the cat, *J. Hirnforsch.*, 31:585, 1990.

14. Cochran, SL, Hackett, JT, and Brown, DL, The anuran Mauthner cell and its synaptic bed, *Neurosci.*, 5:1629, 1980.

15. Cochran, SL, Kasik, P, and Precht, W, Pharmacological aspects of excitatory synaptic transmission to second-order vestibular neurons in the frog, *Synapse*, 1:102, 1987.

16. Cuénod, M, and Streit, P, Neuronal tracing using retrograde migration of labeled transmitter-related compounds, in: *Handbook of Chemical Neuroanatomy*, Vol. 1: Methods in Chemical Neuroanatomy, (Björklund A, Hökfelt T, Eds.), Elsevier, Amsterdam, 1983.

17. Demêmes, D, Raymond, J, and Sans, A, Selective retrograde labeling of neurons of the cat vestibular ganglion with [³H]D-aspartate, *Brain Res.*, 304:188, 1984.

18. Demêmes, D, Wenthold, RJ, Moniot, B, and Sans, A, Glutamate-like immunoreactivity in the peripheral vestibular system of mammals, *Hear Res.*, 46:261, 1990.

19. Demêmes, D, Raymond, J, Atger, P, Grill, C, Winsky, L, and Dechesne, CJ, Identification of neuron subpopulations in the rat vestibular ganglion by calbindin-D 28K, calretinin and neurofilament proteins immunoreactivity, *Brain Res.*, 582:168, 1992.

20. Doi, K, Tsumoto, T, and Matsunaga, T, Actions of excitatory amino acid antagonists on synaptic inputs to rat medial vestibular nucleus: an electrophysiological study in vitro, *Exp. Brain Res.*, 82:254, 1990.

21. Ernfors, P, Lee, K-F, and Jaenisch, R, Mice lacking brain-derived neurotrophic factor develop with sensory deficits. *Nature* 368:147, 1994.

22. Evinger C, and Baker R, Are There Subdivisions of Extraocular Motoneuronal Pools that Can Be Controlled Selectively? in: *Motor Control: Concepts and Issues,* (Humphrey, DR, and Freund H-J, Eds.), John Wiley & Sons, Chichester, 1991.

23. Faber, DS, Kaars, C, and Zottoli, SJ, Dual transmission at morphologically mixed synapses: Evidence from postsynaptic cobalt injections, *Neurosci.*, 5:433, 1980.

24. Furshpan, EJ, "Electrical transmission" at an excitatory synapse in a vertebrate brain, *Science*, 144:878, 1964.

25. Goldberg, JM, and Fernández, C, Conduction times and background discharge of vestibular afferents, *Brain Res.*, 122:545, 1977.

26. Goldberg, JM, Desmadryl, G, Baird, RA, and Fernández, C, The vestibular nerve in the chinchilla. V. Relation between afferent response properties and peripheral innervation patterns in the utricular macula, *J. Neurophysiol.*, 63:781, 1990.

27. Hackett, JT, Cochran, SL, and Brown, DL, Functional properties of afferents which synapse on the Mauthner neuron in the amphibian tadpole, *Brain Res.*, 176:148, 1979.

28. Highstein, SM, Goldberg, JM, Moschovakis, AK, and Fernández, C, Inputs from regularly and irregularly discharging vestibular nerve afferents to secondary neurons in the vestibular nuclei of the squirrel monkey. II. Correlation with output pathways of secondary neurons, *J. Neurophysiol.*, 58:719, 1987.

29. Honrubia, V, Hoffmann, LF, Sitko, S, and Schwartz, IR, Anatomical and physiological correlates in bullfrog vestibular nerve, *J. Neurophysiol.*, 61:688, 1989.

30. Huwe, JA, and Peterson, EH, Differences in brain stem terminations of large- and small-diameter vestibular primary afferents, *J. Neurophysiol.*, 74:1362, 1995.

31. Ito, M, Hongo, T, and Okada, Y, Vestibular-evoked postsynaptic potentials in Deiters' neurons, *Exp. Brain Res.*, 7:214, 1969.

32. Johnson, JW, and Ascher, P, Glycine potentiates the NMDA response in cultured mouse brain neurons, *Nature*, 325:529, 1987.

33. Kasahara, M, and Uchino, Y, Bilateral semicircular canal inputs to neurons in the cat vestibular nuclei, *Exp. Brain Res.*, 20:285, 1974.

34. Kevetter, GA, Immunocytochemistry of glutamate receptors in the vestibular nuclear complex in gerbils, *Soc. Neurosci. Abstr.*, 20:240.3, 1944.

35. Kinney, GA, Peterson, BW, and Slater, NT, The synaptic activation of N-methyl-D-aspartate receptors in the rat medial vestibular nucleus, *J. Neuropysiol.*, 72:1588, 1994.

36. Knöpfel, T, Evidence for N-methyl-D-aspartic acid receptor-mediated modulation of the commissural input to central vestibular neurons of the frog, *Brain Res.*, 426:212, 1987.

37. Korn, H, Sotelo, C, and Crepel, F, Electrotonic coupling between neurons in the rat lateral vestibular nucleus, *Exp. Brain Res.*, 16:255, 1973.

38. Korn, H, Sotelo, C, and Bennett, MVL, The lateral vestibular nucleus of the toadfish Opsanus tau: Ultrastructural and electrophysiological observations with special reference to electrotonic transmission, *Neurosci.*, 2:851, 1977.

39. Lewis, MR, Phelan, KD, Shinnick-Gallagher, P, and Gallagher, JP, Primary afferent excitatory transmission recorded intracellularly in vitro from rat medial vestibular neurons, *Synapse*, 3:149, 1989.

40. Li, H, Godfrey, D, and Rubin, AM,) Receptor binding autoradiography and immunohistochemistry of non-NMDA receptors in rat vestibular nuclear complex, *Soc. Neurosci. Abstr.*, 20:240.6, 1994.

41. Minor, LB, and Goldberg, JM, Vestibular-nerve inputs to the vestibulo-ocular reflex: a functional ablation study in the squirrel monkey, *J. Neurosci.*, 11:1636, 1991.

42. Monaghan, DT, and Cotman, CW, Distribution of N-methyl-D-aspartate-sensitive L-[^{3}H]glutamate-binding sites in the rat brain, *J. Neurosci.*, 5:2909, 1985.

43. Mugnaini, E, Walberg, F, and Hauglie-Hanssen, E, Observations on the fine structure of the lateral vestibular nucleus (Deiters' nucleus) in the cat, *Exp. Brain Res.*, 4:146, 1967.

44. Nakajima, Y, Fine structure of synaptic endings on the Mauthner cell of the goldfish, *J. Comp. Neurol.*, 156:375, 1974.

45. O'Leary, DP, Dunn, RF, and Honrubia, V, Analysis of afferent responses from isolated semicircular canal of the guitarfish using rotational acceleration white-noise inputs. I. Correlation of response dynamics with receptor innervation, *J. Neurophysiol.*, 39:631, 1976.

46. Orrego, F, Criteria for the identification of central neurotransmitters, and their application to studies with some nerve tissue preparations *in vitro*, *Neurosci.*, 4:1037, 1979.

47. Petralia, RS, and Wenthold, RJ, Light and electron immunocytochemical localization of AMPA-selective glutamate receptors in the rat brain, *J. Comp. Neurol.*, 318:329, 1992.

48. Petralia, RS, Wang, Y-X, and Wenthold, RJ, Histological and ultrastructural localization of the kainate receptor subtypes, KA2 and GluR6/7, in the rat nervous system using selective antipeptide antibodies, *J. Comp. Neurol.*, 349:85, 1994.

49. Petralia, RS, Yokotani, N, and Wenthold, RJ, Light and electron microscope distribution of the NMDA receptor subunit NMDAR1 in the rat nervous system using a selective anti-peptide antibody, *J. Neurosci.*, 14:667, 1994.

50. Peusner, KD, Development of vestibular brainstem nuclei. in: *Development of Auditory and Vestibular Brainstem Nuclei 2.* (Romand R, Ed.), Elsevier, Amsterdam, 1992.

51. Peusner, KD, and Giaume, C, The first developing "mixed" synapses between vestibular sensory neurons mediate glutamate chemical transmission., *Neurosci.*, 58:99, 1994.

52. Popper, P, Lopez, I, Alvarez, JC, Rodrigo, JP, Mizevych, PE, and Honrubia, V, Distribution of AMPA receptor subunits in the vestibular nuclei of the chinchilla, *Soc. Neurosci. Abstr.*, 20:240.5, 1994.

53. Raymond, J, Nieoullon, A, Demêmes, D, and Sans, A, Evidence for glutamate as a neurotransmitter in the cat vestibular nerve: radioautographic and biochemical studies, *Exp. Brain Res.*, 56:523, 1984.

54. Raymond, J, Demêmes, D, and Nieoullon, A, Neurotransmitters in vestibular pathways, *Prog. Brain Res.*, 76:29, 1988.

55. Reichenberger, I, and Dieringer, N, Size-related colocalization of glycine- and glutamate-immunoreactivity in frog and rat vestibular afferents, *J. Comp. Neurol.*, 349:603, 1994.

56. Reichenberger, I, Straka, H, Ottersen, OP, Streit, P, Gerrits, NM, and Dieringer, N,) Distribution of GABA, glycine and glutamate immunoreactivities in the vestibular nuclear complex of the frog. in preparation, 1996.

57. Sato, F, and Sasaki, H, Morphological correlations between spontaneously discharging primary vestibular afferents and vestibular nucleus neurons in the cat, *J. Comp. Neurol.*, 333:554, 1993.

58. Sato, K, Mick, G, Kiyama, H, and Tohyama, M,) Expression patterns of a glutamate-binding protein in the rat central nervous system: comparison with N-methyl-D-aspartate receptor subunit 1 in rat, *Neurosci.*, 64:459, 1995.

59. Serafin, M, de Waele, C, Khateb, A, Vidal, PP, and Mühlethaler, M, Medial vestibular nucleus in the guinea-pig. I. Intrinsic membrane properties in brainstem slices, *Exp. Brain Res.*, 84:417, 1991.

60. Shimazu, H, and Precht, W, (1965) Tonic and kinetic responses of cat's vestibular neurons to horizontal angular acceleration, *J. Neurophysiol.*, 28:989, 1965.

61. Sotelo, C, Morphological basis for electrical communication between neurons in the central nervous system of vertebrates. in: *Neuron Concept Today,* (Szentágothai, J, Hámori, J, and Vizi, ES, Eds.), Akadémiai Kiadó, Budapest, 1977.

62. Straka, H, and Dieringer, N, Electrophysiological and pharmacological characterization of vestibular inputs to identified frog abducens motoneurons and internuclear neurons *in vitro, Eur. J. Neurosci.,* 5:251,1993.

63. Straka, H, and Dieringer, N, Disynaptic inhibition of second-order vestibular neurons and its interaction with monosynaptic afferent nerve excitation in the frog, in preparation, 1996.

64. Straka, H, Reichenberger, I, and Dieringer, N, Size-related properties of vestibular afferent fibers in the frog: uptake of and immunoreactivity for glycine and aspartate/glutamate, *Neurosci.,* 70:685,1996a.

65. Straka, H, Debler, K, and Dieringer, N, Size-related properties of vestibular afferent fibers in the frog: differential synaptic activation of NMDA and non-NMDA receptors, *Neurosci.,*70:697,1996b.

66. Straka, H, Biesdorf, S, and Dieringer, N, Canal-specific excitation and inhibition of frog second order vestibular neurons, *J. Neurophysiol.,* 1997.

67. Takahashi, Y, Tsumoto, T, and Kubo, T, N-methyl-D-aspartate receptors contribute to afferent synaptic transmission in the medial vestibular nucleus of young rats, *Brain Res.,* 659:287,1994a.

68. Takahashi, Y, Takahashi, MP, Tsumoto ,T, Doi, K, and Matsunaga, T, Synaptic input-induced increase in intraneuronal Ca^{2+} in the medial vestibular nucleus of young rats, *Neurosci. Res.,* 21:59,1994b.

69. Usami, S, Hozawa, J, Tazawa, M, Jin, H, Matsubara, A, and Fujita, S, Localization of substance P-like immunoreactivity in guinea pig vestibular end organs and the vestibular ganglion, *Brain Res.,* 555:153,1991.

70. Werman, R, Criteria for identification of a central nervous transmitter, *Comp. Biochem. Physiol.,* 18:745,1966.

71. Wilson, Vj, and Felpel, LP, Specificity of semicircular canal input to neurons in the pigeon vestibular nuclei, *J. Neurophysiol.,* 35:253,1972.

72. Won, M-H, Oh, Y-S, and Shin, H-C, Localization of somatostatin-like immunoreactive neurons in the vestibular ganglion of the rabbit, *Neurosci. Lett.,* 217:129,1996.

73. Wylie, RM, Evidence of electrotonic transmission in the vestibular nuclei of the rat, *Brain Res.,* 50:179,1973.

74. Yang, X, Korn, H, and Faber DS, Long-term potentiation of electrotonic coupling at mixed synapses, *Nature ,*348:542,1990.

75. Ylikoski, J, Päivärinta, H, Eränkö, L, Merena, I, and Lehtosalo, J, Is substance P the neurotransmitter in the vestibular endorgans?, *Acta Otolaryngol.,* 97:523,1984.

4

Efferent Vestibular System: Anatomy, Physiology, and Neurochemistry

Jay M.Goldberg, Alan M. Brichta, and Phillip A. Wackym

CONTENTS

0-8493-7679-3/00/$0.00+$.50
© 2000 by CRC Press LLC

4.1. SYNOPSIS

The vestibular organs in all vertebrate classes receive an efferent innervation originating in the brain stem. Although there are relatively few parent efferent neurons, their axons branch to innervate more than one organ and they also ramify extensively in the neuroepithelium of individual organs. By means of this divergent innervation, efferent fibers provide a major source of synaptic input to type II hair cells, afferent calyces, and other nonmyelinated afferent processes. Efferent synapses are characterized by an accumulation of small, clear vesicles and by a small number of large, dense-core vesicles. The synapses on afferent processes have asymmetric membrane thickenings, while a subsynaptic cistern distinguishes those on hair cells.

In many nonvestibular organs, efferent activation reduces afferent activity by inhibiting hair cells. Efferent responses in vestibular organs are more heterogeneous. In fish and mammals, there is an excitation of afferents; whereas in frogs and turtles, both excitation and inhibition are seen. There are variations in the efferent responses of vestibular afferents innervating different parts of the neuroepithelium and differing in their responses to natural stimulation. In the frog posterior crista, efferent synapses on type II hair cells can mediate both excitatory and inhibitory responses. Many of the excitatory responses seen in other animals are likely to involve efferent synapses on afferent processes. Efferent neurons receive a convergent input from several vestibular and non-vestibular receptors and respond in association with active movements. On the basis of the discharge properties of efferent neurons, it is suggested that one of their functions is to switch the vestibular organs from a "postural" mode to a "volitional" mode.

Acetylcholine (ACh) is an efferent neurotransmitter. Most efferent vestibular neurons contain choline acetyltransferase (ChAT) and acetylcholinesterase (AChE). Several nicotinic and musca-rinic receptors have been localized in vestibular organs. Rapid efferent actions, including excitation and inhibition, are likely to involve nicotinic receptors. Slower actions may involve muscarinic receptors. Recent evidence implicates ATP as a possible neurotransmitter involved in efferent excitation of vestibular hair cells. Recordings from solitary hair cells suggest that a variety of purinergic (P_2) receptor subtypes are found in hair cells. Nevertheless, the precise mechanisms of ATP's presumed action have yet to be determined. There is contradictory evidence that GABA is an afferent transmitter and inconclusive evidence that it is an efferent transmitter. Two neuropeptides have been found in efferent vestibular neurons, where they are colocalized with ACh. CGRP is preferentially found in efferent axons synapsing on afferent processes. In lateral lines, application of CGRP results in a slow excitation of hair cells. Its actions on vestibular organs are unknown. Preproenkephalin mRNA is found in the majority of efferent vestibular neurons. Neither the distribution of enkephalinergic efferent axons within the vestibular organs, nor the physiological actions of this innervation have been defined.

4.2. HISTORICAL INTRODUCTION

An efferent innervation of the ear was first described by Rasmussen (1946, 1953),[187,188] who was able to trace fibers from the medial part of the superior olivary complex to the contralateral vestibular nerve and then by way of the vestibulo-cochlear anastomosis to the contralateral cochlea.

Rasmussen[189] identified a second auditory efferent component as originating in the lateral part of the superior olive and running to the ipsilateral cochlea. Rasmussen and Gacek[84,190] defined a vestibular efferent component, arising in the vicinity of the vestibular nuclei and reaching the vestibular organs. In the meantime, ultrastructural techniques had been applied to the mammalian ear.[60,253] Two groups of nerve fibers were observed. One group had poorly vesiculated endings, while the second group's endings were highly vesiculated. Work in other systems had shown that an accumulation of vesicles was characteristic of presynaptic elements.[44,176] This led Engström (in 1958)[58] to suggest that the vesiculated endings in the labyrinth were efferent terminals, and that the nonvesiculated endings were afferent terminals. Support for the dichotomy came from Smith and Sjöstrand's (1961)[222] description of afferent (ribbon) synapses in which hair cells are presynaptic to nonvesiculated boutons. At about the same time, it was found by histochemical methods that efferent nerve fibers in both the cochlea[37,212] and the vestibular organs[51,118] stained positively for AChE. As afferents were not AChE positive, the discovery established that the two sets of fibers were neurochemically distinct. Hilding and Wersäll (1962)[109] connected the histochemical and ultrastructural findings by demonstrating that only the vesiculated endings were AChE positive. Proof of the efferent nature of the vesiculated endings was obtained by showing that they degenerated after central lesions.[123,130,220,221,225]

Rasmussen[187] noted, while tracing the medial group of auditory efferents, that they crossed the midline immediately beneath the floor of the fourth ventricle. This provided a convenient site at which to stimulate the efferent fibers, while cochlear potentials were recorded with gross electrodes. The approach was used by Galambos (1956),[87] Fex (1959),[66] and Desmedt and Monaco (1961).[46] It was found that efferent stimulation reduced the sound-evoked compound action potential[46,66,87] and increased the cochlear microphonic[46,66]. The enhancement of the microphonic could be explained by a postsynaptic inhibition of hair cells.[61] Fex (1962)[67] then studied the effects of efferent stimulation on single cochlear afferents. He observed an inhibitory action, including a reduction of both background and sound-evoked activity. The reduction of background activity was probably a consequence of the experiments being conducted in a noisy environment. In any case, Wiederhold and Kiang (1970)[257,258] showed that electrical stimulation of medial efferents resulted in a decrease in the sensitivity of cochlear afferents to tones near their characteristic frequency, but had almost no influence on the background activity obtained in the quiet. The lack of effect on background activity and the frequency selectivity of the effect on tone-evoked activity would be difficult to explain by a direct inhibition of afferent transmission. It is now known that the myelinated cochlear afferents, which are the only afferents recorded to date,[129] synapse on inner hair cells,[129,224] while the medial efferents synapse on outer hair cells.[224,252] Rather than directly inhibiting afferent transmission, the efferent synapses on outer hair cells reduce active tuning of the cochlear partition and it is the detuning that is reflected in afferent discharge.[29,10,41] The pharmacology of efferent transmission was studied by Bobbin and Konishi (1971, 1974),[18,19] who showed that the effects of medial efferent stimulation on cochlear potentials are mimicked by cholinergic agonists and are blocked by cholinergic antagonists.

Efferent axons also innervate lateral-line sensory organs. As first demonstrated by Russell, (1968)[202,1204] efferent activation resulted in an inhibition of lateral-line afferent activity. Flock and Russell (1973, 1976)[72,73] confirmed these findings and showed that they resulted from an inhibition of hair cells. Evidence was presented that efferent transmission in lateral lines was cholinergic.[203]

Progress on vestibular efferents lagged behind work in the cochlea and in lateral lines. One reason was the uncertainty about the location of efferent neurons and their pathways. Sala (1965),[208] for example, observed a reduction in single-unit activity recorded in the vicinity of the vestibular nerve on one side when the contralateral vestibular nuclei were electrically stimulated. Unfortunately, it is unclear whether Sala was stimulating efferent pathways or whether he was recording from peripheral afferents. Llinás and Precht (1967, 1969)[143,144] used electrophysiological techniques to define an efferent vestibular pathway arising in the frog cerebellum. The conclusion was in error as anatomical results, summarized below, indicate that all efferent pathways have a brain-stem origin. At this point, horseradish peroxidase was being developed as a retrograde tracer.[138] Gacek and Lyon (1974)[85] used this technique to determine that efferent vestibular neurons were bilaterally located in the brain stem just medial to the rostral vestibular nuclei. The labeling of the efferent neurons was an important step that led within a few years to studies of the effects of efferent stimulation on afferent discharge,[50,92] as well as to later studies of the neurochemistry of efferent transmission.

4.3 ANATOMY

4.3.1 EFFERENT CELL-BODY LOCATIONS AND AXONAL TRAJECTORIES

Below we will consider the locations of efferent cell bodies in mammals and compare the results with those obtained in other vertebrate species. The course of the efferent axons from the brain stem to the ear will only be described in mammals.

4.3.1.1 Mammals.

Retrograde-tracer methods have been used to define the locations of vestibular efferent neurons in the cat,[45,85,122,251] rat,[213,237,250] guinea pig,[229] gerbil,[179,205] chinchilla,[120] and squirrel monkey.[32,92] Results are illustrated for the squirrel monkey (Figure 4.1A, B).

The majority of vestibular efferent cell bodies lie between the vestibular and abducens nuclei, in the vicinity of the descending facial nerve. They form a so-called *group e* that in some species is divided into a larger lateral group and a smaller medial group. Retrogradely labeled cells are also found more ventrally in the caudal pontine reticular formation.[45,92,122,213,237] Some of the ventrally located neurons may be labeled as a result of spillage of tracer into the middle ear.[92] Approximately equal numbers of efferent neurons are found on the sides ipsilateral and contralateral to the labyrinth being innervated.

The trajectories of efferent fibers, as revealed by AChE histochemistry, are illustrated in Figure 4.1C. Contralateral vestibular efferent axons cross the midline just underneath the floor of the fourth ventricle, where they are joined by crossing olivocochlear fibers. On the ipsilateral side, the contralateral and ipsilateral axons come together in a fiber bundle underneath group e. Fibers pass over or through the spinal trigeminal tract to reach the vestibular nerve. Vestibular efferents traverse the vestibular ganglion and then run in the various ampullary and macular nerves to reach the individual organs (Figure 4.1D). Auditory afferents travel in the inferior vestibular nerve. On reaching the vestibular ganglion, they enter the vestibulo-cochlear anastomosis and travel to the cochlea.

4.3.1.2 Comparative Anatomy

Labyrinthine sensory organs receive an efferent innervation in all vertebrates with the exception of hagfish.[2,12,149] Efferents and facial motoneurons may differentiate from a common pool of neurons in the developing embryo.[75,77,160] The relation between efferents and the facial nucleus is still evident in the adult brains of fish,[25,42,77,108,136,159,161,193] amphibians,[76,95,178,230,259] reptiles[14,194,227,228] and birds.[56,214,231,256] The locations of efferent cell bodies in various species are depicted in Figure 2.

In fish (Figure 4.2A), efferents are found either partially or entirely within the confines of the facial motor nucleus. While facial motoneurons are located exclusively on the ipsilateral side, some of the efferents are found contralaterally. There is overlap in the distribution of efferent cell bodies innervating the various vestibular and lateral-line organs. Efferent neurons in amphibians (Figure 4.2B) are also closely associated with the facial nucleus; there are relatively few contralateral efferents. In reptiles (Figure 4.2C) and birds (Figure 4.2D), efferent somas are located dorsal to the facial nucleus. Cell bodies innervating the basilar papilla (an auditory organ) and the various .vestibular organs overlap with vestibular efferents tending to have a more dorsal location. Compared to fish and amphibians, a larger fraction of efferent neurons in reptiles and birds are found

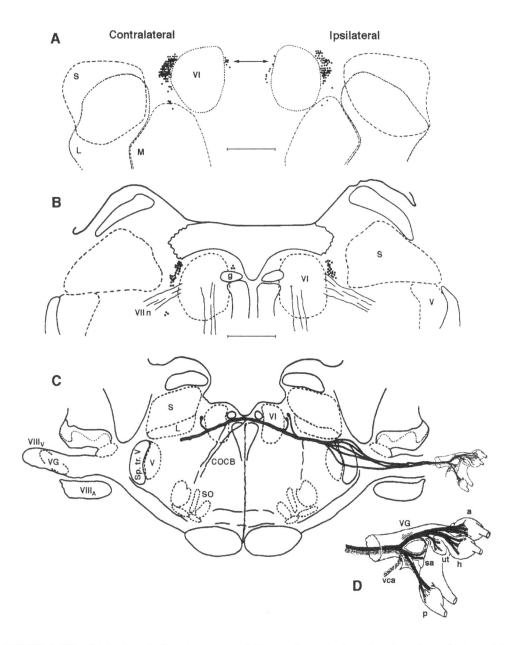

FIGURE 4.1 The distribution of efferent neurons and fibers in the squirrel monkey. Locations of retrogradely labeled neurons following the injection of horseradish peroxidase into the ipsilateral labyrinth are shown in horizontal plan view (A) and in a single transverse section (B). The trajectories of efferent axons are revealed by acetylcholinesterase (AChE) histochemistry (C, D). The olivocochlear tracts are indicated by stippling in D. Abbreviations: a, h, and p: anterior, horizontal and posterior ampullae, respectively; COCB: crossed olivo-cochlear bundle; g: facial genu; L, M, and S: lateral, medial, and superior vestibular nuclei, respectively; LSO and MSO, lateral and medial superior olive; MLF, medial longitudinal fasciculus; MT, medial trapezoid nucleus; sa, sacculus; SO, superior olive; Sp. tr. V: spinal trigeminal tract; ut: utriculus; vca: vestibulo-cochlear anastomosis; VG: vestibular ganglion; V: spinal trigeminal nucleus; VI: abducens nucleus; VII: facial nucleus; VII n, descending facial nerve; VIIIa: auditory nerve; VIIIv: vestibular nerve. (From Goldberg, J.M., and Fernández, *C. J. Neurophysiol.* 43: 986, 1980. With permission.)

on the contralateral. side. Figure 4.2E illustrates the situation in mammals. Now, the vestibular and auditory efferent cell bodies have separate locations. Vestibular efferent somas are found in the dorsolateral tegmentum on both sides of the brain. Auditory efferent somas have a ventral location in the superior olivary complex and can be distinguished into lateral (LOC) and medial (MOC) olivocochlear groups, located respectively lateral and medial to the medial superior olive.[61,96,252] Although both groups project bilaterally, most LOC neurons innervate the ipsilateral cochlea and most MOC neurons supply the contralateral cochlea. LOC fibers contact afferent dendrites innervating inner hair cells, while MOC fibers synapse with outer hair cells.

4.3.2 PERIPHERAL BRANCHING PATTERNS

There are approx. 500 parent efferent neurons innervating the vestibular end organs in mammals,[85,92,251] as compared to >10,000 afferents.[65,86,110] Efferent axons, once they reach the neuro-epithelium, branch several times to provide a divergent innervation. As a result of this divergence, the relatively small number of parent efferent axons give rise to a dense innervation with highly vesiculated efferent boutons being outnumbered by afferent boutons in only a 2 - 3:1 ratio.[93] In recent studies, Purcell and Perachio[182-184] have been able to label efferent neurons and reconstruct their branching patterns in the neuroepithelium. Individual efferent axons are found to contain several hundred boutons and to innervate as much as 250 μm of the neuroepithelium. This may be contrasted with the more restricted branching patterns of afferents. Even the most highly branched afferent terminals contain 50-100 boutons extending over a distance of 50 to 100 μm.[63-65] Despite the divergent nature of efferent innervation, the projection patterns can respect zonal boundaries. Ipsilateral efferent neurons project mainly to the central zone of each cristae, while contralateral efferent neurons project mainly to the surrounding peripheral zone.[183] Separate groups of efferent axons may innervate the striolar, juxtastriolar, and peripheral extrastriolar regions of the utricular and saccular maculae.[182,184]

The divergence in the innervation of individual efferent axons is not confined to single organs. In fish, individual efferent axons can send collaterals to both vestibular and lateral-line organs.[38] Electrophysiological studies in the frog show that efferent axons can branch to innervate two or more vestibular organs on one side.[181,200,201,246] Retrograde-tracer studies suggest that a small pro-portion of efferent neurons can project to both ears.[45,214] Bilaterally projecting efferent neurons may be more common than hitherto suspected as the labeling of a single organ can lead to extensive labeling of efferent terminations in other ipsilateral and contralateral. organs (I.M. Purcell and A.A. Perachio, personal communication).

4.3.3 SYNAPTIC ULTRASTRUCTURE

Vestibular efferent axons terminate as highly vesiculated boutons on vestibular hair cells and on afferent dendrites including calyceal endings.[125,221,254] The innervation of hair cells may be termed presynaptic; that on afferent dendrites, postsynaptic. As is illustrated in Figure 4.3, individual efferent fibers in mammals can provide both presynaptic and postsynaptic endings.[125,221]

In most descriptions of postsynaptic innervation, emphasis has been placed on the efferent terminals found on afferent calyces. This has led to the notion that in fish and amphibians, which lack type I hair cells and calyx endings, efferent innervation is entirely presynaptic.[254] While this may be true in frogs,[151] fish have a postsynaptic efferent innervation of afferent dendrites.[151,209] Both presynaptic and postsynaptic efferent terminals are also found in the peripheral zone of the turtle crista (A. Lysakowski, personal communication), a region only containing type II hair cells and afferent terminals of the bouton variety.[27,126]

The fine structure of efferent terminations is similar in all vertebrate classes (Figure 4.4).[151,251]

Efferent boutons contain an accumulation of round vesicles, which are predominantly small and clear; an occasional larger, dense-core vesicle is seen. Based on studies at other synapses,[43,128]

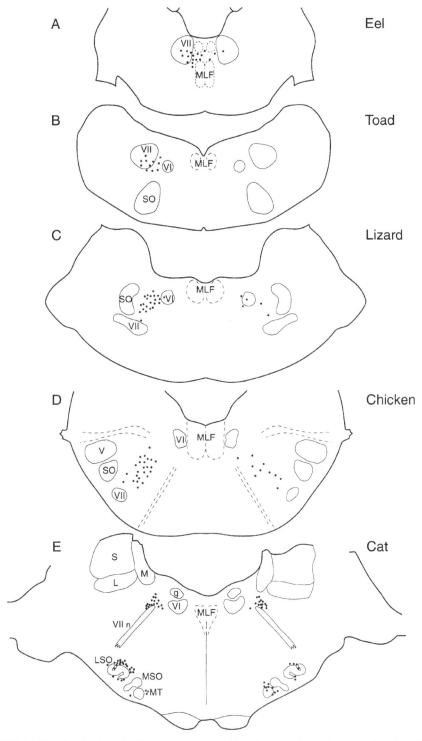

FIGURE 4.2 The distribution of efferent neurons and fibers in several vertebrate species. Based on Meredith and Roberts[161] (A), Pellegrini et al.[178] (B), Barbas-Henry and Lohman[14] (C), Whitehead and Morest[256] (D), and Warr[251] (E). Abbreviations as in Figure 4.1.

the small, clear vesicles presumably contain ACh or other classic neurotransmitters, while the dense-core vesicles are likely to contain neuropeptides. Efferent synapses on afferent terminals (Figure 4.4B) are marked by asymmetric membrane thickenings. Those on hair cells (Figure 4.4A), while they lack obvious membrane specializations, are distinguished by a subsynaptic cistern, a flattened sac in the hair-cell cytoplasm immediately opposite the efferent bouton. The cistern is continuous with the smooth endoplasmic reticulum (A. Lysakowski, personal communication), bearing a resemblance to intracellular Ca^{2+} stores in other tissues.[164] This raises the possibility that one function of efferent synapses on hair cells is to regulate cytosolic $Ca,^{2+}$ especially in the region between the cistern and the hair-cell plasmalemma. A small number of boutons contacting hair cells appear to contain reciprocal synapses.[54,59,93,198] Here, an individual ending has a typical afferent (ribbon) synapse in proximity with an accumulation of vesicles facing a subsynaptic cistern. The origin of the reciprocal synapses is unclear. They could belong to afferent neurons with their cell bodies in the vestibular ganglion or to efferent neurons with their cell bodies in the brainstem

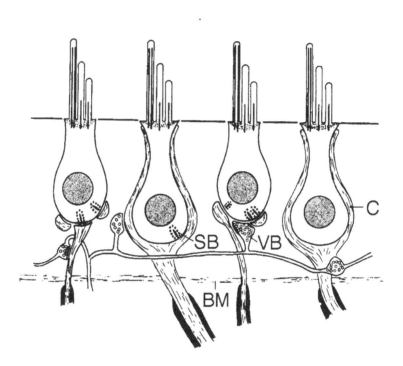

FIGURE 4.3 The branching of a single efferent axon in the neuroepithelium is depicted. Abbreviations: BM, basement membrane; C, calyx ending; SB, synaptic body; VB, vesiculated bouton. (From Smith, C.A. and Rasmussen, G.L. In: Third Symposium on the Role of the Vestibular Organs in Space Exploration, pp 183-201. US Govt Printing Office, Washington D.C., 1968.)

4.4. PHYSIOLOGY

4.4.1. RESPONSES OF AFFERENTS TO EFFERENT ACTIVATION

Electrical activation of efferent fibers results in an inhibition of afferent activity in auditory[6,83] and lateral-line organs.[72,73,202,204,216] The inhibition includes a reduction of the background discharge and a decrease in the responses to sensory stimulation.

Quite a different situation exists in vestibular organs. In mammals, efferent activation excites afferents as evidenced by an increase in background discharge (Figure 4.5).[92,157] The same is true of the great majority of afferents in the toadfish (Figure 4.6).[22,107]

Efferent responses are more heterogeneous in the posterior crista of frogs[16,199-201] and turtles[26]: some afferents are excited, others are inhibited, and still others show a mixed inhibitory-excitatory response (Figure 4.7)

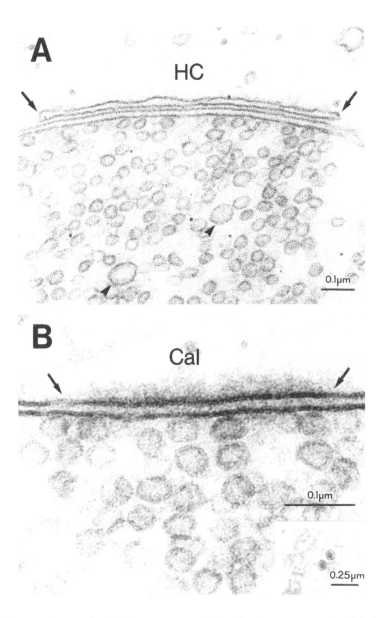

FIGURE 4.4 Electron micrographs of efferent synapses. A) An efferent synapse on a type II hair cell (HC) from a chinchilla posterior crista. Arrows delimit the subsynaptic cistern. Arrowheads point to two large vesicles which, for these fixation conditions, fail to show dense cores. B) An efferent synapse on a calyx ending (Cal) from a chinchilla posterior crista showing asymmetric membrane specializations. Arrows delimit postsynaptic density. Only small, translucent vesicles are seen. *Inset* shows two large dense-core vesicles in an efferent terminal from the horizontal canal of a skate. (Courtesy of Dr. Anna Lysakowski. With permission.)

In the toad, both excitation and inhibition were seen in all vestibular organs, including the three semicircular canals, the utricular macula, and the lagena.[232] The one exception was the saccular

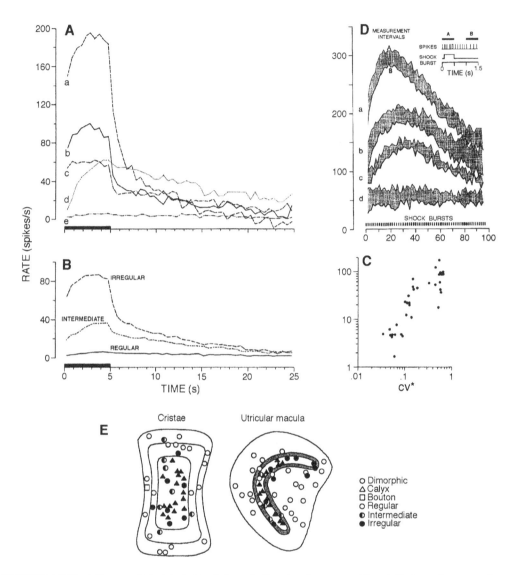

FIGURE 4.5 The influence of electrical stimulation of efferent pathways on afferent discharge in mammals. (A) Increase in firing rate for five vestibular-nerve afferents during and after a shock train of 5-s duration, 333 shocks/s. Bins, 0.5 s. Responses of four irregularly discharging (a - d) and one regularly discharging (e) afferent. (B) Responses averaged for 14 irregular, 10 intermediate and 10 regular vestibular-nerve afferents. Same stimulation parameters as in (A). (C) Relation between the response, averaged over the 5-s shock train, and a normalized coefficient of variation (CV*) for 34 vestibular-nerve afferents. The more irregular the discharge, the higher is the CV*. (D) Firing rates during (A) and after (B) successive shock trains of 0.4-s duration repeated every 1.5 s; shock rate, 200/s. Data from four vestibular-nerve afferents. Counting intervals (A and B) are depicted in the inset. The fast response is illustrated by the difference between the A and B curves. The slow build-up of discharge is evident from either curve. (E) Maps of the cristae and utricular macula in the chinchilla showing the locations of intra-axonally labeled units. The cristae are divided into central, intermediate, and peripheral zones; the macula into a ribbon-shaped striola, a surrounding juxtastriola (shaded), and a large peripheral extrastriola. The type of unit (calyx, dimorphic or bouton) and its discharge regularity (regular, intermediate or irregular) are indicated (see KEY). (A - C from Goldberg, J.M. and Fernández, C.J. Neurophysiol. 43: 986, 1980. D from McCue, M.P., and Guinan J.J. Jr, *J. Neurosci.* 14: 6071-6083. E from Baird, R.A., Desmadryl, G., Fernández, C., and Goldberg, J.M., *J. Neurophysiol.*, 60: 182-203, 1988. With permission.)

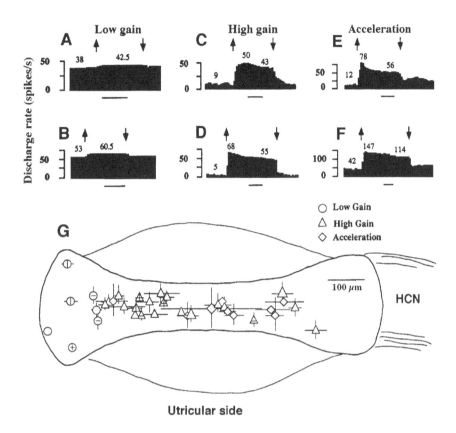

FIGURE 4.6 Firing rates during electrical stimulation of efferent pathways in the toadfish. Six horizontal-canal afferents categorized as low-gain (A, B), high-gain (C, D), or acceleration (E, F) units. Numbers above histograms are firing rates at selected points. Start and end of shock train indicated by arrows. Bars, 10 s. Shock frequency, 100/s. (G): Locations of dye-filled units of various physiological classes (see KEY) are placed on a standard map of the horizontal crista. Lines indicate approximate length of terminal field in longitudinal and transverse directions. The horizontal-canal nerve (HCN) approaches the crista from the right, (A-F from Boyle, R., and Highstein, S.M., *J. Neurosci.,* 10: 1570, 1990. G from Boyle, R., Carey, J.P., Highstein, S.M., *J. Neurophysiol.,* 66: 1504, 1991. With permission.)

macula, where almost all fibers were inhibited. It may be relevant that the saccular organ in anurans functions as a vibratory, rather than as a vestibular receptor.[36] While the effects of electrical stimulation of efferent fibers has not been studied in birds, responses have been recorded from horizontal-canal afferents in response to mechanical stimulation of the contralateral horizontal canal. Both excitatory and inhibitory responses have been recorded.

Besides its effects on background discharge, efferent activation can influence the response of vestibular afferents to natural stimulation. In the squirrel monkey[92] and in the toadfish,[22] efferent excitation is associated with a small (30 to 60 percent) decline in rotational gain without any obvious effect on response dynamics. The reduction in gain can be interpreted in terms of classic synaptic actions: whether these are excitatory or inhibitory or involve presynaptic or postsynaptic inputs, they should be associated with an increase in membrane conductance and a gain decrease. A different result is obtained when audio frequencies are used: efferent stimulation now results in a dramatic (two- to four-fold) increase in the amplitude of the phase-locked response to 800-Hz tones.[157] To explain the gain increases, it might be suggested that membrane conductance is decreased by efferent activation. There are two difficulties with the suggestion: (1) it does not explain the gain decreases seen with more traditional, low-frequency vestibular stimulation; and (2) at 800 Hz, impedance should be so dominated by membrane capacitance that even a large change in membrane conductance should have only a marginal influence on high-frequency gain. Because the high-frequency observations cannot be explained by a classic synaptic action, they raise questions as to the nature of efferent actions.

There are striking variations in the efferent responses of afferents innervating different parts of the neuroepithelium and differing in their responses to natural stimulation. To understand the diversity of efferent responses, we need to consider the regional organization of the endorgans. Since the organization differs across species, separate sections are devoted to mammals, toadfish, and turtles.

4.4.1.1. Mammals

Repetitive electrical stimulation results in an increase in background activity of afferents innervating all five vestibular organs in the squirrel monkey.[92] Typical responses are illustrated for five units in Figure 4.5A. Four of the units had an irregular spacing of action potentials; for the fifth unit, the spacing was regular. Much larger responses are seen in the irregular units. This was a general finding, as is illustrated by the averaged responses for populations of regular, intermediate and irregular units (Figure 4.5B) and by the scatterplot between efferent response magnitude and cv*, a normalized coefficient of variation that serves as a measure of discharge regularity (Figure 5C). Responses consist of two components. A fast component is responsible for the abrupt changes in rate at the onset and termination of the shock train, while a slow component accounts for the gradual build-up of discharge during the train and the gradual decline of the response in the post-stimulus period. A second way of distinguishing fast and slow responses is seen in Figure 4.5D.[157] Here, the fast component reaches a maximum during each 500-ms shock train. In contrast, the slow component continues to increase during the first 20 to 40 seconds of efferent stimulation and then declines over the next 60 to 80 seconds. Fast responses are usually larger than slow responses in irregular units; for regular units the reverse is true. In fact, the efferent responses seen in regular units, are so small and so slow that it seems unlikely that they function to alter sensory transmission on a short time scale.

Recent intra-axonal labeling studies allow us to relate the above results to the morphology and location of afferents. Results are illustrated in Figure 4.5E for the cristae and the utricular macula. There are three kinds of afferents in mammals.[63-65] *Calyx fibers* innervate type I hair cells in the central zone of the crista and the striola of the utricular macula. *Bouton fibers* supply type II hair cells in the peripheral zone of the cristae and in the peripheral extrastriola of the macula. *Dimorphic fibers,* which are the most numerous kind of afferent, provide a mixed innervation to type I and type II hair cells in all zones of the cristae and the macula. The discharge regularity of an afferent is more closely related to its location in the neuroepithelium than to the types and numbers of hair cells it innervates.[13,91,154] Both calyx and dimorphic units innervating the central (striolar) zone are irregularly discharging, while dimorphic fibers innervating the peripheral (peripheral extrastriolar) zone are regularly discharging. It has proved difficult to label bouton fibers. Indirect evidence suggests that the latter units are regularly discharging.[154] On this basis of these findings, it can be inferred that calyx fibers and centrally located dimorphic fibers have large efferent responses with conspicuous fast components. Peripherally located dimorphic and bouton units can be presumed to have small, largely slow responses.

The differences in the efferent responses of afferents innervating the central and peripheral zones cannot be explained by regional differences in the density of efferent innervation. AChE-positive efferent fibers are densest in the peripheral zone,[169] paralleling the density of hair cells[65,140] and afferent fibers.[63-65] Quantitative electron microscopy indicates that the number of efferent boutons supplying individual type II hair cells and individual calyx endings is relatively uniform throughout the crista.[93,152] Different efferent neurons may innervate the separate zones of the neuroepithelium[182-184] and could differ in the kinds or amounts of the neurotransmitters that they release. A regional difference in the receptors present on hair cells and afferent processes must also be considered. In addition, there are variations in the spike-encoding mechanisms of regular and irregular afferents: irregularly discharging afferents are more sensitive to externally applied galvanic currents[94] and the same can be expected for synaptic inputs, including those from efferents.[223]

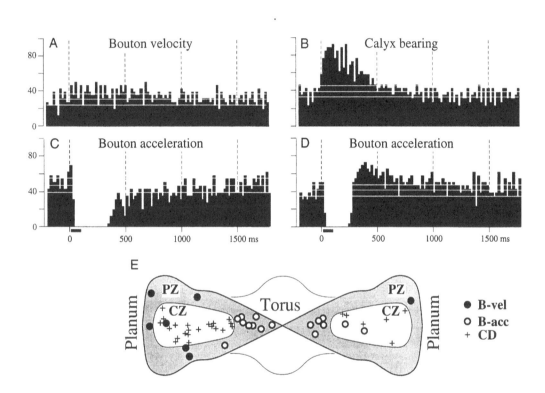

FIGURE 4.7 Firing rates during electrical stimulation of efferent pathways in the turtle. Four posterior-canal afferents categorized as: (A) bouton velocity (B-vel); (B) calyx bearing (CD); or (C, D) bouton acceleration (B-acc) units. Shock trains are 100 ins (bars) with a shock rate of 200/s. (E): Locations of dye-filled units of various physiological classes (see KEY) are placed on a standard map of the posterior crista separated into central (CZ) and peripheral (PZ) zones. (From Brichta, A.M. and Goldberg, *J.M., Ann. New York Acad. Sci.,* 781: 183, 1996. With permission.)

4.4.1.2. Toadfish

Only type II hair cells are found in fish and these are innervated by bouton afferents.[20,151,209,254] Based on their physiology, bouton afferents in the toadfish horizontal crista have been divided into three categories: low-gain, high-gain and acceleration units.[20,21] As their names imply, low-gain units have lower sensitivities to head rotations than do high-gain units; the sinusoidal responses of the former units are nearly in phase with angular head velocity, while the latter units show phase leads. Acceleration afferents have even larger gains and phase leads. The three groups of afferents differ in their branching patterns and their locations in the crista.[20] As illustrated in Figure 4.6G, low-gain afferents have the smallest dendritic fields and are located near the edges of the crista neighboring the planum semilunatum. Acceleration afferents have considerably larger fields located near the center of the organ. High-gain afferents are intermediate in field size and location. Excitatory efferent responses differ in magnitude for the three groups (Figure 4.6A-F). Increases in background discharge are smallest for low-gain afferents and highest for acceleration afferents. Efferent actions are thus largest near the center of the organ. The cellular basis for this regional difference has not been determined.

In one study,22 inhibitory responses were seen in only 4/144 (less than 3 percent) of afferents. The inhibited fibers were classified as high-gain or acceleration and can be presumed to be centrally located.

4.4.1.3 Turtle

The turtle posterior crista is divided into two triangular-shaped hemicristae (Figure 4.7E), which meet at a raised elevation or torus. Each hemicrista has a central zone surrounded by a peripheral zone, the latter extending from the torus to the planum semilunatum. As in mammals, turtles have type I and type II hair cells.[126] Type I and a smaller number of type II hair cells are found in the central zone and are innervated by calyx, dimorphic, and bouton afferents.[27] Only type II hair cells are found in the peripheral zone and are supplied by bouton units. Intra-axonal labeling studies,[26] summarized in Figure 4.7E, indicate that bouton afferents can be divided into two groups. Bouton-velocity (B-vel) units are located near the planum; they are regularly discharging, have low rotational sensitivities, and their sinusoidal responses are nearly in phase with head velocity. Bouton-acceleration (B-acc) units are found near the torus and have an irregular discharge, very large rotational gains, and large phase leads. Calyx and dimorphic units, which are confined to the central zone, have similar response properties and are placed into a single "CD" category; they are irregularly discharging and have gains and phases intermediate between those of B-vel and B-acc units. The few bouton units found in the central zone resemble peripheral bouton units at the same location along the long axis of the crista.

Efferent responses differ for the three afferent groups.[26] B-vel (Figure 4.7A) and CD units (Figure 4.7B) are excited by efferent activation, with the CD units showing much larger responses. B-acc units are inhibited (Figure 4.7C) or show mixed responses consisting of an early inhibition and a late excitation (Figure 4.7D). Once again, the cellular mechanisms responsible for these differences are unknown.

4.4.2 Sites of Efferent Actions: Presynaptic or Postsynaptic

There is a presynaptic efferent innervation of type II hair cells and a postsynaptic innervation of afferent processes (including calyx endings). Which of the variety of efferent responses are mediated by the two kinds of efferent contacts?

In the frog posterior crista, presynaptic efferent synapses can mediate both excitatory and inhibitory responses.[16,199-201] Recordings from afferent terminals show that neither kind of response is associated with monosynaptic efferent-evoked EPSPs or IPSPs; rather, there is a modulation in the rate of miniature excitatory postsynaptic potentials (mEPSPs) produced by synaptic transmission from the hair cell with no change in mEPSP amplitude. Postsynaptic efferent actions have been observed in the toadfish horizontal crista and lagena.[106,107] In both organs, efferent stimulation increases afferent discharge and monosynaptic efferent-evoked EPSPs are recorded from afferent terminals. Studies of presynaptic actions have been confined to organs showing efferent inhibition of afferent discharge. These include lateral-line organs,[73] the goldfish sacculus,[83] and the turtle basilar papilla.[7] In all cases, IPSPs have been recorded from hair cells. Local application of ACh results in IPSPs in outer hair cells isolated from the mammalian cochlea,[52,111] which suggests that an inhibition of these hair cells is responsible for the efferent effects of MOC activation.

Vestibular type I hair cells are only rarely contacted by efferent boutons. Based solely on this anatomical arrangement, the large excitatory efferent responses seen in calyx afferents[26,92,157] should reflect a postsynaptic efferent innervation.

4.4.3 FUNCTIONS OF THE EFFERENT VESTIBULAR SYSTEM

In section 4.4.1, we reviewed the effects of efferent activation on afferent discharge. To understand the function of efferent neurons, we also need to know the conditions that determine their discharge. Most studies of efferent discharge properties have been done in fish and frogs. In many experiments, detaching the particular nerve branch from its peripheral connections eliminated afferent activity. Efferent neurons respond to vestibular stimulation, including activation of semicircular canals[17,33,34,89,102,180,211] and otolith organs.[134] The responses provide evidence that efferent neurons receive a convergent input from several vestibular organs in both ears. Efferent neurons in the toadfish receive a direct input from the vestibular nerve; monosynaptic EPSPs can be recorded from individual efferents on stimulation of nerve branches innervating several organs.[108] An anatomical study suggests that monosynaptic vestibular-nerve inputs to efferent neurons also exist in mammals.[255] Despite the direct vestibular inputs, rotational responses in toadfish efferents are inconsistent; for example, an efferent neuron may respond during several consecutive rotation cycles, but fail to do so on subsequent cycles.[101] Efferent neurons also respond to nonvestibular stimulation, including pressure applied to the skin,[180,211] passive movement of limbs,[180,211] and visual stimulation.[107,135] In the toadfish, responses to sensory stimulation have a long latency (~150 ms) and may outlast the stimulus by 500 ms or more.[106,107] The long latency and prolonged poststimulus responses suggest that polysynaptic pathways are involved.

Several workers have noted that efferents discharge in relation to active body movements.[89,180,211] For a small fraction of vestibular-nerve fibers in the goldfish[47,210] and rabbit[47] discharge was modulated in association with saccadic eye movements. Most (if not all) of the saccade-related fibers in the goldfish were thought to be efferents, while some of the saccade-related fibers in rabbit may have been afferents. Saccade-related activity was not observed in afferent recordings from alert monkeys.[127,148] Efferent activation in the toadfish is most clearly associated with arousal, a stereotyped behavior that can be evoked by touching the snout.[107] The behavior can be a prelude to movement. Arousal leads to efferent activation and an efferent-mediated excitation of afferents.[22,107] Both effects begin before the actual movement.

The results in toadfish[107] and other animals[89,180,211] suggest that vestibular efferents respond in anticipation of movement. Recordings from lateral-line efferents can be interpreted similarly.[195,204,238,239] Taking this as a clue, we can speculate that one function of the efferent vestibular system is to switch the vestibular organs from a "postural" mode to a "volitional" mode.[26] Efferent excitation, by raising the background discharge of afferents and lowering afferent gain, can be viewed as a mechanism for reducing the inhibitory silencing of afferent discharge that would normally take place during large, voluntary movements. The larger efferent responses seen in irregular units in mammals[92,157] and in high-gain and acceleration afferents in toadfish[20,22] can be rationalized by the fact that these afferents are more easily silenced than are low-gain, regular afferents. What is the role of efferent inhibition? At least in the turtle posterior crista, inhibition is specifically targeted to B-acc afferents.[26] These afferents are unusual in that they have very large rotational gains during small (<5 deg/s) head rotations, but become saturated for even modest (20 deg/s) excitatory (as well as inhibitory) head rotations. Presumably, B-acc afferents play an important role in monitoring the small head movements involved in the control of stationary posture, but are of little value in monitoring larger, voluntary movements. The efferent inhibition of B-acc units is powerful and is likely to eliminate their discharge. It is unclear why it might be advantageous to temporarily silence the discharge of these units, rather than ignoring their signals during large movements. Another puzzle relates to acceleration units in the toadfish. These units resemble B-acc units in turtles in that they have very high gains and large phase leads.[21] Rather than being inhibited by efferent activation, the toadfish acceleration afferents show large excitatory efferent responses.[20,22] Obviously, much more needs to be learned about the behaviorally meaningful conditions leading to efferent discharge and about the consequent efferent-mediated effects on afferent discharge.

4.5 NEUROCHEMISTRY

In this section, we review the role of substances that may function as efferent neurotransmitters. Attention is confined to ACh, ATP, GABA, and to three classes of neuropeptides - calcitonin gene-related peptide (CGRP), the enkephalins, and substance P. For each of these potential transmitters, we consider three topics: (1) evidence for its localization in efferent neurons; (2) pharmacological studies suggesting its involvement in efferent actions; and (3) possible cellular mechanisms. In addition, there is enough information on cholinergic receptors to discuss the receptor subtypes present in hair cells and afferents.

4.5.1 ACETYLCHOLINE (ACH)

It is now accepted that acetylcholine (ACh) is an efferent neurotransmitter.[97,133]

4.5.1.1 Cellular Location

Based on the presence of the synthetic enzyme, choline acetyltransferase (ChAT), most efferent vestibular neurons are cholinergic. ChAT has been found associated with efferent neurons in fish,[25,42,193] amphibians,[95,146] reptiles,[194] birds,[163] and mammals.[32,179,213] The presence of ChAT can be taken as a necessary condition for a neuron to be cholinergic. The same is not true for the degradative enzyme AChE, which in the brain is sometimes present in neurons lacking ChAT-immunoreactivity.[70,139] Despite the reservation, AChE is widely distributed among vestibular efferent neurons in mammals[32,85,92,124,179,213,251] and in fish.[159]

4.5.1.2 Receptor Subtypes

The actions of ACh depend on the receptors present in target structures. Classically, cholinergic receptors have been divided into nicotinic (nAChR) and muscarinic (mAChR) categories based on the agonists and antagonists that affect them. Recent cell and molecular studies have indicated that there are fundamental differences in the structure and modes of operation of the two kinds of receptors40 (see also Chapters 11 and 12). In addition, there is diversity within both the nAChR and mAChR families; nAChRs are pentameric ligand-gated ion channels, while mAChRs are single transmembrane proteins coupled to guanine nucleotide-binding G proteins. In turn, the G proteins can act directly on ion channels or on membrane-bound enzyme systems that regulate second-messenger systems. The differences between nAChR and mAChR receptors are reflected in the speed with which they act. The ion channels of nAChRs are activated with kinetics of 1 to 100 ms, while mAChRs operate more slowly (>100 ms).

Nicotinic AChRs can be divided into three categories: (1) nAChRs of skeletal muscle and fish electric organs; (2) neuronal nAChRs that do not bind α–bungarotoxin (α–BTX); and (3) neuronal nAChRs that bind α–BTX.142 Members of each category differ in the combination of subunits they employ. Adult skeletal muscle nAChRs, for example, are composed of $(\alpha1)2(\beta1)\epsilon\delta$ subunits. Neuronal nAChRs are also pentameric; they may assemble as heteromeric channels composed of α and β subunits, or as homomeric channels made up of a single type of a subunit.141,142,158 At present, eight α ($\alpha2$ - $\alpha9$) and three β ($\beta2$ - $\beta4$) neuronal nAChR subunits have been cloned.57,141,142,158 There are differences between the channels involving $\alpha2$-$\alpha6$ subunits and those involving $\alpha7$-$\alpha9$ subunits. The former are heteromers and generally do not bind α–BTX, while the latter are homomers and bind α–BTX.51,142 Each subunit contributes to the pharmacological properties of the fully assembled nAChR.

Five mAChR subtypes (m 1 - m5) have been cloned.115 As is summarized in Table 1, mAChRs can differ in the G proteins and second-messenger pathways they use.40,115

Table 1: Characteristics of *muscarinic acetylcholine receptor subtypes*

	m1	m2	m3	m4	m5
Stimulate PI metabolism	+++	+	+++	+	+++
Increase cAMP levels	+	0	+	0	+
Decrease cAMP levels	0	+++	0	+++	0
Adenyly cyclase inhibition	0	+++	0	+++	0
Release intracellular Ca^{2+}	+++	0	+++	0	+++
Inhibit Ca^{2+} conductances	0	++	0	++	nd
Activate Ca^{2+} dependent K^+ and Cl^- conductances	+++	+	+++	0	+++
G protein subfamily	G_q	G_I	G_q	G_I	G_q

There are subfamilies of G proteins: Gs stimulates and Gi inhibits adenylyl cyclase, while Gq activates phospholipase C;[88] the m1, m3 and m5 mAChR subtypes are linked through Gq and phospholipase C to cause inositol 1,4,5-triphosphate (IP$_3$)-mediated intracellular Ca^{2+} release and diacylglycerol-mediated activation of protein kinase C; m2 and m4 mAChR subtypes inhibit adenylyl cyclase via Gi proteins.

Several nAChR subunits have been localized to hair-cell organs: α9 is found in hair cells of the mammalian cochlea[51,90] and may also be present in vestibular hair cells;[105] α7 has been reported in chick auditory hair cells;[3] and α7 resembles α9 in being blocked by strychnine, but the two receptors can be distinguished by the effects of α–BTX and nicotine.[57,142] Of the two nAChR subtypes, α7 is blocked at much lower concentrations by α–BTX. Nicotine acts as an agonist of α7, but not of α9. *In situ* hybridization studies show that the α4 and β2 subunits are expressed within vestibular ganglion cells, suggesting that postsynaptic efferent effects may be mediated by α4β2 nAChRs.[249] In addition, ganglion cells may express α4 - α7 subunits.[105] In rat[249] and human[121] vestibular end organs, α–BTX binds to afferent calyces and to type II hair cells.

Messenger RNAs (mRNAs) have been found in the vestibular periphery for several mAChRs, including ml, m2 and m5 in humans and ml through m5 in the rat.[248] mRNAs for two Gs and one Gi proteins may also be present (M. Troyanovskaya and P. Wackym, unpublished data). Using reverse transcription polymerase chain reaction (RT-PCR), Tachibana et al[235] found that in mRNAs for several G protein α subunits, a Gi subunit has been localized to inner and outer hair cells.[234]

4.5.1.3 Pharmacology

There have been relatively few studies of efferent pharmacology in vestibular organs. These have been confined to the frog posterior crista, where the major efferent actions are presynaptic.[16,199-201] Bernard et al.[16] studied the effects of various cholinergic agents on background activity and on the responses to efferent activation. Results are illustrated in Figure 4.8 and are consistent with two conclusions: (1) fast efferent actions, including the excitation and the inhibition seen in different afferents, involve nicotinic receptors on hair cells; (2) and although slow responses have not been described following efferent stimulation in the frog, there are muscarinic receptors on hair cells that could mediate a slow excitatory action.

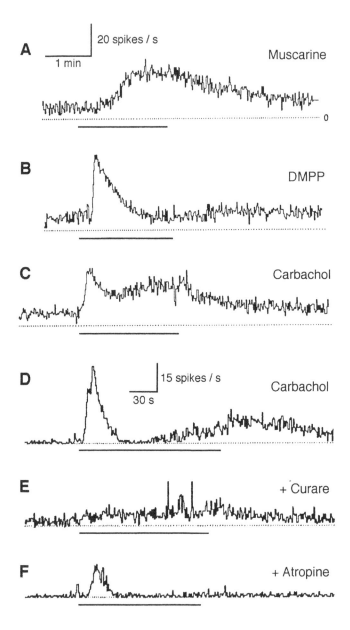

FIGURE 4.8 Cholinergic pharmacology in the frog posterior crista. Responses of different afferents to drug application (bars). (A) a muscarinic agonist (muscarine, 0.2 mM) results in a slow, nondesensitizing response; (B) a nicotinic agonist (DMPP or 1,1-dimethyl-4-phenylpiperazine, 0. 1 mM) leads to a fast, desensitizing response; and (C) a mixed agonist (carbachol, 0.2 mM) produces both a fast and a slow response. D - F: responses of a single afferent. Carbachol (0.2 mM) produces fast and slow responses; only the fast response is eliminated when a nicotinic antagonist (curare, 5 uM) is added to the bath and only the slow response disappears when a muscarinic antagonist (atropine, 5 uM) is added. Top calibration is for A - C; bottom calibration, for D - F. (From Bernard, C., Cochran, S.L., and Precht, W., *Brain Res.,* 338: 225-236, 1985. With permission.)

Similar findings were reported by Rossi et al.[201] for efferent inhibition, but not for efferent excitation. Thus, various nicotinic antagonists were found to abolish inhibitory responses, but had no effect on excitatory responses. A slow excitation was the predominant response to cholinergic agonists observed in whole-nerve recordings from the frog posterior crista.[11,170] ACh evoked both depolarizing and hyperpolarizing currents from frog vestibular hair cells.[114]

More *thorough* pharmacological studies of presynaptic efferent inhibition have been done in two kinds of nonvestibular preparations. The first involves the medial olivocochlear (MOC) system in mammals. MOC fibers are stimulated in the floor of the fourth ventricle and effects are monitored by recording cochlear potentials (see Section 4.2). The second approach involves the use of solitary hair cells, including outer hair cells from the mammalian cochlea,[52,111] short hair cells from the chick cochlea,[80,81,218] and hair cells from the frog sacculus.[233,264] In all cases, application of ACh gives rise to hyperpolarizing IPSPs. Similar results have been obtained from the two kinds of preparations. Effects are blocked by curare and by atropine, usually requiring higher concentrations of the latter.[19,52,81,111,226,26] Strychnine is a potent blocker,[46,81,111,226] as is α–bungarotoxin.[69,81,111,264] Carbachol is an effective agonist, but muscarine and nicotine are ineffective.[81,264] The pharmacological profile is unlike that of any muscarinic receptor,[115] but matches that of the recently cloned α9 nicotinic receptor.[57]

4.5.1.4 Cellular Mechanisms

Studies in chick cochlear hair cells indicate how ACh might mediate presynaptic inhibition. According to Fuchs and Murrow,[80,81] ACh opens a nicotinic channel and this results in a relatively nonspecific cation current. A fraction of the inward (depolarizing) current is carried by $Ca,^{2+}$ which leads to a larger, more prolonged calcium-activated potassium (K_{Ca}) outward (hyperpolarizing) current. A different scheme was envisioned by Shigemoto and Ohmori.[217,218] The latter authors agree that the efferent inhibition is due to a rise in cytosolic free calcium, which activates a K_{Ca} channel. However, they disagree as to the ACh receptor involved believed to be muscarinic; the source of the Ca^{2+} postulated as being intracellular; and the signaling mechanism, which they contend involves a G-protein linked to inositol triphosphate (IP_3). More recent studies favor the Fuchs-Murrow[80,81] interpretation for a variety of hair cells.[52,111,264] In particular, ACh-induced IPSPs are abolished when external Ca^{2+} is removed. An even more persuasive reason is the aforementioned pharmacological resemblance between the efferent actions and the α9 nAChR. At the same time, the influx of Ca^{2+} through a nicotinic channel could be supplemented from intracellular stores, e.g., from the subsynaptic cistern. Based on studies in other systems, such mobilization can be accomplished in two ways: 1) a calcium-mediated Ca^{2+} release triggered by the influx of Ca^{2+} through nicotinic channels,[164,171]; (2) an IP_3-mediated Ca^{2+} release as a consequence of the activation of muscarinic receptors.[115]

Mechanisms similar to those proposed by Fuchs and Murrow[80,81] may operate to produce the presynaptic inhibition seen in the frog posterior crista.[16,199-201] In contrast, a nicotinic action not linked to a K_{Ca} channel would be expected to produce presynaptic excitation. The inward current through nicotinic channels would depolarize the hair cell. Moreover, most nAChR ion channels show a significant permeability to $Ca.^{2+57,141,142,158}$ The consequent rise in cytosolic free Ca^{2+} could enhance excitation by facilitating afferent transmitter release. A pure nicotinic action could lead to the fast excitatory responses seen during efferent stimulation and be mimicked by nicotinic agonists.[16] Given its relatively slow kinetics, the mobilization of Ca^{2+} from internal stores could mediate the slow excitation produced by muscarinic agonists.[16]

The cellular mechanisms of postsynaptic efferent actions have received less attention. Monosynaptic EPSPs recorded from toadfish afferents are fast[106,107] and are likely to be mediated by nicotinic channels. The fast excitation seen in calyx afferents[26,92,157] may also reflect a nicotinic action. Yamaguchi and Ohmori,[260] working with cultured chick cochlear ganglion cells, describe an M current, a voltage-sensitive K^+ current that is suppressed by cholinergic agonists; the cholinergic suppression is blocked by muscarinic antagonists and may involve m1 and/or m3 receptors.[82] In analogy with sympathetic ganglion cells,[28] M current suppression could give rise to a slow EPSP.

4.5.2 ADENOSINE 5'-TRIPHOSPHATE (ATP)

ATP can function as a classic neurotransmitter, acting on ligand-gated ion channels, and also as a neuromodulator, working by way of G proteins to exert several actions, including the release of Ca^{2+} from intracellular stores.[101] Ionotropic receptors are referred to as P_{2X}; metabotropic receptors include P_{2Y} and P_{2U} categories. Several of the receptors have now been cloned.[24,103,113,177,245]

Recently, ATP has been implicated in presynaptic efferent excitation.[200]

4.5.2.1 Cellular Location

There is no evidence for a separate set of purinergic efferent axons. Based on studies in motor-nerve terminals, it can be postulated that ATP is colocalized with ACh in the same neurons and, most likely, in the same synaptic vesicles.[53,215,219]

4.5.2.2 Pharmacology

Rossi et al.200 monitored transmission from hair cells by recording mEPSPs from afferent terminals in the frog posterior crista. The afferents were classified as being excited or inhibited by electrical activation of efferent fibers. Perilymphatic application of ATP increased the rate of hair-cell mEPSPs in afferents that were also excited by efferent stimulation, but had no effect on afferents showing efferent inhibition. The ATP responses were transient; once they disappeared, the afferents could no longer be excited by efferent stimulation. Aubert et al.10,11 found that perilymphatic ATP increased the background activity in whole-nerve recordings from the frog posterior crista; the actions of various agonists and antagonists suggested that a P2Y receptor was involved.

4.5.2.3 Cellular Mechanisms

There are P_{2X} receptors in a variety of hair cells,[9,112,167] including mammalian vestibular type I and type II hair cells.[192] ATP results in a rapid inward current.[9,112,167,192] P_{2X} receptors in hair cells, like those found elsewhere,[15] open a nonselective cation channel, which shows appreciable permeability to $Ca.^{2+}$ [167] The ATP-induced inward current will depolarize the hair cell and is presumably responsible for an observed rapid rise in intracellular free calcium.[9,192,217] A second, slower rise in intracellular free Ca^{2+} takes place even in the absence of extracellular Ca^{2+} [9,217] suggesting the presence of a P_{2Y} or some other metabotropic receptor. ATP is most effective in producing a P_{2X} action when applied to the apical surface of outer hair cells.[9,112] The localization of P_{2X} receptors at the hair-cell apex is supported by binding studies.[165,166] Since efferent synapses are located at the hair-cell base, an apical location would seem to eliminate P_{2X} as an efferent receptor in outer hair cells. The possibility remains that there is a P_{2Y} efferent receptor in the cochlea and that P_{2X} and/or P_{2Y} are efferent receptors in vestibular organs.

If ATP proves to be an efferent neurotransmitter, it could have several actions besides depolarizing hair cells and increasing cytosolic $Ca.^{2+}$ At other cholinergic synapses, ATP can modulate transmission either by acting directly on nicotinic receptors[168] or by phosphorylating such receptors via second-messenger systems.[78,150] The presence of ectonucleotidases[166] will degrade ATP to adenosine, which can act on several classes of G protein-coupled adenosine (P_1) receptors.[175] Among its other actions, adenosine has been found to decrease neurotransmitter release in a variety of systems.[55,100,191] There is evidence that adenosine may have a similar action in the frog posterior crista.[31]

4.5.3 GAMMA-AMINOBUTYRIC ACID (GABA)

The role of GABA in vestibular neurotransmission has proved elusive. That GABA might be an afferent transmitter in hair-cell organs was first proposed by Flock and Lamm,[71] who found that

GABA was synthesized even by organs lacking an efferent innervation and that afferent discharge was blocked by the GABA$_A$ antagonist, picrotoxin. Other workers[5,243] have suggested that GABA is an efferent transmitter in vestibular organs, as is probably the case in the mammalian cochlea.[61] We first consider neurochemical and immunohistochemical evidence, most of which is consistent with the idea that GABA is an afferent neurotransmitter in vestibular organs. We then review pharmacological evidence, which seems inconsistent with this possibility.

4.5.3.1 Cellular Location

Confirmation that the labyrinth contains the synthetic enzyme, L-glutamic acid decarboxylase (GAD), was obtained in biochemical studies in several preparations.[146,162] More recently, immuno-histochemistry has been used to localize GAD, the inactivating enzyme (GABA-transaminase or GABA-T), and subunits of the GABA$_A$ receptor. GAD-IR is found presynaptically in type I and type II hair cells, but not in postsynaptic elements.[147,244] In contrast, GABA-T-IR is found postsynaptically in calyx endings, parent fibers in the stroma, and ganglion cells,[147,244] as well as in some hair cells[147]; there is disagreement whether the degradative enzyme is found intracellularly[147] or extracellularly.[244] GABA$_A$ receptor subunits are localized postsynaptically.[74,131] Assuming a cytoplasmic location of GABA-T, results are consistent with a scheme in which GABA is synthesized and presumably released from hair cells, acts on postsynaptic targets, and is removed and transaminated by hair cells and terminals.

Attempts to localize GABA have used antibodies that were raised to its conjugation to bovine serum albumin with glutaraldehyde. The first such study was done in the squirrel monkey.[243] GABA-like-IR was present in fibers and bouton terminals throughout the cristae and the maculae. No calyx endings were stained. The distribution of the GABA-positive fibers and terminals is consistent with their status as efferent processes. In later studies in the chick[156,244] and pigeon[156] GABA antibodies stained hair cells, as well as presumed efferent fibers. The number of stained fibers in the chick was lower than that seen in the squirrel monkey or pigeon.[156] Contradictory results have been reported in rodents. GABA-like-IR has been variously described as being present only in calyx endings (guinea pig)[49]; being present in hair cells, calyx endings, and nerve fibers (guinea pig)[145]; or not being present (guinea pig and rat)[156].

GABAergic efferents should be capable of synthesizing GABA and consequently should be GAD-IR. Fibers meeting this criterion have not been described.[147,244] Nor have efferent cell bodies been shown to be positive for GABAergic markers.[32,179,243]

4.5.3.2 Pharmacology

Felix and Ehrenberger62 reported that GABA increased afferent discharge in the cat saccular macula, while GABAA antagonists (picrotoxin and bicuculline) had the opposite effect. In contrast, later studies5,39,99 found that GABA had, at most, a small excitatory effect on afferent discharge when compared to the excitatory amino acids and that GABAA antagonists did not alter unitary discharge. Similarly, negative results were obtained in studies of the effects of GABA and GABAA antagonists on the cochlear compound action potential.132 Annoni et al.5 made two observations, which would seem to rule out GABA as an afferent neurotransmitter. First, GABAA antagonists did not reduce the size of mEPSPs recorded from afferent terminals in the frog posterior crista. Second, the slight excitation produced by GABA was eliminated when transmission from the hair cell was blocked by high Mg2+/ low Ca2+ conditions. The last result suggested that the action of GABA was exerted on hair cells and might reflect a role for GABA in efferent transmission.

When the immunohistochemical and pharmacological studies are considered together, it has to be concluded that there is contradictory evidence that GABA is an afferent neurotransmitter and inconclusive evidence that it is an efferent neurotransmitter.

4.5.4 CALCITONIN GENE-RELATED PEPTIDE (CGRP)

The presence of dense-core vesicles in efferent terminals suggests the presence of neuroactive peptides. Of the various neuropeptides that have been considered, evidence for an involvement in efferent function is strongest for CGRP.

4.5.4.1 Cellular Location

aCGRP is a 37-amino-acid neuropeptide that is alternately spliced with calcitonin from a single gene transcript.[196,197] A second gene gives .rise to a closely related neuropeptide, bCGRP.[196] aCGRP is widely distributed in the central and peripheral nervous systems,[8,119,155] including auditory,[61] vestibular,[120,236,237,250] and lateral-line[1] efferent neurons.

In a study by Tanaka et al,[236,237] CGRP-positive vestibular efferent neurons were confined to the lateral part of group e. Wackym et al.[250] found a broader distribution, including the medial part of group e and the caudal pontine reticular (CPR) formation. CGRP-positive neurons project to both ipsilateral and contralateral vestibular organs.[237,250] Most CGRP-positive neurons also contain ChAT, suggesting a colocalization of ACh and the neuropeptide.[174] On reaching the neuroepithelium, the CGRP-positive neurons branch to provide a postsynaptic innervation of afferent chalices,[120,236,237,247,250] and of unmyelinated afferent dendrites[120,247,250]; contacts with type II hair cells, although present, are infrequent.[120, 247,250]

4.5.4.2 Pharmacology

The actions of CGRP in hair-cell systems have only been studied in lateral-line organs.[1,216] Application of CGRP results in an increase in afferent discharge. Recordings from afferent terminals reveal a parallel increase in the rate of hair cell mEPSPs, suggesting that CGRP exerts a presynaptic action. Electrical activation of efferent pathways in lateral lines results in a fast response consisting of an initial inhibition followed by excitation.[204,216] There is, in addition, a slow excitation that can persist for several minutes following efferent stimulation.[216] The slow response, like the excitatory response to CGRP, is unaffected by cholinergic blockers. On this basis, it has been suggested that CGRP may mediate the slow efferent action.

4.5.4.3 Cellular Mechanisms

The mechanisms responsible for the CGRP-induced slow response are far from clear. Shigemoto and Ohmori[217] observed that preincubation of chick cochlear hair cells with CGRP increased the ACh-evoked rise in intracellular Ca^{2+} without having any effect by itself. The neuropeptide has been shown to enhance Ca^{2+} currents in dorsal-root-ganglion neurons.[207] While a similar mechanism might explain the excitatory CGRP response seen in lateral lines,[1,216] it should be noted that the predominant fast efferent action seen in lateral-line hair cells is inhibitory and that the usual effect of increased cytosolic Ca^{2+} in such hair cells is a hyperpolarization.[218] In some neurons, CGRP can reduce voltage-gated K^+ currents,[206,265] which would provide another excitatory mechanism. Many of the actions of CGRP involve cAMP as a second messenger.[35,155,186,265]

The CGRP-mediated excitation seen in lateral lines presumably involves efferent synapses on hair cells.[216] In the case of the mammalian vestibular organs, CGRP-positive efferent fibers would appear to be preferentially targeted to calyx endings and other unmyelinated afferent processes.[120,236,237,250] This raises the possibility that effects may be different in lateral-line and vestibular organs.

Neuropeptides, as well as other neurotransmitters acting by way of second messengers, can have a variety of actions beyond those involved in the short-term modification of sensory transmission (see Chapter 9). Although such long-term actions have not been studied in hair cell organs, they should be kept in mind. To illustrate the versatility of neuropeptide signaling mechanisms, we can consider the effects of CGRP at the neuromuscular junction. This may provide a useful model for hair cell efferent systems, since in both cases ACh and CGRP are co-localized in the presynaptic terminal. As reviewed by Changeux et al.[35] and by Arvidsson et al.,[8] CGRP has the following actions at the neuromuscular junction: (1) it speeds up the desensitization of the postsynaptic nAChR receptor; (2) it enhances transmitter release from the presynaptic terminal; (3) it regulates the synthesis of the a subunit of the nAChR and the total concentration of nicotinic receptors in the postsynaptic membrane; and (4) it suppresses the disuse-induced sprouting of motor nerve terminals. The actions range from the neuromodulation of conventional synaptic transmission to the neurotrophic regulation of synaptic structure.

4.5.5 OTHER NEUROPEPTIDES

Enkephalins have been localized within the cochlear efferent system[61] and are also likely to be present in vestibular efferents. The majority of efferent cell bodies express preproenkephalin mRNA[205] and show met-enkephalin-like immunoreactivity.[179] On the other hand, Ylikoski et al.[263] found no enkephalinergic fibers within the vestibular nerve, the vestibular ganglion, or the neuroepithelium.

Substance P (SP) appears to be involved in afferent, but not in efferent neurotransmission. SP has been localized within a subpopulation of vestibular ganglion cells and afferent terminals,[116,117,156,240,241,261-263] but has not been found in efferent cell bodies[32,179] or in efferent terminals in the neuroepithelium.[241] There is an isolated report that CGRP and SP are colocalized in a subgroup of vestibular fibers found within and beneath the neuroepithelium.[242]

Although not normally detectable in efferent vestibular (EVS) neurons, the neuropeptide galanin and the growth-associated membrane phosphoprotein, GAP-43, become detectable following hemilabyrinthectomy.[172,173] As a similar increase in these substances is seen in other fiber systems following axotomy (for review, see Ref. 79), EVS neurons may be responding to damage of their axons, rather than to an imbalance of vestibular inputs. Consistent with this interpretation, galanin is only found in those EVS neurons that had projected to the damaged labyrinth.[172]

4.6 CONCLUSION

This chapter has reviewed the anatomy, physiology and neurochemistry of the efferent vestibular innervation. While much has been learned about the innervation since it was first described by Rasmussen and Gacek in 1958,[84,190], several questions remain unanswered. It is unclear whether there are several subpopulations of efferent neurons that might differ in their peripheral branching patterns, cellular targets in the end organs, physiological actions, or neurochemistry. In most hair cell organs, efferents influence afferent discharge by inhibiting hair cells. Efferent effects in vestibular organs are more diverse. Both excitation and inhibition are seen and are targeted to specific populations of afferents differing in their response properties and locations in the neuroepithelium. The synaptic mechanisms mediating these diverse actions remain to be elucidated. To understand the functional role of efferents, their discharge properties and the actions they exert have to be studied in behaviorally meaningful circumstances.

ACh is an efferent neurotransmitter. Both nicotinic and muscarinic receptors are involved in efferent actions. Yet, pharmacological studies in vestibular organs have not progressed to the point where they implicate any of the several cholinergic receptor subtypes defined by molecular techniques or localized to the organs by immunohistochemistry or by *in situ* hybridization. ATP may be involved in efferent excitation in the frog posterior crista. Whether ATP works in a similar manner in other species remains to be determined, and the cellular mechanisms of its presumed actions have not been specified, even in the frog. The neuropeptide CGRP has been localized to a subset of efferent vestibular neurons. CGRP has a slow excitatory action in lateral-line organs, but its effects in vestibular organs have not been studied. Preproenkephalin mRNA is expressed in efferent neurons, but the distribution and actions of enkephalinergic fibers in the vestibular organs have yet to be defined.

Speculations as to the function of efferent systems have been limited to their short-term effects on afferent transmission. The diversity of cholinergic receptor subtypes, as well as the presence of neuropeptides, suggests that the efferents are involved in a variety of functions ranging from the modulation of conventional synaptic transmission to the neurotrophic regulation of synaptic structure.

4.7. ADDENDUM

Since this chapter was completed, papers have been published describing the branching patterns of efferent axons within the crist,[185] and the termination of efferent boutons on hair cells and afferent processes.[153] *In-situ* hybridization studies in the rat have clarified a variety of nAChR subunits expressed in hair cells and afferents.[104] $\alpha 4$ - 7 and $\beta 2$ - 3 mRNAs were present in vestibular ganglion cells; $\alpha 6$ and $\beta 2$ mRNAs were found in all ganglion cells, while the mRNAs for the other subunits were restricted to subpopulations. $\alpha 9$ was the only nAChR subunit found to be expressed in hair cells, whether these were type I or type II. The presumed presence of $\alpha 9$ in type II hair cells is intriguing, since in other hair cell systems it is associated with efferent inhibition, whereas the predominant action in mammalian organs is excitation. Why type I hair cells would continue to express $\alpha 9$ is also intriguing, as these hair cells lose their efferent innervation early in development. Evidence that vestibular organs expressed an even wider assortment of nAChR subunits was obtained when RT-PCR was applied to extracted mRNA: $\alpha 3$, $\alpha 5$–7, $\alpha 9$ and $\beta 2$ - 4 in the vestibular epithelium and $\alpha 2$ - 7 and $\beta 2$ - 4 in the vestibular ganglion.[4] mRNAs for P_{2X} purinergic receptors were found in both the endorgan and in the ganglion.[137] In a preliminary report, Boyle et al.[23] described the efferent actions exerted on hair cells in the toadfish horizontal crista. Only inhibitory responses were observed although efferent inhibition is observed in only a small minority of afferents.

REFERENCES

1. Adams, JC, Mroz, EA, and Sewell, WF, A possible neurotransmitter role for CGRP in a hair-cell sensory organ, *Brain Res.,* 419: 347,1987.
2. Amemiya, F, Kishida, R, Goris, RC, Onishi, H, and Kusunoki, T. Primary vestibular projection in the hagfish, *Eptatretus burgeri, Brain Res.,* 337: 73,1985.
3. Anand, R, Fuchs, PA, Cooper, JF, Pend, X, and Lindstrom, J, Evidence that a-bungarotoxin-sensitive acetylcholine receptors in the chick's cochlea contain a7 subunits, *Soc. Neurosci. Abstr.,* 19: 1076,1993.
4. Anderson, AD, Troyanovskaya, M, and Wackym, PA, Differential expression of a2-7, a9 and b2-4 nicotinic acetylcholine receptor subunit mRNA in the vestibular end-organs and Scarpa's ganglia of the rat, *Brain Res.,* 778: 414,1997.
5. Annoni, J-M, Cochran, SL, Precht, W, Pharmacology of the vestibular hair cell afferent fiber synapse in the frog, *J. Neurosci.,* 4: 2106,1984.
6. Art, JJ, Fettiplace, R, Efferent desensitization of auditory nerve fiber responses in the cochlea of the turtle *Pseudemys scripta elegans, J. Physiol.,* 356: 507,1984.
7. Art, JJ, Fettiplace, R, Fuchs, PA, Synaptic hyperpolarization and inhibition of turtle cochlear hair cells, *J. Physiol.,* (London), 356: 525,1984.
8. Arvidsson, U, Piehl, F, Johnson, H, Ulfhake, B, Cullheim, S, Hökfelt, T, The peptidergic motoneuron, *Neuroreport.,* 4: 849,1993.
9. Ashmore, JF, and Ohmori, H, Control of intracellular calcium by ATP in isolated outer hair cells of the guinea-pig cochlea, *J. Physiol.,* (London), 428: 109,1990.
10. Aubert, A, Norris, CH, and Guth, PS, Influence of ATP and ATP agonists on the physiology of the isolated semicircular canal of the frog *(Rana pipiens), Neuroscience,* 62: 963,1995.

11. Aubert, A, Norris, CH, and Guth, PS, Indirect evidence for the presence and physiological role of endogenous extracellular ATP in the semicircular canal, *Neuroscience*, 64: 1153, 1995.

12. Baird, IL, Some aspects of comparative anatomy and evolution of the inner ear in submammalian vertebrates, *Brain Behav. Evol.*, 10: 11, 1974.

13. Baird, RA, Desmadryl, G, Fernández, C, and Goldberg, JM, The vestibular nerve of the chinchilla. II. Relation between afferent response properties and peripheral innervation patterns in the semicircular canals. *J. Neurophysiol*, 60: 182, 1988.

14. Barbas-Henry, HA, and Lohman, AHM, Primary projections and efferent cells of the VIIIth cranial nerve in the monitor lizard, *Varanus exanthematicus, J. Comp. Neurol.*, 277: 234, 1988.

15. Bean, BP, Pharmacology and electrophysiology of ATP-activated ion channels, *Trends Pharmacol. Sci.*, 13: 87, 1992.

16. Bernard, C, Cochran, SL, and Precht, W, Presynaptic actions of cholinergic agents upon the hair cell-afferent fiber synapses in the vestibular labyrinth of the frog, *Brain Res.*, 338: 225, 1985.

17. Blanks, RHI, and Precht, W, Functional characterization of primary vestibular afferents in the frog, *Exp. Brain Res.*, 25: 369, 1976.

18. Bobbin, RP, and Konishi, T, Acetylcholine mimics crossed olivocochlear bundle stimulation, *Nature*, 231: 222, 1971.

19. Bobbin, RP, and Konishi, T, Action of cholinergic and anticholinergic drugs at the crossed olivocochlear bundle-hair cell junction, *Acta Oto-Laryngol.*, 77: 55, 1974.

20. Boyle, R, Carey, JP, and Highstein, SM, Morphological correlates of response dynamics and efferent stimulation in horizontal semicircular canal afferents, *J. Neurophysiol*, 66: 1504, 1991.

21. Boyle, R, and Highstein, SM, Resting discharge and response dynamics of horizontal semicircular canal afferents in the toadfish, *Opsanus tau. J. Neurosci.*, 10: 1557, 1990.

22. Boyle, R, and Highstein, SM, Efferent vestibular system in the toadfish: action upon horizontal semicircular canal afferents, *J. Neurosci.*, 10: 1570, 1990.

23. Boyle, R, Highstein, SM, and Rabbitt, RD, Simultaneous horizontal canal afferent and hair cell recordings during mechanical and efferent stimulation *in vivo* in the toadfish, *Abstracts 21st Midwinter Meeting, Association for Research in Otolaryngology*, p 142, 1998.

24. Brake, AJ, Wagenbach, MJ, and Julius, D, New structural motif for ligand-gated ion channels defined by an ionotropic ATP receptor, *Nature*, 371: 519, 1994.

25. Brantley, RK, and Bass, AH, Cholinergic neurons in the brain of a teleost fish *(Porichtys notatus)* located with a monoclonal antibody to choline acetyltransferase, *J. Comp. Neurol.*, 275: 87, 1988.

26. Brichta, AM, and Goldberg, JM, Afferent and efferent responses from morphological fiber classes in the turtle posterior crista, *Ann. New York Acad. Sci.*, 781: 183, 1996.

27. Brichta, AM, and Peterson, EH, Functional architecture of vestibular primary afferents from the posterior semicircular canal of a turtle, *Psedemys, scripta. J. Comp. Neurol.*, 344: 481, 1994.

28. Brown, DA, M currents: an update. *Trends Neurosci.*, 11: 294, 1988.

29. Brown, MC, and Nuttall, AL, Efferent control of cochlear inner hair cell responses in the guinea-pig, *J. Physiol.*, (Lond), 354: 625, 1984.

30. Brown, MC, Nuttall, AL, and Masta, RI, Intracellular recordings from cochlear inner hair cells: effects of stimulation of the crossed olivocochlear efferents, *Science*, 222: 69, 1983.

31. Bryant, GM, Barron, SE, Norris, CH, and Guth, PS, Adenosine is a modulator of hair cell afferent neurotransmission, *Hearing Res.*, 30: 231, 1987.

32. Carpenter, MB, Chang, L, Pereira, AB, Hersch, LB, Bruce, G, and Wu J-Y, Vestibular and cochlear efferent neurons in the monkey identified by immunocytochemical methods, *Brain Res.*, 408: 275, 1987.

33. Caston, J, L'activite vestibulaire efferente chez le grenouille, *Pfluegers Arch.*, 331: 365, 1972.

34. Caston, J, and Gribenski, A, Responses des fibres vestibulaires efferentes a une rotation dans le plan horizontal chez le grenouille, *(Rana esculenta* L.), *C. R. Soc. Biol.*, 169: 1062, 1975.

35. Changeux, J-P, Duclert, A, and Sekine, S, Calcitonin gene related peptides and neuromuscular interactions, *Ann. New York Acad. Sci.*, 657: 361, 1992.

36. Christensen-Dalsgaard, J, and Narins, PM, Sound and vibration sensitivity of VIIIth nerve fibers in the frogs *Leptodactylus albilabris* and *Rana pipiens pipiens, J. Comp. Physiol.*, A172: 653, 1993.

37. Churchill, JA, Shucknecht, HF, and Doran, R, Acetylcholinesterase activity in the cochlea, *Laryngoscope*, 66: 1, 1956.

38. Claas, B, Fritzcsh, B, and Münz, H, Common efferents to lateral line and labyrinthine hair cells in aquatic vertebrates, *Neurosci. Lett.*, 27: 231,1981.

39. Cochran, S, and Correia, M, Functional support of glutamate as a vestibular hair cell transmitter in an amniote, *Brain Res.*, 670: 321,1995.

40. Cooper, JR, Bloom, FE, and Roth, RH, *The Biochemical Basis of Neuropharmacology*, 7th edition, New York: Oxford Univ Press,1996.

41. Dallos, P, The active cochlea, *J. Neurosci.*, 12: 4575,1992.

42. Danielson, PD, Zottoli, SJ, Corrodi, JG, Rhodes, KJ, and Mufson, EJ, Localization of choline acetyl-transferase to somata of posterior lateral line efferents in goldfish, *Brain Res.*, 448: 158,1988.

43. De, Camilli, P, and Jahn, R, Pathways to regulated exocytosis, *Annu. Rev. Physiol.*, 52: 625,1990.

44. De, Robertis, E, and Bennett, HS, Submicroscopic vesicular component in the synapse, *Federation Proc.*, 13: 35,1954.

45. Dechesne, C, Raymond, J, and Sans, A, The efferent vestibular system in the cat: a horseradish peroxidase and fluorescent retrograde tracer study, *Neuroscience*, 11: 893,1984.

46. Desmedt JE, and Monaco P, Mode of action of the efferent olivocochlear bundle on the inner ear, *Nature*, 193: 1263,1961.

47. Dichgans, J, Schmidt, CL, and Wist, ER, Frequency modulation of afferent and efferent unit activity in the vestibular nerve by oculomotor impulses, Prog *Brain Res.*, 37: 449,1972.

48. Dickman, JD, and Correia, MJ, Bilateral communication between vestibular labyrinths in pigeons, *Neuroscience*, 57: 1097,1993.

49. Didier, A, Dupont, J, and Cazals, Y, GABA immunoreactivity of calyceal nerve endings in the vestibular system of the guinea pig, *Cell Tissue Res* 260: 415,1990.

50. Dieringer, N, Blanks, RHI, and Precht, W, Cat efferent vestibular system: weak suppression of primary afferent activity, *Neurosci. Lett.*, 5: 285,1977.

51. Dohlman, G, Farkashidy, J, and Salonna, F, Centrifugal nerve fibers to the sensory epithelium of the vestibular labyrinth, *J. Laryngol.*, 72: 984,1958.

52. Doi, T, and Ohmori, H, Acetylcholine increases intracellular Ca^{2+} concentration and hyperpolarizes the guinea-pig outer hair cell, *Hearing Res.*, 67: 179,1993.

53. Dowdal, MJ, Boyne, AF, and Whittaker, VP, Adenosine triphosphate, a constituent of cholinergic synaptic vesicles, *Biochem. J.*, 140: 1,1972.

54. Dunn, RF, Reciprocal synapses between hair cells and first-order afferent dendrites in the crista ampullaris of the bullfrog, *J. Comp. Neurol.*, 193: 255,1980.

55. Dunwiddie, TV, The physiological role of adenosine in the central nervous system, *Int. Rev. Neurobiol.*, 27: 11.7,1985.

56. Eden, AR, and Correia, MJ, Identification of multiple groups of efferent vestibular neurons in the adult pigeon using horseradish peroxidase and DAPl, *Brain Res.*, 248: 201,1982.

57. Elgoyhen, AB, Johnson, DS, Boutler, J, Vetter, DE, and Heinemann S, $\alpha 9$ an acetylcholine receptor with novel pharmacological properties expressed in rat cochlear hair cells, Cell 79: 705,1994.

58. Engström, H, On the double innervation of the sensory epithelia of the inner ear, *Acta Oto-Laryngol* 49:109, 1958.

59. Engström, H, Bergström, HB, and Ades, HW, Macula. utriculi and macula sacculi in the squirrel monkey, *Acta. Oto-Laryngol. Suppl.*, 301: 75,1972.

60. Engström, H, and Wersall, J, The ultrastructural organization of the organ of Corti and of the vestibular sensory epithelia, *Exp. Cell Res. Suppl.*, 5: 460,1958.

61. Eybalin, M, Neurotransmitters and neuromodulators of the mammalian cochlea, *Physiol. Rev.*, 73: 309,1993.

62. Felix, D, and Ehrenberger, K, The action of putative neurotransmitter substances in the cat labyrinth, *Acta. Oto-Laryngol.*, 93: 101,1982.

63. Fernández, C, Baird, RA, and Goldberg, JM, The vestibular nerve of the chinchilla. 1. Peripheral innervation patterns in the horizontal and superior semicircular canals, *J. Neurophysiol*,60: 167,1988.

64. Fernández, C, Goldberg, JM, and Baird, RA, The vestibular nerve of the chinchilla. III. Peripheral innervation patterns in the utricular macula, *J. Neurophysiol*,63: 767,1990.

65. Fernández, C, Lysakowski, A, and Goldberg, JM, Hair-cell counts and afferent innervation patterns in the cristae ampullares of the squirrel monkey with a comparison to the chinchilla, *J. Neurophysiol*,, 73:1253,1995.

66. Fex, J, Augmentation of cochlear microphonics by stimulation of efferent fibers to the cochlea, *Acta. Oto-Laryngol.*, 50: 540,1959.

67. Fex, J, Auditory activity in centrifugal and centripetal cochlear fibres in cat, *Acta. Physiol. Scand. Suppl.*, 189: 1,1962.

68. Fex, J, Efferent inhibition of the cochlea related to hair cell dc activity: study of postsynaptic activity of the crossed olivocochlear fibres in the cat, *J. Acoust. Soc. Am.*, 41: 667,1967.

69. Fex, J, and Adams, JC a-bungarotoxin blocks reversibly cholinergic inhibition in the cochlea, *Brain Res.*, 159:440, 1978.

70. Fibiger, HC, The organization and some projections of cholinergic neurons of the mammalian forebrain, *Brain Res.*, 257,1982.

71. Flock, A, and Lam, DMK, Neurotransmitter synthesis in the inner ear and lateral line sense organs, *Nature*, 249,142,1974.

72. Flock, A, and Russell, IJ, The post-synaptic action of efferent fibres in the lateral line organ of the burbot, *J. Physiol., (London)*, 235:591,1973.

73. Flock, A, and Russell, IJ, Inhibition by efferent nerve fibres: action on hair cells and afferent synaptic transmission in the lateral line canal organ of the burbot, *Lota lota. J. Physiol.*, (London) 257: 45,1976.

74. Foster, JD, Drescher, MJ, and Drescher, DG, Immunohistochemical localization of GABA receptors in the mammalian crista ampullaris, *Hearing Res.*, 83: 203,1995.

75. Fritzsch, B, Development of the labyrinthine efferent system, *Ann. New York Acad. Sci.*, in press.

76. Fritzsch, B, Crapon, De and Caprona, D, The origin of centrifugal inner ear fibers of gymnophions (amphibia). A horseradish peroxidase study, *Neurosci. Lett.*, 45: 131,1984.

77. Fritzsch, B, Dubuc, R, Ohta, Y, and Grillner, S, Efferents to the labyrinth of the river lamprey (*Lampetra fluiatilis*) as revealed with retrograde tracing techniques, *Neurosci. Lett.*, 96: 241,1989.

78. Fu, WM, Potentiation by ATP of the postsynaptic acetylcholine response at developing neuromuscular synapses in Xenopus cell cultures, *J. Physiol., (London)*, 477: 449,1994.

79. Fu, SY, and Gordon, T, The cellular and molecular basis of peripheral nerve regeneration, *Mol. Neurobiol.*, 14: 67,1997.

80. Fuchs, PA, and Murrow, BW, Cholinergic inhibition of short (outer) hair cells of the chick's cochlea, *J. Neurosci.*, 12: 800,1992.

81. Fuchs, PA, and Murrow, BW, A novel cholinergic receptor mediates inhibition of chick cochlear hair cells, *Proc. Roy. Soc. B.*, 248: 35,1992.

82. Fukuda, K, Higashida, H, Kubo, T, Maeda, A, Akiba, I, Bujo, H, Mishina, M, and Numa, S, Selective coupling with KI currents of muscarinic acetylcholine receptor subtypes in NG108-15 cells, *Nature*, 335: 355,1988.

83. Furukawa, T, Effects of efferent stimulation on the saccule of goldfish, *J. Physiol., (London)*, 315: 203,1981.

84. Gacek, RR, Efferent component of the vestibular nerve, Neural Mechanisms of the Auditory and Vestibular Systems (Rasmussen GL, and Windle WF, Eds.), pp 276-284, Springfield, IL: Thomas,1960.

85. Gacek, RR, and Lyon, M, The localization of vestibular efferent neurons in the kitten with horseradish peroxidase, *Acta. Oto-Laryngol.*, 77: 92,1974.

86. Gacek, RR, and Rasmussen, GL, Fiber analysis of statoacoustic nerve of guinea pig, cat and monkey, *Anat. Rec.*, 139: 455,1961.

87. Galambos, R, Suppression of the auditory nerve activity by stimulation of efferent fibers to the cochlea, *J. Neurophysiol.*,19: 424,1956.

88. Gilman, AG, G proteins and regulation of adenylyl cyclase, *Biosci. Reports*, 15: 65,1995.

89. Gleisner, L, and Henriksson, NG, Efferent and afferent activity pattern in the vestibular nerve of the frog, *Acta. Oto-Laryngol Suppl.*, 192: 90,1964.

90. Glowatzki, E, Wild, K, Brändle, U, Fakler, G, Zenner, H-P, and Ruppersberg, JP. Cell-specific expression of the a9 n-AChR receptor subunit in auditory hair cells revealed by single-cell RT-PCR, *Proc. Roy. Soc. B*, 262: 141,1995.

91. Goldberg, JM, Desmadryl, G, Baird, RA, and Fernández, C, The vestibular nerve of the chinchilla. V. Relation between afferent discharge properties and peripheral innervation patterns in the utricular macula, *J. Neurophysiol.*,63: 791,1990.

92. Goldberg, JM, and Fernández, C, Efferent vestibular system in the squirrel monkey: anatomical location and influence on afferent activity, *J. Neurophysiol.*, 43: 986,1980.

93. Goldberg, JM, Lysakowski, A, and Fernández, C, Morphophysiological and ultrastructural studies in the mammalian cristae ampullares, *Hearing Res.*, 49: 89,1990.

94. Goldberg, JM, Smith, CE, and Fernández, C, Relation between discharge regularity and responses to externally applied galvanic currents in vestibular nerve afferents of the squirrel monkey, *J. Neurophysiol.*,51: 1236,1984.

95. Gonzalez, A, Meredith, GE, and Roberts, BL, Choline acetyltransferase immunoreactive neurons innervating labyrinthine and lateral line sense organs in amphibians, *J. Comp. Neurol.*, 332: 258,1993.

96. Guinan, JJ Jr, Warr, WB, and Norris, BE, Differential olivocochlear projections from lateral versus medial zones of the superior olivary complex, *J. Comp. Neurol.*, 221: 358,1983.

97. Guth, PS, Norris, CH, and Bobbin, RP, The pharmacology of transmission in the peripheral auditory system, *Phati-nacol Rev.*, 28: 95,1976.

98. Guth, PS, Norris, CH, Guth, SL, Quine, DB, and Williams, WH, Cholinomimetics mimic efferent effects on semicircular canal afferent activity in the frog, *Acta. Oto-Laryngol.*, 102: 194,1986.

99. Guth, SL, and Norris, CH, Pharmacology of the isolated semicircular canal: effect of GABA and picrotoxin, *Hearing Res.*, 22: 235,1984.

100. Hamilton, BR, and Smith, DO, Autoreceptor-mediated purinergic and cholinergic inhibition of motor nerve terminal calcium currents in the rat, *J. Physiol. (London)*, 432:327,1991.

101. Harden, TK, Boyer, JL, and Nicholas, RA, P2-purinergic receptors: subtype-associated signaling responses and structure, *Annu. Rev. Pharamacol. Toxicol.*, 35: 541,1995.

102. Hartmann, R, and Klinke, R, Efferent activity in the goldfish vestibular nerve and its influence on afferent activity, *Pfluegers Arch.*, 388: 123,1980.

103. Henderson, DJ, Elliot, DG, Smith, GM, Webb, TE, and Dainty, IA, Cloning and characterization of a bovine P_{2Y} receptor, *Biochem. Biophys. Res. Comm.*, 212: 648,1995.

104. Hiel, H, Elgoyhen, A.B., Drescher, DG, and Morley, BJ, Expression of nicotinic acetylcholine receptor mRNA in the adult rat peripheral vestibular system, *Brain Res.*, 738: 347,1996.

105. Hiel, H, Drescher, D, Elgoyhen, A, and Morley, BJ, nAChR subunits in the peripheral vestibular system, *Abstracts of a Meeting on the Molecular Biology of Hearing and Deafness*, p 119. Bethesda MD.

106. Highstein, SM, The central nervous system efferent control of the organs of balance and equilibrium, *Neurosci. Res.*, 12: 13,1991.

107. Highstein, SM, and Baker, R, Action of the efferent vestibular system on primary afferents in the toadfish, *Opsanus tau. J. Neurophysiol*, 54: 370,1985.

108. Highstein, SM, and Baker, R, Organization of the efferent vestibular nuclei and nerves in the toadfish, *Opsanus tau, J. Comp. Neurol.*, 243: 309,1986.

109. Hilding, D, and Wersäll, J, Cholinesterase and its relation to the nerve endings in the inner ear, *Acta. Oto-Laryngol.*, 55: 205,1962.

110. Honrubia, V, Kuruvilla, A, Mamikunian, D, and Eichel, JE, Morphological aspects of the vestibular nerve of the squirrel monkey, *Laryngoscope*, 97: 228,1987.

111. Housley, GD, and Ashmore, JF, Direct measurement of the action of acetylcholine on isolated outer hair cells of the guinea pig cochlea, *Proc. Roy. Soc. B*, 244: 161,1991.

112. Housley, GD, Greenwood, D, and Ashmore, JF, Localization of cholinergic and purinergic receptors in outer hair cells isolated from the guinea pig cochlea, *Proc. Roy. Soc. B*, 249:,1992.

113. Housley, GD, Greenwood, D, Bennett, T, and Ryan, AF, Identification of a short form of the P_{2X}R1-purinoceptor subunit produced by alternative splicing in the pituitary and cochlea, *Biochem. Biophys. Res. Comm.*, 212: 501,1995.

114. Housley, GD, Norris, CH, and Guth, PS, Cholinergically-induced changes in outward currents in hair cells isolated from the semicircular canal of the frog, *Hearing Res.*, 43:,1990.

115. Hulme, EC, Birdsall, NJM, and Buckley, NJ, Muscarinic receptor types, *Annu. Rev. Pharamacol. Toxicol.*, 30: 633,1990.

116. Igarashi, S, Koide, C, Sasaki, H, and Nakano, Y, Mercury deposition and its relationship to inner ear function in methylmercury-poisoned rats. A histological and immunohistochernical study, *Acta. Oto-Laryngol.*, 112: 773,1992.

117. Igarashi, S, Sasaki, H, Nakano, Y, and Ino, H, Immunohistochemical examination of S-100 protein and substance P in the inner ear of the rat, *Acta. Oto-Laryngol.*, Suppl 481: 163,1991.

118. Ireland, PE, and Farakashidy, J, Studies on the efferent innervation of the vestibular endorgans, *Trans. Am. Otol. Soc.*, 49: 20,1961.
119. Ishida-Yarnamoto, A, and Tohyama, M, Calcitonin gene-related peptide in the nervous system, *Prog. Neurobiol.*, 33: 335,1989.
120. Ishiyama, A, López, I, and Wackym, PA, Subcellular innervation patterns of the calcitonin gene-related peptidergic efferent terminals in the chinchilla vestibular periphery, *Otolaryngol. Head Neck Surg.*, 111: 385,1994.
121. Ishiyama, A, López, 1, and Wackym, PA, Distribution of efferent cholinergic terminals and a-bungarotoxin binding to putative nicotinic acetylcholine receptors in the human vestibular end-organs, *Laryngoscope*, 105: 1167,1995.
122. Ito, J, Takahashi, H, Matsuoka, I, Takatani, T, Sasa, M, and Takaori, S, Vestibular efferent fibers to ampulla of anterior, lateral and posterior semicircular canals in cats, *Brain Res.*, 259: 293,1983.
123. Iurato, S, Efferent fibers to the sensory cells of Corti's organ, *Exp. Cell Res.*, 27: 162,1962.
124. Iurato, S, Luciano, L, Pannese, E, and Reale, E, Acetylcholinesterase activity in the vestibular sensory areas, *Acta. Oto-Laryngol.*, 71: 147,1971.
125. Iurato, S, Luciano, L, Pannese, E, and Reale, E, Efferent vestibular fibers in mammals: morphological and histochemical aspects, *Prog. Brain Res.*, 37: 429,1972.
126. Jorgensen, JM, The sensory epithelia of the inner ear of two turtles, *Testudo graeca* L. and *Psedemys scripta* (Schoepff), *Acta Zool. (Stockholm)*, 55: 289,1974.
127. Keller, EL, Behavior of horizontal semicircular canal afferents in alert monkey during vestibular and optokinetic stimulation, *Exp Brain Res.*, 24: 459,1976.
128. Kelly, RB, Storage and release of neurotransmitters, *Neuron* 10(Suppl): 43,1993.
129. Kiang, NY, Rho, JM, Northrop, CC, Liberman, MC, and Ryugo, DK, Hair-cell innervation by spiral ganglion cells in adult cats, *Science*, 217: 175,1982.
130. Kimura, R, and Wersall, J, Termination of the olivo-cochlear bundle in relation to the outer hair cells of the organ of Corti in the guinea pig, *Acta. Oto-Laryngol.*, 55: 11,1962.
131. Kitahara, T, Takeda, N, Ohno, K, Araki, T, Kubo, T, and Kiyama, H, Expression of GABA$_A$ receptor g1 and g2 subunits in the peripheral vestibular system of rat, *Brain Res.*, 650: 157,1994.
132. Klinke, R, Neurotransmission in the inner ear, *Hearing Res.*, 22: 235,1986.
133. Klinke, R, and Galley, N, Efferent innervation of vestibular and auditory receptors, *Physiol. Rev.*, 54: 316,1974.
134. Klinke, R, and Schmidt, CL, Efferente Impulse in Nervus vestibularis bei Reizung des kontralateralen Otolithenorgans, *Pfluegers Arch.*, 304: 183,1968.
135. Klinke, R, and Schmidt, CL, Efferent influence on the vestibular organ during active movement of the body, *Pfluegers Arch.*, 318: 325,1970.
136. Koyama, H, Kishida, R, Goris, RC, and Kusunoki, T, Afferent and efferent projection of the VIIIth cranial nerve in the lamprey *Lampetra japonica, J. Comp. Neurol.*, 280: 663,1989.
137. Kreindler, JL, Troyanovskaya, M, and Wackym, PA, Ligand-gated purinergic receptors are differentially expressed in the adult rat vestibular periphery, *Ann. Otol. Rhinol. Laryngol.*, in press, 1998.
138. Kristensson, K, Olsson, Y, and Sjöstrand, J, Axonal uptake and retrograde transport of exogenous proteins in the hypoglossal nerve, *Brain Res.*, 32: 399,1971.
139. Levey, AI, Wainer, BH, Rye, DB, Mufson, EJ, and Mesulam, M-M, Choline acetyltransferase-immunoreactive neurons intrinsic to rodent cortex and distinction from acetylcholinesterase-positive neurons, *Neuroscience*, 13: 341,1984.
140. Lindeman, HH, Studies on the morphology of the sensory regions of the vestibular apparatus, *Ergeb. Anat. Entwicklungsgesch.*, 42: 1,1969.
141. Lindstrom, J, Nicotine acetylcholine receptors. *CRC Handbook of Receptors* (North A, Eds.), pp. 1.53,1994. Boca Raton, FL: CRC Press,1994.
142. Lindstrom, J, Anand, R, Peng, X, Gerzanich, V, Wang, F, and Li, Y, Neuronal nicotinic receptor subtypes, *Ann. New York Acad. Sci.*, 757: 100,1995.
143. Llinás, R, and Precht, W, Cerebellar Purkinje cell Projection to the peripheral vestibular organ in the frog, *Science*, 158: 1328,1967.
144. Llinás, R, and Precht, W, The inhibitory vestibular efferent system and its relation to the cerebellum in the frog, *Exp. Brain Res.*, 9: 16,1969.

145. López, I, Juiz, JM, Altschuler, RA, and Meza, G, Distribution of GABA-like immunoreactivity in guinea pig cristae ampullaris, *Brain Res.*, 530: 170,1990.

146. López, I, and Meza, G, Neurochernical evidence for afferent GABAergic and efferent cholinergic neurotransmission in the frog vestibule, *Neuroscience*, 25: 13,1988.

147. López, I, Wu, J-Y, and Meza, G, Immunocytochemical evidence for an afferent GABAergic neurotransmission in the guinea pig vestibular system, *Brain Res.*, 589: 341,1992.

148. Louie, AW, and Kimm, J, The response of 8th nerve fibers to horizontal sinusoidal oscillations in the alert monkey, Exp *Brain Res.*, 24: 447,1976.

149. Lowenstein, 0, and Thornhill, RA, The labyrinth of *Myxine*: anatomy, ultrastructure and electrophysiology, *Proc. Roy. Soc. B*, 176: 21,1970.

150. Lu, Z, and Smith, DO, Adenosine 5'-triphosphate increases acetylcholine channel opening frequency in rat skeletal muscle, *J. Physiol.*, (London), 436: 45,1991.

151. Lysakowski, A, Synaptic organization of the crista ampullaris in vertebrates, *Ann. New York Acad. Sci.*, 81:164, 1996.

152. Lysakowski, A, and Goldberg, JM, Regional variations in synaptic innervation of the squirrel monkey crista, *Soc. Neurosci. Abstr.*, 19: 1578,1993.

153. Lysakowski, A, and Goldberg, JM, A regional ultrastructural. analysis of the cellular and synaptic architecture in the chinchilla cristae ampullares, *J. Comp. Neurol.*, 389: 419,1997.

154. Lysakowski, A, Minor, LB, Fernández, C, and Goldberg, JM, Physiological identification of morphologically distinct afferent classes innervating the cristae ampullares of the squirrel monkey, *J. Neurophysiol.*,73: 1270,1995.

155. Maggi, CA, Tachykinins and calcitonin gene-related peptide (CGRP) as co-transmitters from peripheral endings of sensory neurons, *Prog. Neurobiol.*, 45: 1,1995.

156. Matsubara, A, Usami, S, Fujita, S, and Shinkawa, H, Expression of substance P, CGRP, and GABA in the vestibular periphery, with special reference to species differences, *Acta. OtoLaryngol. Suppl.*, 519: 248,1995.

157. McCue, MP, and Guinan, JJ, Jr, Influence of efferent stimulation on acoustically responsive vestibular afferents in the cat, *J. Neurosci.*, 14: 6071,1994.

158. McGehee, DS, and Role, LW, Physiological diversity of nicotinic acetylcholine receptors expressed by vertebrate neurons, *Annu. Rev. Physiol.*, 57: 521,1995.

159. Meredith, GE, The relationship of saccular efferent neurons to the superior olive in the eel, *Anguilla anguilla, Neurosci. Lett.*, 68: 69,1986.

160. Meredith, GE, and Roberts, BL, Central organization of the efferent supply to the labyrinth and lateral line receptors of the dogfish, *Neuroscience*, 17: 225,1986.

161. Meredith, GE, and Roberts, BL, Distribution and morphological characteristics of efferent neurons innervating end organs in the ear and lateral line of the European eel, *J. Comp. Neurol.*, 265: 494,1987.

162. Meza, G, Carabez, A, and Ruiz, M, GABA synthesis in isolated vestibular tissue of chick inner ear, *Brain Res.*, 241: 157,1982.

163. Meza, G, and Hinojosa, R, Ontogenetic approach to cellular localization of neurotransmitters in the chick vestibule, *Hearing Res.*, 28: 73,1987.

164. Miller, RJ, The control of neuronal Ca^{2+} homeostasis, *Prog. Neurobiol.*, 37: 255,1991.

165. Mockett, BG, Bo, X, Housley, GD, Thorne, PR, and Burnstock, G, Autoradiographic labeling of P_2 purinoreceptors in the guinea pig cochlea, *Hearing Res.*, 84: 177,1995.

166. Mockett, BG, Housley, GD, and Thorne, PR, Fluorescence imaging of extracellular purinergic receptor sites and putative ecto-ATPase sites on isolated cochlear hair cells, *J. Neurosci.*, 14: 6992,1994.

167. Nakagawa, T, Akaika, N, Kimitsuki, T, Komune, S, and Arima, T, ATP induced current in isolated outer hair cells of the guinea pig cochlea, *J. Neurophysiol*,63: 1068,1990.

168. Nakazawa, K, ATP-activated current and its interaction with acetylcholine-activated current in rat sympathetic neurons, *J. Neurosci.*, 14: 740,1994.

169. Nomura, Y, Gacek, RR, and Balogh, K Jr, Efferent innervation of vestibular labyrinth, *Arch. Otolaryngol.*, 81: 335,1965.

170. Norris, CH, Housley, GD, Williams, WH, Guth, SL, and Guth, PS, The acetylcholine receptors of the semicircular canal in the frog *(Rana pipiens), Hearing Res.*, 32: 197,1988.

171. Ogawa, Y, Role of ryanodine receptors, *Critical Rev. Biochem. Mol. Biol.*, 29: 229,1994.

172. Ohno, K, Takeda, N, Kiyama, H, Kubo, T, and Tohyama, M, Occurrence of galanin-like immunoreactivity in vestibular and cochlear efferent neurons after labyrinthectomy in the rat, *Brain Res.,* 644: 135,1994.

173. Ohno, K, Takeda, N, Kubo, T, and Kiyama, H., Up-regulation of GAP-43 (B50 / Fl) gene expression in vestibular efferent neurons following labyrinthectomy in the rat: in situ hybridization using an alkaline phosphatase-labeled probe, *Hearing Res.,* 80: 123,1994.

174. Ohno, K, Takeda, N, Yamano, M, Matsunaga, T, and Tohyama, M, Coexistence of acetylcholine and calcitonin gene-related peptide in the vestibular efferent neurons in the rat, *Brain Res.,* 566: 103,1991.

175. Olah, ME, and Stiles, GL, Adenosine receptor subtypes: characterization and therapeutic regulation, *Annu. Rev. Pharamacol. Toxicol.,* 35: 581,1956.

176. Palay, S, and Palade, GE, The fine structure of neurons, *J. Biophys. Biochem. Cytol.,* 1: 69,1956.

177. Parr, CE, Sullivan, DM, Paradiso, AM, Lazarowski, ER, Burch, LH, Olsen, JC, Erb, L, Weisman, G, Boucher, RC, and Turner, JT, Cloning and expression of a human P_{2U} nucleotide receptor, a target for cystic fibrosis pharmacotherapy, *Proc. Natl. Acad. Sci. USA.,* 91: 13067,1994.

178. Pellergrini, M, Ceccotti, F, and Magherini, P, The efferent vestibular neurons in the toad *(Bufo bufo* L.): Their location and morphology. A horseradish peroxidase study, *Brain Res.,* 344: 1,1985.

179. Perachio, AA, and Kevetter, GA, Identification of vestibular efferent neurons in the gerbil: histochemical and retrograde labelling, *Exp. Brain Res.,* 78: 315,1989.

180. Precht, W, Llinds, R, and Clarke, M, Physiological responses of frog vestibular fibers to horizontal angular rotation, *Exp. Brain Res.,* 13: 378,1971.

181. Prigioni, I, Valli, P, and Casella, C, Peripheral organization of the vestibular efferent system in the frog: an electrophysiological study, *Brain Res.,* 269: 83,1983.

182. Purcell, IM, and Perachio, AA, Regional distribution of efferent innervation in the utricular maculae of the gerbil, *Soc. Neurosci. Abstr.* 21: 399,1995.

183. Purcell, IM, and Perachio, AA, Three-dimensional analysis of biocytin labeled vestibular efferent neurons in the semicircular canals of the gerbil, *Abstracts 18th Midwinter Meeting, Association for Research in Otolaryngology,* p. 12,1995.

184. Purcell, IM, Perachio, AA, Regional innervation patterns of vestibular efferent and neurons in the saccular macula of the gerbil, *Abstracts, 19th Midwinter Meeting, Association for Research in Otolaryngology,* p 174,1996.

185. Purcell, IM, and Perachio, AA, Three-dimensional analysis of vestibular efferent neurons innervating semicircular canals of the gerbil, *J. Neurophysiol,*78: 3234,1997.

186. Quayle, JM, Bonev, AD, Brayden, JE, and Nelson, MT, Calcitonin gene-related peptide activated ATP-sensitive K^+ currents in the rabbit atrial smooth muscle via protein kinase,*J. Physiol.,* (London), 475: 9,1994.

187. Rasmussen, GL, The olivary peduncle and other fiber connections of the superior olivary complex, *J. Comp. Neurol.,* 84: 141,1946.

188. Rasmussen, GL, Further observations of the efferent cochlear bundle, *J. Comp. Neurol.,* 99: 61,1953.

189. Rasmussen, GL, Efferent fibers of the cochlear nerve and nucleus, *Neural Mechanisms of the Auditory and Vestibular Systems* (Rasmussen GL, and Windle WF, Eds.), pp. 105-115. Springfield, IL: Thomas,1960.

190. Rasmussen, GL, and Gacek, RR, Concerning the question of the efferent fiber component of the vestibular nerve of the cat, *Anat. Rec.,* 130: 361,1958.

191. Redman, RS, and Silinsky, EM, ATP released together with acetylcholine as the mediator of neuromuscular depression of frog motor nerve endings, *J. Physiol.,* (London), 477: 117,1994.

192. Rennie, KJ, and Ashmore, JF, Effects of extracellular ATP on hair cells isolated from the guinea pig semicircular canals, *Neurosci. Lett.,* 160: 185,1993.

193. Roberts, BL, Maslarn, S, Los, I, and Van Der Jagt, B, Coexistence of calcitonin gene-related peptide and choline acetyltransferase in eel efferent neurons, *Hearing Res.,* 74: 231,1994.

194. Roberts, BL, and Meredith, GE, The Efferent Innervation of the Ear: Variations on an Enigma, in *The Evolutionary Biology of Hearing* (Fay, RA, Popper, AN, and Webster, DB, Eds.), pp. 185-210, Berlin: Springer,1992.

195. Roberts, BL, and Russell, IJ, The activity of lateral line efferent neurones in stationary and swimming dogfish, *J. Exp. Biol.,* 57: 433,1972.

196. Rosenfeld, MG, Emeson, RB, Yeakley, JM, Merillat, N, Hedjran, F, Lenz, J, and Delsert, C, Calcitonin gene-related peptide: a neuropeptide generated as a consequence of tissue-specific developmentally regulated alternative RNA processing events, *Ann. New York Acad. Sci.,* 657: 1,1992.

197. Rosenfeld, MG, Mermod, J-J, Amara, SG, Swanson, LW, Sawchenko, PE, Rivier, J, Vale, WW, and Evans, RM, Production of a novel neuropeptide encoded by the calcitonin gene via tissue-specific RNA processing, *Nature,* 304: .1.29,1983.

198. Ross, MD, Rogers, CM, and Donovan, KM, Innervation patterns in rat saccular macula. A structural basis for complex sensory processing, *Acta. Oto-Laryngol.,* 102: 75,1986.

199. Rossi, ML, and Martini, M, Efferent control of posterior canal afferent receptor discharge in the frog labyrinth, *Brain Res.,* 555: 123,1991.

200. Rossi, ML, Martini, M, Pelucchi, B, and Fesce, R, Quantal nature of synaptic transmission at the cytoneural junction in the frog labyrinth, *J. Physiol.,* (London), 478. 1: 17,1994.

201. Rossi, ML, Prigioni, 1, Valli, P, and Casella, C, Activation of the efferent system in the isolated frog labyrinth: effects on the afferent EPSPs and spike discharge recorded from single fibres of the posterior nerve, *Brain Res.,* 185: 125,1980.

202. Russell, IJ, Influence of efferent fibres on a receptor, *Nature,* 219:177, 1968.

203. Russell, IJ, The pharmacology of efferent synapses in the lateral-line system of *Xenopus laevis, J. Exp. Biol.,* 54: 621,1971.

204. Russell, IJ, The role of the lateral-line efferent system in *Xenopus laevis, J. Exp. Biol.,* 54: 621,1971.

205. Ryan, AF, Simmons, DM, Watts, AG, and Swanson, LW, Enkephalin mRNA production by cochlear and vestibular efferent neurons in the gerbil brainstem, *Exp. Brain Res.,* 87: 259,1991.

206. Ryu, PD, Gerber, G, Murase, K, and Randic, M, Actions of calcitonin gene-related peptide on rat spinal dorsal horn neurons, *Brain Res.,* 441: 357,1988.

207. Ryu, PD, Gerber, G, Murase, K, and Randic, M, Calcitonin gene-related peptide enhances calcium current of rat dorsal root ganglion neurons and spinal excitatory synaptic transmission, *Neurosci. Lett.,* 89: 305,1988.

208. Sala, 0, The efferent vestibular system. Electrophysiological research, *Acta. OtoLaryngol. Suppl.,* 197: 1,1965.

209. Sans, A, and Highstein, SM, New ultrastructural features in the vestibular labyrinth of the toadfish, *Opsanus tau . Brain Res.,* 308: 191,1984.

210. Schmidt, CL, and Wist, ER, Dichgans, J, Efferent frequency modulation in the vestibular nerve correlated with eye movements, *Exp. Brain Res.,* 15: 1,1972.

211. Schmidt, RS, Frog labyrinthine efferent impulses, *Acta. Oto-Laryngol.,* 56: 51,1963.

212. Schuknecht, HF, Churchill, JA, and Doran, R, The localization of acetylcholinesterase in the cochlea, *Arch. Otolaryngol.,* 69: 549,1959.

213. Schwarz, DWF, Satoh, K, Schwarz, IE, Hu, K, and Fibiger, HC, Cholinergic innervation of the rat's labyrinth, *Exp. Brain Res.,* 64: 19,1986,

214. Schwarz, IE, Schwarz, DWF, Fredrickson, JM, and Landolt, JP, (1981) Efferent vestibular neurons: a study employing retrograde tracer methods in the pigeon *(Columbia livia), J. Comp. Neurol.,* 196: 1,1981.

215. Schweitzer, E, Coordinated release of ATP and ACh from chotinergic synaptosomes and its inhibition by calmodulin antagonists, *J. Neurosci.,* 7: 2948,1987.

216. Sewell, WF, and Starr, PA, Effects of calcitonin gene-related peptide and efferent nerve stimulation on afferent transmission in the lateral line organ, *J. Neurophysiol,* 65: 1158, 1991.

217. Shigemoto, T, and Ohmori, H, Muscarinic agonists and ATP increase the intracellular Ca^{2+} concentration in chick cochlear hair cells, *J. Physiol.,* (London), 420: 127,1990.

218. Shigemoto, T, Ohmori, H, Muscarinic receptor hyperpolarizes cochlear hair cells of chick by activating Ca^{2+}-activated K^+ channels, *J. Physiol.,* (London), 442:669, 1991.

219. Silinsky, EM, On the association between transmitter secretion and the release of adenine nucleotides from mammalian motor nerve terminals, *J. Physiol.,* (London), 247: 145,1975 .

220. Smit, CA, and Rasmussen GL, Recent observations on the olivocochlear bundle, *Ann. Otol.,* (St Louis) 72: 489,1963.

221. Smith, CA, and Rasmussen, GL, Nerve ending in the maculae and cristae of the chinchilla vestibule, with a special reference to the efferents, in *Third Symposium on the Role of the Vestibular Organs in Space Exploration* Eds., pp. 183-201. Washington D.C.: U.S. Govt Printing Office (SP-152),1968.

222. Smith, CA, and Sjöstrand, FS, A synaptic structure in the hair cells of the guinea pig cochlea, *J. Ultrastruct. Res.,* 5:185,1961.

223. Smith, CE, and Goldberg, JM, A stochastic after-hyperpolarization model of repetitive activity in vestibular afferents, *Biol.Cybern.,* 54: 41,1986.

224. Spoendlin, H, Sensory neural organization of the cochlea, *J. Laryngol. Otol.,* 93: 853,1979.

225. Spoendlin, HH, and Gacek, RR, Electron microscopic study of the efferent and afferent innervation of the organ of Corti in the cat, *Ann. Otol., (St Louis)* 62: 660,1963.

226. Sridhar, TS, Liberman, MC, Brown, MC, and Sewell, WF, A novel cholinergic "slow effect" of efferent stimulation on cochlear potentials in the guinea pig, *J. Neurosci.,* 15: 3667,1995.

227. Strutz, J, The origin of centrifugal fibers to the inner ear in *Caiman crocodilus,* A HRP study, *Neurosci. Lett.,* 27: 65,1981.

228. Strutz, J, The origin of efferent fibers to the inner ear in a turtle *(Terrapene ornata).* A horseradish peroxidase study, *Brain Res.,* 244:165,1982.

229. Strutz, J, The origin of efferent vestibular fibers in the guinea pig, *Acta. Oto-Laryngol.,* 94: 299,1982.

230. Strutz, J, Bielenberg, K, and Spatz, WB, Location of efferent neurons to the labyrinth of the green tree frog (*Hyla cinerea):* A horseradish peroxidase study, *Arch. Otolaryngol.,* 234: 245,1982 .

231. Strutz, J, and Schmidt, CL, Acoustic and vestibular efferent neurons in the chicken *(Gallus domesticus), Acta. Oto-Laryngol.,* 94: 45.

232. Sugai, T, Sugitani, M, and Ooyama H, Effects of activation of the divergent efferent fibers on the spontaneous activity of vestibular afferent fibers in the toad, *Jap. J. Physiol.,* 41: 217,1991.

233. Sugai, T, Yano, J, Sugitani, M, and Ooyama, H, Actions of cholinergic agonists and antagonists on the efferent synapse in the frog saccule, *Hearing Res.,* 61: 56,1992.

234. Tachibana, M, Asano, T, Wilcox, E, Yokotani, N, Rivolta, MN, and Fex J, G protein G_{i2} alpha in the cochlea: cloning and selective occurrence in receptor cells, *Mol. Brain Res.,* 21: 355,1994.

235. Tachibana, M, Wilcox, E, Yokotani, N, Schneider, M, and Fex, J, Selective amplification and partial sequencing of cDNAs encoding G protein subunits from cochlear tissues, *Hearing Res.,* 62: 82,1992.

236. Tanaka, M, Takeda, N, Senba, E, Tohyama, M, Kubo, T, and Matsunaga, T, Localization of calcitonin gene-related peptide in the vestibular end-organs in the rat: an immunohistochemical study, *Brain Res.,* 447: 175,1988.

237. Tanaka, M, Takeda, N, Senba, E, Tohyama, M, Kubo, T, and Matsunaga, T, Localization, origin and fine structure of calcitonin gene related peptide-containing fibers in the vestibular end-organs of the rat, *Brain Res.,* 504: 31,1989.

238. Tricas, TC, and Highstein, SM, Visually mediated inhibition of lateral line primary activity by the octavolaterahs efferent system during predation in the free-swimming toadfish, *Opsanus tau, Exp. Brain Res.,* 83: 233,1990.

239. Tricas, TC, and Highstein, SM, Action of the octavolateralis efferent system upon the lateral line of free-swimming toadfish, *Opsanus tau. J. Comp. Physiol.,* A, 169, 25,1991.

240. Usami, S, Hozawa, J, Shinkawa, H, Tazawa, M, Jin, H, Matsubara, A, Fujita, S, and Ylikoski, J, Immunocytochemical localization of substance P and neurofilament proteins in the guinea pig vestibular ganglion, *Acta. Oto-Laryngol. Suppl.,* 503: 127,1993.

241. Usami, S, Hozawa, J, Tazawa, M, Jin, H, Matsubara A, and Fujita, S, Localization of substance P like immunoreactivity in guinea pig vestibular endorgans and the vestibular ganglion, *Brain Res.,* 555: 1,53,1991.

242. Usami, S, Hozawa, J, and Yhkoski J, Coexistence of substance P and calcitonin generelated peptide like immunoreactivities in the rat vestibular endorgans, *Acta. Oto-Laryngol. Suppl.,*481:166,1991.

243. Usami, S, and Igarashi, M, GABA-like immunoreactivity in the squirrel monkey vestibular endorgans, *Brain Res.,* 417: 367,1987.

244. Usami, S-I, Hozawa, J, Tazawa, M, Igarashi, M, Thompson, GC, Wu, J-Y, and Wenthold, RJ, Immunocytochemical study of the GABA system in chicken vestibular endorgans and the vestibular ganglion, *Brain Res.,* 503: 214,1989.

245. Valera, S, Hussy, N, Evans, RJ, Adami, N, North, RA, Surprenant, A, and Buell, G, A new class of ligand-gated ion channel defined by P_2 receptor for extracellular ATP, *Nature,* 371: 516,1994.

246. Valli, P, Botta, L, Zucca, G, and Casella, C, Functional organization of the peripheral efferent vestibular system in the frog, *Brain Res.,* 362: 92,1986.

247. Wackyrn, PA, Ultrastructural organization of calcitonin gene-related peptide immunoreactive efferent axons and terminals in the rat vestibular periphery, *Am. J. Otol.*, 14: 41,1993.

248. Wackym, PA, Chen, C. Ishiyama, A. Pettis, R, López, 1, and Hoffman, L, Muscarinic acetylcholine receptor subtype mRNAs in the human and rat vestibular periphery, *Cell Biol. Int.*, 20: 187,1996.

249. Wackym, PA, Popper, P, López, 1, Ishiyama, A, and Micevych, PE, Expression of a4 and b2 nicotinic acetylcholine receptor subunit mRNA and localization of a-bungarotoxin binding proteins in the rat vestibular periphery, *Cell Biol. Int.*, 19: 291,1995.

250. Wackym, PA, Popper, P, Ward, PH, and Micevych, PE, Cell and molecular anatomy of nicotinic acetyl-choline receptor subunits and calcitonin gene-related peptide in the rat vestibular system, *Otolaryngol. Head Neck Surg.*, 105: 493,1991.

251. Warr, WB, Olivocochlear and vestibular efferent neurons of the feline brain stem: their location, morphology and number determined by retrograde axonal transport and acetylcholinesterase histochemistry, *J. Comp. Neurol.*, 161: 159,1975.

252. Warr, WB, and Guinan, JJ Jr, Efferent innervation in the organ of Corti: two separate systems, *Brain Res.*, 173: 660,1979.

253. Wersäll, J, Studies on the structure and innervation of the sensory epithelium of the cristae ampullaris in the guinea pig. A light and electron microscopic investigation, *Acta. Oto-Laryngol. Suppl.*, 126: 1,1956.

254. Wersäll, J, and Bagger-Sjöbäck D, Morphology of the vestibular sense organ, in *Handbook of Sensory Physiology, Vestibular System, Basic Mechanisms*, Vol. 6, part I (Kornhuber H, Eds.), pp. 123-170. Berlin: Springer -Verlag,1974.

255. White, JS, Fine structure of vestibular efferent neurons in the albino rat. *Soc. Neurosci. Abstr.* 11: 322.

256. Whitehead, MC, and Morest, DK, Dual populations of efferent and afferent cochlear axons in the chicken, *Neuroscience*, 6: 2351,1981.

257. Wiederhold, ML, Variations in the effects of electrical stimulation of the crossed olivocochlear bundle on single auditory nerve-fiber responses to tone bursts, *J. Acoust. Soc. Am.*, 48: 966,1970.

258. Wiederhold, ML, and Kiang, NYS, Effects of electrical stimulation of the crossed olivocochlear bundle on single auditory nerve-fibers in the cat, *J. Acoust. Soc. Am.*, 48: 950, 1970.

259. Will, U, Efferent neurons of the lateral-line system and the VIIIth cranial nerve in the brainstern of anurans: a comparative study using retrograde tracer methods, *Cell Tissue Res.*, 225: 673,1982.

260. Yamaguchi, K, and Ohmori, H, Suppression of the slow K^+ current by chohne agonists in cultured chick cochlear ganglion cells, *J. Physiol. (London)*, 464: 213,1993.

261. Ylikoski, J, Eranko, L, and Paivarinta, H, Substance P-like immunoreactivity in the rabbit inner ear, *J. Laryngol. Otol.*, 98: 759,1984.

262. Ylikoski, J, Paivarinta, H, Eranko, L, Merena, I, and Lehtosalo, J, Is substance P the neurotransmitter in the vestibular endorgans?, *Acta. Oto-Laryngol.*, 97: 523,1984.

263. Ylikoski J, Pirvola U, Happola O, Panula P, and Virtanen 1, Immunohistochemical demonstration of neuroactive substances in the inner ear of rat and guinea pig, *Acta. OtoLaryngol.*, 107: 417,1989.

264. Yoshida, N, Shigemoto, T, Sugai, T, and Ohmori, H, The role of inositol trisphosphate on ACh-induced outward currents in bullfrog saccular hair cells, *Brain Res.*, 644: 90,1994.

265. Zona, C, Farini, D, Palma, E, and Eusebi, F, Modulation of voltage-activated channels by calcitonin gene-related peptide in cultured rat neurones, *J. Physiol.*, (London), 433: 631,1991.

Section III

Neurochemistry of the Vestibular Nuclear Complex

5

Excitatory Amino Acids and Nitric Oxide in the Vestibular Nuclei

Golda Anne Kevetter, Dale W. Saxon, and Alvin J. Beitz

CONTENTS

0-8493-7679-3/00/$0.00+$.50
© 2000 by CRC Press LLC

5.1. INTRODUCTION

5.1.1 GLUTAMATE AS A TRANSMITTER

The simple amino acid glutamate and perhaps the structurally related amino acid, aspartate156 are
the most prevalent excitatory transmitters in the nervous system.30,139   Glutamate fulfills all the
major criteria of a neurotransmitter. It is found in special stores in certain neuron;, it is released
in a membrane depolarized dependent fashion; it acts on a select group of glutamate receptors; its
actions on the postsynaptic cell mimic the effects of electrical stimulation of the appropriate
presynaptic neurons; and glutamate transporters exist in the presynaptic neuron and glia to transport
glutamate released in the synaptic cleft back into the presynaptic cell or into surrounding glia.30,139

5.1.2. GLUTAMATE RECEPTORS

Glutamate can act through a variety of glutamate postsynaptic and presynaptic receptors, including
the cation-specific ion channels AMPA, NMDA and kainate receptors; and G-protein-coupled metabotropic glutamate receptors.14,78,110,139,142,147   At glutamatergic synapses, neurotransmission
depends on the different kinds of receptor being concentrated at appropriate postsynaptic sites where
they can respond to synaptically released glutamate. Native glutamate receptor channels clustered at
these postsynaptic sites are thought to be oligomeric complexes of different subunits. Over the past
decade, metabotropic, AMPA, kainate and NMDA receptors have all been characterized by molecular
cloning.27,36,42,48,70,83,98,141 Molecular cloning and expression experiments have demonstrated
that the subtype diversity of glutamate receptors is much larger than expected from pharmacological
studies. In fact an additional degree of subunit diversity is generated by alternative splicing and RNA
editing.44,68,120,121 Amino acid changes generated by alternative splicing or RNA editing alters
functional properties of the glutamate-activated channels.44,112 Characteristics of each of the four
main classes of glutamate receptors are summarized below.

5.1.2.1 AMPA Receptors

The AMPA-type glutamate receptors are the principal mediators of fast excitatory neurotransmission
in the mammalian CNS. They are composed of various combinations of four subunit proteins,
[GluR1, GluR2, GluR3 and GluR4 (or GluRA-D)] that exist in two versions (flip and flop) generated
by an alternative splicing.44,110,142   In addition certain subunit transcripts are subject to RNA
editing, a mechanism that strongly influences properties such as $Ca2+$ permeability and rectification
characteristics.118 Although expressed ubiquitously, AMPA receptor subunits show differential
expression patterns in rodent brains.69,91,128 Recent work has shown that AMPA receptors are
both physically and functionally coupled to a variety of cytoskeletal and signalling molecules via
an adaptor protein called GRIP (glutamate receptor interacting protein25). Grip contains seven
protein-interaction domains (PDZ domains) and appears essential for synaptic clustering of AMPA
receptors.[115]

Based on the published cDNA sequences, it has been possible to synthesize messenger RNAs (mRNAs) specific for GluR subunits, to make mixtures of various combinations of these GluR subunits, to express these receptor channel mixtures in oocytes and to study physiologically the receptor properties of the resulting channels.[44,80,82,110] This has also been done for combinations of AMPA and NMDA receptor subunits. Studies of the AMPA receptor response have shown that it can be dominated by the GluR2 response when present.[44,82,110] GluR2 is present predominantly in the edited form, which, when incorporated into AMPA receptor channels, leads to an inhibition in channel permeability to Ca^{2+} and specifies distinct current-voltage relationships.[121,122] The addition and combination of the other subunits can result in varied response characteristics.[44,82,110] For example, it has been shown that desensitization kinetics of AMPA receptors are regulated by the expression of GluR4 spice variants.[112] Therefore, by varying the composition of the glutamate receptor channel, different physiological properties can be expressed in specific neurons and perhaps even change under different physiological states, such as after tetanic stimulation.[80,82,110] All four AMPA receptor subtypes are present in the vestibular nuclei as detailed below, suggesting that AMPA receptors play an important role in vestibular function.

5.1.2.2. Kainate Receptors

The first cloned mammalian kainate receptor subunit, distantly related to the AMPA receptor family, has been termed GluR5.[3] Two additional kainate receptor subtypes, termed GluR6 and GluR7, have been cloned.[4,27] Further candidates for a high affinity kainate receptor subunit, KA1 and KA2, were isolated by Werner et al.[141] and Herb et al.[38] GluR5 and GluR6 can form functional homomeric channels.[3,4] The other three subunits (GluR7, KA1, and KA2) apparently do not form homomeric ion channels, but can contribute to heteromeric assemblies with GluR5 and GluR6.[38]

5.1.2.3 NMDA Receptors

NMDA receptors are modifiable channels that appear to be involved in plastic changes in the nervous system, including changes that occur in the vestibular nuclei during vestibular compensation.[118,119] These channels are voltage-dependent, meaning that they remain closed until a certain amount of depolarization of the neuron is achieved. The NMDA receptor is widely distributed and possesses a monovalent and divalent cation conducting channel, which regulates the survival, differentiation, plasticity, and activity of a large number of neurons within the central nervous system. Pharmacological studies have led to a picture of the native NMDA receptor as a complex, ligand-gated channel composed of multiple subunits with several allosteric modulatory sites. Recent cloning experiments have confirmed this picture and have shown that the assembly of various combinations of subunits from two major families, termed NR1 and NR2, (GluRz and GluRe in the mouse) results in functional NMDA receptors when co-expressed in Xenopus oocytes or transfected into mammalian cells.[82,97] The NR1 subunit is widely expressed in the CNS and displays at least seven isoforms that influence the receptor's sensitivity to $Zn,^{2+}$ protons, polyamines, and protein kinase C.[82] In particular, the presence or absence of exon 5 near the N-terminus is important for influencing the size of evoked currents, potentiation by zinc, and sensitivity to extracellular pH.[7,131] In contrast, expression of the NR2 subunits is spatially restricted to particular brain regions[7,62,81,82] and the NR2 subtype appears more directly related to particular functional and pharmacological properties of NMDA receptors.[143] Heterologous expression systems in particular have established that the kinetic and pharmacological properties of NMDA receptors are determined by their subunit composition.[59,81,142,143] It is worth noting that a novel NR3A subunit has recently been discovered which appears to inhibit the conductance of the NMDA channel (S. A. Lipton, personal communication). While the subunit composition of individual NMDA receptors in the vestibular nuclei have not been examined in detail, it is interesting to note that the majority of NMDA receptor complexes in the rat cerebral cortex have been shown to contain at least three different subunits (NR1/NR2A/NR2B).[67]

Glutamate autoreceptors have been shown to be present on presynaptic terminals and they can thus alter the amount of transmitter (including glutamate) released from the terminal with which they are associated.[103] In this regard NMDA autoreceptors have been found on the central terminals of spinal ganglion neurons[65] and are also present on the central terminals of the vestibular nerve.[31,64,136] Thus if glutamate is released from the central terminals of the vestibular nerve at termination sites in the brain stem, subsequent release could be modified by glutamate acting in a feedback fashion at these NMDA autoreceptors. Finally, with respect to the functional role of NMDA receptors in the vestibular nuclei, work from Vidal's laboratory has shown that NMDA receptors are present on central vestibular neurons and appear to be crucial to the maintenance of the resting discharge of these neurons.[136] This same group has also shown that at a 5 hr timepoint following vestibular labyrinthectomy, there is a significant decrease in the mRNA for the NMDAR1 receptor subunit in the medial vestibular nuclei, which returns to normal by 3 days post labyrinth-ectomy.[20] Furthermore, based on administration of NMDA antagonists in labyrinthectomized animals, Vidal and co-workers have hypothesized that a denervation supersensitivity resulting from an increase in the sensitivity of NMDA receptors is a factor responsible for the recovery of the static syndrome following unilateral labyrinthectomy.[136] These data clearly indicate that NMDA receptors play an important role in the function of the vestibular nuclei and further suggest that alterations in NMDA receptors occur in the vestibular nuclei following vestibular labyrinthectomy.

5.1.2.4 d Glutamate Receptors

The d glutamate receptors (d1 and d2) constitute a separate family of proteins that, on the basis of their amino acid sequence, has been positioned between the NMDA and AMPA receptor families.[1,66,153] The d2 subunit is predominantly expressed in cerebellum and mutant mice that lack this subunit and exhibit defects that could be attributed to cerebellar dysfunction, such as impaired motor coordination.[46] This is accompanied by a strong reduction in cerebellar long term depression (LTD), confirming the results of previous studies using antisense technology.[41] With regard to the role of the d2 receptor in vestibular function, Funabiki and co-workers[33] have shown that following a unilateral labyrinthectomy, compensation of the righting response in mice deficient in the d2 protein is retarded. This indicates that compensation of the dynamic symptoms of vestibular dysfunction is delayed in mice that fail to show LTD as a result of the absense of this particular glutamate receptor subtype, and suggests that this receptor subtype may play an inportant role in central vestibular function. Recent work has shown the presence of the d1 receptor subtype in both types of hair cells in the vestibular endorgan and also in vestibular ganglion neurons.[102] The prominent expressions of the delta 1 receptor in type I and type II vestibular hair-cells suggests a functional role in hair cell neurotransmission.

5.1.2.5 Metabotropic Glutamate Receptors

The cloning of a family of genes coding for metabotropic glutamate receptors, which include eight subtypes (mGluR1 - mGluR8) and several splice variants, has revealed that they show sequence similarity to $GABA_B$ receptors,[47] but that they are not homologous with other known G-protein-coupled receptors.[95,110] Metabotropic glutamate receptors are widely distributed throughout the CNS and are divided into the following three major groups: (1) group I includes mGluR1 (splice variants a, b, c, d, e) and mGluR5 (spice variants a and b) receptors, which are coupled to the stimulation of phosphoinositol turnover; (2) group II includes mGluR2 and mGluR3; and (3) group III includes mGluR4a, mGluR4b, mGluR6, mGluR7a, mGluR7b and mGluR8. The members of groups II and III are coupled to the inhibition of adenylyl cyclase. Additional studies have provided evidence that mGluRs expressed in the CNS are also coupled via G-proteins to the regulation of voltage-gated ion channels.[95] Ultrastructural studies suggest that the mGluR1a receptor subtype appears to be localized at the periphery of the postsynaptic membrane, as opposed to ionotropic

Glu receptors which occupy the core of the synapse. Moreover, the mGluR2, -4 and -7 receptor subtypes have been found to be preferentially localized to presynaptic terminals,[110,115,116] and their activation inhibits glutamate release (see Pin and Duvoisin[95] for review). In addition, an mGlu receptor coupled to PI hydrolysis (the identity of which is unknown) is located at the presynaptic level and its activation enhances glutamate release in the presence of arachidonic acid.[39] Finally, electrophysiological data from brainstem slices that contained the MVN demonstrate that many MVN neurones have functional metabotropic receptors and suggest that this glutamate receptor subclass plays an important role in the function of the MVN.[16]

5.2 GLUTAMATE AS A TRANSMITTER IN THE VESTIBULAR NUCLEI

Not surprisingly, glutamate has been proposed to be a major transmitter in vestibular nuclei and in pathways terminating and arising in vestibular nuclei.[13,21,53,56,57,64,119] The localization and function of glutamate in vestibular nuclei has been studied utilizing a variety of anatomical, pharmacological, and electrophysiological techniques. These have included immunocytochemistry for localization of glutamate and glutamate receptor protein, receptor binding autoradiography for localization of glutamate binding sites, and *in situ* hybridization for the localization of messenger RNAs involved in the manufacture of glutamate synthetic enzymes, glutamate receptors, etc. The following sections discuss findings from each of these approaches.

Most of the data related to excitatory amino acid and glutamate receptor localization has been obtained in rodents; therefore, comparisons with other mammalian and non-mammalian species is limited.

5.2.1 IMMUNOHISTOCHEMICAL LOCALIZATION OF EXCITATORY AMINO ACIDS (EAAs) IN VESTIBULAR NUCLEI

One of the first attempts to anatomically define neurotransmitters in vestibular nuclei involved immunocytochemical studies with antibodies raised against the amino acids glutamate and aspartate, conjugated through glutaraldehyde to thyroglobulin. The presence of a prospective neurotransmitter in a presynaptic neuron is an essential requirement that is used to establish that a neurochemical is a bonafide neurotransmitter, although presence alone is insufficient to absolutely determine the transmitter status in the neuron. This is a particular problem with glutamate, since it is also a major building block of proteins, and is also found in the metabolic pool of all neurons.[30] Thus, a major issue that must be addressed when examining glutamate localization in neurons is the problem of distinguishing the metabolic pool of glutamate from the transmitter pool. Several studies argue that immunocytochemical staining for excitatory amino acids results in relatively selective staining, indicative of staining transmitter pools and not metabolic stores. For example, in one study of amino acid transmitter localization within the vestibular nuclei, procedures were designed to optimize the capture of all amino acid stores, and few neurons were found that contained both excitatory and inhibitory amino acid staining.[135] Rather, excitatory and inhibitory amino acids were found to be complementary and the staining for glycine and GABA was also observed to be generally non-overlapping. On the other hand, Zhang and colleagues[158] have shown that in the olivocerebellar system, glutamate-immunoreactivity was sevenfold higher in climbing fiber terminals than in the olivary neuronal perikarya of origin, while aspartate immunoreactivity was similar in the climbing fiber terminals compared to the olivary neurons of origin. Based on these results, Zhang and coworkers concluded that glutamate—and not aspartate —is the transmitter of olivocerebellar climbing fibers. Thus, caution must be exercised in interpreting immunocytochemical data related to glutamate and aspartate neuronal localization, and in drawing conclusions related to transmitter function based solely on the presence of an excitatory amino acid in neuronal perikarya.

Many neurons in all four vestibular nuclei and Scarpa's ganglion are immunocytochemically labeled with antibodies specific for glutamate or aspartate.[2,51,52,61,160] Most of the positive neurons contain dense staining in the nucleus and the cytoplasm, often extending from the perikaryon into the dendrites of the multipolar neurons. Many small- to medium-size perikarya are labeled in the medial (MVN) and superior (SVN) vestibular nuclei. Neurons are also labeled in the inferior (IVN) vestibular nuclei. Most striking is the presence of immunolabel in the giant 'Deiters' neurons in the lateral (LVN) vestibular nuclei. Virtually all of these neurons stain positively for both glutamate- and aspartate-immunoreactivity.[51] The distribution of cells that stain for EAAs is similar in gerbil,[49] guinea pig,[61] cat,[9,138] and monkey.[9]

Many neurons containing glutamate-immunoreactivity also stain with antibodies for aspartate.[138] While the work of Zhang et al.[157] suggests that aspartate staining in the cytoplasm and nucleus of a neuron may not be indicative of increased concentrations of aspartate in the nerve terminal, neurons that stain densely with glutamate-immunoreactivity appear to contain high levels of glutamate in their terminals.[15,138,157] In addition, experiments showing the depletion of terminal stores of glutamate under depolarizing conditions,[88,89,152,158] provide further evidence for glutamate as a transmitter in selected axonal terminals. These studies suggest that aspartate- and glutamate-like immunocytochemical staining in the perikarya may be indicative of a population of neurons that release glutamate from their terminals.[157,158]

The identification of perspective transmitters associated with afferent inputs to the vestibular nuclei have also been suggested by immunohistochemical studies. For example, neurons in the vestibular ganglion stain with glutamate-like immunoreactivity (Glu-lir) in frogs, rodents, and cats,[9,19,99] suggesting that the input from the vestibular nerve may be glutamatergic (cf. also chapters 2 and 3). This has been confirmed in the medial vestibular nuclei using microdialysis techniques. Yamanaka et al.[152] have shown that repetitive electrical stimulation of the vestibular nerve causes a significant frequency-dependent increase in the release of glutamate in the MVN, while the levels of other amino acids, (including aspartate, glycine, and GABA) remained unaltered. In addition to the vestibular nerve, the inferior olivary input to the vestibular nuclei may be glutamatergic, similar to that proposed for the climbing fiber endings in the cerebellum.[157,158] Finally, immuno-cytochemical evidence suggests that the input from the cerebellar nuclei to the vestibular nuclei may contain a glutamatergic component.[79]

With respect to efferent projections from the vestibular nuclei, immunocytochemical evidence suggests that vestibulospinal (VST) neurons may use an excitatory amino acid as a transmitter. In the LVN and magnocellular MVN, for example, most retrogradely-labeled VST neurons stain with Glu-lir. However, in the parvocellular portion of the MVN less than half of the VST neurons are double-labeled.[49,51] Unlike the lateral VST, the medial VST has both excitatory and inhibitory actions on neurons in the spinal cord.[32,144,145] Therefore, neurons that give rise to the medial VST may utilize one neurotransmitter for excitation and a different neurotransmitter for inhibition.

In addition to vestibulospinal neurons, many vestibular neurons that project to the cerebellum and to the occulomotor nuclei (those involved with the vestibulo-ocular reflex, VOR neurons) are also glutamate-immunoreactive, suggesting that they may be glutamatergic. Injections of retrograde tracers have been made into the trochlear nucleus, a midbrain nucleus that receives an excitatory input from second-order neurons in the contralateral MVN and an inhibitory input from second order neurons in the ipsilateral SVN.[40,73,132,133,151] Most of the retrogradely labeled VOR neurons in MVN were found to be GLU-lir, while far fewer of VOR neurons in the SVN were GLU-lir.[52] In addition, following injections of the anterograde tracer PHA-L into the vestibular complex, anterogradely labeled mossy fiber terminals in the cerebellar cortex have high levels of glutamate-immunogold labeling (unpublished observations), suggesting that some vestibulo-cerebellar neurons may utilize glutamate as a primary transmitter. The above data suggest that several major inputs to the vestibular complex utilize glutamate as their primary transmitter and that glutamate may also serve as a putative transmitter for several efferent projections from the vestibular nuclei, including the vestibulo-spinal, vestibulo-oculomotor, vestibulo-trochlear, and vestibulo-cerebellar pathways.

5.2.2 LOCALIZATION OF GLUTAMATE RECEPTORS

Glutamate receptors can be identified and studied with several techniques. Few of these techniques have been applied systematically to the vestibular nuclei, thus, it is often difficult to compare the results of the various methodologies. The following section summarizes what is known regarding the localization of glutamate receptors in the vestibular nuclei.

5.2.2.1 Localization of Glutamate Receptors Using Receptor Binding Autoradiography

There have been no published studies to date that have determined the binding properties of glutamate receptors and receptor subtypes among the vestibular nuclei. However, one study has examined glutamate binding specifically in the MVN.[129] This study reported a single population of binding sites with a $K_D = 126$ nM.[129] The MVN had a high number of glutamate binding sites (2.3 plus/minus 0.2 pmol/mg protein). Unfortunately, this study looked at general glutamate binding, using glutamate as a displacer, and few conclusions can be drawn regarding the localization of glutamate receptor subtypes.

Survey reports of the presence of pharmacological subgroups of glutamate receptors indicate the presence of all receptor subgroups in the vestibular nuclei. AMPA receptors[11] (using [³H] CNQX binding; 31.2 plus/minus 6.6 fmol/mg) and NMDA receptors[77] (158 plus/minus 23 fmol/mg protein²) have been found in the vestibular nuclei in rat. Quisqualate binding, indicative of both AMPA and metabotropic receptor subgroups, has also been identified in the vestibular nuclei of monkey.[154] Recent quantitative autoradiographic receptor binding analysis in the rat indicates the distribution of glutamate receptors among the vestibular nuclei shown in Table 5.1.

TABLE 5.1
Distribution of Glutamate Receptor Binding Sites Among the Four Vestibular Nuclei and the Nucleus Prepositus Hypoglossi in the Rat

Vestibular Nucleus	Receptor Subtypes [Specific Binding, fmol/mg protein]			
	NMDA	AMPA	Kainate	Metabotropic
Lateral Vestibular N.	35 ± 2.8	360 ± 20	4430 ± 560	120 ± 92
Medial Vestibular N.	125 ± 1.6	530 ± 21	4320 ± 431	280 ± 40
Inferior Vestibular N.	28 ± 3.9	330 ± 27	2770 ± 436	240 ± 37
Superior Vestibular N.	84 ± 1.9	380 ± 32	0	0
Prepositus Hypoglossi	57 ± 1.8	90 ± 15	940 ± 190	20 ± 13

These results clearly show that the highest concentration of NMDA, AMPA, and metabotropic receptors are in the medial vestibular nucleus, while kainate receptors are highest in the lateral vestibular nucleus. The superior vestibular nucleus contains high levels of NMDA and AMPA receptors, but contains very few, if any, kainate or metabotropic receptors. Finally, the inferior vestibular nucleus contains high levels of AMPA and metabotropic receptors and low levels of NMDA receptors, while the prepositus hypoglossi contains moderate levels of NMDA and kainate receptors, but low levels of AMPA and metabotropic receptors. These findings clearly suggest that glutamate receptor subtypes are differentially distributed among the vestibular nuclei and that certain subtypes predominate in each nucleus. These findings have been augmented by recent molecular studies that have examined the distribution of the mRNAs for glutamate receptor subtypes and immunocytochemical studies which have mapped the distribution of the receptor subtypes in the brain, as discussed below.

5.2.2.2 Localization of Glutamate Receptors Using *In Situ* Hybridization and RT-PCR

Using molecular biological approaches, a large number of ionotropic glutamate receptor (GluR) and metabotropic glutamate receptor (mGluR) subunits have been characterized as reviewed above.[44,80,82,110,111] Based on the cloned sequences of these subunits, *in situ* hybridization and reverse transcription-polymerase chain reaction (RT-PCR) studies have been performed to determine if the mRNA for a specific subunit is present in a given brain region. The presence of an mRNA for a specific subunit of glutamate receptor would suggest that the cell is capable of manufacturing that receptor protein. Such data can then be combined with immunocytochemical data and physiological data to determine the potential role that each receptor subtype plays in neurons located in different brain regions.

5.2.2.2.1 Glutamate *receptors in the vestibular ganglion.*

With respect to the vestibular ganglion, expression of mRNA for various glutamate ionotropic receptors is found in the vestibular ganglion.[24,31,85] Results of studies with 14 different subunit-specific oligonucleotides to ionotropic glutamate receptors revealed conspicuous labeling for several AMPA, kainate, and NMDA subunits.[85] Using *in situ* hybridization only two subunits (AMPA subunit GluR1 and kainate subunit GluR7) were not detected.[85] However, RT-PCR analysis of the vestibular ganglion has shown that mRNAs for the GluR1-4 subunits in two alternative spliced versions (flip and flop) are all present in the vestibular ganglion;[24] thus, all AMPA receptor subtypes appear to be present in the vestibular ganglion. Low levels of other kainate and NMDA subunits indicate that specific combinations of receptor channels may exist in the ganglion.[85] These studies support the concept of glutamate as the transmitter between hair cells and the vestibular nerve and further indicate the possibility of NMDA autoreceptors on central processes of the vestibular nerve, as has been found for dorsal root ganglion processes.[65]

5.2.2.2.2 Glutamate *receptors in the vestibular nuclei*

In situ hybridization studies have shown that the mRNA for all four AMPA receptor subtypes are distributed throughout the vestibular nuclei.[96] Quantitative analysis of the levels of hybridization show a high degree of diversity in the levels of expression of the GluR2 subunit mRNA, with the highest levels in the giant Deiter's cells of the LVN and the lowest levels in the small neurons throughout the vestibular nuclei. GluR1, GluR3, and GluR4 mRNA levels were less numerous than GluR2, with Glu4 mRNA being present in the fewest number of neurons. Since the subunit composition of AMPA receptors determines their physiological properties as discussed above, the differential distribution and levels of expression of AMPA subunits in the vestibular nuclei may relate to the characteristics of information processing by the vestibular system.

Neurons throughout the vestibular nuclei express mRNA for the NMDA NR1 receptor subtype.[20,104,135] This includes small and medium neurons in all four nuclei and the giant neurons in the LVN.[20] Within the MVN, the number of labeled neurons as a percentage of the total number of MVN neurons varies between 51 and 80 percent, depending on the level of the nucleus;

whereas 78 to 85 percent of the cells in the LVN are labeled.[135] These results are consistent with receptor binding data showing that NMDA receptors are distributed ubiquitously throughout the vestibular complex. A small proportion of neurons in the vestibular nuclei contain the NMDA R2 receptor subunit mRNAs.[104] NMDA has a role in maintaining the resting discharge of MVN neurons[22,113,136] and is involved in the alterations in response characteristics of MVN neurons to tetanic stimulation of the nerve.[8] NMDA receptors are involved in the synaptic response of second-order vestibular cells in young rats[123] and in particular contribute to the monosynaptic transmission from the first order vestibular afferents.[136] NMDA receptors also play a tangible role in neuronal responses to commissural stimulation in frogs[56] and young rats.[123]

Metabotrophic receptor subunit mRNA has also been identified in the vestibular nuclei; however, these data come primarily from studies involved in mapping the distribution throughout the entire brain and thus details of localization of receptor subtypes among the individual vestibular nuclei are often neglected. The vestibular nuclei express moderate amounts of message for the metabotropic glutamate receptor mGluR1[117] (but see de Waele[20]) and low, but detectable levels of mRNA for the mGluR2 receptor, which appears to be most prominent in the MVN.[87] The mRNA for the mGluR4 receptor is present within the vestibular complex, but mRNA for mGluR3 is not detectable within the vestibular complex.[125] Message for the mGluR7 subunit is present in the vestibular complex where it appears most prominent in the spinal vestibular nucleus.[54] mGluR8 mRNA is also present in the vestibular nuclei,[105] but its exact distribution requires further study.

5.2.2.3 Immunocytochemical Localization of Glutamate Receptors

In situ hybridization techniques can determine if a cell contains the mRNA necessary to synthesize a particular protein subunit of a selected ionotropic or metabotropic glutamate receptor, but this procedure cannot provide information on whether the actual protein has been synthesized and is present in the cell. In contrast, immunocytochemical techniques can be employed to determine if the protein has indeed been manufactured by the cell. Since antibodies have been made to a majority of the glutamate receptor subunits,[85,91-94] it is now possible to study the distribution, segregation and possible overlap of glutamate receptors in the vestibular nuclei as well as gain insight into their role in sensorimotor integration and plasticity.

5.2.2.3.1 AMPA Receptor Localization

Antibodies against the GluR1, GluR2/3, and GluR4 subtypes of AMPA receptors have been used to localize staining for this glutamate receptor subclass.[50,91,96] In the rat, the four vestibular nuclei stained intensely with an antibody to GluR2/3 and moderately with an antibody to GluR1.[91] Immunocytochemical staining for GluR4 was intermediate in intensity between the other two subtypes.[91] A similar pattern of labeling was observed in the chinchilla with a majority of the neurons labeled with an antiserum specific for the GluR2/3 receptor subtypes.[96] GluR1-immunoreactive neurons were fewer in number than GluR2/3, and GluR4-immunoreactivity was found in the fewest number of neurons.[96] It is interesting to note that GluR4-immunostaining was also present in some astrocyte-like structures.

An antibody to GluR1 produces little staining in the vestibular nuclei of the gerbil compared to the intense staining observed in other areas, such as the external cuneate nucleus, spinal nucleus of V, inferior olive, area postrema, and the molecular layer of the cerebellar cortex. However, widely scattered, positively stained cells are present in the vestibular nuclear complex, especially in the lateral portions of IVN and LVN, adjacent to the restiform body. These cells are fusiform in shape with long dendrites (Figure 5.1A).

A group of smaller multipolar cells is located at the border between MVN and PH (Figure 5.2A). A cluster of small labeled neurons is also observed dorsal to the genu of the seventh nerve. In the

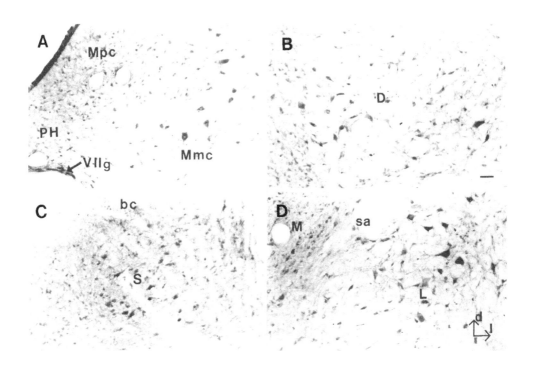

FIGURE 5.1 Photomicrographs through the lateral vestibular nucleus in the same gerbil: (A). staining with antibody to AMPA receptor GluR1, (B). Staining with antibody to AMPA receptor GluR2/3; (C). staining with antibody to AMPA receptor GluR4 and (D). staining with antibody to kainate receptor GluR5/6/7. Scale bar = 50 mm. Abbreviations: rb, restiform body; Ld, dorsal division; Lv, ventral division; sa, stria acoustica.

LVN, the NR1 antibody lightly labels large neurons. In addition neurons in Scarpa's ganglion are lightly stained.

The pattern of GluR2&3 immunostaining is very different from that observed for GluR1. Many large neurons with elaborate dendritic profiles stain in the vestibular nuclei. The largest labeled neurons are observed in the IVN and LVN (Figure 5.1B). In these nuclei, staining is also present in the neuropil surrounding the neurons, but absent from the fiber fascicles. Medium to large neurons in the MVN (Figure 5.1C), SVN, and PH are also immunostained for GluR2/3. The MVN and PH are the most densely stained nuclei of the vestibular nuclear complex. In addition, Scarpa's ganglion cells and Purkinje cells are densely stained.

Numerous brainstem regions, including the vestibular complex and almost all motor nuclei, contain immunostaining specific for the GluR4 subtype. In the vestibular nuclei, neuropil staining resembles that of GluR1; moderately stained, widely scattered neurons are observed throughout the vestibular nuclear complex. Many of the large neurons in LVN and DVN are immunolabeled for GluR4, as are smaller neurons in the SVN and MVN. The neuropil staining in the MVN is extremely dense, and labeled cells are found predominately in the lateral portion of this nucleus. A small group of neurons is also labeled in the area just dorsal to the genu of the seventh nerve. In addition neurons in Scarpa's ganglion show GluR4-immunostaining.

5.2.2.3.2. NMDA Receptor Localization

Immunocytochemical studies have begun to resolve the distribution of neurons that express NMDAR1 (NR1) and the specific location in the neuron where NR1 channels exist.[94] NR1 immunocytochemical staining is located in neuronal perikarya and dendrites in all four vestibular nuclei and the adjacent

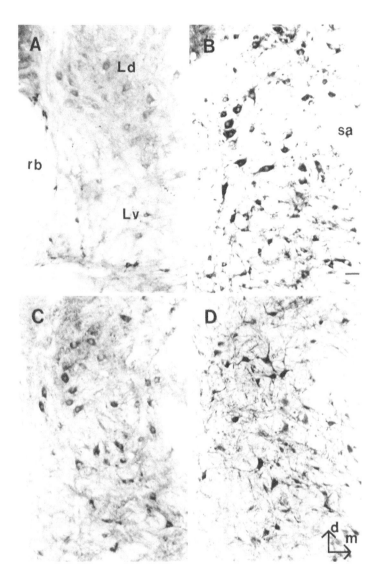

FIGURE 5.2 Photomicrographs through adjacent sections of the medial vestibular nucleus and PH (compare to figure 5.3A): (A). AMPA receptor GluR1 antibody; (B) AMPA receptor GluR2/3 antibody, (C). AMPA receptor GluR4 antibody; and (D). kainate receptor GluR5/6/7 antibody. Scale bar = 50 mm.

nucleus prepositus hypoglossi.[50,94] Labeling for NR1 is similar in the vestibular complex of both the rat[94] and gerbil.[50] Data from Kevetter's laboratory indicate that many uniformly distributed, small to medium-sized neurons are immunolabeled in the MVN (Figure 5.3A). The neuropil of the MVN is also homogeneously labeled. In the IVN (Figure 5.3B), labeled neurons larger than those labeled in MVN or SVN are present. Within the LVN, the perikarya and long dendrites of both large neurons and giant Deiter's neurons are labeled (Figure 5.3D). In addition, distinct labeled fibers are scattered throughout the neuropil of LVN. While the staining pattern in the rostral ventral portion of the LVN is similar to that observed throughout the rest of the nucleus, there are fewer

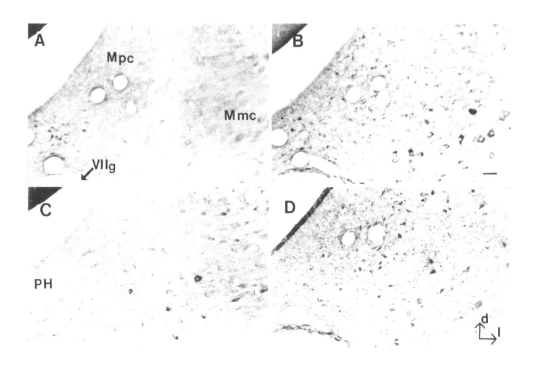

FIGURE 5.3 Photomicrographs of coronal sections through gerbil brain stained with antibodies to NMDAR1. (A). Section through the medial vestibular nucleus (Mpc, parvocellular portion; Mmc, magnocellular portion) and the nucleus prepositus hypoglossi (PH), (B). section through the descending vestibular nucleus (D). C. Section through the superior vestibular nucleus (S, bc brachium conjunctivum), (D). and section through the lateral vestibular nucleus (L, sa stria acoustica). Orientation the same in all photos (d, dorsal, l, lateral). Scale bar is the same for all photos. Scale bar = 50 mm.

neurons labeled in this region and the neurons are smaller in size. Within the SVN, many small- to medium-sized neurons are also labeled (Figure 5.3C).

Many fibers from the vestibular nerve travel through this area en route to the cerebellum. These unstained fibers pierce through the nucleus from its ventrolateral extent to the dorsomedial extent. The neuropil in other parts of the SVN is stained more intensely than in the LVN, but not as intensely as in the MVN. In addition to neurons in the vestibular nuclei, ganglion cells in Scarpa's ganglion also stained with the NR1 polyclonal antibody.[73,94] In the neural areas studied, NR1 staining is localized in postsynaptic densities of synapses and in the soma and dendrites.[94]

While NR1 is ubiquitous in neurons that have NMDA receptors, the NR2 subunits (NR2A, NR2B, NR2C, NR2D-1, NR2D-2[80,82]) are not. As indicated above, the physiological properties of the NMDA channel vary, depending on whether NR1 is homogeneously present and the number and type of NR2 receptors present. NR1 and NR2A/B are colocalized in many regions of the nervous system.[92] NR2A/B immunocytochemical staining is present in neurons of the vestibular nuclei and ganglion.[92] Staining is most intense in neurons of the LVN, while moderate staining is found in the other vestibular nuclei and the nucleus prepositus hypoglossi (PH).

Although pharmacological, physiological, and behavioral effects will be dealt with in detail in other chapters (Chapters 2, 3, 6, 10, and 18), the dense immunocytochemical localization of NMDA receptors in the vestibular nuclei provides anatomical substrates for the role of NMDA in vestibular compensation. Perfusion of NMDA antagonists in the fourth ventricle will cause a decompensation in hemilabyrinthectomized animals.[22,118,136] NMDA antagonists can also block LTP (long-term potentiation) and LTD (long-term depression) in MVN neurons after tetanic stimulation of the vestibular nerve.[8]　NMDA receptors also play a role in normal gaze mechanisms controlled by neurons present in MVN and the PH.[74,75]　In addition NMDA receptors play a role in the neuronal control of normal maintenance of skeletal musculature, as well as eye musculature.[22]　Thus it is clear that NMDA receptors are present on both central and peripheral vestibular neurons and that these receptors play an important role in the normal functioning of the vestibular system.

5.2.2.3.3. Kainate *Receptor Localization*

Kainate receptors include subtypes GluR5 - 7, KA1, and KA2.[111]　These ionotropic receptors mediate fast-rising EPSPs that might be involved in fast synaptic activation of the post synaptic receptor.　Immunohistochemical studies have identified KA2 and GluR6/7-positive neurons in the rat vestibular complex[93] and GluR5/6/7-stained neurons in the vestibular nuclei of the gerbil (Kevetter). Immunostaining for both the KA2 and GluR6/7 subtypes is present in all four vestibular nuclei and intense staining is also present in cell group y.[93]　In addition, moderate levels of KA2 and GluR6/7 are found in Scarpa's ganglion neurons in the rat.[93]

Using an antibody against GluR5/6/7, vestibular ganglion neurons of the gerbil do not stain, but there is an abundance of labeling in the vestibular nuclei.　Neurons are labeled in all four vestibular nuclei and in the PH. Neuropil staining is most intense in the MVN and PH (Figure 5.2D), but immunostained dendrites and fibrous processes are present in all four vestibular nuclei. In the SVN and MVN, many small- to medium-sized neurons are labeled, while in the IVN, LVN (Figure 5.1D), and the magnocellular portion of MVN primarily large and giant cells are labeled.

5.2.2.3.4 Glutamate *Transporters*

The termination of glutamatergic transmission and the clearance of the excessive, neurotoxic concentrations of glutamate is ensured by a high affinity glutamate uptake system associated with specific glutamate transporters in the plasma membranes of both neurons and glial cells.　Four L-glutamate neurotransmitter transporters —the three Na+- dependent transporters, glutamate/aspartate transporter 1 (GLAST-1), glutamate transporter 1 (GLT-1), and excitatory amino acid carrier 1 (EAAC-1), and the Cl- dependent EAAT-4—form a new family of structurally related integral plasma membrane proteins with different distribution in the central nervous system.　The genomic organization of these transporters is still under investigation.　To date, very little is known about the nature of the factors and molecular mechanisms that regulate developmental, regional, and cell type-specific expression of the glutamate transporters and their aberrant functioning in neurodegenerative diseases.　Recent work has shown that glutamate transporters are present in the vicinity of most central glutamate synapses, including those associated with the vestibular nuclei.　However, the role of postsynaptic, neuronal glutamate transporters in terminating signals at central excitatory, vestibular synapses is currently not known.

GLAST-1 and GLT-1 are two high affinity glutamate transporters that are associated exclusively with glial cells.[134]　Takumi et al.[124] have shown that GLAST-1 and the glutamate- synthesizing enzyme, glutamine synthetase, are colocalized in supporting cells apposed to hair cells in the vestibular endorgan and in non-neuronal cells in the vestibular ganglion.　These findings suggest that glutamate released at the afferent synapse of vestibular hair cells or by neurons in the vestibular ganglia are taken up by adjacent supporting cells and converted to glutamine.　This data strengthens the view that glutamate is a transmitter released by vestibular hair cells and further suggests that supporting cells in the vestibular endorgan and in the vestibular ganglion may carry out functions similar to those of glial cells in the CNS.

5.2.2.3.5 Functional Significance

Immunocytochemical localization of glutamate receptors complement physiological and pharmacological studies of glutamate effects on vestibular neurons.[13,53,57,64]　The response of second-order neurons to stimulation of the vestibular nerve can be blocked by AMPA antagonists[53] and can be partially blocked by NMDA antagonists.[136] The functional significance of AMPA and NMDA receptors in vestibular function has been reviewed by Vidal.[21,136] In addition to AMPA and NMDA receptors, kainate receptors are present on second-order neurons.[93] While the role of AMPA and NMDA receptors in vestibular function are beginning to be defined,[21,136] very little is currently known regarding the role of kainate receptors. Nonetheless, the fact that the GluR6/7 and KA2 subtypes are localized to the vestibular nuclei suggests that the kainate glutamate receptor subclass plays some role in central vestibular processing.

5.3 NITRIC OXIDE, NITRIC OXIDE SYNTHASE, AND THE VESTIBULAR SYSTEM

5.3.1 INTRODUCTION

In recent years, it has become apparent that in addition to the traditional neurotransmitters like acetylcholine, glutamate, and GABA, etc., there are also small molecules of an inorganic nature which impart unique signaling capabilities to neurons. Of particular note is the gaseous molecule, nitric oxide (NO), which is known to be produced not only by various cells of the immune system and vascular endothelium, but also by select neurons of the central and peripheral nervous systems.[6,12,17,58,84,114] Neurons producing NO contain the enzyme, nitric oxide synthase (NOS), from which NO is derived in a co-factor-dependent manner in response to increased levels of intracellular calcium.[58,84,114] As a highly reactive free radical with a short half-life and a propensity to diffuse indiscriminately across lipid membranes, NO is released into the intercellular space where it has access to a large variety of neighboring cells.[17,84] However, NO can only affect target cells that contain an appropriate substrate upon which NO can act. NO has been implicated in a variety of functional roles in different brain regions.[17] The current broad interest in NO and NOS/NADPH-d has not only been driven by the important roles that NO is proposed to play in normal brain function, but also by the implication that NO is likely to be involved in a number of neuropathological syndromes and neurodegenerative diseases,[6,12,17,127] as well as by studies that have presented evidence that would suggest that NO may play a neuroprotective role.[12,17,28,35,108,109,127,146]

In light of the inherent properties of NO, including its extemely short half-life, its direct measurement and localization is problematic even under the most controlled conditions. In partial solution to this problem, researchers have cloned and subsequently produced antibodies against the NOS enzyme, making it possible to identify NO-producing cells.[17] In what might be considered a twist of good fortune, it was also determined that the reaction product deposited following the relatively simple histochemical methodology for localizing nicotinamide adenine dinucleotide phosphate-diaphorase (NADPH-d) colocalized for the most part with the constitutive NOS enzyme in brain neurons.[18,45,71,137] Thus, NADPH-d histochemical staining is being used as an indicator of NO-producing neurons.

Since the identification of NO as a potentially important messenger in the nervous system many reports have appeared in the literature that expound on the distribution of neurons containing constitutive NOS/NADPH-d, both in the peripheral and central nervous systems.[26,63,76,137] Although such reports provide valuable information regarding the location of neurons containing constitutive NOS/NADPH-d, they are often of a general nature and frequently lack the detail in individual nuclei that would permit anything more than remedial comparisons with maps based on specific anatomical or neurochemical studies. The vestibular complex and associated structures are no exception in this respect. The paucity of information in scientific literature regarding the specifics of constitutive NOS/NADPH-d in the individual vestibular nuclei will be at least partially addressed in the following sections. Since there is very little information regarding NOS/NADPH-d in the vestibular complex of any species, the following description is based primarily on the authors' personal observations in the rat and mouse. We will attempt where possible to compare and contrast features from other species where information is available.

5.3.2 CONSTITUTIVE NOS/NADPH-D IN NEURONS OF THE MVN

Of the four major nuclei comprising the vestibular complex, only the MVN exhibits concentrations of moderately to intensely stained neuronal perikarya following NADPH-d histochemistry or NOS immunocytochemistry (Figure 5.4A-D,F).

Attempts to decipher the location of NOS/NADPH-d-containing neurons in the vestibular complex from the data presented in the literature is somewhat frustrating. In one previous study that mapped the distribution of NOS/NADPH-d throughout the rat brain,[137] the presence of stained neurons in the MVN was noted, while a similar study in the cat[76] reported stained neurons in the vestibular complex without identifying which specific vestibular nucleus contained the labeled neurons. On the other hand Leigh et al.,[63] made no mention of neurons containing NOS/NADPH-d in the vestibular nuclei of the rat. In a recent publication related to the distribution in the human brain, moderate numbers of positively stained neurons were identified in the MVN, although specific distributional information was lacking.[26] Our own results are in general agreement with those of Vincent and Kimura,[137] but in addition, we have noted that there is a distinct longitudinal organization of NOS/NADPH-d neurons along the caudorostral axis of the MVN. Beginning at its most caudal pole, where the MVN occupies a dorsal triangular strip of tissue overlying the dorsal vagal complex, there are small numbers of positively stained neurons (Figure 5.4A). The morphology of MVN neurons containing constitutive NOS/NADPH-d is fairly homogeneous and they appear to form a continuous stream along the caudal two-thirds of the nucleus. The largest concentration of these neurons can be found in the intermediate one-third of the MVN (Figure 5.4B). The rostral 1/3 of the nucleus typically contains very few NOS/NADPH-d neurons and in many sections, no stained perikarya are found. The homogeneity in morphology of constitutive NOS/NADPH-d neurons throughout the MVN might suggest that they belong to the same population and perhaps have similar function(s). This hypothesis is further supported by the fact that many of these MVN neurons project to the oculomotor nucleus, although a few also project to the thalamus (see Section 5.3.2.3 below).

According to the rat brain atlas of Paxinos and Watson,[90] the MVN can be divided into dorsal and ventral subdivisions. However, the distribution of constitutive NOS/NADPH-d neurons in the MVN does not appear to adhere to these subdivisional boundaries. In fact both the dorsal and ventral subdivisions of the MVN contain appreciable numbers of neurons that stain positive for NOS/NADPH-d. In the dorsal part of the nucleus, stained neurons are restricted almost exclusively to the lateral aspect of the dorsal subdivision along the border with the IVN. In transverse brain sections, these neurons extend in a continuous band from the dorsal edge of the section down along the border with the IVN, and into the ventral part of the MVN (Figure 5.4B). The medial or periventricular part of the dorsal MVN is easily distinguished from the lateral portion by the absence of perikarya stained for NOS/NADPH-d. In the ventral MVN, stained perikarya can be found extending from the lateral aspect of the nucleus toward the midline, where they become less plentiful at the border with the prepositus hypoglossal nucleus.

5.3.2.1 NOS/NADPH-d and the Neuropil of the MVN

The identification of stained cell processes (i.e., dendrites and axons) in the vestibular complex is best revealed using NADPH-d histochemistry. Both the dorsal and ventral portions of the MVN contain marked fiber staining. Dendritic branching of resident NOS/NADPH-d neurons is extensive, with the dendritic domains of individual neurons often overlapping with the dendritic arbors or perikarya of neighboring NOS/NADPH-d neurons, as well as non-NOS neurons. In general, stained dendrites are primarily localized to the region of the nucleus where positively stained cell bodies

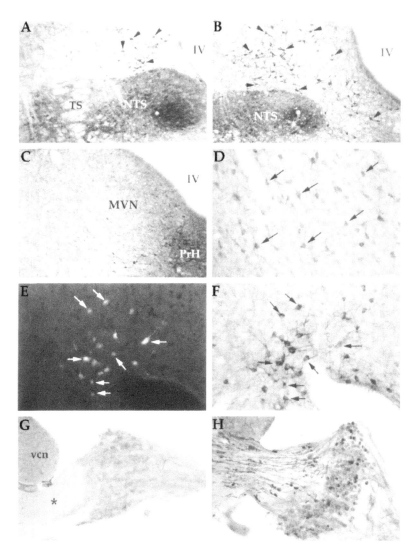

FIGURE 5.4 Photomicrographs illustrating the distribution of neurons containing NOS/NADPH-d in the MVN at different rostrocaudal levels (A - D and F) and representative sections through the vestibular ganglion (G and H). A - C. Transverse sections through the MVN showing neurons containing NADPH-d (arrowheads) in the caudal (A), intermediate (B) and rostral 1/3 of the nucleus. Note that in (B) the stained perikarya are concentrated in the lateral and ventral portion of the MVN, while in (C), there is a conspicuoious absence of stained perikarya in the MVN. (D). A sagittal section through the intermediate 1/3 of the MVN showing numerous perikarya (arrows) stained positive for NOS immunoreactivity. (E and F). The same section through the MVN showing vestibulo-oculomotor neurons (arrows) that are double labeled with the retrograde tracers Fluoro-Gold (E) and NADPH-d (F). (G and H). Sections through the vestibular ganglion taken from a control animal (G) and from a rat 28 days post cerebellar lesion (H). Note that following NADPH-d histochemistry the control ganglion and its central root (*) are devoid of staining while the ipsilateral experimental ganglion contains numerous induced primary sensory neurons. Additionally, there are many stained axonal profiles (some with swellings) in the central root of the vestibular nerve. Abbreviations: PrH (prepositus hypoglossal nucleus), NTS (nucleus of the tractus solitarius), TS (tractus solitarius), vcn (ventral cochlear nucleus), and IV (fourth ventricle).

are numerous, but occasionally extend into the periventricular region of the nucleus where stained perikarya are scarce. Overall, the NADPH-d staining intensity of the neuropil of the MVN is moderate in comparison to the intensely stained neuropil in the subjacent nucleus of the tractus solitarius (NTS) (Figure 5.4A and B). At higher magnifications, a fine network of positively stained processes are revealed in the MVN. These processes are remarkably uniform in size and often have small swellings along their length. Their uniformity and fine morphology would suggest that these processes are axons. This network appears equally represented in all parts of the MVN. A similar network of fibers is present in the MVN of the adult mouse, but this network is negligible in day-old mouse pups, although characteristically stained perikarya are present.

5.3.2.2 Functional Considerations of NOS/NADPH-d in the MVN

Since NOS/NADPH-d is found within a significant subpopulation of MVN neurons (which presumably are producing and releasing NO), the question arises as to the functional significance of this NOS/NADPH-d subpopulation. If one concedes that these neurons are in fact producing and releasing NO into the surrounding intercellular space (where it is free to diffuse into nearby cells), it is reasonable to suggest that neurons and glia in the vicinity of these neurons are potential targets. Although very little is known about the details of NO interactions at the target cells in the vestibular complex, it is known that NO is capable of stimulating the formation of the second messenger cGMP via interaction with soluble guanylate cyclase[28,35,58,114] and that cGMP has been demonstrated in both neurons and glia.[28,35,137] In fact, a study by Furuyama et al.[34] has shown that there are moderate levels of soluble guanylate cyclase in the vestibular complex. Once formed, cGMP has the potential to act through protein kinases, phosphodiesterases, or perhaps directly on ion channels to influence cellular activity.[17,114,137] Since NO has a very short half-life, and thus is unlikely to have effects outside the immediate area of its release, the localization of NOS/NADPH-d neurons and their processes in specific regions of the MVN suggests that these are the areas where NO exerts its effects. However, the presence of a network of NOS fibers permeating areas of the MVN which lack stained somata should not be overlooked, since significant, as of yet unknown, functional implications could be mediated by the release of NO from these fibers. Nonetheless, while the role of NO at the cellular level in the vestibular complex requires further study, the fact that NOS antagonists have an effect on vestibular compensation[29] would suggest that NO in the vestibular complex plays an important role in the compensation process.

5.3.2.3 Efferent Projections of MVN Neurons Containing NOS/NADPH-d

In light of the paucity of information concerning the function(s) of constitutive NOS/NADPH-d neurons in the vestibular complex, it is useful to establish in the initial phase of such investigation(s) the efferent and afferent connectivity of these neurons. From such investigations it is possible to gain some insight into the pathways and systems on which NO may have an effect. A series of double-labeling experiments in the authors' laboratory[107,108] have provided details on the efferent connections of MVN neurons containing constitutive NOS/NADPH-d. Using previously established vestibular efferent targets such as the cerebellum, ventral basal thalamus, cervical spinal cord and oculomotor complex[60,72,73,100,101,130,137] as the sites for the injection of the retrograde tracer Fluoro-Gold (FG), sections were subsequently processed for NADPH-d histochemistry to reveal double-labeled vestibular neurons.

Multiple injections of FG into the cerebellar cortex - including the vermis, paravermis, and hemisphere on one side - resulted in many retrogradely labeled neurons in the ipsilateral MVN, and although the distribution of precerebellar neurons partially overlapped with the distribution of constitutive NOS/NADPH-d neurons, no double-labeled neurons were found. This result would suggest that vestibulocerebellar neurons in the MVN do not produce NO, but perhaps by virtue of their proximity to NO producing neurons, may be influenced by NO. Following FG injections into the ventral basal complex of the thalamus, a small number of retrogradely labeled neurons are consistently found in the contralateral caudal MVN and some of these neurons also contain NOS/NADPH-d. This result provides evidence that NO probably plays a role in vestibulo-thalamic connections. The most dramatic result was observed following injection of FG into the oculomotor nucleus. Such injections resulted in large numbers of retrogradely labeled neurons throughout the MVN, and many of these neurons colocalized with NADPH-d (Figures 5.4E and 5.4F). In some instances virtually all the neurons containing constitutive NOS/NADPH-d also contained FG. It is noteworthy that both the thalamus and oculomotor nucleus have been shown to contain high levels of soluble guanylate cyclase.[34] Furthermore, the authors' own studies indicate that the oculomotor nucleus contains a moderately dense network of fibers that stain positive for NOS/NADPH-d, perhaps in part originating from neurons in the MVN. In light of these results, it would appear that there is a specific subpopulation of vestibulo-oculomotor neurons in the MVN that release NO at the level of the oculomotor complex.

In addition to the oculomotor nucleus, the MVN is known to have efferent connections to other ocular motor nuclei (i.e., abducens and trochlear) and we cannot rule out the possibility that MVN neurons containing NOS/NADPH-d also project to these regions. However, it is worth noting that of these two nuclei, only the abducens contains appreciable amounts of soluble guanylate cyclase.[34] Previous reports have also identified the MVN as having significant connections with the spinal cord.[130] However, most of these vestibulospinal neurons appear to reside in the rostral part of the MVN,[86] an area typically lacking NOS/NADPH-d neurons. In confirmation of these reports, Fluoro-Gold was injected into the cervical spinal cord and it was determined that vestibulospinal neurons in the MVN do not contain constitutive NOS/NADPH-d.[107] In addition to the efferent targets outside the vestibular complex, the vestibular nuclei are highly interconnected by commissural pathways.[100,101] In particular, there is a major commissural projection from the MVN to the contralateral MVN and IVN. Whether NO-producing neurons are involved in these commissural connections remains to be determined.

5.3.2.4 NOS/NADPH-d Colocalization with Other Neurotransmitters in MVN

NOS/NADPH-d has been colocalized in the CNS with other neurotransmitters[17,26,137] and this has lead to the suggestion that NO may play a neuromodulatory role in neurotransmission.[17,26,84] Although there are no definitive studies regarding the colocalization of NOS/NADPH-d and other transmitters in the vestibular complex, comparison of the distribution of NOS/NADPH-d with various neurochemical mapping studies is possible. When compared to the distribution of Chat-positive neurons,[2] the distribution of constitutive NOS/NADPH-d appears to be very similar, although it remains to be determined if these two substances colocalize in any of these neurons. In light of the fact that the efferent connections from the MVN to the oculomotor and trochlear nuclei are considered to be exclusively excitatory and mediated through either glutamate or aspartate neurotransmission,[140] these two excitatory amino acids are likely candidates for colocalization in NOS/NADPH-d neurons.

5.3.2.5 NOS/NADPH-d and Development

In addition to functions in the adult nervous system, NO has been linked to the pre- and postnatal development of neurons. Unfortunately, at present, there is very little known about the development of the constitutive NOS/NADPH-d system in the rodent. The authors' own preliminary observations in mice suggest that at as early as postnatal day 1, there are neuronal profiles in the MVN that stain positive for NADPH-d. Typically, these neurons do not stain as intensely as those in the adult, but the numbers and morphology of neurons in the 1-day-old pups are similar to that in the adult. This is an interesting finding, especially since most of the constitutive NOS/NADPH-d neurons in the adult rat (and presumably the same holds true in the mouse) project to the oculomotor nucleus. Since the eyes do not open in the mouse until several days after birth, the relevance of these NO- producing vestibulo-ocular neurons at birth is unclear with respect to oculomotor function. However, there is evidence to suggest that NO in the developing nervous system promotes and perhaps guides certain developmental events, such as axonal migration and the establishment of synaptic contacts.[17] However, it remains to be determined if NO in neurons and/or fibers of the VC play any role in the development of the vestibular system.

5.3.3 NOS/NADPH-d LOCALIZATION IN THE IVN, LVN, AND SVN

In the rat, the IVN is largely co-extensive caudorostral with the MVN and lies lateral to the MVN over most of its length. At its rostral pole, the IVN lies ventral to the LVN. Typically, the IVN contains very few and in most instances no NOS/NADPH-d-positive neurons. In the small number of cases where positively stained neurons did appear to be in the IVN, these neurons were exclusively found in the caudal half of the nucleus near the border with the MVN. The lack of consistency with regard to the presence of these neurons - combined with their location near the MVN /IVN border and similarity in morphology to NOS/NADPH-d neurons in the MVN - suggests that these neurons may only be displaced MVN neurons rather than a separate IVN population. The absence of constitutive NOS/NADPH-d-containing neurons in the IVN is consistent with other reports in the rat[63,137] but differs from the human, where stained neurons are present in the IVN.[26] This difference probably represents species differences much like that noted between human Purkinje cells[26] which contain constitutive NOS/NADPH-d, and rat[63,106,137] and cat,[76] where Purkinje cells lack NOS/NADPH-d. Although lacking in stained perikarya, the neuropil of the IVN, much like that of the MVN, contains a network of fine processes that do stain for NADPH-d. The density of this network is similar to that found in the MVN and when the fibers are viewed at high magnification they are indistinguishable from those found in the MVN.

In the rat, the remaining two subdivisions of the vestibular complex—the LVN and the SVN —appear approximately at the level of the caudal attachment of the cerebellum to the brainstem where the inferior cerebellar peduncle enters the cerebellum. Both nuclei share similar staining following NOS immunocytochemistry and NADPH-d histochemistry. As in the human,[26] neither of these nuclei in the rat contains NOS/NADPH-d-stained perikarya. There is, however, a fine network of stained processes throughout the neuropil. Although these fibers are morphologically similar to those found in the MVN and IVN, they are in a relative sense less plentiful in the LVN and SVN. As with the other vestibular nuclei, the precise origin of this sparse fiber network is currently unknown.

5.3.4 POSSIBLE ORIGINS OF THE NOS/NADPH-d FIBER NETWORK

The presence of a network of fibers containing constitutive NADPH-d throughout the VC implicates NO in the function of each of the vestibular nuclei. The degree to which NO is involved is not known, although it might be inferred from the varying network densities in the individual nuclei that NO is more significant in those regions where the density is higher. Thus, it is likely that NO plays a more prominent role in the MVN and IVN relative to the LVN and SVN. Although the morphology of the fibers that make up the network in each of the vestibular nuclei is indistin-guishable and might indicate a common origin, this origin is at present unknown. From the authors' own studies and information taken from those of others, some useful comments can be made about the possible origin of the NOS/NADPH-d fibers innervating the VC. From the authors' own experiments, it can be said with certainty that the vestibular ganglion (the source of many primary afferents innervating all parts of the VC) is not a likely source of the NOS/NADPH-d fibers, since ganglion neurons are not found to contain constitutive NOS/NADPH-d (Figure 5.4G). In mammals, many axons originating from Purkinje cells (PC) in the cerebellar cortex innervate the VC, but constitutive NOS/NADPH-d is not found in PC of the rodent. On the other hand, in humans, there is evidence that some PC contain NOS/NADPH-d,[26] and thus, these neurons might contribute to the network found in the VC of that species. Other potential sites of origin include spinal cord, upper cervical ganglia, deep cerebellar nuclei (medial), reticular formation, inferior olive, trigeminal system, commissural neurons, diencephalon, and cerebral cortex.[101] Of these sites, only the diencephalon, cerebral cortex, and commissural connections are likely candidates. This is primarily based on the fact that these other regions either lack neurons containing constitutive NOS/NADPH-d, or alternatively, the known connectivity of a particular region with the VC is inconsistent with the pattern of NOS/NADPH-d fiber labeling in the vestibular nuclei. In the future, it may be possible through double-labeling experiments to determine from where and to what degree different regions contribute to the constitutive network of NOS/NADPH-d fibers in the VC.

5.3.5 Induction of NOS in the Vestibular Ganglion and Nuclei

A number of recent reports have presented evidence that NOS/NADPH-d can be induced in neurons of the peripheral and central nervous systems.[10,23,28,37,55,106-109,137,148,150,155] Especially relevant to the present discussion is the report that neurons in most precerebellar nuclei showed marked induction of NOS following lesion of the cerebellar cortex.[106,107] These studies reported that the distribution of induced neurons was consistent with previously reported neuroanatomical evidence for neurons projecting to the cerebellum.

Primary vestibular neurons in Scarpa's ganglion, as well as secondary relay neurons in the vestibular nuclei, are known contributors to the mossy fiber input to the cerebellar cortex.[5,101] Experiments designed to investigate the induction of NOS/NADPH-d in neurons of the vestibular ganglion and complex were carried out in the authors' lab (unpublished results). In these experiments, either unilateral puncture or large aspiration lesions were made in the cerebellar cortex. Following such lesions, varying numbers of neurons (not only in the ipsilateral VC, but also vestibular ganglion) were found to contain induced NOS/NADPH-d. The time course for the detection of induced NOS/NADPH-d was approximately 72 hours in both ganglion cells and secondary vestibular neurons. Large aspiration lesions produce greater numbers of induced neurons in the ganglion and VC.

5.3.5.1 Induction of NOS/NADPH-d in the Vestibular Ganglion

Unilateral lesion of the cerebellar cortex and the consequential damage to primary afferent axons originating in the vestibular ganglion result in the induction of NOS/NADPH-d in ipsilateral ganglion cells and also in axonal segments found in the vestibular nerve root[107] (Figure 5.4G). The cell bodies of induced ganglion cells do not show obvious morphological distortion that might indicate degenerative changes. However, some stained axons, particularly segments proximal to the ganglion, do show unusual swellings that could indicate cellular changes due to insult. Whether or not all, some, or none of the induced neurons undergo degeneration is still in question. Two interesting side notes to this issue; (1) during embryonic life, at the time when axonal migration and synapse formation is ongoing, virtually all neurons of sensory ganglia express NOS/NADPH-d,[17] and, (2) NOS/NADPH-d is rapidly induced in regenerating olfactory receptor neurons and particularly in outgrowing axons.[17]

5.3.5.2 Induction of NOS/NADPH-d in the Vestibular Complex

The same unilateral cerebellar lesions that cause induction of NOS/NADPH-d in ganglion cells also result in induced NOS/NADPH-d in neurons of the vestibular complex, and, more precisely, in the ipsilateral IVN and SVN but not MVN. Typically, IVN neurons containing induced NOS/NADPH-d are large and multipolar, although smaller neurons are also present in smaller numbers. Induced neurons in the SVN are usually smaller relative to the large induced neurons in the IVN and they are fusiform or pyramidal in shape. The apparent lack of induced neurons in the MVN is surprising, especially when neurons of the MVN are known to project to the cerebel-lum.[101] This result is not readily explained, especially in light of the induction in vestibulocerebellar neurons elsewhere in the complex, as well as in most other precerebellar nuclei.[107,108] However, in two animals where the aspiration lesion clearly involved the SVN, and lateral aspect of the MVN, induced neurons were found in the contralateral MVN. This would suggest that there are neurons in the MVN that are capable of being induced, but that these neurons have contralateral internuclear projections. Although it remains to be determined why there is an apparent lack of induction in vestibulocerebellar neurons in the MVN, it is clear that the response of neurons in the vestibular complex to target specific lesions is variable.

5.3.5.3 Possible Implications and Consequences of Induced NOS/NADPH-d

Although it is far from clear what the consequence(s) of induced NOS/NADPH-d is (are) to vestibular neurons, it would seem that damage to vestibular axons projecting to the cerebellum or contralateral VC is at least in part responsible for the induction. This result is consistent with the hypothesis put forth in previous studies in which it was reported that the induction of NOS/NADPH-d in some neurons occurs following axonal insult.[28,37,55,137,148,149] Unfortunately, there is no real consensus with regard to the effect of NOS/NADPH-d induction on neurons following axonal damage. Some reports indicate that NOS/NADPH-d induction signals the impending demise of induced neurons[10,46,148] while others have proposed that long-term induction might perform some neuroprotective role in damaged neurons.[28,37,106,107,108,109,137,146,150] Such polarized conclusions have yet to be explained adequately but may be related to differences in the systems studied or to factors as yet undetermined. The authors' own studies of NOS/NADPH-d induction in precerebellar neurons indicate that some induced neurons (including vestibular) can be detected at times out to 120 days, a time at which it could be argued that degeneration might be expected to have been complete. Furthermore, these neurons do not show an obviously distorted morphology, perhaps bolstering the argument that induced NOS/NADPH-d has a neuroprotective or a least beneficial role in some, but perhaps not all induced neurons. With regard to a possible neuroprotective role for induced NOS/NADPH-d and presumably NO, several interesting suggestions have been put forth as to how neuroprotection might be affected. A review paper by Dawson and Dawson[17] has cited evidence to support the contention that NO can control blood pressure through effects on sympathetic tone. This being the case, NO produced by damaged (i.e., induced) neurons may influence local blood flow by a direct paracrine action on nearby blood vessels in a self-preservative manner. Along this same line, it is well known that glial cells can exert strong trophic effects on normal as well as damaged neurons, and that NO can stimulate glial cGMP.[28] In this manner the induction of NOS/NADPH-d and the accompanying NO production may induce glial cells to respond with trophic factors that might assist in the neuroprotective and/or neuroregenerative activities of damaged neurons. Another suggestion (also by Fiallos-Estrada et al.[28]) was that NO derived from induced NOS/NADPH-d might act back upon the induced neuron in a self-stimulating manner so as to partially offset the loss of functional connections due to axonal damage, a scenario that might provide the neuron with much-needed time to initiate regenerative processes. Unfortu-nately, at this juncture in time, there are no techniques to track or determine in a definitive manner the fate of all induced neurons. As each new study concerning the induction of NOS/NADPH-d reveals more about this phenomenon, it is becoming clear that induction of NOS/NADPH-d is not only variable but increasingly more complex. Whatever the precise implications, certain neurons of the vestibular complex can now be added to the expanding list of neuronal types that can be induced following various cerebellar insults.

5.4 CONCLUSIONS

There is abundant anatomical, molecular biological, biochemical, and physiological evidence to indicate that glutamate and glutamate receptors are present in the vestibular nuclei and that excitatory amino acids play a fundamental role in vestibular neurotransmission as well as in central vestibular processing. In addition to the vestibular nerve, several other vestibular afferents likely use glutamate or aspartate as a transmitter. All major glutamate receptor subtypes—including AMPA, kainate, NMDA, and metabotropic subtypes—are present in the vestibular complex, and pharmacological data derived from administration of glutamate receptor agonists and antagonists suggest that these receptors play critical roles in normal vestibular function. In addition, pharmacological and molecular biological data suggest that NMDA, d-2, and metabotropic glutamate receptor subtypes are involved in the process of vestibular compensation. A variety of glutamate receptor subtypes appear to be present in the vestibular nuclei, but the exact combinations of subtypes that comprise AMPA, NMDA, and kainate receptor channels associated with vestibular neurons requires further elucidation, as do the combinations of glutamate receptors associated with identified vestibular efferent, commissural, and intrinsic neurons.

In the above sections, we have also reviewed and commented on what is currently known regarding nitric oxide in the vestibular complex. Neuroanatomical data indicates that constitutive NOS/NADPH-d is found in a specific population of MVN neurons, the majority of which have efferent connections to the oculomotor nucleus. In addition, a small number of NOS/NADPH-d-containing neurons contribute to the vestibulothalamic pathway. The presence of soluble guanylate cyclase (a substrate with which NO can interact) in both the oculomotor nucleus and thalamus is consistent with these authors' hypothesis that a subpopulation of efferent neurons in the MVN can effect neurons in these two regions via the release of NO. While the remaining vestibular nuclei do not contain resident NOS/NADPH-d neuronal perikarya, they do contain a fine network of NOS/NADPH-d-containing fibers. This NOS/NADPH-d-positive fiber plexus innervates each of the vestibular nuclei to varying degrees, perhaps representing the relative importance that NO plays in each of these nuclei. Chronic inhibition of NOS has been shown to prevent functional recovery following vestibular lesions, thus, NO appears to play an important role in acquistion of a compensated state following hemilabyrinthectomy. Finally, we demonstrate that neurons found in the vestibular ganglion and in the vestibular complex are susceptible to the induction of NOS/NADPH-d following target- specific damage. At the present time, the underlying mechanisms responsible for NOS/NADPH induction in the vestibular system and the functional consequences of this induction remain to be elucidated.

ACKNOWLEDGMENTS

The authors would like to thank Dr. Robert B. Leonard for critical comments and suggestions. Thanks to Amber Neff for photographic assistance. Supported in part by the Deafness Research Foundation and by NIH grants DC00052 and NS31318.

REFERENCES

1. Araki, K, Meguro, H, Kushiya, E, Takayama, C, Ionoue, Y, and Mishina, M., Selective expression of the glutamate receptor channel delta 2 subunit in cerebellar Purkinje cells, *Biochem. Biophys. Res. Commun.,* 197:1267,1993.
2. Barmack, NH, Baughman, RW, Eckenstein, FP, Shojaku, H, Secondary vestibular cholinergic projection to the cerebellum of rabbit and rat as revealed by choline acetyltransferase immunocytochemistry, retrograde and orthograde tracers, *J. Comp. Neurol.,* 317: 250,1992.

3. Bettler, B, Boulter, J, Hermans-Borgmeyer, I, O'Shea-Greenfield, A, Deneris, ES, Moll, C., Borgmeyer, U, Hollman, M, and Heinemann, S, Cloning of a novel glutamate receptor subunit, GluR5: Expression in the nervous system during development, *Neuron,* 5:583,1990.

4. Bettler, B, Egebjerg, J, Sharma, G, Pecht, G, Hermans-Borgmeyer, I, Moll, C, Stevens, CF, and Heinemann, S, Cloning of a putative glutamate receptor: a low affinity kainate-binding subunit, *Neuron,* 8:257,1992.

5. Brodal, A, and Hovik, B, Site and mode of termination of primary vestibulocerebellar fibers in the cat. An experimental study with silver impregnation methods, Arch. *Ital. Biol.,* 102:1, 1964.

6. Bruhwlyer, J, Chileide, E, Liegeois, JF, Carreer, F Nitric oxide: A new messenger in the brain., *Neurosci. Neurobehav.,* 17:373, 1993.

7. Buller, AL, Larson, HC, Schneider, BE, Beaton, JA, Morrisett, RA, and Monaghan, DT The molecular basis of NMDA receptor subtypes: Native receptor diversity is predicted by subunit composition, *J. Neurosci.,* 14:5471,1994.

8. Capocchi, G, Della Torre, G, Grassi, S, Petorrossi, VR, and Zampolini, M, NMDA receptor-mediated long term modulation of electrically evoked field potentials in the rat medial vestibular nuclei., *Exp. Brain Res.,* 90:546, 1992.

9. Carpenter, MB, Huang, Y, Pereira, AB, and Hersh, LB Immumocytochemical features of the vestibular nuclei in the monkey and cat, *J. Hirnforsch.,* 5:585,1990.

10. Chen, S, and Aston-Jones, G, Cerebellar injury induces NADPH diaphorase in Purkinje and inferior olivary neurons in the rat, *Exp. Neurol.,* 126:270,1994.

11. Chinnery, RM, Shaw, PJ, Ince, PG, and Johnson, M, Autoradiographic distribution of binding sites for the non- NMDA receptor antagonist [H^{3}]CNQX in human motor cortex, brainstem and spinal cord, *Brain Res.,* 630:75,1993.

12. Choi, DW, Nitric oxide: Foe or friend to the injured brain? *Proc. Natl. Acad. Sci.,* USA, 90:9741,1993.

13. Cochran, SL, Kasik, P, and Precht, W, Pharmacological aspects of excitatory synaptic transmission to second-order vestibular neurons in the frog, *Synapse,* 1:102,1987.

14. Collingridge, GL, and Lester, RA, Excitatory amino acids in the vertebrate nervous system, *Pharmacol. Rev.,* 40:143, 1989.

15. Conti, F, Rustioni, A, Petruz, P, Towle, AC, Glutamate-positive neurons in the somatic sensory cortex of rats and monkeys, *J. Neurosci.,* 7:1887, 1987.

16. Darlington, CL, and Smith, PF, Metabotropic glutamate receptors in the guinea-pig medial vestibular nucleus in vitro, *Neuroreport.,* 6:1799, 1995.

17. Dawson, TM, and Dawson, VL, Nitric Oxide: Actions and pathological roles, *The Neuroscientist,* 1:7, 1995.

18. Dawson, TM., Bredt, DS, Fotuhi, M, Hwang, PM, and Snyder, SH, Nitric oxide synthase and neuronal NADPH-diaphorase are identical in brain and peripheral tissues, *Proc. Natl. Acad. Sci. USA.,* 88:7797, 1991.

19. Demêmes, D, Wenthold, RJ, Moniot, B, and Sans, A, Glutamate-like immunoreactivity in the peripheral vestibular system of mammals, *Hearing Res.,* 46:261, 1990.

20. de Waele, C, Abitbol, M, Chat, M, Menini, C, Mallet, J, and Vidal, PP, Distribution of glutamatergic receptors and GAD mRNA- containing neurons in the vestibular nuclei of normal and hemilabyrinth-ectomized rats, *Eur. J. Neurosci.,* 6:565, 1994.

21. de Waele, C., Mühlethaler, M, and Vidal, PP, Neurochemistry of the central vestibular pathways, *Brain Res. Rev.,* 20:24,1995.

22. de Waele, C, Vibert, N, Baudrimont, M, and Vidal, PP, NMDA receptors contribute to the resting discharge of vestibular neurons in the normal and hemilabyrinthectomized guinea pig, *Exp.Brain Res.,* 81:125,1990.

23. Divac, I, Ramirez-Gonzalez, JA, Ronn, LCB, Jahnsen, H, and Regidor, J, NADPH-diaphorase (NOS) is induced in pyrimidal neurons of hippocampal slices, *Neuroreport.,* 5: 325,1993.

24. Doi, K, Ohno, K, Iwakura, S, Takahasi, Y and Kubo, T, Glutamate receptor gene family expressed in vestibular Scarpa's ganglion of rat. *Acta Otolaryngol. Suppl.* 520:334,1995.

25. Dong, H, O'Brien, RJ, Fung, ET, Lanahan, AA, Worley, PF, and Huganir, RL. GRIP: a synaptic PDZ domain-containing protein that interacts with AMPA receptors,*Nature,* 386:279,1997.

26. Egberongbe, YI, Gentlman, SM, Falkai, P, Bogerts, B, Polak, JM, and Roberts, GW, The distribution of nitric oxide synthase immunoreactivity in the human brain, *Neuroscience,* 59: 561,1994.

27. Egebjerg, J, Bettler, B, Hermans-Borgmeyer, I, and Heinemann, S, Cloning of a cDNA for a glutamate receptor subunit activated by kainate but not AMPA, *Nature,* 351:745,1991.

28. Fiallos-Estrada, CE, Kummer, W, Mayer, B, Bravo, R, Zimmermann, M, and Herdegen, T, Long-lasting increase of nitric oxide synthase immunoreactivity, NADPH-diaphorase reaction and c-JUN co-expression in rat dorsal root ganglion neurons following sciatic nerve transection, *Neurosci. Lett.*, 150: 169,1993.

29. Flügel, G, Holm, S, and Flohr, H, Chronic inhibition of nitric oxide synthase prevents functional recovery following vestibular lesions. in: *Biology of Nitric Oxide*, London: Portland Press, pp. 381,1994.

30. Fonnum, F, Glutamate: A neurotransmitter in mammalian brain, *J. Neurochem.*, 42:1,1984.

31. Fujita, S, Usami, S, Shinkawa, H, Sato, K, Kiyama, H, and Tohyama,, M, Expression of NMDA receptor subunit mRNA in the vestibular ganglion of the rat and guinea-pig, *Neuroreport.*, 5:862,1994.

32. Fukushima, Y, Igusa, Y, and Yoshida, D, Characteristics of responses of medial brainstem neurons to horizontal head angular acceleration and electrical stimulation of the labyrinth in the cat, *Brain Res.*, 120:564,.

33. Funabiki, K, Mishina, M, and Hirano, T, Retarded vestibular compensation in mutant mice deficient in d2 glutamate receptor subunit, *NeuroReport*, 7:189,1995.

34. Furuyama, T, Inagaki, S, and Takagi, H, Localization's of a1 and b1 subunits of soluble guanylate cyclase in the rat brain, *Mol. Brain Res.*, 20: 335,1993.

35. Garthwaite, J, Glutamate, nitric oxide and cell-cell signalling in the nervous system, *Trends Neurosci.*, 14: 60,1991.

36. Gasic, GP, and Hollmann, M, Molecular neurobiology of glutamate receptors. *Annu. Rev. Physiol.*, 54:507,1992.

37. Gonzalas, MF, Sharp, FR, Sagar, SM, Axotomy increases NADPH-diaphorase staining in rat vagal motor neurons, *Brain Res. Bull.*, 18: 417,1987.

38. Herb, A, Burnashev, N, Werner, P, Sakmann, B, Wisden, W, and Seeburg, PH, The KA-2 subunit of excitatory amino acid receptors shows widespread expression in brain and forms ion channels with distantly related subunits, *Neuron*, 8:775,1992.

39. Herrereo, I, and Miras-Portugal, T, Sanchez-Prieto J, *Nature*, 360:163,1992.

40. Highstein, SM, Ito, M, Differential localization within the vestibular nuclear complex of the inhibitory and excitatory cells innervating IIIrd nucleus oculomotor neurons in rabbit, *Brain Res.*, 3:306,1971.

41. Hirano, T, Kasono, K, Araki, K, Shinozuka, K, and Mishina, M., Involvement of the glutamate receptor delta 2 subunit in the long term depression of glutamate responsiveness in cultured rat Purkinje cells, *Neurosci. Lett.*, 182:172,1994.

42. Hollmann, M, O'Shea-Greenfield, A, Rogers, SW, Heinemann, S, Cloning by functional expression of a member of the glutamate receptor family, *Nature*, 342:643,1989.

43. Hollmann, MA, Hartley, M, and Heinemann, S Calcium permeability of KA-AMPA-gated glutamate receptor channels depends on subunit composition, *Science*, 252:851,1991.

44. Hollmann, M, and Heinemann, S, Cloned glutamate receptors. *Annu. Rev. Neurosci.* 17:31,.

45. Hope BT, Micheal GJ, Knigge KM, Vincent SR Neuronal NADPH-diaphorase is a nitric oxide synthase, *Proc. Natl. Acad. Sci.*, (USA) 88: 2811,1991.

46. Kashiwabuchi, N, Ikeda, K, Araki, K, Hirano, T, Shibuki, K, Takayama, C, Inoue, Y, Kutsuwada, T, Yagi, T, Kang, Y, Aizawa, S, and Mishina, M, Impairment of motor coordination, Purkinje cell synapse formation, and cerebellar long term depression in GluRd2 mutant mice, *Cell*, 81:245,1995.

47. Kaupmann, K, Huggel, K, Heid, J, Flor, PJ, Bischoff, S, Mickel, SJ, McMaster, G, Angst, C, Bittiger, H, Froestl, W, and Bettler, B, Expression cloning of GABA_B receptors uncovers similarity to metabotropic glutamate receptors, *Nature*, 386:239,1997.

48. Keinänen, K, Wisden, W, Sommer, B, Werner, P, Herb, A, Verdoorn, TA, Sakmann, B, and Seeburg, P, A family of AMPA-selective glutamate receptors, *Science*, 249:556,1990.

49. Kevetter, GA, Some excitatory transmitters in the central vestibular pathways in the gerbil, *Ann.NY.Acad.Sci.*, 656:940,1993.

50. Kevetter, GA, Immunocytochemistry of glutamate receptors in the vestibular nuclear complex in gerbils, *Society for Neuroscience Abstracts*, 20:569,1995.

51. Kevetter, GA, and Coffey, AR, Aspartate-like immunoreactivity in vestibulospinal neurons in gerbils, *Neurosci.Lett.*, 123:273,1991.

52. Kevetter, GA, and Hoffman, RD, Excitatory amino acid immunoreactivity in vestibulo-ocular neurons in gerbils, *Brain Res.*, 554:348,1991.

53. Kinney, GA, Peterson, BW, and Slater, NT, The synaptic activation of N-Methyl-D-Aspartate receptors in the rat medial vestibular nucleus, *J. Neurophys.*, 72:1588,1994.

54. Kinzie, JM, Saugstad, JA, Westbrook, GL, and Segerson, TP, Distribution of metabotropic glutamate receptor 7 messenger RNA in the developing and adult rat brain, *Neuroscience*, 69:167,1995.

55 .Kitchener, PD, Van der Zee, CEEM, and Diamond, J, Lesion-induced NADPH-diaphorase reactivity in neocortical pyramidal neurons, *NeuroReport*, 4: 487,1993.

56. Knöpfel,, T, Evidence for N-methyl-D-aspartic acid receptor-mediated modulation of the commissural input to central vestibular neurons of the frog, *Brain Res.*, 426:212,.

57. Knöpfel, T, Dieringer, N, The role of NMDA and non-NMDA receptors in the central vestibular synaptic transmission, Adv. *Oto-Phino. Laryng.*, 42:229,1988.

58. Knowles, RG, and Moncada, S, Nitric oxide synthase in mammals, *Biochem,* J. 298:249,1994.

59. Kohr, G, and Seeburg, PH, Subtype-specific regulation of recombinant NMDA receptor channels by protein tyrosine kinases of the src family, *J. Physiol (Lond)*, 492:445,1996.

60. Kotchabhakdi, N, Rinvik, E, Walberg, F, and Yingchareon, K, The vestibulo-thalamic projections in the cat studied by retrograde axonal transport of horseradish peroxidase, *Exp. Brain Res.*, 40: 405,1980.

61. Kumoi, K, Saito, N, and Tanaka, C, Immunohistochemical localization of gama-aminobutyric acid- and aspartate- containing neurons in the guinea pig vestibular nuclei, *Brain Res.*, 416:22,1987.

62. Kutsuwada, T, Kashiwabushi, N, Mori, K, Sakimua, K, Kushiya, E, Araki, K, Meguro, H, Masaki, H, Kumanishi, T, Arakawa, M, and Mishina, M, Molecular diversity of the NMDA receptor channel, *Nature*, 358:36,1992.

63. Leigh, PN, Connich, JH, and Stone, TW, Distribution of NADPH-diaphorase-positive cells in the rat brain, *Comp. Biochem. Physiol.*, 97: 259,1990.

64. Lewis, MR, Phelan, KD, Shinnick-Gallagher, P, and Gallagher, JP, Primary afferent excitatory transmission recorded intracellularly in vitro from rat medial vestibular neurons, *Synapse* 3:149,1989.

65. Liu, H, Wang, H, Sheng, M, Jan, LY, Jan, YN, and Basbaum, AI, Evidence for presynaptic N-methyl-D-aspartate autoreceptors in the spinal cord dorsal horn, *Proc. Natl. Acad. Sci. USA* 91:8383,1994.

66. Lomeli, H, Sprengel, R, Laurie, DJ, Kohr, G, Herb, A, Seeburg, PH, and Wisden, W, The rat delta 1 and delta 2 subunits extend the excitatory amino acid receptor family, *FEBS Lett.*, 315:318, 1993.

67. Luo, JH, Wang, Y, Yasuda, RP, Dunah, AW, and Wolfe, BB, The majority of NMDA receptor complexes in adult rat cerebral cortex contain at least three different subunits (NR1/NR2A/NR2B), *Mol. Pharmacol.*, 51:79,1997.

68. Maas, S, Thorsten, M, Herb, A, Seeburg, PH, Keller, W, Krause, S, Higuchi, M, and O'Connell, MA, Structural requirements for RNA editing in glutamate receptor pre-mRNAs by recombinant double-stranded RNA adenosine deaminase, *J Biol Chem.*, 271:12221,1996.

69. Martin, LJ, Blackstone, CD, Levey, AI, Huganir, RL, and Price, DL, AMPA receptor subunits are differentially distributed in rat brain, *Neuroscience*, 53:327,1993.

70. Masu, M, Tanabe, Y, Tsuchida, K, Shigemoto, R, and Nakanishi, S, Sequence and expression of a metabotropic glutamate receptor, *Nature*, 349:760,1991.

71. Matsumoto, T, Nakane, M, Pollack, JS, Kuk, JE, and Forstermann, U, A correlation between soluble nitric oxide synthase and NADPH-diaphorase activity is only seen after exposure of the tissue to fixative, *Neurosci. Lett.*, 155: 61,1993.

72. McCrea, RA, Strassman, A, May, E, and Highstein, SM, Anatomical and physiological characteristics of vestibular neurons mediating the horizontal vestibulo-ocular reflex in the squirrel monkey, *J. Comp. Neurol.*, 264:547,1987a.

73. McCrea, RA, Strassman, A, May, E, and Highstein, SA, Anatomical and physiological characteristics of vestibular neurons mediating the vertical vestibulo-ocular reflexes of the squirrel monkey, *J. Comp. Neurol.*, 264:571,1987b.

74. Mettens, P, Cheron, G, and Godaux, E, Involvement of the N-methyl-D-aspartate receptors of the vestibular nucleus in the gaze-holding system of the cat, *Neurosci.Lett.*, 174:209,1994a.

75. Mettens, P, Cheron, G, and Godaux, E, NMDA receptors are involved in temporal integration in the oculomotor system of the cat., *Neuroreport*, 5:1333,1994b.

76. Mizukawa, K, Vincent, SR, McGeer, PL, and McGeer, EG, Distribution of reduced-nicotinamide-adenine-dinucleitide-phosphate diaphorase-positive cells and fibers in the cat central nervous system, *J. Comp. Neurol.*, 279: 281,1989.

77. Monaghan, DT, and Cotman, CW, Distribution of N-Methyl-D-aspartate-sensitive L-[³H]glutamate-binding sites in rat brain, *J. Neurosci.,* 5: 2909,1985.

78. Monaghan, DT, Bridges, RJ, and Cotman, CW, The excitatory amino acid receptors: Their classes, pharmacology, and distinct properties in the function of the central nervous system, *Ann. Rev. Pharmacol. Toxicol.,* 29:91,1989.

79. Monaghan, PL, Beitz, AJ, Larson, AA, Altschuler, RA, Madl, JE, and Mullett, MA, Immunocytochemical localization of glutamate-, glutamainase- and aspartate aminotransferase-like immunoreactivity in the rat deep cerebellar nuclei, *Brain Res.,* 363:364,1986.

80. Monyer, H, Sprengel, HR, Schoepfer, R, Herb, A, Higuchi, M, Lomeli, H, Burnashev, N, Sakmann, B, and Seeburg, PH, Heteromeric NMDA receptors: Molecular and functional distinction of subtypes, *Science,* 256:1217,1992.

81. Monyer, H, Burnashev, N, Laurie, DJ, Sakmann, B, and Seeburg, PH, Developmental and regional expression in rat brain and functional properties of four NMDA receptors, *Neuron,* 12:529,1994.

82. Nakanishi, S, Molecular diversity of glutamate receptors and implications for brain function, *Science,* 258:597,1992.

83. Nakanishi, N, Shneider, N, and Axel, R, A family of glutamate receptor genes: evidence for the formation of heteromultimeric receptors with distinct channel properties, *Neuron,* 5:569,1990.

84. Nathan, C, Nitric oxide as a secretory product of mammalian cells, *J. Biol. Chem.,* 268: 2231,1992.

85. Niedzielski, AS, and Wenthold, RJ, Expression of AMPA, Kainate, and NMDA receptor subunits in cochlear and vestibular ganglia, *J. Neurosci.,* 15(3):2338,1995.

86. Nomura, I, Senba, E, Kubo, T, Shiraishi, T, Matsuraga, T, Tohyama, M, Shiotani, Y, and Wu, J-Y, Neuropeptides and GABA in the vestibular nuclei of the rat: An immunohistochemical analysis, *I. Distribution. Brain Res.,* 311: 109,1984.

87. Ohishi, H., Shigemoto, R., Nakanishi, S. and Mizuno, N. Distribution of the messager RNA for a metabotrophic glutamate receptor, mGluR2, in the central nervous system, *Neuroscience,* 53:1009,1993.

88. Ottersen, OP, Laake, JH, and Storm-Mathisen, J, Demonstration of a releasable pool of glutamate in cerebellar mossy and parallel fibre terminals by means of light and electron microscopic immunocytochemistry, *Arch. Ital. Biol.,* 128:111,1990a.

89. Ottersen, OP, Storm-Mathisen, J, Bramham, CL, Trop, R, Laake, J, and Gundersen, V, A quantitative electron microscopic immunocytochemical study of the distribution and synaptic handling of glutamate in rat hippocampus, *Progresss in Brain Res.,* 83:99,1990b.

90. Paxinos, G, C, and Watson, *The Rat Brain in Stereotaxic Coordinates,* 2nd Edition, 1986.

91. Petralia, RS, and Wenthold, RJ, Light and electron immunocytochemical localization of AMPA-selective glutamate receptors in the rat brain, *J. Comp Neurol.,* 318:329,1992.

92. Petralia, RS, Wang, YX, and Wenthold, RJ, The NMDA receptor subunits NR2A and NR2B show histological and ultrastructural localization patterns similar to those of NR1, *J. Neurosci.,* 14:6102,1994a.

93 . Petralia, RS, Wang, YX, and Wenthold, RJ, Histological and ultrastructural localization of the kainate receptor subunits, KA2 and GluR6/7, in the rat nervous system using selective antipeptide antibodies, *J.Comp Neurol.,* 349:85,1994b.

94. Petralia, RS, Yokotani, N, Wenthold, RJ, Light and electron microscope distribution of the NMDA receptor subunit NMDAR1 in the rat nervous system using a selective anti-peptide antibody, *J. Neurosci.,* 14:667,1994c.

95. Pin, JP, and Duvoisin, R, The metabotropic glutamate receptors: structure and functions, *Neuropharmacology,* 34:1-26, 1995.

96. Popper, P, Rodrigo, JP, Alvarez, JC, Lopez, I, and Honrubia, V, Expression of the AMPA-selective receptor subunits in the vestibular nuclei of the chinchilla, *Mol. Brain Res.,* 4:22,1997.

97. Priestley, T, Laughton, P, Myers, J, Le Bourdelles, B, Kerby, J, and Whiting, PJ, Pharmacological properties of recombinant human N-methyl-D-aspartate receptors comprising NR1a/NR2A and NR1a/NR2B subunit assemblies expressed in permanently transfected mouse fibroblast cells, *Molec. Pharmacol.,* 48:841,1995.

98. Puckett, C, Gomez, CM, Korenberg, JR, Tung, H, Meier, TJ, Chen, XN, and Hood, L, Molecular cloning and chromosomal localization of one of the human glutamate receptor genes, *Proc. Natl. Acad. Sci. U.S.A.,* 88:7557,1991.

99. Reichenberger, I, and Dieringer, N, Size-related colocalization of glycine and glutamate-immunoreactivity in frog and rat vestibular afferents, *J. Comp. Neurol.,* 349:603,1994.

100. Rubertone, JA, Mehler, WR, Cox, GE, The intrinsic organization of the vestibular complex: Evidence for internuclear connectivity, *Brain Res.,* 263: 137,1983.

101. Rubertone, JA, Mehler, WR, Voogd, J, The vestibular nuclear complex, in: *The Rat Nervous System,* G. Paxinos Ed., pp. 773,1995.

102. Safieddine, S, and Wenthold, RJ, The glutamate receptor subunit delta-1 is highly expressed in hair cells of the auditory and vestibular systems, *J. Neurosci.,* 17:7523,.

103. Sanchez-Prieto J, Budd DC, Herrero I, Vazquez E and Nicholls DG, Presynaptic receptors and the control of glutamate exocytosis, *Trends Neurosci.*, 19:235,1996.
104. Sans, N, Sans, A, and Raymond, J, Regulation of NMDA receptor subunit mRNA expression in the guinea pig vestibular nuclei following unilateral labyrinthectomy, *Eur J Neurosci.*, 9:2019,1997.
105. Saugstad, JA, Kinzie, JM, Shinohara, MM, Segerson, TP, and Westbrook, GL, Cloning and expression of rat metabotropic glutamate receptor 8 reveals a distinct pharmacological profile, *Mol. Pharmacol.*, 51:119,1997.
106. Saxon, DW, and Beitz, AJ, Cerebellar injury induces NOS in Purkinje cells and cerebellar afferent neurons, *Neuroreport*, 5: 809,1994.
107. Saxon, DW, and Beitz, AJ, Constitutive and induced nitric oxide synthase in the vestibular complex and ganglion, *Neurosci. Abst.*, 26:720.8,1995.
108. Saxon, DW, and Beitz, AJ, An experimental model for the non-invasive transynaptic induction of nitric oxide synthase in Purkinje cells of the rat cereballum, *Neuroscience*, 72: 157,1996.
109. Saxon, DW, and Beitz, AJ, Induction of NADPH-diaphorase/nitric oxide synthase in the brainstem trigeminal system resulting from cerebellar lesions, *J. Comp. Neurol.*, 371: 41,1996.
110. Schoepp, D, Bockaert, J, and Sladeczek, F, Pharmacological and functional characteristics of metabotropic excitatory amino acid receptors, *Trends Pharmacol. Sci.* 11:508,1990.
111. Seeburg, PH, The TINS/TIPS Lecture - The molecular biology of mammalian glutamate receptor channels, *Trends.Neurosci.*, 16:359,1993.
112. Seeburg, PH, The role of RNA editing in controlling glutamate receptor channel properties, *J. Neurochem.*, 66,1,1996.
113. Serafin, M, Khateb, A, de Waele, C, Vidal, PP, and Mühlethaler, M, Medial vestibular nucleus in the guinea pig. NMDA induced oscillations, *Exp. Brain Res.*, 88:187,1992.
114. Schmidt, HHHW, Lohmann, SM, and Walter, U, The nitric oxide and cGMP signal transduction system: Regulation and mechanism of action, *Biochimica et Biophysica Acta.*, 1178: 153,1993.
115. Sheng, M, Glutamate receptors put in their place, *Nature*, 386:221,1997.
116. Shigemoto, R, Kulik, A, Roberts, JD, Ohishi, H, Nusser, Z, Kaneko, T, and Somogyi, P, Target cell—specific concentration of a metabotropic glutamate receptor in the presynaptic active zone, *Nature*, 381:523,1996.
117. Shigemoto, R, Nakanishi, S, and Mizuno, N, Distribution of the mRNA for a metabotropic glutamate receptor (mGluR1) in the central nervous system: An *in situ* hybridization study in adult and developing rat, *J. Comp. Neurol.*, 322:121,1992.
118. Smith, PF, Darlington, CL, The NMDA antagonists MK801 and CPP disrupt compensation for unilateral labyrinthectomy in the guinea pig, *Neurosci. Lett.*, 94:309,1988.
119. Smith, PF, de Waele, C, Vidal, PP, and Darlington, CL, Excitatory amino acid receptors in normal and abnormal vestibular function, *Mol. Neurobiol.*, 5:369,1991.
120. Sommer, B, Kohler, M, Sprengel, R, and Seeburg, PH, RNA editing in brain controls a determinant of ion flow in glutamate-gated channels, *Cell* 67:11,1991.
121. Sommer, B, and Seeburg, PH, Glutamate receptor channels: novel properties and new clones, *TiPS*, 13:291,1992.
122. Takahashi, Y, Takahashi, MP, Tsumoto, T, Doi, K, and Matsunaga, T, Synaptic input-induced increase in intraneuronal Ca^{2+} in the medial vestibular nucleus of young rats, *Neurosci. Res.*, 21:59,1994a.
123. Takahashi, Y, Tsumoto, T, and Kubo, T, N-Methyl-D-aspartate receptors contribute to afferent synaptic transmission in the medial vestibular nucleus of young rats, *Brain Res.*, 659:287,1994b.
124. Takumi, Y, Matsubara, A, Danbolt, NC, Laake, JH, Storm-Mathisen, J, Usami, S, Shinkawa, H, and Ottersen, OP, Discrete cellular and subcellular localization of glutamine synthetase and the glutamate transporter GLAST in the rat vestibular end organ, *Neuroscience*, 79:1137,1997.
125. Tanabe, Y, Masu, M, Ishii, T, Shigemoto, R, and Nakanishi, S, A family of metabotropic glutamate receptors. *Neuron* 8:169,1992.
126. Tanabe, Y, Normua, A, Masu, M, Shigemoto, R, Mizuno, N, and Nakanishi, S, Signal transduction pharmacological properties, and expression patterns of two rat netabotropic glutamate receptors, *J. Neurosci.*, 13:1372,1993.
127. Tatter, SB, Galpern, WR, and Isacson, O, Neurotrophic factor protection against excitotoxic neuronal death, *The Neuroscientist*, 1: 286,1995.

128. Tolle, TR, Berthele, A, Zieglgansberger, W, Seeburg, PH, and Wisden, W, The differential expression of 16 NMDA and non-NMDA receptor subunits in the rat spinal cord and in the periaqueductal gray, *J. Neurosci.,* 13:5009,1993.

129. Touati, J, Raymond, J, and Dememes, D, Quantitative autoradiographic characterization of L-[³H] glutamate binding sites in rat vestibular nuclei, *Exp.Brain Res.,* 76:646,1989.

130. Tracey, DJ, Ascending and descending pathways in the spinal cord, in: *The Rat Nervous System,* 2nd edition, G. Paxinos, Ed., pp. 67-80, 1995.

131. Traynelis, SF, Hartley, M, and Heinemann, S, Control of proton sensitivity of the NMDA receptor by RNA splicing and polyamines, *Science,* 268:873,1995.

132. Uchino, Y, Hirai, N, Suzuki, S, and Watanabe, S, Properties of secondary vestibular neurons fired by stimulation of ampullary nerve of the vertical, anterior or posterior semicircular canals in the cat, *Brain Res.,* 223:273,1981.

133. Uchino, Y, Hirai, N, Suzuki., S, Branching pattern and properties of vertical- and horizontal-related excitatory vestibulo-ocular neurons in the cat, *J. Neurophysiol.,* 48:891,1982.

134. Ullensvang, K, Lehre, KP, Storm-Mathisen, J, and Danbolt, NC, Differential developmental expression of the two rat brain glutamate transporter proteins GLAST and GLT, *Eur. J. Neurosci.,* 9:1646,1997.

135. Verge, VMK, Khang, X, Xu, X-J, Wiesenfeld-Hallin, Z, and Hokfelt, T, Marked increase in nitric oxide synthase mRNA in rat dorsal root ganglia after peripheral axotomy: *in situ* hybridization and functional studies, *Proc. Natl. Acad. Sci. USA,* 89:11617,1992.

136. Vidal, PP, Babalian, A, deWaele, C, Serafin, M, Vibert, N, and Muhlethaler, M, NMDA receptors of the vestibular nuclei neurons. *Brain Res. Bull.,* 40:347,1996.

137. Vincent, SR, and Kimura, H, Histochemical mapping of nitric oxide synthase in the rat brain, *Neuroscience,* 46: 755,1992.

138. Walberg, F, Ottersen, OP, and Rinvik, E, GABA, glycine, aspartate, glutamate and taurine in the vestibular nuclei: an immunocytochemical investigation in the cat, *Exp. Brain Res.,* 79:547,1990.

139. Watkins, JC, Krogsgaard-Larsen, P, and Honor, T, Structure-activity relationships in the development of excitatory amino acid receptor agonists and competitive antagonists, *Trends Pharmacol. Sci.,* 11:25,1990.

140. Wentzel, PR, DeZeeuw, CI, Holstege, JC, and Gerrits, NM, Inhibitory synaptic inputs to the oculomotor nucleus from vestibulo-ocular-reflex-related nuclei in the rabbit, *Neuroscience,* 65: 161,1995.

141. Wenzel, A, Fritschy, JM, Mohler, H, and Benke, D, NMDA receptor heterogeneity during postnatal development of the rat brain: differential expression of the NR2A, NR2B, and NR2C subunit proteins, *J. Neurochem.,* 68:469,1997.

142. Werner, P, Voigt, M, Keinanen, K, Wisden, W, and Seeberg, PH, Cloning of a putative high affinity kainate receptor expressed predominantly in hippocampal cells, *Nature,* 351:742,1991.

143. Westbrook, GL, Glutamate receptor update, *Curr. Opin. Neurobiol.,* 4:337,1994.

144. Williams, K, Pharmacological properties of recombinant NMDA receptors containing the NR2D subunit, *Neurosci. Lett.,* 184:181,1995.

145. Wilson, VJ, Maeda, M, Connections between semicircular canals and neck motoneurons in the cat, *J. Neurophysiol.,* 37:346,1974.

146. Wilson, VJ, Peterson, BW, Vestibulospinal and reticulospinal systems, in *Handbook of Physiology - The Nervous System II,* Rockefeller University, New York, pp. 667,1981.

147. Wink, DA, Hanbauer, I, Krishna, MC, DeGaff, W, Gamson, J, and Mitchell, JB, Nitric oxide protects against cellular damage and cytotoxicity from reactive oxygen species, *Proc. Natl. Acad. Sci., (USA),*90: 9813,1993.

148. Wisden, W, and Seeburg, PH, Mammalian ionotropic glutamate receptors, *Cur. Opin. Neurobiol.,* 3:291,1993.

149. Wu, W, Expression of nitric oxide synthase (NOS) in injured CNS neurons as shown by NADPH diaphorase histochemistry, *Exp. Neurol.,* 120: 153,1993.

150. Wu, W, and Li, L, Inhibition of nitric oxide synthase reduces motoneuron death due to spinal root avulsion, *Neurosci. Lett.,* 153: 121,1993.

151. Wu, W, Scott, DE, Increased expression of nitric oxide synthase in hypothalamic neuronal regeneration, *Exp. Neurol..* 121: 279,1993.

152. Yamamoto, M, Shimoyama, I, Highstein, SM, Vestibular nucleus neurons relaying excitation from the anterior canal to the oculomotor nucleus, *Brain Res.,* 148:31,1978.

153. Yamanaka, T, Sasa, M, Matsunaga, T, Glutamate as a primary afferent neurotransmitter in the medial vestibular nucleus as detected by *in vivo* microdialysis, *Brain Res.,* 762:243,1997.

154. Yamazaki, M., Mori, H, Araki, K., Mori, KJ, and Mishina, M, Cloning, expression and modulation of a mouse NMDA receptor subunit, *FEBS Lett.,* 300:39,1992.

155. Young, AB, Dauth, GW, Hollingsworth, Z, Penney, JB, Kaatz, K, and Gilman, S, Quisqualate- and NMDA-sensitive [^{3}H]Glutamate binding in primate brain, *J. Neurosci. Res.,* 27:512,1990.

156. Yu, W-H, A Nitric oxide synthase in motor neurons after axotomy, *J. Histochem, Cytochem,* 42:451, 1994.

157. Yuzaki, M, Forrest, D, Curran, T, and Conner, JA, Selective activation of calcium permeability by aspartate in Purkinje cells, *Science,* 273:1112,1996.

158. Zhang, N, and Ottersen, OP, In search of the identity of the cerebellar climbing fiber transmitter: Immunocytochemical studies in rats, *Can. J. Neurol. Sci.,* 20 (Suppl. 3), S36,1993.

159. Zhang, N, Walberg, F, Laake, JH, Meldrum, BS, and Ottersen, OP, Aspartate-like and glutamate-like immunoreactivities in the inferior olive and climbing fibre system: A light microscopic and semiquantitative electron microscopic study in rat and baboon (*Papio anubis*), *Neuroscience,* 1:61,1990.

160. Zhongqi, J, Aas, J-E., Laake, JH, Walberg, F, and Ottersen, OP, Aspartate and glutamate-like immunoreactivities in the vestibular nuclei: A semiquantitative study, *J. Comp. Neurol;* 307:296,1991.

6 Glutamatergic Transmission in the Medial Vestibular Nucleus

N.T. Slater, L.N. Eisenman, G.A. Kinney, B.W. Peterson, and J.C. Houk

CONTENTS

6.1 INTRODUCTION

The signal transformations that occur at synapses between first-order vestibular afferent fibers and second-order neurons of the vestibular nuclear complex represent a critical step in the events which underlie vestibulo-ocular and vestibulo-collic reflexes. Studies of the pharmacology and biophysics of transmission at these synapses have been largely restricted to an examination of transmission within the medial vestibular nucleus (MVN). Such studies have been done using *in vitro* brain slice preparations, as coronal slices can be cut which retain both the nVIII afferent fibers and the second-order neurons within the MVN (Figure 6.1A, Refs. 18,40).

These studies, together with an examination of the firing properties of MVN neurons, are an important step in the analysis of the cellular mechanisms by which signal transformations within the vestibular system occur. They should ultimately provide predictive information which will be of value in the dissection of the complex circuitry mediating vestibular reflexes. In this chapter, the available data concerning the pharmacological and biophysical properties of transmission is reviewed, and incorporated in a model that relates these features to the performance of the primary afferent second order neuron pathway during vestibular stimulation.

6.2 GLUTAMATERGIC TRANSMISSION IN THE MEDIAL VESTIBULAR NUCLEUS

A considerable body of evidence now exists demonstrating that transmission at synapses between vestibular primary afferents and second-order neurons of the MVN is glutamatergic in nature (cf. Chapters 3 and 5 and Refs. 41, and 61 for reviews). This was first demonstrated in lower vertebrate preparations *in vitro*, where it was shown that glutamate receptor antagonists block EPSPs evoked by ipsi- or contralateral nVIII stimulation.[9,34,35] Gallagher and colleagues also demonstrated that EPSPs evoked by ipsilateral nVIII stimulation in mammalian MVN were blocked by broad-spectrum glutamate receptor antagonists; but paradoxically, the selective NMDA receptor antagonist AP5 potentiated EPSPs or was without effect.[41] This would suggest that AMPA receptors mediate the slow EPSP evoked by low frequencies of stimulation of nVIII under conditions where GABAergic pathways within the MVN are not impaired. In other brain regions, the visualization of an underlying NMDA receptor-mediated EPSP is frequently masked both by feedforward and recurrent GABA receptor-mediated inhibition, and by the voltage-dependent blockade of NMDA receptor-channels by external magnesium.[10,45] A blockade of NMDA receptor-mediated excitation of feed-forward GABAergic inhibitory interneurons may, therefore, explain the potentiation of nVIII-evoked EPSPs reported by Lewis et al.[41]

A shunting of the NMDA-receptor-mediated EPSP could originate from the feedforward activation of GABAergic neurons within the MVN. Immunocytochemical studies have demonstrated the presence of GABA or glutamic acid decarboxylase (GAD) immunoreactive neurons in the vestibular nuclear complex.[4,12,36,47,68] In the MVN, these cells are generally small, and immuno-stained fibers and puncta are found throughout the vestibular nuclei. These fibers and puncta originate both from cells within the vestibular nuclei and from the projections of cerebellar Purkinje cells.[22,24] Half of these intrinsic GABAergic neurons within the MVN contribute to a brainstem pathway[4] which likely forms the substrate for the observed inhibition of oculomotor neurons by vestibular nuclei mediated by $GABA_A$ receptors.[23,48] Individual MVN cells also possess both $GABA_A$ and $GABA_B$ receptors located at both pre- and postsynaptic sites.[18,21] Thus, within the MVN there exist GABAergic afferent projections to the nucleus and GABAergic efferents to oculomotor neurons. Evidence has been obtained demonstrating that activation of GABA receptors does shunt the NMDA component, as in the presence of $GABA_A$ receptor antagonists, the nVIII-evoked synaptic response is comprised of both AMPA and NMDA receptor-mediated compo-nents,[33,64] as is the case at the majority of glutamatergic synapses.[10,45]

6.3 BIOPHYSICAL PROPERTIES OF TRANSMISSION

Patch-clamp studies of synaptic transmission in MVN neurons have focused primarily on the properties of the NMDA receptor-mediated component.[35,64] The latter study employed very young rats (P 4-6) in which MVN neurons were visually identified and patch-clamped using both whole-cell or perforated patch-clamp methods, while the former employed somewhat older animals (P 7-25) and blind patch recording methods to study cells below the slice surface.[33] The results from both laboratories were qualitatively similar. In the presence of $GABA_A$ receptor antagonists, the EPSP evoked by nVIII stimulation is a comprised of both AMPA and NMDA receptor-mediated components (Figures 6.1D and 6.2). In current-clamp, these two components can be visualized in I - V curves as a short latency AMPA receptor-mediated component with linear I - V characteristics, and a longer latency component that displays the non-linear I - V characteristics of magnesium channel block of the NMDA receptor (Figure 6.2).

The combined application of both $GABA_A$ and AMPA receptor antagonists allows the NMDA receptor-mediated component to be clearly visualized (Figure 6.3). In the presence of external magnesium, it can be seen that the voltage-dependent rectification of the late component of the composite EPSP derives from the voltage-dependent block by magnesium of the NMDA receptor-mediated component (Figure 6.3).

The time course of synaptically evoked EPSPs recorded under current-clamp are sculpted by factors other than the underlying synaptic current alone, such as the membrane time constant and voltage-dependent conductances. Thus, voltage-clamp studies are required to accurately measure the time course and voltage dependence of the synaptic current. Voltage-clamp studies of the NMDA receptor-mediated EPSC evoked by nVIII stimulation have been performed.[33,64] The NMDA component was pharmacologically isolated by GABA and AMPA receptor antagonists in a mag-nesium-free saline, using patch pipettes filled with cesium methanesulphonate and QX-314 to block voltage-gated K^+ and Na^+ conductances, respectively (Figure 6.4).

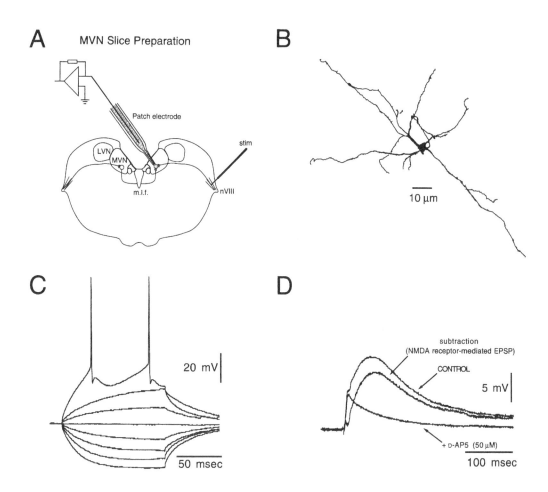

FIGURE 6.1 Whole-cell recording from rat medial vestibular nucleus (MVN) neurons *in vitro*. (A) Schematic illustration of the *in vitro* preparation of rat medial vestibular nucleus. Stimulating electrodes (stim) can be placed on the vestibular nerve root (nVIII) or in the nVIII fiber pathway at the lateral margin of the slice, and patch-clamp recording electrodes placed in the MVN. (B) Reconstruction of an MVN neuron patch-clamped with a Lucifer Yellow-filled pipette. (C) The passive and active properties of an MVN neuron recorded with K gluconate-filled patch electrodes. Note the prominent biphasic afterhyperpolarization and the lack of inward rectification. Membrane potential = -80 mV. (D) EPSPs evoked by nVIII stimulation in the presence of the GABA$_A$ receptor antagonist bicuclline (10 µM) before (Control) and after (+D-AP5) the application of the NMDA receptor antagonist D-AP5 (50 µM); digital subtraction of these traces (faint line) illustrates the time course of the NMDA receptor-mediated component of the nVIII-evoked EPSP in the presence of bicuculline. **A**, **C** and **D** are from Kinney et al., 1994, with permission.

Under these conditions an nVIII-evoked slow EPSC could be observed with a relatively slow rise time (10-90%: 5.8 msec) and linear *I - V* relations over the range -90 to +50 mV (Figure 6.5C and D; Ref 33). These synaptic currents were evoked with relatively short latencies (2.7 m, Ref 33; 3.4 ms, Ref 64), indicative of a monosynaptic response recorded at lower temperatures. The decay of the synaptic current could be best fit as the sum of two exponentials (t_f = 27.6 ms; t_s = 147 ms), which surprisingly did not display any evidence of voltage dependence (Figure 6.4C).

The slow risetime of the NMDA receptor-mediated synaptic current and the biexponential decay of the EPSC is consistent with other measurements of this conductance made using patch-clamp recording methods in slices of other brain regions.[5,11,19,20,30,54,58,59] The time course of the NMDA-mediated synaptic current in MVN neurons was unlikely to have been significantly affected by dendritic RC filtering, as the slow time course of this component would largely preclude such an effect.[62] Moreover, MVN neurons at the postnatal ages used in patch-clamp experiments are relatively unbranched by comparison with cortical pyramidal neurons (Figure 6.1B). Electron microscopic studies of primary afferent terminals on second-order MVN neurons also indicate that a significant proportion (29%) of these synapses are axo-somatic,[31] which provides for optimal resolution of the time course of the response, as in CA3 hippocampal neurons.[28] We have constructed a dendrogram of an MVN neuron using methods similar to those of Jonas et al.,[28] and the results of such modeling confirm that little filtering of the NMDA receptor-mediated EPSC is observed in these cells, whereas AMPA receptor-mediated synaptic currents display considerable filtering for sites located at distal dendritic regions.

In studies of other brain regions in which the voltage-dependence of the decay of the synaptic conductance was measured,[11,30] both components were reported to be voltage-dependent, increasing with membrane depolarization, whereas no significant voltage dependence of decay kinetics were observed in MVN neurons (Figure 6.4C, Ref. 33). The origin of this difference is not clear; it was suggested[33] that this lack of voltage dependence might result from a lack of contribution of a voltage-dependent component of desensitization to the time course of decay. Desensitization of NMDA receptors is regulated by a number of factors (e.g., see Refs. 3,55,66), some of which may modulate desensitization in a voltage-dependent manner[2]. Since desensitization may contribute to the decay phase of NMDA receptor-mediated synaptic currents,[37,38] the voltage-dependence of the

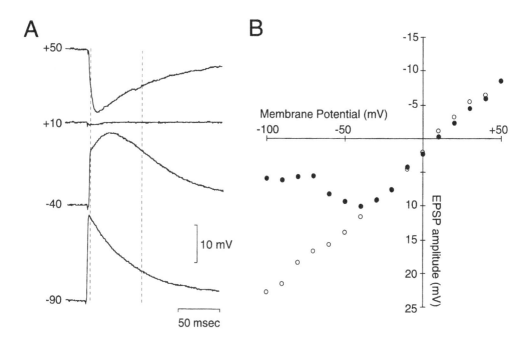

FIGURE 6.2 Voltage-dependent properties of the vestibular nerve (nVIII)-evoked EPSP. The amplitude of the EPSP was recorded over a range of membrane potentials at two latencies (indicated by dashed lines, 4 and 75 m). **(A)** Examples of EPSPs recorded at four representative membrane potentials; **(B)** the amplitudes of the early (open circles) and late (filled circles) components of the composite EPSP evoked by nVIII stimulation over a range of membrane potentials (+50 to -100 mV). (From Kinney et al. (1994) with permission.)

decay phase at other synapses could result, in part, from a voltage-dependent regulation of desensitization. However, primary vestibular afferent fibers fire at very high rates during head rotation (up to 250 Hz), thus, the slow time course of the NMDA component would result in a summating depolarization plateau at high frequencies, which would be impaired by significant desensitization at this synapse. The absence of voltage sensitivity of the NMDA receptor-mediated synaptic current decay may, therefore, result from a lack of voltage-dependent regulation of desensitization. Voltage-independent forms of desensitization would contribute, however, but the extent of contribution may vary between physiologic cell classes within the MVN, depending on the particular subunits expressed (see below).

The AMPA receptor-mediated synaptic current has also been studied to some extent in MVN neurons.[64] In the presence of external magnesium at hyperpolarized membrane potentials (-65 mV), the AMPA component can be observed as a very rapidly activating and deactivating current with linear I - V relations and is abolished by the AMPA receptor antagonist CNQX.[64] An example of an nVIII-evoked AMPA receptor-mediated EPSC recorded in the presence of bicuculline and D-AP5 is shown in Fig 6.4B. As in other preparations studied, the rise time is very rapid, with a simple exponential decay. To study the properties of the AMPA component, it is necessary to pharmacologically isolate this component by the use of GABA and NMDA antagonists. Under these conditions, however, nVIII stimulation may also result in a considerable amount of polysynaptic activation of MVN neurons which obscures the analysis of the monosynaptic EPSC. Indeed, polysynaptic activation of MVN neurons mediated by both NMDA and AMPA receptors has been demonstrated in patch-clamp studies.[33]

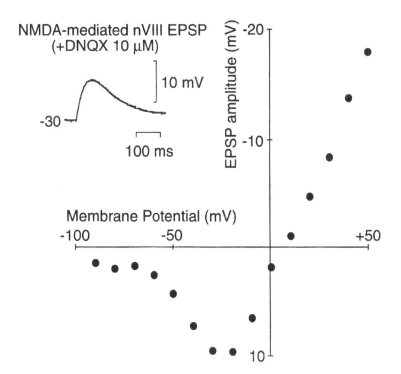

FIGURE 6.3 The voltage dependence of NMDA receptor-mediated EPSPs in a rat MVN neuron in the presence of external magnesium (1 mM). Inset shows an example of an EPSP recorded at -30 mV. Graph illustrates the voltage dependence of the EPSPs evoked by nVIII stimulation in the presence of DNQX and bicuculline.

The existence of recurrent excitatory collaterals between MVN neurons that are modulated by GABAergic inhibition is supported by extracellular single-unit studies of the tonic firing rate of MVN neurons *in vitro*, which is enhanced by $GABA_A$ receptor antagonists.[14,43] This effect can be blocked by the broad-spectrum glutamate receptor antagonist kynurenate.[43] NMDA receptor antagonists also directly depress the rate of spontaneous firing.[43,60] These results would suggest that a portion of the spontaneous firing of MVN neurons is modulated via both glutamate and $GABA_A$ receptors within recurrent pathways of the MVN. Thus, the endogenous firing rate of MVN neurons

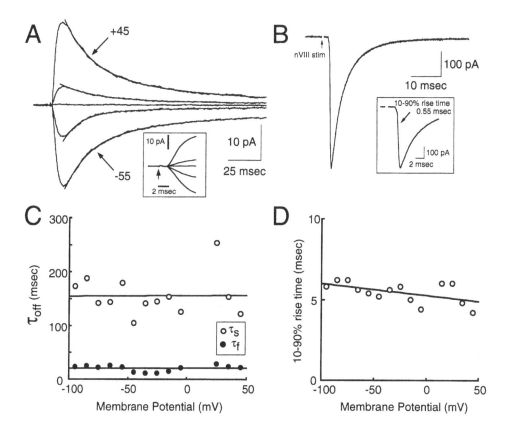

FIGURE 6.4   The time course of the NMDA receptor-mediated postsynaptic current in voltage-clamped MVN neurons is not voltage dependent.   (A) NMDA receptor-mediated EPSCs recorded at a range of membrane potentials in the presence of bicuculline (10 μM) and DNQX (10 μM).   Traces are averaged responses (n = 6) to nVIII stimulation recorded at five membrane potentials (-55, -15, +15 (reversal potential), +25 and +45 mV). In each trace, the time course of decay of the synaptic current was fit with a biexponential decay that is superimposed on the current response. The inset is an enlargement of the onset of the EPSCs to illustrate the latencies and risetimes of the EPSCs. The arrow indicates the onset of the nVIII stimulus.   (B) An nVIII-evoked EPSC mediated via AMPA/KA receptors in a patch-clamped rat MVN neuron. The response was pharmacologically isolated by the application of bicuculline (10 μM) and D-AP5 (50 μM). The inset shows the same response at an expanded time base. The rise time of these EPSCs was relatively fast (mean = 0.65 ms, n = 5), suggesting that the MVN neuron may be electrotonically compact, and/or the majority of nVIII terminals are close to the soma. (C,D) The relationship between the membrane potential and the fast (τ_f) and slow (τ_s) time constants of decay (B) and the rise time of the EPSC (10 - 90% of peak amplitude; C) for the cell shown in (A). Note the lack of obvious voltage-dependence of the kinetics of the rise time or decay. Lines in (C) and (D) are best-fit linear regression lines. (A, C and D are from Kinney et al. (1994) with permission.)

results from both the inherent pacemaker properties of these cells, and amino-acid receptor-mediated collateral pathways within the nucleus. A large proportion of neurons in the MVN are immuno-reactive to glutamate[32] and these neurons may represent the source of the intrinsic glutamatergic pathways. The potential contribution of these intrinsic glutamatergic pathways within the MVN to vestibular signal processing has not been explored.

6.4 KINETIC MODELS OF TRANSMISSION

The complex circuitry of the MVN makes analysis of the synaptic physiology of individual neurons somewhat problematic. This inherent complexity suggests that computational modeling may provide an important contribution to enhancing our understanding of information processing in the MVN. Although a cellular model of MVN neurons has recently been presented,[53] the nonlinear dynamics of synaptic transmission that derive from receptor kinetics were not explored. As a preliminary step to constructing a network model, the authors have constructed models of the NMDA component of the primary afferent input to the MVN. NMDA receptor-mediated synaptic currents were studied for several reasons. Attenuation of currents in the dendritic arbor of a neuron is frequency-dependent, and slow currents such as the NMDA-mediated current are much less attenuated than the more rapid AMPA-mediated currents.[62] Furthermore, anatomical evidence suggests that a significant fraction of the primary afferent synapses terminate at somatic or proximal dendritic sites (see above), which would also minimize attenuation due to the passive properties of the MVN neurons.

Patch-clamp studies of glutamate receptor-mediated currents have provided a detailed picture of the behavior of the NMDA receptor-channel complex to be utilized in the construction of a kinetic model of the NMDA receptor-mediated EPSC in MVN neurons. The following account summarizes the factors considered in the construction of such a model. After binding two molecules of glutamate in the presence of glycine, the NMDA receptor-channel complex exhibits multiple transitions between open and closed states. The state transitions continue until glutamate dissociates from the receptor and is removed from the synaptic cleft by rapid diffusion. In order to explain their observations on glycinergic modulation of NMDA receptor activity, Benveniste et al.[3] proposed a model of the NMDA receptor that required binding of both glutamate and glycine before it can enter an open state. In this model, glycine-sensitive desensitization occurs as a result of the dissociation of glycine. It was further postulated that the model could account for glycine-insensitive desensitization with the inclusion of an additional desensitized state entered from the fully liganded, closed state. Benveniste and Mayer[1] later demonstrated that models with two binding sites each for glutamate and glycine provided the best fit to whole-cell patch-clamp data from concentration jump experiments. In concentration jump experiments on outside-out patches, Clements and Westbrook[6] observed that, in the presence of saturating concentrations of glycine, the time course of NMDA channel activation became concentration-insensitive at high glutamate concentrations, suggesting that channel opening was the rate-limiting step. These responses could be fit by a simple kinetic scheme in which a liganded, closed state could switch either to an open state or to a desensitized state (see below). At sufficiently low concentrations of glutamate, binding becomes the rate-limiting step. The best fit of the responses to low glutamate was achieved using a model with two glutamate binding sites, confirming the results obtained by Benveniste and Mayer[1] and providing an estimate of 4.9 $\mu M^{-1} s^{-1}$ for the glutamate binding rate. Model-fitting results also suggest that glutamate and glycine binding are independent. Therefore, in the presence of a saturating concentration of glycine, it can be assumed that the glycine binding sites on the NMDA receptor are always occupied, resulting in a simplified model of the NMDA receptor consisting of two binding sites, an open state and a desensitized state, as shown below.

in NMDA receptor subunit expression. The MVN is a heterogeneous structure, and different cell classes are likely to express different combinations of subunits, which will result in subtle differences in the physiological properties of transmission at these synapses.

Simulations of responses to 10-Hz stimulation using parameters derived from fits to hippo-0campal data suggested that the average NMDA current would not significantly increase as stimulus frequency increases beyond 20 Hz, because desensitization builds up quickly and reaches a steady-state level after very few stimuli.[15] However, in the MVN, desensitization builds up more slowly, allowing the average current to increase as stimulus frequency increases well above 20 Hz. One possible explanation for a reduced contribution of desensitization in the MVN is the expression of different subunit combinations of NMDA receptors in the two brain regions. Monyer et al.[46] have shown that hippocampal pyramidal neurons express predominantly the NR2A and NR2B subtypes of the NMDA receptor, whereas neurons in the brainstem and cerebellum also express the NR2C and NR2D subtypes, which have differing kinetic properties.[46,63] Recombinant NR2A receptors show the greatest degree of desensitization and the fastest offset decay time constant. NR2B and NR2C receptors display slow kinetics of desensitization and offset decay time constants, and NR2D subunit combinations have a particularly slow deactivation time constant (4.8 s). Therefore, greater expression of the NR2A subtype of receptors in hippocampal neurons relative to MVN neurons could explain the observed differences in desensitization.

6.5 SYNAPTIC INPUTS AND FIRING BEHAVIOR DURING VESTIBULAR STIMULATION

The full development of cellular models of glutamatergic transmission that will be of value for higher network considerations of vestibular function must also incorporate other factors, such as the kinetics of the AMPA current, knowledge of the intrinsic pathways of the MVN (including GABAergic and glycinergic pathways) that regulate transmission of the primary afferent-second-order synapse, and details of the cable properties of these neurons. A growing number of studies have also provided evidence for a heterogeneity of the electrophysiological characteristics of neurons within the MVN,[17,27,56,57] which must also be considered in the implementation of such models (e.g., Ref. 53). While most neurons within the MVN appear to be tonic pacemaker cells, two physiological classes of neurons have been identified by their characteristic action potential shapes: type A cells with a monophasic AHP, and type B cells which also display a slow apamin-sensitive component of the AHP.[13,27,56,57] Differences in the biophysical properties of synaptic transmission in these cells have not been identified, nor has evidence been presented that these two physiological classes of neurons can be further differentiated by their connectivity or responsiveness to behavioral stimulation. Using linear exponential functions to emulate synaptic input, Quadroni and Knöpfel[53] have constructed models of these cells that demonstrate that the differing firing patterns will produce markedly different input-output functions, but this approach will not simulate the nonlinear dynamics of synaptic transmission.

Multicompartment models of type A and B MVN neurons incorporating voltage-gated currents, internal calcium buffering and passive membrane properties similar to that described by Quadroni and Knöpfel[53] have been implemented in NEURON software in which the synaptic input was derived from a kinetic model of the AMPA receptor[28] fit to AMPA receptor-mediated synaptic currents (Figure 6.4B), and the NMDA current in MVN neurons (see above). This multicompartment model accurately simulates the spontaneous firing behavior of MVN neurons and responses to injected current; an example of a model of a type B neuron is illustrated in Figure 6.7A. The relationship between current input and firing rate was linear. However, when a sinusoid of afferent synaptic activity with a cycle duration of 0.5 Hz (range of afferent activation rates: 0 - 100 Hz) was applied to this model neuron, a more complex pattern of neural discharge was evoked, as shown in Figure 6.7B. Activity first increased gradually, then more steeply as the input activation rate increased. Such conspicuous non-linearity is not observed in the MVN *in vivo*, and other factors that could linearize neural behavior including sculpting of excitatory currents by GABA-mediated

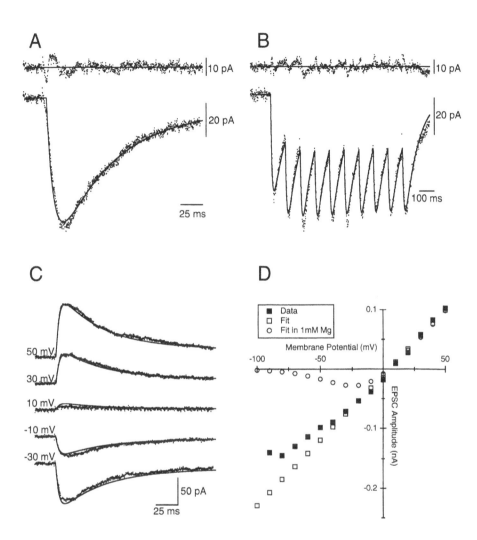

FIGURE 6.5 The fit of a kinetic model of NMDA receptor-mediated currents to experimental data. Traces in (A) and (B) illustrate the NMDA receptor-mediated synaptic current of a rat MVN neuron (dots) and the fit of the model (solid line) for responses to nVIII stimulation evoked at 1 Hz (A) or 10 Hz (B). Residuals plots of the difference between the model and the data are shown above each trace. (C) Experimental data and models fits at varying holding potentials in a nominally magnesium-free saline. (D) Plots of the experimental data (filled diamonds) and model fits (open diamonds) from the cell shown in (C). Open circles show the simulated voltage-dependent rectification of the model in the presence of external magnesium (1mM).

pre- or postsynaptic inhibition, must be considered. The present study focused on differences in modulation of discharge of MVN neurons in which the NMDA receptor-mediated currents exhibited differing amounts of desensitization (Figures 6.5 and 6.6). Changes in membrane potential varied greatly between model cells as a consequence of the relative contribution of NMDA currents during high frequencies of afferent activity. Model neurons with weakly desensitizing NMDA currents exhibited the incrementing activity shown in Figure 6.7B, whereas model neurons with strong

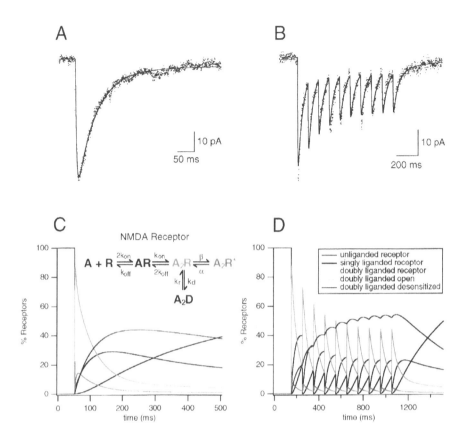

FIGURE 6.6 Time-dependent changes in the fraction of receptors in each of the kinetic states predicted by the model. Traces in (A) and (B) illustrate the NMDA receptor-mediated synaptic current of a rat MVN neuron (dots) and the fit of the model (solid line) for responses to nVIII stimulation evoked at 1 Hz (A) or 10 Hz (B). The temporal dynamics of the models states during synaptic stimulation at 1 and 10 Hz are illustrated in (C) and (D). Following transmitter release, NMDA receptors become rapidly saturated and the fraction of unliganded receptors falls to zero (assuming complete saturation). The fraction of receptors in the doubly liganded, desensitized state builds with time during repetitive stimulation to a peak of approximately 50% at 10 Hz. Thus, during periods of high-frequency activation, the fraction of receptors in the desensitized state is relatively stable, and the time course of the current (reflecting the fraction of open channels) will be primarily determined by transitions between other states and dissociation of glutamate.

desensitization exhibited a saturating plateau of discharge rate as the afferent input rate increased (not shown).

Other properties of the cells, such as the slow AHP, which has been used to define the type A and B neurons, had less effect on the firing rate at physiologically relevant frequencies of afferentactivity (>50 Hz). These results would suggest that the input-output relations of any given primaryafferent-MVN synapse will be principally shaped by the subunit composition of the NMDA receptors at the synapse, rather than differences in voltage-gated channels expressed. These results emphasize the need to accurately model the nonlinear dynamics of the synaptic currents. It should be further noted that these models only consider the monosynaptic primary afferent second-order MVN synapse, and do not incorporate the presently unknown polysynaptic circuitry within the MVN, which will doubtless contribute to information processing.

Further advances in understanding the cellular basis of signal transformations within the vestibular system will benefit from multidisciplinary approaches. For instance, the dynamic prop-

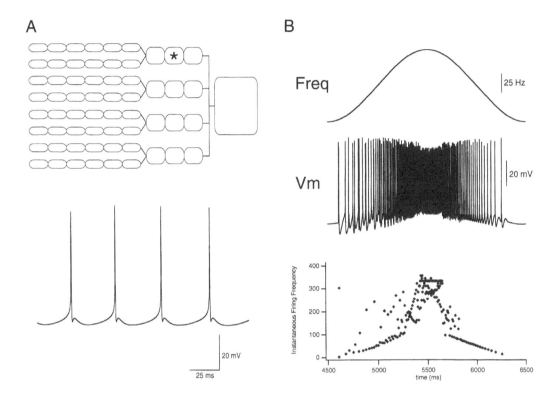

FIGURE 6.7   Simulations of membrane potential changes of a model type B MVN cell to a sinusoidal pattern of afferent synaptic input emulating head rotation at 0.5-Hz. (A) Multicompartment model of a type B MVN neuron (upper diagram) and the spontaneous firing behavior of this model (lower trace). The cell was modelled with four dendrites of identical geometry with proximal dendrites 99 μm in length and 3 μm in diameter, terminating in two distal dendrites 198 μm in length and 1 μm in diameter, with a soma diameter of 31 μm, and an axial resistance of 150?cm. The asterisk indicates the location of the synaptic input to the model. (B), The pattern of membrane potential oscillations in the model cell to a 0.5 Hz sinusoidal change in afferent synaptic activity of the range 0 to 100 Hz. An external concentration of 1 mM Mg^{2+} was included in the simulation. Upper trace shows the change in afferent synaptic firing frequency, middle trace shows the membrane potential of the model cell to this simulated synaptic drive, and the lower graph is is plot of the instantaneous firing frequency for the middle trace.

erties of synaptic transmission revealed by cellular studies can be incorporated into the neural networks that have been defined by anatomical and physiological studies such as intrinsic and commissural vestibulo-vestibular pathways, cerebello-vestibular pathways, and pathways interconnecting the vestibular nuclei with the nucleus prepositus hypoglossi and interstitial nucleus of Cajal. The differing dynamics of synaptic transmission provided by the kinetic analysis of AMPA and NMDA receptor-mediated transmission will be especially interesting to explore in the context of the latter pathways, which are involved in integrating velocity-related signals provided by vestibular receptors to obtain position-related signals required to control horizontal and vertical eye movements.[16] It will also be of great interest to explore the cellular correlates of adaptive plasticity[44,52] and context-specific modulation[50] of vestibular reflexes. As yet, the cellular analysis of vestibular signal transformations represents a relatively unexplored territory likely to provide significant contributions to the understanding of vestibular reflex physiology.

ACKNOWLEDGMENTS

We would like to thank Dr. Mayank B. Dutia for reading the manuscript and Drs. S. Alford, D. Johnston, B. Sakmann, and N. Spruston for helpful discussions. The authors' work is supported by the United States Public Health Service and the Office of Naval Research.

REFERENCES

1. Benveniste, M, and Mayer, ML, Kinetic analysis of antagonist action at N-methyl-D-aspartic acid receptors, *Biophys. J*, 59:560,1991.
2. Benveniste, M, and Mayer, ML, Multiple effects of spermine on N-methyl-D-aspartic acid receptor responses of rat cultured hippocampal neurones, *J. Physiol. (London)*, 464,131,1993.
3. Benveniste, M, Clements, J, Vyklicky, LJ, and Mayer, ML, A kinetic analysis of the modulation of N-methyl-D-aspartic acid receptors by glycine in mouse cultured hippocampal neurones, *J. Physiol. (London)*, 428,333,1990.
4. Blessing, WW, Hedger, SC, and Oertel, WH, Vestibulospinal pathway in rabbit includes GABA-synthesizing neurons, *Neurosci. Lett.*, 80,158,1987.
5. Carmignoto, G, and Vicini, S, Activity-dependent decrease in NMDA receptor responses during development of the visual cortex, *Science*, 258,1007,1992.
6. Clements, JD, and Westbrook, GL, Activation kinetics reveal the number of glutamate and glycine binding sites on the N-methyl-D-aspartate receptor, *Neuron.*, 7,605,1991.
7. Clements, JD, and Westbrook, GL, Kinetics of AP5 dissociation from NMDA receptors: Evidence for two identical cooperative binding sites, *J. Neurophysiol.*, 71,2566,1994
8. Clements, JD, Lester, RAJ, Tong, G, Jahr, CE, and Westbrook, GL, The time course of glutamate in the synaptic cleft, *Science*, 258,1498,1992.
9. Cochran, SL, Kasik P, and Precht, W, Pharmacological aspects of excitatory synaptic transmission to second-order vestibular neurons in the frog, *Synapse*, 1,102,1987.
10. Collingridge, GL, Lester, RAJ, Excitatory amino acids in the vertebrate central nervous system, *Pharmacol. Rev.*, 40,143,1990.
11. D'Angelo, E, Rossi, P, and Taglietti, V, Voltage-dependent kinetics of N-methyl-D-aspartate synaptic currents in rat cerebellar granule cells, *Eur. J. Neurosci.*, 6:640, 1994.
12. Dupont, J, Geffard, M, Calas, A, and Aran, J-M, Immunohistochemical evidence for GABAergic cell bodies in the medial nucleus of the trapezoid body and in the lateral vestibular nucleus in the guinea pig brainstem, *Neurosci. Lett.*, 111,263,1990.
13. Dutia, MB, Membrane properties and firing behaviour of rat medial vestibular nucleus neurones *in vitro:* Modulation of tonic activity by step and sinusoidal current injection, *J. Physiol*, in press,1995.
14. Dutia, MB, Johnston, AR, and McQueen, DS, Tonic activity of rat medial vestibular nucleus neurones in vitro and its inhibition by GABA, *Exp. Brain Res.*, 88,466,1992.
15. Eisenman, LN, Alford, ST, and Houk, JC, Simulation of temporal summation in kinetic models of glutamate receptors, *Soc. Neurosci. Abstr.*, 20,1994.
16. Fukushima, K, Kaneko, CR, and Fuchs, AF, The neuronal substrate of integration in the oculomotor system, *Prog. Neurobiol.*, 39,609,1992.
17. Gallagher, JP, Lewis, MR, and Shiinick-Gallagher, P, An electrophysiological investigation of the rat medial vestibular nucleus *in vitro*, in: *Progress in Clinical and Biological Research*, vol. 176, (Correia MJ, and Paerachio, AA, Eds.,) p. 293-304, Alan Liss, New York,1985.
18. Gallagher, JP, Phelan, KD, and Shinnick-Gallagher, P, Modulation of excitatory transmission at the rat medial vestibular nucleus synapse, *Ann. NY Acad. Sci.*, 656,630,1992.
19. Hestrin, S, Developmental regulation of NMDA receptor-mediated synaptic currents at a central synapse, *Nature*, 357,686,1992.

20. Hestrin, S, Sah, P, and Nicoll, RA, Mechanisms generating the time course of dual component excitatory synaptic currents recorded in hippocampal slices, *Neuron*, 5,247,1990.

21. Holstein, GR, Martinelli, GP, and Cohen, B, Baclofen-sensitive $GABA_B$ binding sites in the medial vestibular nucleus localized by immunocytochemistry, *Brain Res.*, 581,175,1992.

22. Houser, CR, Barber, R.P. and Vaughn, JE Immunocytochemical localization of glutamic acid decarboxylase in the dorsal lateral vestibular nucleus: evidence for an intrinsic and extrinsic GABAergic innervation. *Neurosci. Lett.* 47,413,1984.

23. Ito, M., Highstein, S.M. and Tsuchiya, T. The post-synaptic inhibition of rabbit oculomotor neurons by secondary vestibular impulses and its blockade by picrotoxin. *Brain Res.* 17,520,1970.

24. Ito, M, Raven Press: New York, *The Cerebellum and Neural Control*, 1984.

25. Jahr, CE, High probability opening of NMDA receptor channels by L-glutamate, *Science*, 255,470,1992.

26. Jahr, CE, and Stevens, CF, Voltage dependence of NMDA-activated macroscopic conductances predicted by single-channel kinetics, *J. Neurosci.*, 10,3178,1990.

27. Johnston, AR, MacLeod, NK, Dutia, MB, Ionic conductances contributing to spike repolarization and after-potentials in rat medial vestibular nucleus neurones, *J. Physiol. (London)*, 481,61,1994.

28. Jonas, P, Major, G, and Sakmann, B, Quantal components of unitary EPSCs at the mossy fibre synapse on CA3 pyramidal cells of rat hippocampus, *J. Physiol.*, 472,615,1993.

29. Jonas, P, and Spruston, N, Mechanisms shaping glutamate-mediated excitatory postsynaptic currents in the CNS, *Curr. Opin. Neurobiol.*, 4,366,1994.

30. Keller, BU, Konnerth, A, and Yaari, Y, Patch clamp analysis of excitatory synaptic currents in granule cells of rat hippocampus, *J. Physiol. (London)*, 435,275,1991.

31. Kevetter, GA, Morphology and ultrastructure of vestibular afferent fibers in the medial vestibular nucleus of the gerbil, *Anat. Record*, 218,72A,1987.

32. Kevetter, GA, and Hoffman, RD, Excitatory amino acid immunoreactivity in vestibulo-ocular neurons in gerbils, *Brain Res.*, 554,348,1991.

33. Kinney, GA, Peterson, BW, and Slater, NT, The synaptic activation of N-methyl-D-aspartate receptors in the rat medial vestibular nucleus, *J. Neurophysiol.*, 72,1588,1994.

34. Knöpfel, T, Evidence for N-methyl-D-aspartic acid receptor-mediated modulation of the commisural input to central vestibular neurons of the frog, *Brain Res.*, 426,212,1987.

35. Knöpfel, T, and Dieringer, N, Lesion-induced vestibular plasticity in the frog: are N-methyl-D-aspartate receptors involved? *Exp. Brain Res.*, 72,129,1988.

36. Kumoi, K, Saito, N, and Tanaka, C, Immunohistochemical localization of a-aminobutyric acid- and aspartate-containing neurons in the guinea pig vestibular nuclei, *Brain Res.*, 416,22,1987.

37. Lester, RAJ, and Jahr, CE, NMDA channel behavior depends on agonist affinity, *J. Neurosci.*, 12,635,1992.

38. Lester, RAJ, Clements, JD, Westbrook, GL, and Jahr, CE, Channel kinetics determine the time course of NMDA receptor-mediated synaptic currents, *Nature*, 346,565,1990.

39. Lester, RAJ, Tong, G, and Jahr, CE, Interactions between the glutamate and glycine binding sites of the NMDA receptor, *J. Neurosci.*, 13,1088,1993.

40. Lewis, MR, Gallagher, JP, and Shinnick-Gallagher, P, An *in vitro* brain slice preparation to study the pharmacology of central vestibular neurons, *J. Pharmacol. Meth.*, 18,267,1987.

41. Lewis, MR, Phelan, KD, Shinnick-Gallagher, P, and Gallagher, JP, Primary afferent excitatory transmission recorded intracellularly *in vitro* from rat medial vestibular neurons, *Synapse*, 3,149,1989.

42. Lin, F, and Stevens, CF, Both open and closed NMDA receptor channels desensitize, *J. Neurosci.*, 14,2153,1994.

43. Lin, Y, and Carpenter, DO, Medial vestibular neurons are endogenous pacemakers whose discharge is modulated by neurotransmitters, *Cell. Molec. Neurobiol.*, 13,601,1993.

44. Lisberger,, S The neural basis for the learning of simple motor skills, *Science*, 242,728,1988.

45. Mayer, ML, and Westbrook, GL, The physiology of excitatory amino acids in the vertebrate central nervous system, *Prog. Neurobiol.*, 28,197,1987.

46. Monyer, H, Burnashev, N, Laurie, DJ, Sakmann, B, and Seeburg, PH, Developmental and regional expression in the rat brain and functional properties of four NMDA receptors, *Neuron*, 12,529,1994.

47. Nomura, I, Senba, E, Kubo, T, Shiraishi, T, Matsunaga, T, Tohyama, M, Shiotani, Y, and Wu, J-Y Neuropeptides and a-aminobutyric acid in the vestibular nuclei of the rat: an immunocytochemical analysis. I. Distribution, *Brain Res.*, 311,109,1984.

48. Obata, K, and Highstein, SM, Blocking by picrotoxin of both vestibular inhibition and GABA action on rabbit oculomotor neurons, *Brain Res.*, 18,538,1970.

49. Olverman, HJ, Jones, AW, Mewett, KN, and Watkins, JC, Structure/activity relations of *N*-methyl-D-aspartate receptor ligands as studied by their inhibition of [³H]D-2-amino-5-phosphonopentanoic acid binding in rat brain membranes, *Neuroscience*, 26,17,1988.

50. Paige, GD, and Tomko, DL, Eye movement responses to linear head motion in the squirrel monkey. II. Visual-vestibular interactions and kinematic considerations, *J. Neurophysiol.*, 65,1183,1991.

51. Patneau, DK, and Mayer, ML, Structure-activity relationships for amino acid transmitter candidates acting at *N*-methyl-D-aspartate and quisqualate receptors, *J. Neurosci.*, 10,2385,1990.

52. Peterson, BW, Kinney, GA, Quinn, KJ, and Slater, NT, Potential mechanisms of plastic adaptive changes in the vestibulo-ocular reflex, *Ann. NY Acad. Sci.*, 781:499,1996.

53. Quadroni, R, and Knöpfel, T, Compartmental models of type A and type B guinea pig medial vestibular neurons, *J. Neurophysiol.*, 72,1911,1994.

54. Randall, AD, Schofield JG, Collingridge, GL Whole-cell patch-clamp recordings of an NMDA receptor-mediated synaptic current in rat hippocampal slices, *Neurosci. Lett.*, 114,191,1990.

55. Sather, W, Dieudonné, S, MacDonald, JF, and Ascher, P, Activation and desensitization of *N*-methyl-D-aspartate receptors in nucleated outside-out patches from mouse neurones, *J. Physiol., (London)* 450,643,1992.

56. Serafin, M, de Waele, C, Khateb, A, Vidal, PP, Mühlethaler, M, (1991a) Medial vestibular nucleus in the guinea-pig. I. Intrinsic membrane properties in brainstem slices, *Exp. Brain. Res.*, 84,417,1991a.

57. Serafin, M, de Waele, C, Khateb, A, Vidal, PP, and Mühlethaler, M, (1991b) Medial vestibular nucleus in the guinea-pig. II. Ionic basis of the intrinsic membrane properties in brainstem slices, *Exp. Brain Res.*, 84,426,1991b.

58. Silver, RA, Traynelis, SF, and Cull-Candy, SG, Rapid-time-course miniature and evoked excitatory currents at cerebellar synapses *in situ*, *Nature*, 355,163,1992.

59. Slater, NT, Rossi, DJ, (1996). Functional expression of NMDA receptors in developing neurons. In *Excitatory Amino Acids and The Cerebral Cortex*, F. Conti and T.P. Hicks, Eds., M.I.T. Press, p. 215,1996.

60. Smith, PF, Darlington, CL, and Hubbard, JI, Evidence that NMDA receptors contribute to synaptic function in the guinea pig medial vestibular nucleus, *Brain Res.*, 513,149,1990.

61. Smith, PF, de Waele, C, Vidal, P-P, and Darlington, CL, Excitatory amino acid receptors in normal and abnormalvestibular function, *Molec. Neurobiol.*, 5,369,1991.

62. Spruston, N, Jaffe, DB, Williams, SH, and Johnston, D, Voltage- and space-clamp errors associated with the measurement of electrotonically remote synaptic events, *J. Neurophysiol.*, 70,781,1993.

63. Stern, P, Behe, P, Schoepfer, R, and Colquhoun, D, Single-channel conductances of NMDA receptors expressed from cloned cDNAs: comparison with native receptors, *Proc. R. Soc. Lond.*, B 250,271,1992.

64. Takahashi, Y, Tsumoto, T, and Kubo, T, *N*-methyl-D-aspartate receptors contribute to afferent synaptic transmission in the medial vestibular nucleus of young rats, *Brain Res.*, 659,287,1994.

65. Tong, G, Shepherd, D, and Jahr, CE, Synaptic desensitization of NMDA receptors by calcineurin, *Science*, 267,1510,1995.

66. Vyklicky, LJ, Benveniste, M, and Mayer, ML, Modulation of *N*-methyl-D-aspartic acid receptor desensitization by glycine in mouse cultured hippocampal neurones. J Physiol *(London)* 428,313,1990.

67. Vyklicky LJ, Patneau DK, Mayer ML Modulation of excitatory synaptic transmission by drugs that reduce desensitization at AMPA/kainate receptors, *Neuron*, 7,971,1991.

68. Walberg, F, Ottersen, OP, and Rinvik, E, GABA, glycine, aspartate, glutamate and taurine in the vestibular nuclei: an immunocytochemical investigation in thecat, *Exp. Brain Res.*, 79: 547,1990.

7

Inhibitory Amino Acid Transmitters in the Vestibular Nuclei

G. R. Holstein

CONTENTS

7.1 INTRODUCTION

Gamma-aminobutyric acid (GABA) and glycine are recognized as the major inhibitory neurotransmitters in the vertebrate central nervous system. In general, the rostro-caudal densities of these two short-chain amino acids are inversely distributed through the neuraxis, with glycine prevalent caudally in the spinal cord and lower brainstem, and GABA predominant in rostral diencephalic and telencephalic regions. The central vestibular and oculomotor systems, and in particular the vestibular nuclear complex (VNC), are areas in which these two neurotransmitters co-exist and may even colocalize. The following review focuses on current evidence available concerning the presence, mechanisms of action, and significance of inhibitory amino acid neurotransmission in the four main nuclear groups of the mammalian VNC.

7.2 CONNECTIVITY, NEUROCYTOLOGY, AND ULTRASTRUCTURE OF VESTIBULAR NEURONS

The superior vestibular nucleus (SVN) is composed of a central core of multipolar, round or fusiform neurons, 25 to 50 mμ in diameter, partially encapsulated by smaller cells that are also of variable shape. The larger neurons include the excitatory and inhibitory neurons of the angular vestibulo-ocular reflex (aVOR) pathway, which send projections to contralateral and ipsilateral oculomotor and trochlear nucleus neurons, respectively.[42,72] The dendrites of all SVN cell types tend to arborize near the cell body, except for the extensive bushy dendritic trees of neurons sending projections to the cerebellum; and in all cases, these dendritic processes tend to remain within the borders of SVN.

The dorsal portion of the lateral vestibular nucleus (DLVN) is notable for the presence of large (40 to 70 μm diameter) Deiter's projection cells, although smaller neurons are present throughout the region, and cells of all sizes contribute axons to the lateral vestibulospinal tract. In contrast, the ventral part of LVN (VLVN) is comprised of medium-sized neurons, important for conveying vertical aVOR information to the extraocular motor neurons. There is some indication of a soma-totopic organization in the spinal projections from LVN, such that cells projecting to lumbar spinal cord segments are located dorsally, and cells projecting to the cervical cord are found primarily in the ventral region.[110]

The medial and descending vestibular nuclei (MVN and DVN, respectively) are composed of small- and medium-sized neurons of triangular, multipolar, pear, or round shape (for reviews, see Reference 14). The dendritic processes of these cells tend to be long and slender, with few branches, no obvious pattern of radiation, and restricted but overlapping dendritic fields. In rat and cat, most efferent axons from MVN cells course ventromedially or dorsolaterally, often without issuing local collaterals. Some of the smaller neurons, however, have axons that divide near the soma, providing collateral branches in the region of the parent cell.

Using ultrastructural criteria, two types of neurons have been differentiated in the border zone between MVN and DVN of the rat.[98] One type is a small cell with little cytoplasm but a relatively large nucleus, which receives many axosomatic synapses. The other is a larger cell, identical to the medium-sized neurons of primates, having rich cytoplasm and receiving no synapses directly on the perikaryon. Both cell types display nuclear indentations. In addition, three morphologically distinct terminals have been described: small boutons that lack neurofilaments, contain spherical vesicles, and typically contact cell bodies and proximal dendrites at asymmetric synapses; large boutons that have many neurofilaments, spherical vesicles, and form asymmetric contacts with finer dendritic branches; and boutons with elongated vesicles that form symmetric synapses with all regions of the postsynaptic perikarya and dendrites. The small boutons with spherical vesicles degenerate after vestibular nerve transection, suggesting that they arise from primary vestibular afferents. Occasional axo-axonic synapses are reported to involve boutons with spherical vesicles presynaptic to those with elongated vesicles. This synaptology suggests a morphologic basis for presynaptic inhibition[47].

Experimental investigations of the cat SVN have identified vestibulo-ocular neurons as medium size with round, unindented nuclei[32-34]. Eight weeks and 1 year following vestibular nerve transection, such cells in ipsilateral SVN decrease 17 percent in size, and lose 74 percent of the synaptic input to the soma. While there is no evidence for sprouting, the authors suggest that reactive synaptogenesis accounts for the reported changes to rounder synaptic vesicles and more asymmetric synapses involving the remaining boutons forming axosomatic contacts with vestibulo-ocular neurons. These residual boutons are thought to originate from commissural fibers. In the contralateral SVN, neurons with both indented and unindented nuclei suffer a 35 percent decrease in axosomatic boutons, but only those cells with indented nuclei are significantly smaller in size than controls. It is hypothesized that these neurons represent the cells with commissural projections[32].

7.3 GABA AND GLYCINE IN THE VESTIBULAR NUCLEI

7.3.1 GABA

Five functional types of GABAergic neurons have been described or implicated in one or more parts of the VNC. These include cells with efferent projections to extra-ocular motor neuron pools, neurons mediating disynaptic commissural inhibition, vestibulo-olivary and vestibulo-spinal projection neurons, and local circuit neurons. GABAergic terminals in various VNC regions are derived from these commissural and intra-VNC connections, as well as from Purkinje cell afferents.

7.3.1.1 The Cytology of GABAergic VNC Neurons

Electrophysiological[80,81] and biochemical[82] studies provided early experimental evidence that GABA might serve as a neurotransmitter in some cells of the VNC. Subsequent immunocytochemical investigations have been conducted using antibodies directed against GABA or its synthesizing

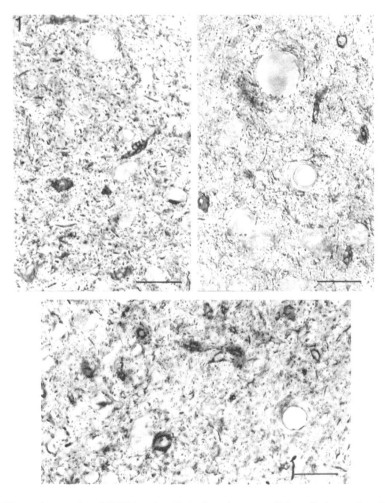

FIGURE 7.1 Photomicrographs of GABAergic cells in the primate medial vestibular nucleus (MVN). Small- and medium sized round and fusiform GABA-immunostained cells are scattered throughout the nucleus. GABAergic fibers and puncta provide a background reticulum of immunoreactivity. Peroxidase immunostain, Nomarski optics. Scale bars: 50 μm.

enzyme glutamic acid decarboxylase (GAD). These studies have revealed labeled neurons in the vestibular nuclei of a variety of species, including mouse,[83] guinea pig,[26,63] rat,[24, 52, 75, 79] rabbit,[8] cat,[103, 105, 110] and monkey [50] (see Figures 7.1 – 7.3). Estimates of the number or density of stained cells are somewhat variable, undoubtedly depending upon species, antibody, and/or methodological considerations. The most conservative estimates indicate that such neurons represent less than ten percent of the total cell population in each of the four main vestibular nuclei, and are distributed uniformly through the VNC.[110] However, most immunocytochemical studies, as well as oligonu-cleotide probes used to visualize mRNA for the GABA synthetic enzyme glutamic acid decarbox-ylase (GAD),[22] have found intense labeling throughout nucleus prepositus hypoglossi (NPH), MVN and SVN, where these elements comprise an estimated 33 percent to 43 percent of the total cell populations. Somewhat fewer cells are reportedly labeled in the DVN and LVN, and none of these are the large Deiters' neurons.

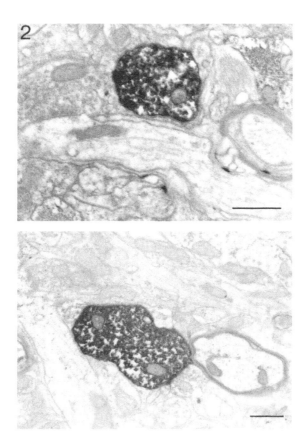

FIGURE 7.2 GABA-immunoreactive dendrites in Rhesus monkey (top) and rat (bottom) MVN. At the ultrastructural level, GABAergic dendrites in both species contain a dark cytoplasmic matrix and loosely scattered, small dark mitochondria. Peroxidase immunostain. Scale bars: 0.5 μm.

In the rat, small round neurons are the predominant GAD-positive cell type in MVN, whereas small- to medium- sized triangular or pear-shaped labeled neurons are scattered throughout DVN.[79] In the guinea pig,[63] GABA-like immunoreactivity is localized in small round, oval, or fusiform vestibular cells. Similarly, *in situ* hybridization studies in rat provide histochemical evidence for the presence of GAD mRNA in many small to medium size MVN neurons, but not in large diameter cells.[29,77] In most studies, regardless of technique, the giant Deiters' neurons in LVN are not

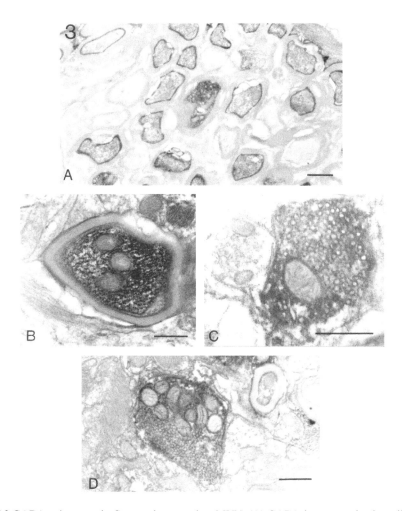

FIGURE 7.3 GABAergic axons in Cynomolgus monkey MVN. (A) GABA-immunostained myelinated axons are not segregated in fiber bundles coursing through MVN. Scale bar: 1 μm. (B) One type of GABA-labeled myelinated axon contains dark, scattered mitochondria, and numerous neurofilaments. (C, D). One type of GABAergic bouton contains a high packing density of spherical synaptic vesicles, and numerous mitochondria that are often clustered on one side of the bouton. Scale bars in B - D: 0.5 μm.

themselves immunoreactive, although labeled puncta surround these somata, sometimes so thoroughly as to give the appearance of perikaryal staining.[52] Such an interpretation is supported by classical degeneration studies that show Purkinje cell axons forming synapses with the cell bodies and proximal dendrites of DLVN neurons.[76] Physiological studies also indicate that the major efferent projection pathway of these cells, the lateral vestibulospinal tract, is excitatory, and immunocytochemical observations suggest that the pathway is glutamatergic.[110] However, one study in the guinea pig has reported GABA immunostaining of Deiters' neurons in LVN.[26]

7.3.1.2 Commissural Inhibition

A commissural system interconnects all parts of MVN, as well as SVN, DVN, and NPH[28]. This fiber system was first demonstrated between homonymous areas of SVN and DVN using axonal degeneration, and then subsequently shown using horseradish peroxidase injections to involve MVN neurons as well. It has since been suggested that MVN constitutes the single most important source of crossing axons. Some commissural cells receive direct input from the labyrinth, although physiological studies have found little evidence for convergence of afferents to these neurons from different semicircular canal nerves. This absence of convergence suggests that the commissural pathway may produce push-pull reactions in the aVOR from reciprocal canal pairs, thereby increasing the sensitivity of second-order vestibular neurons during head movements.[58]

Physiologic recordings from second-order MVN neurons during contralateral vestibular nerve stimulation have demonstrated disynaptic as well as polysynaptic inhibition of these commissural targets. The disynaptic inhibition of contralateral neurons is suppressed by application of either picrotoxin and bicuculline, or strychnine, but not by both types of antagonists.[86] Since picrotoxin and bicuculline block postsynaptic GABA receptors, and strychnine is a specific glycine receptor antagonist, these data suggest that there are separate populations of GABA-receptive and glycine-receptive postsynaptic MVN neurons. Since the primary afferents terminate ipsilateral to their somata, it can be inferred that there is an inhibitory portion of the vestibular commissure, and that it contains distinct GABAergic and glycinergic components. Support for these physiological observations is derived from the author's recent immunocytochemical studies, in which a portion of the indirect aVOR commissural pathway mediating velocity storage is demonstrated to be GABAergic.[48] Since transection of rostral medullary commissural fibers causes functionally-discrete damage to the velocity storage pathway, but leaves the direct aVOR pathway intact,[59,112,113] neurons that degenerate after this midline section can be interpreted as direct participants in the velocity storage network.[48] The GABAergic cells of this pathway are small- and medium-sized neurons located laterally in rostral and rostro-intermediate MVN.[48] They have large nuclei, loose strands of endoplasmic reticulum, occasional cisterns and vacuoles, and round or tubular mitochondria. Their dendrites do not appear to receive GABAergic innervation. The terminals of GABAergic commissural axons contain a moderate density of round/pleomorphic synaptic vesicles, and often form axo-axonic synapses with non-GABAergic terminals.[48]

7.3.1.3 Projections to Extraocular Motor Neuron Pools

GABAergic second-order vestibular neurons have been demonstrated using a variety of approaches to form inhibitory synaptic connections with ipsilateral oculomotor and trochlear motoneurons.[19,103,104,114] These cells receive anterior and posterior semicircular canal-related input and are critical for mediating the vertical aVOR.[72] Most of the second-order cells of this type are located in SVN, with a smaller contingent in rostral MVN. Their axons course in the ipsilateral MLF, and terminate on the somata and proximal dendrites of recipient motor neurons.[43,86] SVN lesions, or unilateral section of the MLF, reduce the concentration of GABA in the ipsilateral trochlear nucleus.[94,103] Similarly, iontophoretic GABA injections in the region of the oculomotor nucleus depress the antidromic field potentials elicited by oculomotor nerve stimulation, as well as spike generation in these motor neurons.[73] In the cat, the concentration of GABA is highest in the oculomotor and trochlear nuclei, and low in the abducens nucleus. Although not the subject of this review, it should be noted that complementary excitatory contralateral projections to the appropriate oculomotor neurons balance these ipsilateral inhibitory vertical aVOR projections. In contrast, inhibitory inputs from second-order vestibular neurons to the abducens nucleus appear to utilize glycine,[103-105] and are discussed below.

7.3.1.4 Vestibulo-olivary and Vestibulo-spinal Neurons

GABAergic vestibular neurons in MVN, as well as NPH, have been shown to project to the ß nucleus of the inferior olivary complex.[4,5,25] In MVN, these neurons are primarily found scattered near the surface of the fourth ventricle. It has also been suggested that some vestibulospinal neurons may utilize GABA for neurotransmission.[8] Many of the neurons that send projections in the medial vestibulospinal tract via the descending MLF are located in the rostral portion of MVN and the VLVN, areas that are rich in GABAergic neurons. In fact, in the rabbit, almost half of the retrogradely-labeled vestibulospinal neurons present in MVN and DVN are GAD-immunostained[8]. However, since labyrinthine-evoked inhibition in neck motor neurons is strychnine-sensitive, and bicuculline and picrotoxin-insensitive, glycine may also play a role in vestibular innervation of the neck. Support for this speculation derives from the observation that presumed glycinergic inhibitory vestibular neurons that project to the ipsilateral abducens nucleus issue collaterals that descend in the ipsilateral MLF toward the spinal cord.[74] Unfortunately, at present, there is only an incomplete understanding of the neurotransmitters involved in vestibulospinal projections and synaptic inter-actions, particularly since these tracts are traditionally considered to be excitatory pathways.

7.3.1.5. Interneurons.

There have been few reports and no consensus regarding the existence or characteristics of inhibitory interneurons in the VNC. Traditional Golgi-impregnation studies[38,68,90] failed to reveal a significant population of true Golgi Type II cells. The results of lesion studies[15] and physiological recordings[100] in the vestibular nuclei have led to the opposite conclusion: that intercalated neurons are present and participate in commissural pathway interactions. Moreover, a small population of intrinsic GAD-containing neurons has been identified immunocytochemically in the dorsal portion of LVN.[52] Pharmacologic studies further support this observation, and indicate that GABAergic interneurons in MVN are critical for the production of long term depression following high-frequency stimulation of primary vestibular afferents.[36]

7.3.1.6 Fibers and Terminals

In all parts of the VNC, GABA- or GAD-immunolabeled fibers, usually beaded and of variable diameter, course multidirectionally.52,63,75,79,110 In MVN, most of the largest-caliber stained fibers project medially toward NPH, while bundles of immunostained fibers course directly through DVN. Large-diameter GAD-immunoreactive fibers are apparent in LVN, with a higher density observed in DLVN.79,110 As described above, these fibers are often seen in close proximity to the somata of Deiters' neurons. The lowest densities of GABAergic fibers are reported in DVN and SVN.

Immunolabeled puncta are apparent throughout the vestibular nuclei.8,52,63,75,79,110. Many of these are attributable to GABAergic VNC neurons that provide axonal arborizations within the nucleus of origin or to other regions of the VNC. In addition, most extrinsic GABAergic terminals are undoubtedly derived from cerebellar Purkinje cells.[24] In fact, following extensive lesioning of cerebellar Purkinje cell afferents, only 30 percent of the GAD activity remains in LVN.[52] In this context, it is not surprising that DLVN, which receives direct Purkinje cell afferents[109] and has a higher GABA content,[30] also contains more GABA-immunoreactive fibers than VLVN.[110] Immu-nostained puncta are observed throughout the neuropil of the VNC, often in close proximity to the proximal dendrites and perikarya of vestibular cells. In LVN, there is a somewhat greater density of labeled puncta in the ventral region,[110] suggesting that GABAergic Purkinje cell afferents course through DLVN but terminate ventrally.

7.3.2 Calcium Binding Proteins

Changes in intracellular calcium concentration have been correlated with a variety of neuronal functions, including signal transduction, shaping of the action potential, neurotransmitter release, and synaptic alterations,[39,118] Calcium binding proteins may modulate these activities using two alternative approaches. Buffer proteins, such as spot-35 protein (S-35), calbindin D-28k (CB), and calretinin, may directly regulate the Ca^{2plus} concentration inside specific cells. In contrast, trigger proteins such as parvalbumin (PV), calmodulin, and proponin-C, change their conformation after binding to free $Ca.^{2plus}$ This alteration exposes regions of the protein's surface that interact with nearby target molecules and modify activity.

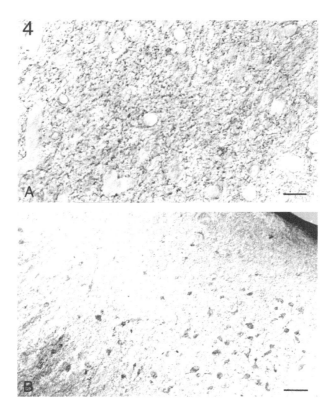

FIGURE 7.4 Photomicrographs of Parvalbumin-immunoreactive fibers (A) and Calbindin D-28k- immuno-stained cells and fibers (B) in the vestibular nuclei of the Cynomolgus monkey. Peroxidase immunostain, Nomarski optics. Scale bars: 50 µm (A) and 100 µm (B).

More than 200 calcium-binding proteins have been reported to date. Some of these, such as calmodulin, are present in all cells, although most are expressed in a tissue- and cell-type-specific fashion.[62] In particular, CB and PV have been associated with specific functional systems and pathways.[40] In distribution studies, these molecules are often found in the same region, but rarely coexist in the same neuron.[16] CB has been viewed as a marker for several neurodegenerative disorders, including Huntington's,[61] Parkinson's,[117] and Alzheimer's[53] diseases. For example, lower CB mRNA expression has been reported in Alzheimer's brain hippocampus.[106] In the normal brain, CB has been postulated to play a role in neuronal protection against neurotoxic levels of free intracellular Ca.$^{2+}$ [16,101] PV, instead, has often been associated with GABAergic, metabolically-active neurons,[17] although some reports indicate an additional association with glutamatergic cells.[18] The co-existence of PV with GABA is well-documented for the subset of GABAergic neurons having fast electrical and high metabolic activity.[12,13] However, a direct relationship with GABAergic neurotransmission itself is unlikely.[85] The association of the calcium-binding proteins with neurotransmission is further suggested by the observations in hippocampus that CaBP mRNA levels are increased after electrical stimulation,[69] and immunoreactivity is altered in histologically-normal tissue after kindling.[3,57] Deafferentation in the visual system has also been reported to affect calcium-binding protein levels,[97] suggesting their sensitivity to altered sensory input.

In the adult rat cerebellar cortex, CB immunoreactivity is present in Purkinje cells of all folia.[2,6,16,35,62] In addition, two distinct morphologic types of mossy fiber axons appear immunostained, although climbing fibers do not. Some labeled axons appear beaded, and distribute collaterals in the infraganglionic plexus, particularly of lobule X. One report also provides evidence

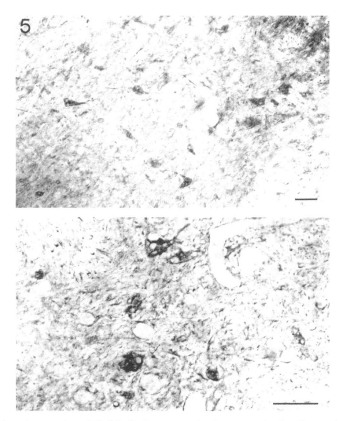

FIGURE 7.5 Light micrographs of Calbindin-immunostained neurons in the Cynomolgus monkey VNC. Peroxidase immunostain, Nomarski optics. Scale bars: 50 µm.

for CB-immunoreactive cerebellar Golgi cells,[35] especially in lobules I and X. Dense CB immunore-activity is found in fibers of the mouse DLVN.[7] These fibers are interpreted as Purkinje cell afferents, since they are not present in Purkinje cell-deficient mutants, but are present in normal mice and Weaver mutants.

In contrast, PV is present to varying degrees in basket cells, stellate cells, and the somata, nuclei, dendrites, spines, axons and terminals of Purkinje cells.[16,62] In newborn rats, climbing fibers from the dorsal cap of the inferior olive are identified by PV-immunoreactivity, which distributes to lobules IX and X of the vermis, and to the flocculus.[111] In lobule X, PV-positive fibers form two sagittal bands on each side of the midline. In the adult rat, some individual Purkinje cells co-localize CB and PV.[16] It is noteworthy that Purkinje, basket, and stellate cells, which are PV-positive, are all rapidly firing GABAergic neurons, whereas GABAergic Golgi cells are slow-firing and PV-negative. In the region of the VNC, a small amount of PV-cell body immunostaining is reported in rodent and cat, although a moderate density of fiber staining is found in MVN and DVN, a moderate number of fibers are immunoreactive in LVN, and few fibers stain in NPH.[21,60] We have observed dense PV fiber staining in the primate MVN (Figure 7.4), but no perikaryal label is apparent. It is possible that colchicine injection would reveal a population of PV-containing neurons, since some CB and PV immunoreactive fibers in the rat paraflocculus have been interpreted as afferents from the vestibular nuclei.[16]

The presence of calretinin has also been evaluated by immunohistochemistry in rodents,[1,60,92] and by *in situ* hybridization using a digoxigenin-labeled oligonucleotide probe to study calretinin mRNA localization in rat and guinea pig.[95] These studies demonstrate the presence of calretinin in cells throughout the VNC, particularly in medium and large size multipolar neurons near the ventricular wall of MVN. These cells are similar in appearance to the CB-immunoreactive neurons apparent in primate MVN (Figures 7.4 and 7.5).

7.3.3 GLYCINE

Behavioral, physiological and biochemical evidence supports the idea that glycinergic neurons are present in the vestibular nuclei.[31,86,105,110] For example, electrical stimulation of the horizontal canal nerve causes inhibition of abducens motor neurons. This effect is abolished by strychnine, but is unaffected by picrotoxin or bicuculline. These recordings suggests that second order vestibular neurons that mediate the horizontal aVOR do so through glycinergic projections to the abducens nucleus. In fact, glycine immunoreactive cells have been reported in MVN, DVN, and LVN of cat.[105,110] The stained cells are small or medium-size, and have a distribution and density similar to that of GABAergic neurons. Additionally, glycinergic afferents to the abducens nucleus from second-order neurons in the vestibular nuclei, as well as from neurons in NPH and the dorsolateral medullary reticular formation, have been demonstrated using immunocytochemistry in the cat.[103] These premotor neurons are also selectively labeled by retrograde transport following [³H]glycine injection in the abducens nucleus.

Labeled fibers and puncta are observed in all parts of VNC, albeit infrequently.[103,110] There are glycine-labeled fibers present in the penetrating bundles of the DVN, and some stained axons in the hook bundle traversing LVN.[110] Glycine is localized predominantly in descending axons in the MLF that project to the abducens nucleus and the spinal cord.[103] In addition, the abducens nuclei contain a high density of glycine-immunoreactive boutons, some which are observed in close proximity to neuronal somata.[105]

7.4 COLOCALIZATION AND COEXISTENCE OF THE INHIBITORY AMINO ACID NEUROTRANSMITTERS

The coexistence or colocalization of GABA and glycine has been reported in several brain regions, including the cerebellum,[84] vestibular nuclei,[110] and oculomotor nuclei.[114] It has been suggested that cerebellar Golgi cells release both amino acids at their terminals, and biochemical studies of GABA and glycine transporters (see below) are also compatible with the possibility of terminals containing both GABA and glycine.[107]

In the vestibular system, electrophysiological data suggests that vestibular Type I neurons may be controlled by both GABA and glycine.[31] Extensive colocalization of GABA and glycine in the rabbit VNC is reported in LVN neurons, and in some axons and boutons of MVN and DVN.[114] In general, most glycinergic cells and myelinated axons appear to co-localize GABA,[110] and conversely, few glycine-immunostained profiles are GABA-immuno-negative.[114] In contrast, GABA-immunolabeled fibers of all sizes, and most of the GABA-immunoreactive perisomatic puncta in LVN, are not glycinergic. Coexistence of GABA and glycine is observed less frequently in DVN, where glycine-only neuronal elements were common. Last, although GABA apparently does not coexist with glutamate or aspartate in the neurotransmitter pool, some glycinergic cells of DVN also show aspartate and glutamate immunoreactivity.[114] However, since glycine, aspartate, and glutamate are all present as metabolites in neurons, these observations must be viewed with some caution. In addition, despite the colocalization of GABA and glycine in single vestibular neurons, the ultimate impact of synaptic neurotransmission is dependent on the presence and localization of specific postsynaptic receptor sites.

7.5 GABA AND GLYCINE RECEPTORS IN THE VESTIBULAR NUCLEI

Three principal types of GABA receptors have been localized in the mammalian CNS. $GABA_A$ receptors belong to the ionotropic family of receptor-channel complexes. They are activated by muscimol, blocked by bicuculline, and are associated with Cl⁻ ion channels in the membrane. $GABA_B$-binding sites are bicuculline insensitive, modulate Ca^{2+} and/or K^+ ion conductances, and belong to the family of metabotropic receptors that regulate adenylate cyclase activity through linkage with inhibitory guanosine triphosphate-binding proteins. In addition, $GABA_B$ receptors are sensitive to baclofen (γ-amino-b-(p-chlorophenyl)butyric acid; Lioresal), a lipid-soluble GABA analog which is inactive at $GABA_A$ sites . $GABA_C$ receptors have been identified in the retina, and may be present in other parts of the CNS,[9] but the presence, localization, and function of these receptors in the central vestibular system is not known, so they will not be reviewed here.

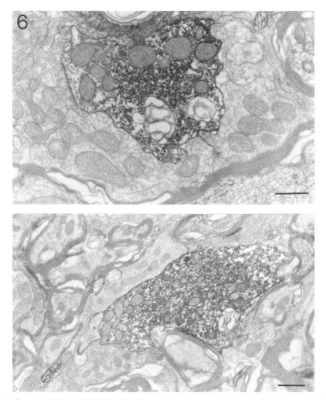

FIGURE 7.6 L-Baclofen-sensitive $GABA_B$ binding sites on dendrites in Rhesus monkey MVN, providing evidence for postsynaptic $GABA_B$ receptor sites. Scale bars: 0.5 μm (top); 1 μm (bottom).

The regional distributions of $GABA_A$ and $GABA_B$ receptor sites were originally identified by autoradiographic methods using a variety of agonist and antagonist ligands (for reviews, see ref. [11] and [55]), and more recently have been visualized by immunocytochemistry[20,71, 93, 99] and *in situ* hybridization histochemistry.[99,115,116] In general, the autoradiographic studies suggest that $GABA_A$ receptor sites are found in greatest density in structures of the forebrain, whereas $GABA_B$ binding sites are concentrated in specific nuclei distributed throughout the neuraxis. The cerebellum is one region where both $GABA_A$ and $GABA_B$ receptor subtypes are found in high concentrations. In the cerebellar cortex, $GABA_B$ receptors are most highly concentrated in the molecular layer, while $GABA_A$ sites are denser in the granular layer. Reports suggest that $GABA_B$ sites may be located on the dendritic processes, (but not the somata), of Purkinje cells, as well as on parallel fibers and/or climbing fiber terminals. Immunocytochemical observations of L-baclofen binding sites in cerebellar cortex indicate that most baclofen-sensitive $GABA_B$ binding sites are localized on parallel fibers and their terminals, and few are present on the other elements of the neuropil.[49,51,71]

In the vestibular nuclei, $GABA_A$ receptor autoradiography has failed to reveal binding sites in MVN, DVN or LVN. In contrast, mRNA encoding the a_1 subunit of the $GABA_A$ receptor has been detected by *in situ* hybridization histochemistry in the rat vestibular nuclei.[44] Heavy labeling is present in the giant Deiter's neurons of LVN, moderate grain densities are apparent in MVN and DVN, and no specific labeling is apparent in SVN. Since ligand-binding studies indicate the presence of a receptor protein, while *in situ* hybridization reveals the somata containing a particular subunit mRNA, this discrepancy may be attributable to a difference between the presence of mRNA in the perikaryon and the presence of binding sites on axon terminals.[44] However, since GABAergic synapses onto the cell bodies and proximal dendrites of VNC neurons are not uncommon, this discrepancy is more likely due to methodological differences between the two type of studies.

$GABA_B$ receptor localization studies in the vestibular nuclei have primarily been restricted to observations of L-baclofen-sensitive binding sites,[71] since the receptor has not yet been sequenced and cloned. These localization studies are based on the specificity of the agonist for the $GABA_B$ receptor, and the development of a monoclonal antibody directed against the agonist.

In these experiments, L-baclofen immunoreactivity is observed in discrete neuronal elements in the MVN of L-baclofen-injected animals[45-47,71] (Figures 6 and 7).

Immunoreactive myelinated axons, mostly of small caliber, are frequently observed. In addition, densely packed axonal profiles with small mitochondria and spherical or pleomorphic synaptic vesicles are markedly immunostained. These profiles form synaptic contacts with unlabeled dendrites, which also receive synapses from unlabeled boutons. This synaptology represents one possible substrate for presynaptic inhibition. In addition, there are many immunoreactive dendrites present in the neuropil, often participating in symmetric synapses. The presynaptic elements in these synapses usually contain loosely packed spherical or ellipsoid vesicles. These elements may represent the morphologic basis for postsynaptic inhibition. Thus, there is immunocytochemical evidence for both pre- and postsynaptic GABAergic inhibition in MVN mediated by $GABA_B$ receptors.

Taken together, these observations provide some insight into the specific subregions of the VNC that are more likely to contain GABA- and glycine-recipient vestibular neurons. Behaviorally, administration of L-baclofen causes a reduction of the dominant time constant of the aVOR and optokinetic after-nystagmus, and a decrease in the nystagmus produced by off-vertical axis rotation.[41] These are all functions of velocity storage,[91] suggesting that inhibitory control of velocity storage is mediated by GABA effects on $GABA_B$ receptors. Consistent with this, an increased response to baclofen follows nodulo-uvulectomy, indicating $GABA_B$ receptor supersensitivity. Functionally, extracellular recordings from MVN cells in an *in vitro* slice preparation indicate that GABA, muscimol, and baclofen all inhibit the discharge of tonically-active MVN neurons.[27] Picrotoxin and bicuculline antagonize the GABA and muscimol-induced inhibition, and phacolfen and 2-OH-saclofen antagonize the baclofen effect. The observation that this antagonism persists in the presence of picrotoxin, Co^{2+} and high Mg^{2+} media provides yet another line of evidence supporting the claim of postsynaptic $GABA_B$ receptors on tonically active MVN cells, in addition to the presynaptic receptor sites.

In the context of inhibitory neurotransmission, glycine acts through a strychnine-sensitive receptor that belongs to the family of ligand-gated Cl^- channel membrane proteins (for review, see Ref. [64]). Glycine also functions as a co-agonist with glutamate to activate postsynaptic *N*-methyl-D-aspartate (NMDA) receptors. In general, endogenous levels of glycine are highest in the spinal cord, somewhat lower in the brainstem, and markedly decreased more rostrally. This is also true of specific strychnine binding, as assessed by autoradiography. In the human vestibular nuclear complex, high densities of strychnine-sensitive glycine receptors are present in MVN and NPH, but a low density is reported in DVN. In the rat, densities are two-to-threefold fold higher than in humans, and moderate grain counts are obtained in MVN.[87] It is unclear whether this represents a

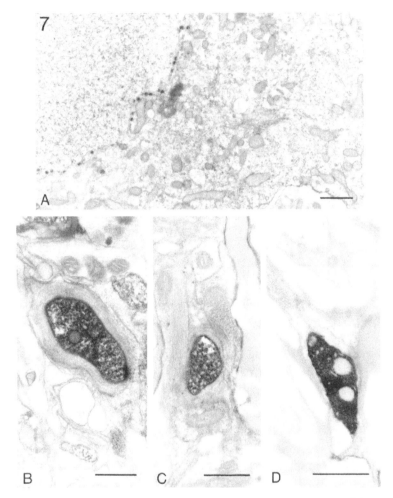

FIGURE 7.7 (A) L-Baclofen-sensitive GABA$_B$ binding sites are present on portions of the nuclear envelope. Scale bar: 0.5 μm. (B - D) L-Baclofen-sensitive GABA$_B$ binding sites on myelinated axons (B, C) and terminals (D) in Rhesus monkey MVN, providing evidence for presynaptic GABA$_B$ receptor sites. Scale bars (B - D): 0.5 μm.

methodological difference, or an actual phylogenetic difference in glycine receptor density. Physiologic recordings from MVN neurons *in vitro* indicate that almost all of these cells are inhibited in a dose-dependent manner by exogenous glycine, and this inhibition is blocked with strychnine pretreatment.[23] In addition, many glycine receptor γ_1 subunit mRNA-containing neurons are present in all regions of the VNC.

7.6 TRANSPORTERS FOR THE INHIBITORY AMINO ACID NEUROTRANSMITTERS

GABAergic and glycinergic synaptic transmission involves the release of neurotransmitter molecules into the cleft, interaction of the neurotransmitters with postsynaptic receptors, and removal of these molecules from the cleft through a re-uptake mechanism. Re-uptake provides a substrate for the process of molecular recycling, and also prevents diffusion of neurotransmitter molecules away from the region of the synapse, thereby decreasing the likelihood of chemical crosstalk between adjacent synapses. In addition, rapid inactivation of released neurotransmitters by transport back into the presynaptic elements results in temporal sharpening, in contrast to the prolonged time course of passive diffusion.

Re-uptake is catalyzed by two general types of neurotransmitter-specific transporters.[67,78,108] Plasma membrane transporters, or "symporters," move their specific substrate along with sodium and chloride ions, following the membrane sodium gradient. Members of this family include the transporters for the inhibitory amino acids, monoamines, and catecholamines. As an example, it has been demonstrated for one of the GABA transporters that one molecule of GABA co-transports with one chloride and two sodium ions.[88] Vesicle transporters, or "antiporters," are driven by the electrochemical proton gradient produced by a H^+-translocating ATPase associated with the vesicle membrane. The antiporter carries a substrate such as glutamate into, and protons out of, the synaptic vesicle.

The first to be purified and cloned was a rat brain GABA transporter, designated GAT-1.[37] Subsequent expression cloning of the norepinephrine, serotonin, and then dopamine transporters revealed the existence of a gene family with a high degree of sequence homology. Thus, the GABA and monoamine transporters have certain domains that are nearly identical. At present, cDNAs encoding four GABA transporters (GAT 1 - GAT 4) and two glycine transporters (GLYT1a, b, and c, and GLYT2) have been identified in several species, albeit using alternative nomenclatures (for review, see ref. [96]). All of these transporters are Na^{2+}- and Cl⁻-dependent glycoproteins that share a putative structure of 1-transmembrane helices with short N- and C-termini on the cytoplasmic side of the membrane and with a single large loop in the external face of the membrane (between helices III and IV).[56]

The regional distribution of transporter activity is often consistent with the distributions of the associated neurotransmitter, suggesting that transporters may be expressed in a neurotransmitter-specific manner. Using antibodies produced against the least homologous portions of the four mouse GABA transporters, or hybridization histochemistry, the regional distributions of the GABA transporters have been assessed in the CNS of several species.[54,56,65] Of the four, only GAT1 and GAT4 prove to be CNS-specific, whereas GAT2 and GAT3 are also expressed in peripheral tissues such as liver and kidney. In the nervous system, all four are present in the neuropil, and are not expressed in cell bodies.[89] While GAT1 appears to be distributed rather uniformly throughout the brain, GAT4 is strongly expressed in the spinal cord, brainstem, thalamus, hypothalamus, and basal forebrain, and is weakly expressed in the cerebellum, hippocampus, striatum, and cerebral cortex.[54,70,89] In the rat, it appears that GAT4 is strongly expressed, and GAT1 is moderately expressed in the vestibular nuclei, particularly MVN (data from ref. [54]). Comparison of these overall distribution patterns with the widespread localization of GABA in the CNS leads to the suggestion that GAT1 is the more likely transporter at established GABAergic synapses.[56,65] It is interesting to note that GAT3 mRNA is plentiful in neonatal rodents, reaching a maximum several days after birth before falling to a low level that persists in adult brain, suggesting that this transporter may be involved in, and/or regulated by, postnatal developmental processes.[65]

At present, two glycine transporters have been cloned. One, GLYT1,[37,67,102] shows three splice variants with different amino termini, designated GLYT1a, GLYT1b, and GLYT1c,[10, 66,119] the second identified glycine transporter is designated GLYT2.[66] GLYT1b is expressed exclusively in the nervous system, including glial cells, whereas GLYT1a is found in gray matter regions of the nervous system expressing glycine receptor subunits, but also in non-neural tissue. GLYT2 differs from GLYT1 by its extended intracellular amino terminus, its insensitivity to N-methyl-aminoacetic acid, and its localization primarily in the hindbrain.[56] These attributes suggest that GLYT2 serves as the transporter at strychnine-sensitive inhibitory glycinergic synapses, that are also concentrated in the spinal cord and brainstem. However, in addition to activating Cl-channel receptors, glycine also co-activates glutamatergic excitatory synapses by binding to sites associated with the NMDA receptor. The high levels of GLYT1 mRNA that have been reported in hippocampus and neocortex suggest the colocalization of this transporter with the NMDA receptor.[10]

In the rat, the highest levels of GLYT1 mRNA are observed in the midbrain, brainstem, hypothalamus, hippocampus, and cerebellum; and the highest levels of GLYT2 mRNA are reported in the hypothalamus, midbrain, and brainstem.[10] Using histochemical probes, GLYT1 hybridizes almost exclusively to white matter, while GLYT2 hybridizes only to gray matter and shows the rostro-caudal gradient characteristic of glycinergic synapses. The GLYT2-specific probe strongly labels many nuclei of the brainstem, especially the pontine, vestibular, and trigeminal nuclei.

To further study the localization of these proteins, sequence-specific antibodies have been raised against the cloned glycine transporters, and used for immunocytochemical visualization at the light and electron microscopic levels.[119] In these studies, GLYT1 immunoreactivity appears strongest in spinal cord, brainstem, diencephalon, and retina. Immunostain is present in glial cell processes around both glycinergic and non-glycinergic neurons, in all regions except the retina. GLYT2 immunoreactivity is restricted to the spinal cord, brainstem, and cerebellum, with low levels apparent in the diencephalon. The GLYT2 protein is found in presynaptic, putative glycinergic boutons. In the vestibular nuclei, strong glial labeling is observed with GLYT1 antibody, and GLYT2 immunoreactivity is high in boutons apparent throughout the VNC. In the cerebellar cortex, GLYT1 immunolabel is densest in the molecular layer, primarily associated with Bergmann glia. GLYT2 is observed in terminal boutons, mainly appearing as rings of Golgi cell axon terminals surrounding the glomeruli of the granular layer, although glial elements also show some immunostain. The correlation between neurotransmitter, receptor, and transporter is reinforced by the observation that GLYT1 is distributed throughout the cranial nerve brainstem motor nuclei, including the oculomotor and abducens, whereas GLYT2 immunoreactivity is dense in the abducens nucleus, but not visible in the oculomotor complex.

In summary, GABA is associated predominantly with ascending second-order vestibular axons that project to the oculomotor and trochlear nuclei, and that participate in the production of the vertical aVOR. Glycine is localized primarily in descending axons of second-order vestibular neurons that project to the abducens nuclei and the spinal cord, and that participate in the horizontal aVOR.[103] This functional and neurochemical distinction may also correlate with the differential roles of GABA and glycine in vestibular commissural inhibition. In addition, there is evidence for the presence of $GABA_A$, $GABA_B$, and glycine receptors in the vestibular nuclei. Studies of GABA and glycine transporter mechanisms are just beginning to clarify the presence, localization, and function of these molecules in inhibitory neurotransmission in the vestibular nuclear complex.

ACKNOWLEDGMENTS

Supported in part by National Institutes of Health research grants DC 01705 and DC 02451 from the National Institute on Deafness and Other Communication Disorders, and NASA Research Grant NAG 2.

REFERENCES

1. Arai, R, Winsky, L, Arai, M, and Jacobowitz, DM, Immunohistochemical localization of calretinin in the rat hindbrain, *J. Comp. Neurol.,* 310,21,1991.
2. Baimbridge, KG, and Miller, JJ, Immunohistochemical localization of calcium-binding protein in the cerebellum, hippocampal formation, and olfactory bulb of the rat, *Brain Res.,* 245,223,1982.
3. Baimbridge, KG, and Miller, JJ, Hippocampal calcium-binding protein during commissural kindling-induced epileptogenesis; Progressive decline and effects of anticonvulsants, *Brain Res.,* 324,85,1984.
4. Barmack, NH, Fagerson, M, and Errico, P, Cholinergic projection to the dorsal cap of the inferior olive of the rat, rabbit, and monkey, *J. Comp. Neurol.,* 328,263,1993.

5. Barmack, NH, Mugnaini, E, and Nelson, BJ, Vestibularly-evoked activity of single neurons in the B-nucleus of the inferior olive, in *The Olivocerebellar System in Motor Control.*, P. Strata, Editor, 1989, Springer-Verlag, Berlin. p. 313,1989.

6. Batini, C, Cerebellar localization and colocalization of GABA and calcium binding protein-D28K, *Arch. Ital. Biol.*, 128,127,1990.

7. Baurle, J, and Grusser-Cornehls, U, Calbindin D-28k in the lateral vestibular nucleus of mutant mice as a tool to reveal Purkinje cell plasticity, *Neurosci. Letts.*, 167,85,1994.

8. Blessing, WW, Hedger, SC, and Oertel, WH, Vestibulospinal pathway in rabbit includes GABA-synthesizing neurons, *Neurosci. Lett.*, 80,158,1987.

9. Bormann, J, and Feigenspan, A, GABAc Receptors, *Trnds. Neurosci.*, 18,515,1995.

10. Borowsky, B, Mezey, E, and Hoffman, BJ, Two glycine transporter variants with distinct localization in the CNS and peripheral tissues are encoded by a common gene, *Neuron,* 10,851,1993.

11. Bowery, NG, Hill, DR, Hudson, AL, Price, GW, Turnbull, MJ, and Wilkin, GP, Heterogeneity of mammalian GABA receptors, in *Actions and Interactions of GABA and Benzodiazepines*, N.G. Bowery, Ed., New York, Raven Press, p. 81,1984.

12. Braun, K, Scheich, H, Schachner, M, and Heizmann, CW, Distribution of parvalbumin, cytochrome oxidase activity and 14C-2-deoxyglucose uptake in the brain of zebra finch I. Auditory and vocal motor systems, *Cell Tiss. Res.*, 240,101,1985.

13. Braun, K, Scheich, H, Schachner, M, and Heizmann, CW, Distribution of parvalbumin, cytochrome oxidase activity and 14C-2-deoxyglucose uptake in the brain of zebra finch. II. Visual system, *Cell. Tiss. Res.*, 240,117,1985.

14. Brodal, A, The vestibular nuclei in the macaque monkey, *J. Comp. Neurol.*, 227,252,1984.

15. Brodal, A, *Organization of the commissural connections: Anatomy*, in *Basic Aspects of Central Vestibular Mechanisms*, A. Brodal and O. Pompeiano, Eds. Elsevier: Amsterdam, p. 167,1972.

16. Celio, MR, Calbindin D-28k and parvalbumin in the rat nervous system, *Neuroscience*, 35,375,1990.

17. Celio, MR, GABA neurons contain the calcium binding protein parvalbumin, *Science*, 232,995,1986.

18. Cote, P-Y, Sadikot, AF, and Parent, A, Complementary distribution of calbindin D-28k and parvalbumin in the basal forebrain and midbrain of the squirrel monkey, *Eur. J. Neurosci.*, 3,1316,1991.

19. de la Cruz, RR, Pastor, AM, Martínez-Guijarro, FJ, López-García, C, and Delgado-García, JM, Role of GABA in the extraocular motor nuclei of the cat, A postembedding immunocytochemical study, *Neuroscence*, 51,911,1992.

20. deBlas AL, Victorica, J, and Friedrich, P, Localization of the $GABA_A$ receptor in the rat brain with a monoclonal antibody to the 57,000 MW peptide of the $GABA_A$ receptor/benzodiazepine receptor/Cl- channel complex, *J. Neurosci.*, 8,602,1988.

21. DeLeon, M, Covenas, R, Narvaez, JA, Aguirre, JA, and Gonsalez-Baron, S, Distribution of parvalbumin immunoreactivity in the cat brain stem, *Brain Res. Bull.*, 32,639,1993.

22. deWaele, C, Abitbol, M, Chat, M, Menini, C, Mallet, J, and Vidal, PP, Distribution of glutamatergic receptors and GAD mRNA-containing neurons in the vestibular nuclei of normal and hemilabyrinth-ectomized rats, *Eur. J. Neurosci.*, 6,565,1994.

23. deWaele, C, Muhlethaler, M, and Vidal, PP, Neurochemistry of the central vestibular pathways, *Brain Res. Rev.*, 20,24,1995.

24. DeZeeuw, CI, and Berrebi, AS, Postsynaptic targets of Purkinje cell terminals in the cerebellar and vestibular nuclei of the rat. *Eur. J. Neurosci.*, 7,2322,1995.

25. DeZeeuw, CI, Wentzel, P, and Mugnaini, E, Fine structure of the dorsal cap of the inferior olive and its GABAergic and non-GABAergic input from the nucleus prepositus hypoglossi in rat and rabbit *J. Comp. Neurol.*, 327,63,1993.

26. Dupont, J, Geffard, M, Calas, A, and Aran, J-M, Immunohistochemical evidence for GABAergic cell bodies in the medial nucleus of the trapezoid body and in the lateral vestibular nucleus in the guinea pig brainstem, *Neurosci. Letts.*, 111,263,1990.

27. Dutia, MB, Johnston, AR, and McQueen, DS, Tonic activity of rat medial vestibular nucleus neurones in vitro and its inhibition by GABA, *Exp. Brain Res.*, 88,466,1992.

28. Epema, AH, Gerrits, NM, and Voogd, J, Commissural and intrinsic connections of the vestibular nuclei in the rabbit: a retrograde labeling study, *Exp. Brain Res.*, 71,129,1988.

29. Ferraguti, F, Zoli, M, Aronsson, M, and Agnati, LF, Goldstein, M, Filer, D, Kjell, F, Distribution of glutamic acid decarboxylase messenger RNA-containing nerve cell populations of the male rat brain, *J. Chem. Neuroanat.,* 3,377,1990.

30. Fonnum, F, Storm-Mathisen, J, and Walberg, F, Glutamate decarboxylase in inhibitory neurons. A study of the enzyme in Purkinje cell axons and boutons in the cat, *Brain Res.,* 20,259,1970.

31. Furuya, N, Yabe, T, and Koizumi, T., *Neurotransmitters regulating the activity of the vestibular commissural inhibition in the cat,* in *Sensing and Controlling Motion: Vestibular and Sensorimotor Function,* Palo Alto, CA: NY Acad. Sci.,1991.

32. Gacek, R, Lyon, M, and Schoonmaker, J, Ultrastructural changes in vestibulo-ocular neurons following vestibular neurectomy in the cat, *Ann. Otol. Rhin. Laryngol.,* 97,42,1988.

33. Gacek, R, Lyon, M, and Schoonmaker, J, Morphologic correlates of vestibular compensation in the cat, *Acta Otolaryngol. (Stockh.) Suppl.,* 462,1,1989.

34. Gacek, R, Lyon, M, and Schoonmaker, J, Ultrastructural changes in contralateral vestibulo-ocular neurons following vestibular neurectomy in the cat, *Acta Otolaryngol. (Stockh.) Suppl.,* 477,3,1991.

35. Garcia-Segura, LM, Baetens, D, Roth, J, Norman, AW, and Orci, L, Immunohistochemical mapping of calcium-binding protein immunoreactivity in the rat central nervous system, *Brain Res.,* 296,75,1984.

36. Grassi, S, Della Torre, G, Capocchi, G, Zampolini, M, and Pettorossi, VE, The role of GABA in NMDA-dependent long term depression (LTD) of rat medial vestibular nuclei, *Brain Res.,* 699,183,1995.

37. Guastella, J, Nelson, N, Nelson, H, Czyzyk, L, Keynan, S, Miedel, MC, Davidson, N, Lester, HA, and Kanner, BI, Cloning and expression of a rat brain GABA transporter, *Science,* 249,1303,1990.

38. Hauglie-Hanssen, E, Intrinsic neuronal organization of the vestibular nuclear complex in the cat. A Golgi study, *Adv. Anat. Embryol. Cell Biol.,* 40,1,1968.

39. Heizmann, CW, Parvalbumin, an intracellular calcium-binding protein: Distribution, properties and possible roles in mammalian cells, *Experientia,* 40,910,1984.

40. Heizmann, CW, and Hunziker, W, Intracellular calcium-binding proteins: more sites than insights, *Trends. Biochem. Sci.,* 16,98,1991.

41. Helwig, D, and Cohen, B, L-baclofen and the VOR before and after nodulo-uvulectomy; GABA receptor hypersensitivity in the vestibular nuclei? *Soc. Neurosci. Abstr.,* 14,334,1988.

42. Highstein, SM, Goldberg, JM, Moschovakis, AK, and Fernandez, C, Secondary neurons. Inputs from regularly and irregularly discharging vestibular nerve afferents to secondary neurons in the vestibular nuclei of the squirrel monkey: II. Correlation with output pathways of secondary neurons, *J. Neurophysio.,* 58,714,1987.

43. Highstein, SM, and Ito, M, Differential localization within the vestibular nuclear complex of the inhibitory and excitatory cells innervating IIIrd nucleus oculomotor neurons in rabbit., *Brain Res.,* 29,355,1971.

44. Hironaka, T, Morita, Y, Bagihira, S, Tateno, E, Kita, H, and Tohyama, M, Localization of GABA_B receptor alpha-1 subunit mRNA-containing neurons in the lower brainstem of the rat, *Mol. Brain Res.,* 7,335,1990.

45. Holstein, GR, Martinelli, GP, and Cohen, B, Immunocytochemical visualization of L-baclofen-sensitive GABAB binding sites in the medial vestibular nucleus, in *Sensing and controlling motion: Vestibular and sensorimotor function,* B. Cohen, D.L. Tomko, and F. Guedry, Eds., New York: The New York Academy of Sciences, p. 933,1992.

46. Holstein, GR, Martinelli, GP, and Cohen, B, Visualization of L-baclofen sensitive binding sites in the medial vestibular nucleus, *Ann. NY Acad. Sci.,* in press.,1991.

47. Holstein, GR, Martinelli, GP, and Cohen B, L-Baclofen-sensitive GABA_B binding sites in the medial vestibular nucleus localized by immunocytochemistry, *Brain Res.,* 581,175,1992.

48. Holstein, GR, Martinelli, GP, and Cohen, B, The ultrastructure of GABA-immunoreactive vestibular commissural neurons related to velocity storage in the monkey, *Neuroscience,* 93,171,1999.

49. Holstein, GR, Martinelli, GP, and Cohen, B, mmunocytochemical localization of L-baclofen-sensitive GABA_B binding sites in cerebellar cortex, *Soc. Neurosci. Abstr.,* 18,853,1992.

50. Holstein, GR, Martinelli, GP, Degen, JW, and Cohen, B, GABAergic neurons in the primate vestibular nuclei, *Ann. N.Y. Acad. Sci,* 781,443,1996.

51. Holstein, GR, Martinelli, GPT, Reis, ED, and Pasik, P, Immunogold localization of probable GABA_B receptor sites by monoclonal anti-baclofen antibodies, *Soc. Neurosci. Abstr.,* 15,771,1989.

52. Houser, CR, Barber, RP, and Vaughn, JE, Immunocytochemical localization of glutamic acid decarboxylase in the dorsal lateral vestibular nucleus: Evidence for an intrinsic and extrinsic GABAergic innervation, *Neurosci. Lett.,* 47,413,1984.

53. Ichimiya, Y, Emson, PC, Mountjoy, CQ, Lawson, DEM, and Iizuka, R, Calbindin-immunoreactive cholinergic neurons in the nucleus basalis of Meynert in Alzheimer-type dementia, *Brain Res.,* 499,402,1989.

54. Ikegaki, N, Saito, N, Hashima, M, and Tanaka, C, Production of specific antibodies against GABA transporter subtypes (GAT1, GAT2, GAT3) and their application to immunocytochemistry, *Mol. Brain Res.,* 26,47,1994.

55. Johnston, GAR, Multiplicity of GABA receptors, in *Receptor Biochemistry and Methodology,* J.C. Venter and L.C.Harrison, Eds. 1986, *New York: Liss.,* p. 57,1986.

56. Jursky, F, Tamura, S, Tamura, A, Mandiyan, S, Nelson, H, and Nelson, N, Structure, function and brain localization of neurotransmitter transporters, *J. Exp. Biol.,* 196,283,1994.

57. Kamphuis, W, Wadman, WJ, Huisman, E, Heizmann, CW, and Lopes da Silva, FH, Kindling induces changes in parvalbumin immunoreactivity in rat hippocampus and its relation to long-term decrease in GABA immunoreactivity, *Brain Res.,* 479,119,1989.

58. Kasahara, M, and Uchino, Y, Selective mode of commissural inhibition induced by semicircular canal afferents on secondary vestibular neurones in the cat, *Brain Res.,* 34,366,1971.

59. Katz, E, DeJong, JMBV, Buettner-Ennever, J, and Cohen, B, Effects of midline medullary lesions on velocity storage and the vestibulo-ocular reflex, *Exp. Brain Res.,* 87,505,1991.

60. Kevetter, GA, Pattern of selected calcium-binding proteins in the vestibular nuclear complex of two rodent species, *J. Comp. Neurol.,* 365,575,1996.

61. Kiyama, H, Seto-Ohshima, A, and Emson, PC, Calbindin D28k as a marker for the degeneration of the striatonigral pathway in Huntington's disease, *Brain Res.,* 525,209,1990.

62. Kosaka, T, Kosaka, K, Nakayama, T, Hunziker, W, and Heizmann, CW, Axons and axon terminals of cerebellar Purkinje cells and basket cells have higher levels of parvallbumin immunoreactivity than somata and dendrites: Quantitative analysis by immunogold labeling, *Exp. Brain Res.,* 93,483,1993.

63. Kumoi, K, Saito, N, and Tanaka, C, Immunohistochemical localization of g-aminobutyric acid- and aspartate-containing neurons in the guinea pig vestibular nuclei, *Brain Res.,* 416,22,1987.

64. Langosch, D, Becker, CM, and Betz, H, The inhibitory glycine receptor: a ligand-gated channel of the central nervous system, *Eur. J. Biochem.,* 194,1,1990.

65. Liu, Q-R, Lopez-Corcuera, B, Mandiyan, S, Nelson, H, and Nelson, N, Molecular characterization of four pharmacologically distinct GABA transporters in mouse brain, *J. Biol. Chem.,* 268,2106,1993.

66. Liu, Q-R, Lopez-Corcuera, B, Mandiyan, S, Nelson, H, and Nelson, N, Cloning and expression of a spinal cord- and brain-specific glycine transporter with novel structural features, *J. Biol. Chem.,* 268,22802,1993.

67. Liu, Q-R, Mandiyan, S, Nelson, H, and Nelson, N, A family of genes encoding neurotransmitter transporters, *Proc. Natl. Acad. Sci.,* USA 89,6639,1992.

68. Lorente de Nó, R, Vestibulo-ocular reflex arc, *Arch. Neurol. Psychiat.,* (Chic.) 30,245,1933.

69. Lowenstein, DH, Miles, MF, Hatam, F, and McCabe, T, Up-regulation of calbindin-D28k mRNA in the rat hippocampus following focal stimulation of the perforant path, *Neuron,* 6,627,1991.

70. Mabjeesh, N, Frese, M, Rauen, T, Jeserich, G, and Kanner, BI, Neuronal and glial gamma-aminobutyric acid transporters are distinct proteins, *FEBS,* 299,99,1992.

71. Martinelli, GP, Holstein, GR, Pasik, P, and Cohen, B, Monoclonal antibodies for ultrastructural visualization of L-baclofen-sensitive GABA$_B$ receptor site, *Neuroscience,* 46,23,1992.

72. McCrea, RA, Strassman, A, and Highstein, SM, Anatomical and physiological characteristics of vestibular neurons mediating the vertical vestibulo-ocular reflexes of the squirrel monkey, *J. Comp. Neurol.,* 264,571,1987.

73. McCrea, RA, Strassman, A, May, E, and Highstein, SM, Anatomical and physiological characteristics of vestibular neurons mediating the horizontal vestibulo-ocular reflex in the squirrel monkey, *J. Comp. Neurol.,* 264,547,1987.

74. McCrea, RA, Yoshida, K, Berthoz, A, and Baker, R, Eye movement related activity and morphology of second order vestibular neurons terminating in the cat abducens nucleus, *Exp. Brain Res.,* 40,468,1980.

75. Mugnaini, E, and Oertel, WH, An atlas of the distribution of GABAergic neurons and terminals in the rat CNS as revealed by GAD immunohistochemistry, in *Handbook of Chemical Neuroanatomy,* A. Bjorklund and T. Hokfelt, Eds. Amsterdam: Elsevier/North Holland, p. 436,1985.

76. Mugnaini, E, and Walberg, F, An experimental electron microscopical study on the mode of termination of cerebellar cortico-vestibular fibers in the cat lateral vestibular nucleus (Deiters' nucleus), *Exp. Brain Res.*, 4,212,1967.

77. Najlerahim, A, Harrison, PJ, Barton, AJL, Heffernan, J, and Pearson, RCA, Distribution of messenger RNAs encoding the enzymes glutaminase, aspartate aminotransferase and glutamic acid decarboxylase in rat brain, *Mol. Brain Res.*, 7,17,1990.

78. Nelson, N, Lill, H, Porters and neurotransmitter transporters, *J. Exp. Biol.*, 196,213,1994.

79. Nomura, I, Senba, E, Kubo, T, Shiraishi, T, Matsunaga, T, Matsunaga, T, Tohyama, M, Shiotani, Y, and Wu, J.Y, Neuropeptides and a-aminobutyric acid in the vestibular nuclei of the rat: an immunocytochemical analysis, I. Distribution, *Brain Res.*, 311,109,1984.

80. Obata, K, Ito, M, Ochi, R, and Sato, N, Pharmacological properties of the postsynaptic inhibition by Purkinje cell axons and the action of a-aminobutyric acid on Deiters neurones, *Exp. Brain Res.*, 4,43,1967.

81. Obata, K, and Takeda, K, Release of g-aminobutyric acid into the fourth ventricle induced by stimulation of the cat's cerebellum, *J. Neurochem.*, 16,1043,1969.

82. Otsuka, M, Obata, K, Miyata, Y, and Tanaka, Y, Measurement of g-aminobutyric acid in isolated nerve cells of cat central nervous system, *J. Neurochem.*, 18,287,1971.

83. Ottersen, OP, and Storm-Mathisen, J, Glutamate- and GABA-containing neurons in the mouse and rat brain, as demonstrated with a new immunocytochemical technique, *J. Comp. Neurol.*, 229,374,1984.

84. Ottersen, OP, Storm-Mathisen, J, and Somogyi, P, Colocalization of glycine-like and GABA-like immunoreactivities in Golgi cell terminals in the rat cerebellum: a postembedding light and electron microscopic study, *Brain Res.*, 450,342,1988.

85. Plogman, D, and Celio, MR, Parvalbumin does not influence the activity of glutamate decarboxylase, *Brain Res.*,1990.

86. Precht, W, Schwindt, PC, and Baker, R, Removal of vestibular commissural inhibition by antagonists of GABA and glycine, *Brain Res.*, 62,222,1973.

87. Probst, A, and Cortes, R, palacios, JM, The distribution of glycine receptors in the human brain. A light microscopic autoradiographic study using [3H]strychnine, *Neuroscience,* 17,11,1986.

88. Radian, R, and Kanner, BI, Stoichiometry of sodium- and chloride-coupled-gamma-aminobutyric acid transport by synaptic membrane vesicles isolated from rat brain, *Biochemistry*, 22,1236,1983.

89. Radian, R, Otterson, OP, Storm-Mathisen, J, Castel, M, and Kanner, BI, Immunocytochemical localization of the GABA transporter in rat brain, *J. Neurosci.*, 10,1319,1990.

90. Ramón y Cajal, S, *Histologie du Systeme Nerveux de l`Homme et des Vertébrés.* Vol. 1, Paris: Maloine,1909.

91. Raphan, T, Matsuo, V, and Cohen, B, Velocity storage in the vestibulo-ocular reflex (VOR), *Exp. Brain Res.*, 35,229,1979.

92. Resibois, A, and Rogers, JH, Calretinin in rat brain: an immunohistochemical study, *Neurosci.*, 46,101,1992.

93. Richards, JG, Schoch, P, Hartig, P, Takacs, B, and Mohler, H, Resolving $GABA_A$/benzodiazepine receptors: cellular and subcellular localization in the CNS with monoclonal antibodies, *J. Neurosci.*, 7,1866,1987.

94. Roffler-Tarlov, S, and Tarlov, E, Reduction of GABA synthesis following lesions of inhibitory vestibulotrochlear pathway, *Brain Res.*, 91,326,1975.

95. Sans, N, Moniot, B, and Raymond, J, Distribution of calretinin mRNA in the vestibular nuclei of rat and guinea pit and the effects of unilateral labyrinthectomy: a non-radioactive *in situ* hybridization study, *Mol. Brain Res.*, 28,1,1995.

96. Schloss, P, Püschel, AW, and Betz, H, Neurotransmitter transporters: new members of known families, *Curr. Opin. Cell Biol.*, 6,595,1994.

97. Schmidt-Kastner, R, Meller, D, and Eysel, UT, Immunohistochemical changes of neuronal calcium-binding proteins parvalbumin and calbindin-D-28k following unilateral deafferentation in the rat visual system, *Exp. Neurol.*, 117,230,1992.

98. Schwarz, DWF, Schwarz, IE, and Fredrickson, JM, Fine structure of the medial and descending vestibular nuclei in normal rats and after unilateral transection of the vestibular nerve, *Acta Otolaryngol*, 84,79,1977.

99. Sequier, JM, Richards, JG, Maiherbe, P, Price, GW, Mathews, S, and Mohler, H, Mapping of brain areas containing RNA homologous to cDNAs encoding the a and B subunits of the rat $GABA_A$ gamma-aminobutyrate receptor, *Proc. Natl. Acad. Sci.,* U.S.A. 85,7815,1988.

100. Shimazu H, and Precht W Inhibition of central vestibular neurons from the contralateral labyrinth and its mediating pathway, *J. Neurophysiol.,* 29,467,1966.

101. Sloviter, RS, Calcium binding (calbindin D-28k) and parvalbumin immunocytochemistry: localization in the rat hippocampus with specific reference to the selective vulnerability of hippocampal neurons to seizure activity, *J. Comp. Neurol.,* 280,183,1989.

102. Smith, KE, Borden, LA, Hartig, PR, Branchek, T, and Weinshank, RL, Cloning and expression of a glycine transporter reveal colocalization with NMDA receptors, *Neuron,* 8,927,1992.

103. Spencer, RF, and Baker, R, GABA and glycine as inhibitory neurotransmitters in the vestibulo-ocular reflex, in *Sensing and Controlling Motion: Vestibular and Sensorimotor Function,* B. Cohen, F. Guedry, and D. Tomko, Eds. 1992, The New York Academy of Sciences, p. 602,1992.

104. Spencer, RF, Wang, S-F, and Baker, R, The pathways and functions of GABA in the oculomotor system, *Prog. Brain Res.,* 90,307,1992.

105. Spencer, RF, Wenthold, RJ, and Baker, R, Evidence for glycine as an inhibitory neurotransmitter of vestibular, reticular, and prepositus hypoglossi neurons that project to the cat abducens nucleus, *J. Neurosci,* 9,2718,1989.

106. Sutherland, MK, Wong, L, Somerville, MJ, Yoong, LK, Bergeron, C, Parmentier, M, and McLachlan, DR, Reduction of calbindin-28k mRNA levels in Alzheimer as compared to Huntington hippocampus, *Mol. Brain Res.,* 18,32,1993.

107. Taal, W, and Holstege, JC, GABA and glycine frequently colocalize in terminals on cat spinal motoneurons, *Neuroreport,* 5,2225,1994.

108. Uhl, GR, and Hartig, PR, Transporter explosion: update on uptake, *Trends Pharmacol. Sci.,* 13,421,1992.

109. Walberg, F, and Jansen, J, Cerebellar corticovestibular fibers in the cat, *Exp. Neurol.,* 3,32,1961.

110. Walberg, F, Ottersen, OP, and Rinvik, E, GABA, glycine, aspartate, glutamate and taurine in the vestibular nuclei: An immunocytochemical investigation in the cat, *Exp. Brain Res.,* 79,547,1990.

111. Wassef, M, Cholley, B, Heizmann, CW, and Sotelo, C, Development of the olivocerebellar projection in the rat. II. Matching of the developmental compartmentations of the cerebellum and inferior olive through the projection map, *J. Comp. Neurol.,* 323,537,1992.

112. Wearne, S, Raphan, T, and Cohen, B, Contribution of vestibular commissural pathways to velocity storage and spatial orientation of the angular vestibulo-ocular reflex, *J. Neurophys.,* 78,1193,1997.

113. Wearne, S, Raphan, T, and Cohen, B, Nodulo-uvular control of central vestibular dynamics determines spatial orientation of the angular vestibulo-ocular reflex, in *New Directions in Vestibular Research,* S.M. Highstein, B. Cohen, and J.A. Büttner-Ennever, Eds., New York Academy of Sciences, New York,1996.

114. Wentzel, PR, deZeeuw, CI, Holstege, JC, and Gerrits, NM, Colocalization of GABA and glycine in the rabbit oculomotor nucleus, *Neurosci. Letts.,* 164,25,1993.

115. Wisden, W, Morris, BJ, Darlison, MG, Hunt, SP, and Bernard, EA, Localization of $GABA_A$ receptor a-subunit mRNAs in relation to receptor subtypes, *Mol. Brain Res.,* 5,305,1989.

116. Wisden, W, Morris, BJ, Darlison, MG, Hunt, SP, and Bernard, EA, Distinct $GABA_A$ receptor a-subunit mRNAs show differential patterns of expression in bovine brain, *Neuron,* 2,937,1988.

117. Yamada, T, McGeer, PL, Baimbridge, KG, and McGeer, EG, Relative sparing in Parkinson's disease of *substantia nigra* dopamine neurons containing calbindin-D28k, *Brain Res.,* 526,303,1990.

118. Yamagishi, M, Ishizuka, YF, M., Nakamura, H, Igarashi, S, Nakano, Y, and Kuwano, R, Distribution of calcium binding proteins in sensory organs of the ear, nose and throat, *Acta Otolaryngol. (Stockholm) Suppl.,* 506,85,1993.

119. Zafra, F, Aragon, C, Olivares, L, Danbolt, NC, Gimenez, C, and Storm-Mathisen, J, Glycine transporters are differentially expressed among CNS cells, *J. Neurosci.,* 15,3952,1995.

8 Modulatory Effects of Monoamines on Central Vestibular Neurons: Possible Functional Implications

Nicolas Vibert, Mauro Serafin, Catherine De Waele, Alexander Babalian, Michel Mühlethaler, and Pierre-Paul Vidal

CONTENTS

8.1 INTRODUCTION

8.1.1 GOALS OF THE CHAPTER

The vestibular nuclei, located in the caudal brainstem, are known to be strongly involved in the sensorimotor transformations underlying gaze and posture stabilization in vertebrates.[6] Indeed, the central vestibular neurons receive most of the sensory vestibular inputs coming from the ipsilateral labyrinth, and are also fed with visual and proprioceptive information. They are moreover the targets of numerous cortical, cerebellar, and reticular projections, and also receive commissural fibers coming from the other vestibular complex. The second-order vestibular neurons integrate all these signals to elaborate an internal, tridimensional representation of the position and movements of the head in space. This representation is then used to stabilize gaze and posture through various vestibulo-oculomotor and vestibulo-spinal synergies. Therefore, in order to understand the neural mechanisms of these sensorimotor transformations, it was necessary to get some precise knowledge of the physiological and pharmacological properties of central vestibular neurons.

During the last 20 years, several groups have used *in vivo* preparations to demonstrate that these neurons were sensitive to most monoaminergic transmitters. We and others have recently confirmed these data, using both *in vitro* and *in vivo* techniques. The aim of this chapter is twofold: first, to summarize the present state of knowledge on this particular topic and, second, to try to evaluate the functional implications of the sensitivity of central vestibular neurons to monoamines.

8.1.2. THE MONOAMINES ARE MODULATORY TRANSMITTERS IN THE CENTRAL NERVOUS SYSTEM

Apart from the various neuroactive peptides, classical neurotransmitters of low molecular weight can be divided in two main groups. The excitatory and inhibitory amino acids, which include aspartate, glutamate, GABA, and glycine, seem to be responsible for about 90 percent of chemical neurotransmission in the brain.[59] They would be involved in most of the fast synaptic events, since they mainly activate postsynaptic, ionotropic receptors. The five monoamines (histamine, dopamine, serotonin, noradrenaline, and adrenaline) are included in the second group with acetyl-choline. In contrast to amino acids, they generally have more diffuse and moderate effects on central neurons. Most of them can only activate metabotropic receptors acting through second messenger systems, and have therefore much slower actions on the neuronal activity.[29]

Recent anatomical and physiological studies have confirmed that this distinction was also valid in the vestibular system. The main transmitter released by the sensory vestibular afferents, as well as the excitatory second-order vestibular neurons, is an excitatory amino acid.[73,102] The cerebellar and commissural inhibitory fibers reaching vestibular neurons seem to mainly release GABA and glycine.[102] Electrophysiological studies have furthermore shown that central vestibular neurons were generally much more sensitive to amino acids than to any other type of receptor agonists.[23,98]

In spite of their rather weak effects on neuronal discharges, the physiological importance of monoamines should not be underestimated. The monoaminergic neuronal systems are essential, in that they are able to precisely modulate and coordinate the physiological activities of large cerebral regions according to the behavioral state of the animal. Most of the common neurological disorders (such as schizophrenia or Parkinson's disease) are deeply related to dysfunctions in one or more of the aminergic systems. In the vestibular system, various agonists and antagonists of the monoaminergic receptors are also commonly used to treat vertigos and motion sickness,[40,72,105] or to accelerate vestibular compensation following surgical sections of the vestibular nerve.[85] A recent biochemical study has furthermore shown that, at least in the medial vestibular nucleus, turnover rates of monoaminergic metabolites were compatible with significant monoaminergic activity.[16] A better knowledge of the modulatory effects of monoamines on vestibular neurones should supply a better understanding of vestibular-related pathologies, and could lead to the elaboration of more efficient pharmaceutical treatments.

8.1.3 TWO MAIN FUNCTIONAL CLASSES OF NEURONS IN THE MEDIAL VESTIBULAR NUCLEUS

This chapter includes the results of *in vitro* studies on slices, designed in order to precisely define the actions of monoaminergic agonists on the different cell types present in the medial vestibular nucleus (MVN).[81,98,99] Using intracellular recordings in guinea pig coronal brain slices containing the vestibular nuclei, we have demonstrated two major cell types among MVN neurons (MVNn) differing by their intrinsic membrane properties (Figure 8.1).[79,80] Type A MVNn (about 30 percent) were characterized by a large, single afterhyperpolarization (AHP) and a transient rectification due

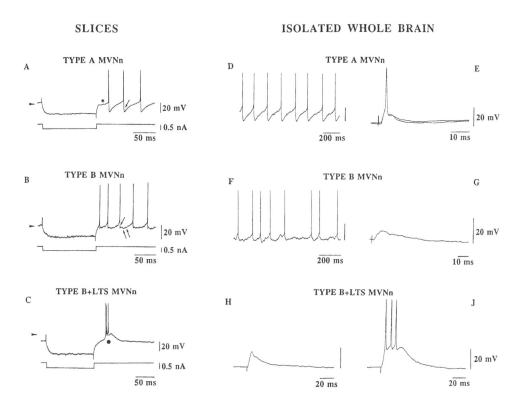

FIGURE 8.1 Intracellular recordings of the three main types of MVNn in slices (left side) and in the IWB (right side).

(A – C). Intracellular recordings showing the responses of MVN neurons to hyperpolarizing current pulses in slices. From top to bottom are demonstrated: first a type A MVNn (A) displaying transient rectification due to the activation of an I_A-like current (asterisk) and a single, deep AHP (arrow); second, a type B MVNn (B) with its biphasic AHP (arrow and double arrow); third, a type B+LTS neuron (C) characterized by the presence of a low-threshold calcium spike (dot). In this part of the figure, arrowheads indicate the level of the cell's resting membrane potential.

(D – J). Intracellular recordings of identified, second-order MVNn of all three main cell types in the IWB. From top to bottom are displayed:
- a regular spontaneous discharge (D), and an excitatory monosynaptic potential induced by stimulation of the ipsilateral vestibular nerve (E), recorded in a type A second-order neuron ;
- an irregular spontaneous discharge (F), and an excitatory monosynaptic potential induced by stimulation of the ipsilateral vestibular nerve (G), recorded in a type B second-order neuron ;
- an irregular spontaneous discharge (H), and excitatory monosynaptic potentials induced by stimulations of the ipsilateral vestibular nerve (J), in a type B+LTS second-order neuron. When the neuron is hyperpolarized, the same vestibular stimulation can induce either a single synaptic potential (J, lower trace) or a low-threshold calcium spike (LTS) which superimposes on the synaptic potential (J, upper trace).

to an I_A-like potassium current, whereas type B MVNn (about 50 percent) displayed a biphasic AHP and sodium-dependent subthreshold plateau potentials, but no transient rectification. In addition, about one quarter of type B MVNn, namely type B+LTS MVNn, displayed calcium-dependent, low-threshold spikes (LTS). A third, non-homogeneous class of cells (type C MVNn), could not be fitted into either one of the two main classes (about 20 percent). Whatever their type, most MVNn exhibited a spontaneous, regular resting discharge in slices, and their functional identification is very complex in this kind of preparation. It was therefore difficult to link our slice-based classification of MVNn with the results of *in vivo* extracellular recordings segregating horizontal canal-related vestibular neurons in tonic, regular and kinetic, irregular neurons.[49,82,83]

Following this first description of the physiological properties of MVNn, we have therefore begun to study the vestibulo-oculomotor pathways on a new, *in vitro* preparation of isolated, perfused whole brain (IWB) of guinea pig.[4,55] This technique allows stable intracellular recordings of MVNn, while the observed cells can be functionally identified as in acute *in vivo* preparations. We were able to retrieve in the whole brain (Figure 8.1) MVNn of all the cell types that were defined in slices (A, B, B+LTS, and C MVNn). Contrary to what happens in slices where spontaneous synaptic drive is lacking, the regularity of MVNn resting discharges is highly variable in the IWB. Interestingly, most of the type A MVNn had a regular firing rate, while type B MVNn generally had a more irregular one. This observation is important to bridge the gap between the results obtained on slices and on the alert animal: the type A MVNn apparently correspond to the tonic cells identified *in vivo*, while type B MVNn would be the kinetic ones. This correlation between the *in vitro*- and *in vivo*-based classifications was reinforced by the fact that in the isolated whole brain, numerous neurons of both cell types were monosynaptically activated following eletrical stimulations of the ipsilateral vestibular nerve, and could thus be identified as second-order MVNn (Figure 8.1).

The functional meaning of this segregation of MVNn is still a matter of debate. *In vivo*, the kinetic, irregular neurons have a higher sensitivity to acceleration than the tonic neurons. The coexistence of these two types of neurons could therefore underlie a functional separation of the networks controlling gaze and posture in frequency-tuned channels.[5,26,48] Type A MVNn would only be activated by low-frequency stimuli and would mainly be involved in static postural and oculomotor control, whereas type B MVNn seem to be more able to follow high-frequency head movements and would implement the dynamic responses. An alternative hypothesis states that, due to the different biomechanical properties of the eyes compared with the other segments of the body (neck, trunk, limbs), the tonic MVNn would mainly be restricted to the oculomotor system, while the kinetic ones would be more involved in vestibulo-spinal control.[11,28,34] It is noteworthy, however, that irregular, kinetic MVNn could also be used to input the velocity storage integrator included in the vestibulo-oculomotor system.[2] Indeed, this proposition nicely fits with the presence in B MVNn of a subthreshold, persistent sodium conductance. In the isolated whole brain, this conductance can produce long plateau potentials in response to a single shock applied on the vestibular nerve,[4,80] which could partly underlie the integrative processes.

8.2 THE BRAIN MONOAMINERGIC SYSTEMS IN VERTEBRATES: A BRIEF SUMMARY

Only five main monoaminergic molecules are functioning as neurotransmitters in vertebrates. The three catecholamines (dopamine, noradrenaline, and adrenaline) are all synthesized from the amino acid tyrosine. In contrast, serotonin derives from tryptophane, whereas histamine is produced by a decarboxylation of histidine. Since noradrenaline and adrenaline act through the same set of specific receptors, data concerning them will be grouped together.

Each one of the monoamines is produced by small, specific subsets of neurons with very restricted anatomical locations. In each case, however, these few cells give rise to very diffuse axonal arborizations which altogether extend to almost every structure of the central nervous system. On the other hand, each monoaminergic transmitter can activate a whole set of various, specific receptors on target neurons (Table 8.1).

Most of them are metabotropic receptors, and can be located either postsynaptically or on presynaptic terminals. Therefore, each monoamine can have very different effects on neuronal activity in any particular brain structure, depending on the nature of the activated receptors and on their localization at the cellular level.

TABLE 8.1 Summary of the Main Classes of Monoaminergic Receptors with Prominent Characteristics

Name of Receptor	Type of Receptor	Main Transduction Mechanism	Main Localization
Histamine	H1metabotropic H2metabotropic H3metabotropic	activation of phospholipase C activation of adenylate-cyclase inhibition of phospholipase C?	postsynaptic postsynaptic presynaptic
Serotonin	5-HT 1metabotropic 5-HT 2metabotropic 5-HT 3ionotropic 5-HT 4metabotropic	inhibition of adenylate-cyclase activation of phospholipase C receptor includes a cationic channel activation of adenylate-cyclase	pre- and postsynaptic pre- and postsynaptic pre- and postsynaptic mostly postsynaptic
Dopamine	"D 1-like" metabotropic (D1 and D5) "D 2-like" (D2, D3 andmetabotropic D4)	activation of adenylate-cyclase inhibition of adenylate-cyclase,activation of phospholipase C or phospholipase A	pre- and postsynaptic pre- and postsynaptic (D3 mostly presynaptic)
Noradrenal	α1metabotropic α2metabotropic βmetabotropic	activation of phospholipase C inhibition of adenylate-cyclase activation of adenylate-cyclase	pre- and postsynaptic pre- and postsynaptic pre- and postsynaptic

8.2.1 THE CENTRAL HISTAMINERGIC SYSTEM

In Vertebrates, all histaminergic neurons of the central nervous system are grouped in the tubero-mammillary nucleus of the posterior hypothalamus. These histaminergic cells innervate almost every structure in the brain, except the cerebellum. Histamine released from diffuse, sparsed axon terminals modulates neural activities in very large cerebral regions. Most of its specific receptors are located on neurons, but some are also present on astrocytes and on blood vessels (where they would regulate the blood flow).[100] Behavioral studies have shown that the histaminergic system was mainly involved in the daily regulation of vigilance according to a well-defined, species-specific circadian rhythm. Administration of anti-histaminergic drugs decreases the level of vigilance, and often induces somnolence. Histamine is also involved in the control of neuroendocrine systems, and in the regulation of internal temperature and cerebral blood flow. Since all these activities are regulated according to the day-night alternance, histamine could play an essential role in the definition of intrinsic biological rhythms.[61]

Three types of metabotropic, histaminergic receptors have been identified in Table 8.1. The H_1 and H_2 receptors are postsynaptic receptors, whereas the H_3 receptors generally are presynaptic ones. H_1 receptors are positively coupled to phospholipase C and mostly induce phospholipid hydrolysis, whereas H_2 receptors are in most cases positively coupled to adenylate-cyclase. Both postsynaptic receptors generally induce neuronal excitation. H_3 receptors are often located on histaminergic terminals, where they exert a negative control on the synthesis and release of histamine. They may also inhibit the release of other transmitters at non-histaminergic axon terminals.[3]

8.2.2 THE CENTRAL SEROTONINERGIC SYSTEM

The serotoninergic neurons of the central nervous system are located within the brainstem reticular formation. These cells are grouped in eight distinct clusters, which include the different raphe nuclei. Their diffuse projections, and particularly those originating in the dorsal raphe nucleus, reach every part of the central nervous system. The serotoninergic system is involved in the control of various physiological functions, including arousal, feeding behavior, nociception, thermoregulation, and sexual activity. It seems to be also important for the regulation of emotional states. Indeed, clinical studies in man have shown that modifications of serotoninergic transmission were associated with various psychiatric and/or neurological pathologies, including depression, anxiety, and dementias. In numerous cases, aggressive behaviors and suicidal tendencies were apparently linked to persistent high serotonin concentrations in the brain[8].

The recent use of molecular biology techniques has led to the identification of at least 10 distinct, specific serotoninergic receptors. They have been classified in four main groups (Table 8.1), which include three types of metabotropic receptors ($5-HT_1$, $5-HT_2$, and $5-HT_4$) and one ionotropic receptor ($5-HT_3$). Several subtypes of both $5-HT_1$ and $5-HT_2$ receptors have been cloned and pharmacologically identified. Other genes corresponding to new serotoninergic binding sites have been identified, but their physiological characteristics are still unknown.[31] $5-HT_1$ receptors are characterized by their high affinity for serotonin (< 10 nM), and can be divided in three subtypes (the $5-HT_{1A}$, $5-HT_{1B}$, and $5-HT_{1D}$). Generally, all of them are negatively coupled with adenylate-cyclase, and can be located both pre- and postsynaptically. Presynaptic $5-HT_1$ receptors often inhibit the release of various neurotransmitters, including serotonin itself. $5-HT_2$ receptors also include three different receptor subtypes ($5-HT_{2A}$, $5-HT_{2B}$, and $5-HT_{2C}$). They are positively coupled with phospholipase C, and their activation generally increases the intracellular calcium concentration. They can also be located either pre- or postsynaptically. Postsynaptic $5-HT_2$ receptors generally induce a neuronal excitation through inactivation of potassium channels. The $5-HT_4$ receptors are positively coupled to adenylate-cyclase, and are mostly postsynaptic receptors. Their effects seem to be quite similar to those of $5-HT_2$ receptors. In contrast with the other monoaminergic receptors, the $5-HT_3$ binding sites include a cation-selective channel, and their activation results in short postsynaptic depolarizations. Most of them are presynaptically located, and might stimulate the release of various neurotransmitters, including serotonin itself.[111]

8.2.3 THE CENTRAL DOPAMINERGIC SYSTEM

Numerous studies have been devoted to the physiology of central dopaminergic pathways. Indeed, dopaminergic disorders underlie several well-known neurological pathologies, including Parkinson's disease, Huntington's chorea, and progressive supranuclear palsy. Three main dopaminergic pathways have been described in the central nervous system.[14]

- The nigro-striatal fibers originate in the substantia nigra and play an essential role in the control of locomotion and movement.

- The meso-cortico-limbic pathway includes dopaminergic neurons of the ventral tegmental area, and innervates all limbic structures (hippocampus, entorhinal cortex). These dopaminergic cells appear to be essentially involved in the regulation of emotional states.

- The tubero-infundibular pathway originates from dopaminergic neurons located in the hypothalamus, and participates in the control of hypophyseal activities. This dopaminergic system regulates prolactine concentration in the blood, and would strongly influence the hormonal control of reproductive activities.

Some smaller dopaminergic cell groups, with more restricted projection sites, have been furthermore identified in various brain structures. Dopaminergic neurons have been localized, for instance, in the olfactive bulb, the retina, the thalamus,[89] and the dorsal motor nucleus of the vagus nerve.[39]

Since 1979, two main types of dopaminergic receptors (D_1 and D_2) have been classically described[37]. Recent studies, however, have shown that dopamine could actually activate at least five distinct subtypes of metabotropic receptors[14]. These five receptors can be grouped into two classes according to their pharmacological and structural homologies with the prototypical D_1 and D_2 binding sites defined in 1979 (Table 8.1). The "D_1-like" receptors include the D_1 and D_5 subtypes. They are, in general, positively coupled with adenylate-cyclase, and can be both pre- and postsynaptically localized. In presynaptic position, they mostly stimulate the release of various transmitters. The "D_2-like" receptors include the D_2, D_3, and D_4 subtypes. In most cases, they seem to be negatively coupled with adenylate-cyclase, and can be both pre- or postsynaptically located. The presynaptic ones apparently inhibit the release of various neurotransmitters in many different brain regions. On the other hand, postsynaptic "D_2-like" receptors generally hyperpolarize neurons by activating some of their potassium conductances.[97]

8.2.4. THE CENTRAL ADRENERGIC SYSTEM

Like the other monoamines, noradrenaline has quite diffuse effects on large neuronal populations, and acts through a whole set of specific metabotropic receptors. Most noradrenergic fibers in the brain originate from the locus coeruleus, located in the dorsal pons. The axons of these neurons innervate most of the forebrain, the cerebellum, and the dorsal half of the brainstem. More ventrally located noradrenergic cells innervate the rest of the brainstem and the hypothalamus. Noradrenaline can have multiple pre- and post-synaptic effects, and one of its main actions is to increase the signal-to-noise ratio of amino acid-mediated synaptic transmission;[106] therefore, it might play a key-role in selective attention processes. Noradrenaline also seems to be involved in the regulation of the sleep-wake cycle; it is one of the transmitters mediating thalamo-cortical activation processes, together with acetylcholine and serotonin.[53]

Recent biochemical and pharmacological studies have demonstrated that adrenergic receptors could be classified into three main groups: namely, the α_1, α_2 and β receptors (Table 8.1).[13,57] Alpha$_1$-Receptors are generally considered to include three or four distinct subtypes (named α_{1A} to α_{1D}). They all induce an increase of the intracellular calcium concentration through activation of phospholipase C. They can be presynaptically located, but are also known to induce direct neuronal excitation by inactivating voltage-dependent potassium channels.[53] Alpha$_2$-Receptors would include at least three distinct receptor subtypes. Each of them is generally negatively coupled with adenylate-cyclase, and can also be pre- or post-synaptically located. Presynaptic α_2 receptors generally decrease transmitter release, whereas activation of postsynaptic α_2 receptors generally hyperpolarizes neurons through activation of potassium conductances. Beta-receptors can also be divided in three distinct subypes (β_1 to β_3). They are all positively coupled with adenylate-cyclase. Presynaptic β-receptors are known to stimulate transmitter release, whereas the postsynaptic ones would mainly induce neuronal depolarizations.[13,86]

8.3 HISTAMINERGIC MODULATION OF CENTRAL VESTIBULAR NEURONS

Various behavioral data suggest that the central histaminergic system can strongly modulate vestibular function. First, histamine is known to be highly involved in the control of arousal and sleep, and several authors have shown that the horizontal vestibulo-ocular reflex (HVOR) was very sensitive to the state of alertness.[15,21] Second, both histamine itself and histaminergic ligands have been succesfully used in humans for the symptomatic treatment of vertigo and motion sickness. Because of its vasodilatating properties, histamine has been widely used in the past for the treatment of inner ear dysfunctions of vascular origin. In contrast, other histaminergic ligands like betahistine (which is both a partial H_1 agonist and an H_3 antagonist) seem to have a direct action on central vestibular structures.[20] Following this line of thought, the author's group has recently shown[108] that in the guinea pig, the gain of the HVOR was depressed following intraperitoneal injection of the pure H_3 antagonist thioperamide. Classical anti-histaminergic drugs like cinnarizine have also been extensively used to treat vestibular-related disorders, despite their sedative properties.[105]

Recent anatomical data have confirmed that the vestibular neurones were probably subjected to strong histaminergic influences. The histaminergic cells of the posterior hypothalamus send direct projections toward the whole vestibular complex.[88,91] The distribution of these histaminergic fibers shows spatial variations,with significantly heavier labeling in the medial and superior subnuclei.[96] Accordingly, autoradiographic and *in situ* hybridization studies have detected numerous H_1 and H_2 binding sites in all vestibular nuclei,[9,75] including the MVN.

Several *in vivo* electrophysiological studies have shown that the lateral and medial vestibular nuclei neurones could be both inhibited or excited by histamine or histaminergic agonists,[76,90] but the use of microiontophoretic techniques could not rule out the possibility of indirect effects. In slices, histamine mostly depolarizes medial vestibular neurons.[66,104] Using intracellular recordings, we have demonstrated that the three types of MVNn (A, B, and B+LTS neurons) previously identified in slices[79] were equally sensitive to this histaminergic action.[81] In the guinea pig slices, this depolarization was apparently due to the direct activation of postsynaptically located H_2 receptors on MVNn; indeed, the selective H_1 antagonist mepyramine, and the H_3 agonist and antagonist α-methyl-histamine and thioperamide were ineffective in this condition. However, according to Wang and Dutia[104] who worked on rat slices, the excitatory responses of MVN to histamine could be partially antagonized by the specific H_1 antagonist triprolidine. These excitatory effects, therefore, could be mediated by both H_1 and H_2 receptors. This discrepancy may be due to species differences in either the anatomical repartition or the pharmacological properties of H_1 histaminergic receptors.

Concerning H_3 receptors, which are mainly located on histaminergic terminals, *in vitro* recordings could not exclude their presence within the MVN. Indeed, the histaminergic axons reaching vestibular neurons were cut during the slicing procedure. The authors have therefore studied in alert animals the effect of selective perfusion of one of the vestibular complex with α-methyl-histamine or thioperamide.[108] Perfusion with the H_3 agonist induced postural and oculomotor changes similar to those observed following hemilabyrinthectomy, whereas thioperamide had no effect. This suggests that the histaminergic fibers reaching the vestibular nuclei indeed carry presynaptic H_3 receptors whose activation inhibits a tonic histamine release, thus decreasing the spontaneous discharge of vestibular neurons on the injected site. In contrast, the mechanism underlying the vestibuloplegic effects of systemic injection of thioperamide[108] (see above) remain to be elucidated, since it was impossible to determine the neuronal population involved.

Histamine seems to have a general excitatory influence on most medial vestibular neurons by activating H_1 and/or H_2 postsynaptic receptors, but would also control its own release through H_3 presynaptic receptors. The whole histaminergic system seems to be activated both by disturbances from the afferent inflow from the peripheral vestibular receptors (like unilateral vestibular deafferentation),[30] and, in case of central sensory conflicts, inducing motion sickness.[92] This activation may be a normal physiological response to the stress associated with such conditions, and would explain the clinical efficiency of histaminergic ligands in vestibular-related pathologies. The discovery of the vestibuloplegic effects of betahistine and thioperamide, which are both H_3 antagonists, opens the possibility for new, more efficient treatments of these disorders ;[108] in fact, in contrast to standard anti-histaminergic agents, thioperamide does not seem to induce drowsiness.[47]

8.4. SEROTONINERGIC MODULATION OF CENTRAL VESTIBULAR NEURONS

Unlike those conducted for histamine, very few clinical or behavioral studies have carefully examined serotoninergic actions on vestibular function. Only Ternaux and Gambarelli[95] have shown that intra-cerebroventricular injection of serotonin actually increased the gain of the vestibulo-ocular reflex in the rat. In fact, serotonin was initially considered as a modulator of central vestibular neurons entirely on the basis of anatomical data. Immunocytochemical studies have demonstrated the existence of a dense network of serotoninergic fibers innervating all parts of the vestibular complex,[27,87] which originate from the serotoninergic neurons of the dorsal raphe nucleus.[25] These data have been confirmed and extended by several autoradiographic and *in situ* hybridization studies that have shown the presence of 5-HT_{1A}, 5-HT_{1B} and 5-HT_2 receptors (or the corresponding mRNAs) in all subnuclei of the vestibular complex.[63,64,107]

In vivo electrophysiological studies have demonstrated that in the rat, the effects of iontophoretically applied serotonin varies among the different vestibular nuclei. Neurons of the lateral vestibular nucleus are mostly depolarized by 5-HT, but this depolarization is often preceded by a short hyperpolarization.[43] In contrast, in the medial and superior vestibular nuclei serotonin has been shown to have purely excitatory, purely inhibitory, or biphasic actions.[44,45] The excitatory responses are apparently mediated through 5-HT$_2$ receptors, whereas the inhibitory ones are due to the activation of 5-HT$_{1A}$ binding sites. Using *in vitro* extracellular recordings in slices, Johnston et al.[35] have confirmed that serotonin has both excitatory and inhibitory effects on the spontaneous discharge of MVNn, but they found a greater proportion of excitatory responses than reported in the corresponding *in vivo* study of Licata et al.[45]

Using intracellular recordings in guinea pig brainstem slices, the authors have recently studied the sensitivity of the three main, previously defined,[79] types of MVNn to bath applications of serotonin and various serotoninergic ligands.[98] Most MVNn (about 80 percent) were depolarized by serotonin, but the type B and B+LTS neurons were significantly more sensitive to this action than the type A neurons. Perfusion of serotonin in synaptic-uncoupling conditions (for instance, in the presence of tetrodotoxin) revealed that serotonin directly activated postsynaptic receptors on B MVNn whereas, in contrast, its excitation of type A MVNn was indirectly induced (Figure 8.2).

As described by Johnston et al.,[35] the depolarizing effects of serotonin could be reproduced with α-methyl-serotonin, known as a specific agonist of 5-HT$_2$ receptors, but were only partly

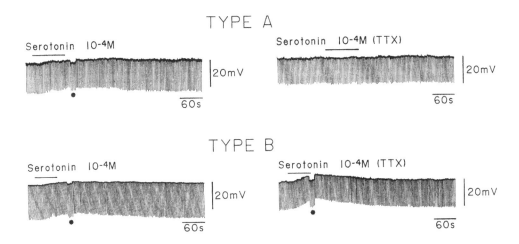

Figure 8.2 On slices, MVN neurons could be either directly or undirectly depolarized by serotonin.

Type A MVNn: In most of these neurons, the depolarization induced by serotonin in normal medium (left trace) disappears following addition of 1 μM tetrodotoxin (which blocks the spontaneous discharge of all neurons in the slice) in the perfusion solution (right trace). Type A MVNn are therefore only undirectly affected by serotonin. The asterisk on the left trace indicates where the cell has been transiently, manually clamped back to its original membrane potential, to show the decrease of the cell membrane resistance induced by serotonin.

Type B MVNn: In contrast, the depolarization induced by serotonin in normal medium (left trace) persists on most B MVNn in presence of tetrodotoxin (right trace). These cells seem to be the only one to express postsynaptic, serotoninergic receptors. In both cases, the serotonin-induced depolarization was associated with a decrease of the cell's membrane resistance (see the asterisks showing the localization of the manual clamps).

blocked by the specific antagonist ketanserin. Furthermore, these responses were mostly associated with a decrease of the recorded cells' membrane resistance, whereas depolarizations due to typical

5-HT$_2$ receptors generally induce a resistance increase by inactivating potassium conductances.[7,60] It is therefore tempting to link these depolarizing responses with several similar, serotonin-induced depolarizations which have been observed in various central nervous system structures, and whose pharmacological characteristics do not correspond to any one of the currently well-defined serotoninergic binding sites.[1] In addition to these excitatory effects, the author has also observed hyperpolarizing responses to serotonin in about 15 percent of both type A and type B cells. Due to the lower occurrence of these inhibitory effects, the pharmacological specificity of these direct postsynaptic responses could not be tested. According to the literature (see above), they were most likely due to the activation of 5-HT$_{1A}$ receptors.

The majority of neurons in the MVN appear to be primarily depolarized and excited by serotonin, following activation of "5-HT$_2$-like" receptors. This activating effect will be more efficient on type B or B+LTS neurones, which are the only ones to carry the corresponding serotoninergic binding sites. If indeed "tonic" and "kinetic" vestibular neurons are part of distinct, frequency-selective networks (see the third paragraph of the introduction), serotonin could be more involved in the regulation of dynamic, vestibular-related synergies than in static postural and oculomotor control. It could increase the dynamic sensitivity of the vestibular system, allowing the animal to react rapidly to external stimulations.

8.5 CENTRAL VESTIBULAR NEURONS ARE SENSITIVE TO DOPAMINE

The sensitivity of central vestibular neurons to dopamine has only been recently demonstrated. During the past 10 years, data from several behavioral and clinical studies have actually suggested that the dopaminergic system could be involved in the regulation of vestibular function. For instance, Piribedil (Trivastal®), a specific agonist of "D$_2$-like" receptors,[17] has been shown to reduce most of the classical symptoms associated with some of the age-related cochleo-vestibular syndromes affecting elderly people.[24] It is now commonly used in the treatment of these cochleo-vestibular pathologies. Nardo et al.[56] have shown that dihydroergocristine, a non-specific dopaminergic agonist, reduces the spontaneous nystagmus following hemilabyrinthectomy in the guinea pig. Their findings were confirmed by Petrosini and Dell'Anna,[65] who demonstrated that vestibular compensation was enhanced following systemic injections of "D$_2$-like" dopaminergic agonists.

The problem is, however, that the vestibular nuclei have not been shown to receive any specific dopaminergic innervation.[41,54] Despite these negative results, Bouthenet et al.[10] have demonstrated the presence of dopaminergic D$_2$ receptor mRNA in the medial vestibular nucleus using *in situ* hybridization techniques. A recent autoradiographic study has furthermore detected a weak density of D$_2$ receptors in the vestibular complex of the rat, mostly concentrated in the MVN.[110] This discrepancy could be due to some technical limitations of former immunohistochemical studies, and/or to the fact that the rather weak dopaminergic innervation of the brainstem and spinal cord was generally given little attention.

The presence of specific dopaminergic binding sites on central vestibular neurones has been confirmed by several electrophysiological works. As early as 1973, Matsuoka et al.[50] had reported that the spontaneous discharge of vestibular neurons was increased following systemic injections of L-DOPA (a metabolic precursor of dopamine). A more recent study has shown that in the cat, the activity of central vestibular neurons was modulated by microiontophoretic ejections of dopamine.[32] *In vitro*, Gallagher et al.[23] have furthermore described depolarizing effects of dopamine on intracellularly recorded medial vestibular neurons in the rat.

In view of these data, the author designed a study in order to determine the precise effects of dopaminergic agonists on the different MVN cell types that we have identified in slices (A, B, and B+LTS neurons). In a normal medium or in the presence of tetrodotoxin (which blocks the firing of action potentials and evoked synaptic transmission in the slice), dopamine induces a membrane depolarization of about 75 percent of MVNn, associated with an increase of their membrane resistance; this holds true for all three cell types. We checked the pharmacological specificity of this depolarizing response using selective ligands of the various dopaminergic binding sites, and found that this depolarization was due to the activation of "D_2-like" receptors. The fact that dopamine depolarizes MVNn by acting on "D_2-like" receptors was somewhat surprising since, according to the literature, the activation of postsynaptic "D_2-like" receptors generally induces membrane hyperpolarizations.[97] Actually, the dopamine-induced depolarization corresponded to an indirect, presynaptic effect since, when synaptic transmission was blocked, in the presence of a high Mg^{2+} and low Ca^{2+}-containing solution, dopamine had in contrast a weak postsynaptic, hyperpolarizing action on all types of MVNn. This hyperpolarizing action also resulted from the activation of "D_2-like" receptors (Figure 8.3).

What is the mechanism underlying the depolarizing action of dopamine in normal medium? Our original hypothesis was that dopaminergic agonists, by acting on presynaptic "D_2-like" receptors, could inhibit in the slice a spontaneous, tetrodotoxin-resistant, tonic release of an inhibitory transmitter like GABA. Indeed, the vestibular complex is known to contain an important population of local GABAergic interneurons, which mediate the commissural inhibition coming from the contralateral vestibular nuclei.[71] In addition, a large number of GABAergic terminals are present in the vestibular nuclei.[42,101] GABA would be used as a neurotransmitter by the Purkinje cell fibers reaching vestibular neurons from the vestibulo-cerebellum,[12,33] and by some inhibitory axons coming from the contralateral inferior olive.[51] Spontaneous GABA release has been demonstrated to occur in the vestibular nuclei *in vitro*,[18,84] and we have confirmed its existence by applying bicuculline (a specific $GABA_A$ antagonist) on the slices, both in normal medium and in the presence of tetrodotoxin. We found in both cases that bicuculline induced a strong membrane depolarization of all MVNn, associated with an increase of their membrane resistance. In the continuous presence of bicuculline, the depolarizing effects of dopamine were no longer seen, and the responses were always hyperpolarizing (as in synaptic uncoupling conditions). These results are in agreement with our initial hypothesis: the dopamine-induced depolarizations were probably due to the activation of presynaptic "D_2-like" receptors located on GABAergic terminals, and to the subsequent removal of a tonic GABA release on MVNn (Figure 8.3). Such presynaptic, inhibitory effects of dopaminergic agonists on the release of various neurotransmitters have already been reported in several structures of the central nervous system.[86]

Briefly, dopamine can act on most MVNn in two distinct ways: by activating both pre- and postsynaptically-located "D_2-like" receptors. This sensitivity of MVNn could underlie a functional modulation of vestibular function by dopaminergic systems, and could explain why some dopaminergic agonists are efficient in the treatment of cochleo-vestibular pathologies. In particular, the dopaminergic inhibition of GABA release in the MVN will probably decrease the efficiency of inhibitory commissural connections linking together the two vestibular complexes. According to several authors,[22,36] this process would induce a decrease of both the gain and the time constant of the horizontal vestibulo-ocular reflex. The presynaptic, excitatory effect of dopamine would thus not contradict its direct, hyperpolarizing action on MVNn. Altogether, dopamine would mainly decrease the dynamic sensitivity of vestibularly driven synergies, as was suggested by the clinical actions of dopaminergic agonists. In line with the strong involvement of dopamine in motor control, such an effect would be useful in reducing the vestibular-related perturbations during organized motor behavior such as locomotion.

8.6 EFFECTS OF NORADRENALINE ON CENTRAL VESTIBULAR NEURONS

As for histamine, numerous behavioral studies have shown that noradrenaline was largely involved in the control of vestibular function. In particular, the noradrenergic system would play an important role in the regulation of the vestibulo-spinal[68] and vestibulo-ocular reflexes.[94] Unilateral lesion of

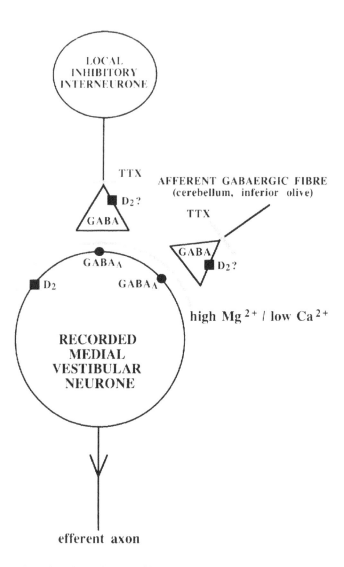

Figure 8.3 Summary of possible dopaminergic effects in the MVN.
An intracellularly-recorded MVNn is shown to receive GABAergic inputs either by a local inhibitory interneuron or by afferent GABAergic fibers coming, for instance, from the cerebellum or the inferior olive. In a high Mg^{2+}/low Ca^{2+} solution the recorded MVNn is synaptically isolated, as indicated by the grey half-circle. In the presence of tetrodotoxin, the firing of action potentials and evoked synaptic transmission are blocked, as indicated by the grey bars on the axons, but the spontaneous release of GABA by terminals is not.

the noradrenergic neurons of the locus coeruleus induces asymmetric postural responses in decerebrate animals. *In vivo*, the activity of these noradrenergic neurons is modulated by both vestibular and cervical proprioceptive stimulations.[69] One of the specific roles of noradrenaline would be to facilitate adaptive processes in both the vestibulo-spinal and the vestibulo-ocular systems.[52,70]

Until recently, very few noradrenergic neuronal endings had been found in the vestibular nuclei,[88] and noradrenaline was thought to affect the vestibular system mainly by acting on Purkinje cells in the vestibulo-cerebellum. However, a recent immunohistochemical study[78] has now described two separate fiber bundles originating from the locus coeruleus which project to the four subnuclei of the vestibular complex. The highest densities of noradrenergic fibers were found in the superior and lateral nuclei. These noradrenergic projections may directly affect information processing by vestibular neurons. Indeed, autoradiographic and immunocytochemical studies have revealed the presence of β and α_2 noradrenergic receptors in the vestibular complex[19,62,103]. In accordance with the differential repartition of noradrenergic terminals, β receptors were particularly abundant in the lateral and superior subnuclei. In contrast, α_2 receptors were mostly concentrated in the MVN[74,93]. Finally, mRNAs coding for the α_1, α_{2A} and α_{2C} receptor subtypes have been recently detected in all vestibular subdivisions by *in situ* hybridization techniques.[58,67,77]

Electrophysiological data confirm that noradrenaline can directly modulate the activity of central vestibular neurons. *In vivo* microiontophoretic injections in the cerebellectomized cat have shown that noradrenaline generally increased the resting activity of lateral vestibular nucleus neurons, whereas it mostly decreased the activity of MVNn.[38,109] In contrast, in the rat, Licata et al.[46] found that most lateral and superior vestibular nuclei neurons were inhibited by noradrenaline, mainly acting through α_2 receptors. Even if this discrepancy could be due to the different types of *in vivo* preparations used in these studies, we were interested in clarifying these rather contradictory data.

We have, therefore, decided to study the effects of noradrenaline and of several specific agonists of the different noradrenergic receptor subtypes, on the three main types of neurons defined in the medial vestibular nucleus (Figure 8.4).[98]

In normal medium, about 55 percent of MVNn are depolarized and excited by noradrenaline, while their membrane resistance decreases. Type A MVNn (which are excited in about 40 percent of the cases) are generally much less sensitive to this depolarizing action than type B or B+LTS MVNn (excited in about two-thirds of the cases). In contrast, about 20 percent of both A and B MVNn have their discharge rate decreased following bath-application of noradrenaline. Using selective agonists of the three main subtypes of noradrenergic receptors, we demonstrated that α_1, α_2, as well as β receptors were present in the MVN. Selective activations of each one of these receptor subtypes induced distinct responses of MVNn. Isoproterenol, a specific β agonist, mostly reproduced the depolarizing responses to noradrenaline; it depolarized about 60 percent of MVNn while decreasing their membrane resistance. Perfusion of L-phenylephrine, a selective α_1 agonist, also depolarized about 60 percent of the tested neurons, but this depolarization was associated with an increase of their membrane resistance. Finally, bath-application of clonidine, a specific α_2 agonist, hyperpolarized most MVNn, while decreasing their membrane resistance. While the hyperpolarizing effects of clonidine always persisted in conditions of synaptic uncoupling, a large proportion of the other responses were apparently indirectly mediated, and further studies should be done to clarify these points. What is clear is that, at least in the *in vitro* conditions, noradrenaline seems to affect MVNn mainly by activating either α_2 or β noradrenergic receptors (Figure 8.4), in accordance with autoradiographic and immunocytochemical studies.[19,74,103] The fact that, on slices, bath-applied noradrenaline never induced clear α_1-related effects suggests a presynaptic localization of α_1 receptors in the vestibular nuclei (as for H_3 histaminergic receptors, see above).

Bath application of noradrenaline can have various effects on MVNn through simultaneous activation of the three main subtypes of noradrenergic receptors. This direct modulation of the activity of central vestibular neurons could play an important role in the noradrenergic control of vestibular function revealed by behavioral studies. If "tonic" (putative type A) and "kinetic" (putative type B) vestibular neurons really correspond to distinct frequency-selective channels, the higher sensitivity of type B and B+LTS MVNn to noradrenaline would explain why this monoamine has been essentially involved in the control of dynamic vestibular synergies. As in other structures of the central nervous system,[106] noradrenaline could increase the signal-to-noise ratio of synaptic connections in the MVN. Such an effect could underlie the facilitatory effects of noradrenaline on vestibular adaptive processes, by increasing the sensitivity of central vestibular neurons to conflicting visual or proprioceptive signals.

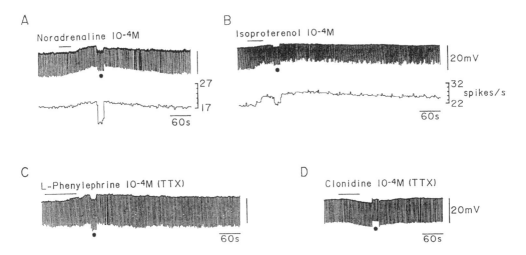

FIGURE 8.4 Typical responses of MVNn to the main noradrenergic agonists.

(A). Whatever their type, a majority of MVNn are slightly depolarized by noradrenaline in normal medium. This effect is generally associated with an increase of their spontaneous firing rate, and a decrease of the cell's membrane resistance. The asterisk indicates where the neuron has been transiently, manually clamped back to its initial membrane potential, to evaluate the change in membrane resistance induced by noradrenaline.

(B). As does noradrenaline itself, the specific b agonist, isoproterenol, mainly depolarizes all types of MVNn. This action is accompanied by a decrease of the cell's membrane resistance (see the manual clamp indicated by the asterisk).

(C). The specific a_1 agonist, L-phenylephrine, applied here in presence of 1 μM tetrodotoxin, generally induces a slight depolarization of MVNn. In contrast with noradrenaline, this effect is associated with an increase of the cell's membrane resistance (see the asterisk marking the transient, manual clamp).

(D). The specific a_2 agonist, clonidine, has a clear hyperpolarizing effect on all types of MVNn, associated with a decrease of their spontaneous discharge rate. This hyperpolarization probably results from the activation of postsynaptic receptors, since it always persists in presence of tetrodotoxin. The clonidine-induced action is generally concomitant with a decrease of the neuron's membrane resistance (see asterisk).

8.7 CONCLUSION

As pointed out in the introduction, the modulatory effects of monoamines on central vestibular neurons are much weaker than those of amino acids, or even of acetylcholine.[98] Due to the diffuse repartition pattern of their terminals, activation of monoaminergic neurons will probably simultaneously influence neuronal activity throughout the whole vestibular complex.

Histamine would regulate both the tonic activity and excitability of central vestibular neurons according to the main circadian rhythms of the animal, by acting on both type A and type B neurons. Dopamine seems to similarly affect all types of MVNn, but have a global inhibitory effect on vestibular function. On the other hand, serotonin and noradrenaline mainly act on type B and B+LTS neurons, and could be more involved in the control of dynamic vestibular responses. While serotonin apparently increases the responsiveness of the whole vestibular system to external stimulations, noradrenaline would have more subtle effects by facilitating the adaptive processes and plastic changes which continuously maintain the functional efficiencies of vestibulo-ocular and vestibulo-spinal reflexes.

The monoaminergic sensitivity of central vestibular neurons definitely opens new perspectives for further clinical research. We hope that some of the new, more selective agonists or antagonists of the various monoaminergic receptors detected in the vestibular nuclei will prove to be useful in the treatment of cochleo-vestibular syndromes.

ACKNOWLEDGMENTS

Our collaboration with the group of Dr. Mühlethaler was supported by a "Programme International de Coopération Scientifique" awarded by the Centre National de la Recherche Scientifique (CNRS), the Fond National Suisse pour la Recherche Scientifique (FNRS), and the French Ministère des Affaires Etrangères. We also received support from the Centre National d'Etudes Spatiales (CNES). We thank Mrs. D. Machard for her excellent technical assistance.

REFERENCES

1. Andrade, R, and Chaput, Y, 5-hydroxytryptamine$_4$-like receptors mediate the slow excitatory response to serotonin in the rat hippocampus, *J. Pharmacol. Exp. Therap.*, 257:930,1990.
2. Angelaki, DE, and Perachio, AA, Contribution of irregular semicircular canal afferents to the horizontal vestibuloocular response during constant velocity rotation, *J. Neurophysiol.* 69:996,1993.
3. Arrang, JM, Pharmacological properties of histaminergic receptor subtypes, *Cell Mol. Biol.*, 40: 273,1994.
4. Babalian, A, Vibert, N, Assié, G, Serafin, M, Mühlethaler, M, and Vidal, PP, Central vestibular networks in the guinea pig: functional characterization in the isolated whole brain in vitro, *Neuroscience,* 81: 405,1997.
5. Baker, R, Evinger, C, McCrea, RA, Some thoughts about the three neurons in the vestibuloocular reflex, *Ann. NY Acad. Sci.*, 374:171,1981.
6. Berthoz, A, Cooperation and substitution between the saccadic system and "reflexes" of vestibular origin: should we revise the notion of "reflex" ?, *Rev. Neurol. (Paris),* 145: 513,1989.
7. Bobker, DH, A slow excitatory postsynaptic potential mediated by 5-HT$_2$ receptors in nucleus prepositus hypoglossi, *J. Neurosci.,* 14:2428,1994.
8. Bonate, PL, Serotonin receptor subtypes: functional, physiological, and clinical correlates, *Clin. Neuropharmacol.,* 14: 1,1991.
9. Bouthenet, ML, Ruat, M, Sales, N, Garbarg, M, and Schwartz, JC, A detailed mapping of histamine H$_1$ receptors in guinea pig central nervous system established by autoradiography with [^{125}I]iodobolpyramine, *Neuroscience,* 26: 553,1988.
10. Bouthenet, ML, Souil, E, Martres MP, Sokoloff, P, Giros B, and Schwartz, JC, Localization of dopamine D$_3$ receptor mRNA in the rat brain using *in situ* hybridization histochemistry: Comparison with dopamine D$_2$ receptor mRNA, *Brain Res.,* 564:203,1991.
11. Boyle, R, Goldberg, JM, and Highstein, SM, Inputs from regularly and irregularly discharging vestibular nerve afferents to secondary neurons in the vestibular nuclei of the squirrel monkey. III. Correlation with vestibulospinal and vestibuloocular output pathways, *J. Neurophysiol.,* 68:471,1992.
12. Brodal, A, Anatomy of the vestibular nuclei and their connections, in: *Handbook of Sensory Physiology,* Vol. I (Kornhuber HH, Ed.), pp 240-351. Berlin-New York: Springer,1974.
13. Bylund, DB, Eikenberg, DC, Hieble, JP, Langer, SZ, Lefkowitz, RJ, Minneman, KP, Molinoff, PB, Ruffolo, RR, Jr, and Trendelenburg, U, International Union of Pharmacology nomenclature of adrenergic receptors, *Pharmacol. Rev.,* 46:121,1994.
14. Civelli, O, Bunzow, JR, and Grandy, DK, Molecular diversity of the dopamine receptors, *Annu. Rev. Pharmacol. Toxicol.,* 32:281,1993.
15. Collins, WE, and Guedry, FE, Arousal effects and nystagmus during prolonged constant angular acceleration, *Acta. Otolaryngol.,* 54:349,1962.
16. Cransac, H, Cottet-Emard, JM, Pequignot, JM, and Peyrin, L, Monoamines (norepinephrine, dopamine, serotonin) in the rat medial vestibular nucleus: endogenous levels and turnover, *J. Neural. Transm.,* 103:391,1996.
17. Dourish, CT, Piribedil: behavioral, neurochemical, and chemical profile of a dopamine agonist, *Prog. Neuro-Psychopharmacol. Biol. Psychiatr.,* 7:3,1983.

18. Dutia, MB, Johnston, AR, and McQueen, DS, Tonic activity of rat medial vestibular nucleus neurons in vitro and its inhibtion by GABA, *Exp. Brain Res.*, 88:466,1992.

19. Fernandez-Lopez, A, del Arco, C, Gonzales, AM, Gomez, T, Calvo P, and Pazos A, Autoradiographic localization of a_2-adrenoceptors in chick brain, *Neurosci. Lett.*, 120:97,1990.

20. Fisher, AJEM, Histamine in the treatment of vertigo, *Acta. Otolaryngol. (Suppl)*, 479:24,1991.

21. Flandrin, JM, Courjon, JH, Jeannerod, M, Schmid, R, Vestibulo-ocular responses during the states of sleep in the cat, *Electroenceph.Clin. Neurophysiol.*, 46:521,1979.

22. Galiana, HL, and Outerbridge, JS, A bilateral model for central vestibular pathways in vestibuloocular reflex, *J, Neurophysiol.*, 51:210,1984.

23. Gallagher, JP, Phelan, KD, and Shinnick-Gallagher, P, Modulation of excitatory transmission at the rat medial vestibular nucleus synapse, *Ann. NY Acad. Sci.*, 656:630,1992.

24. Gallet, B, Ané P, Efficacité de Trivastal 50 retard dans les syndromes cochléo-vestibulaires, *Act. Ther. J. ORL.*, special issue:50,1989.

25. Giuffrida, R, Licata, F, Li, Volsi, G, Santangelo, F, and Sapienza, S, Immunocytochemical localization of serotoninergic afferents to vestibular nuclei in the rat, *Eur. J. Physiol.*, 419:R33,1991.

26. Godaux, E, Halleux, J, and Gobert, C, Adaptive changes of the vestibulo-ocular reflex in the cat: effects of a long-term frequency-selective procedure, *Exp. Brain Res.*, 49:28,1983.

27. Harvey, JA, McMaster, SE, and Romano, AG, Methylenedioxyamphetamine: neurotoxic effects on serotonergic projections to brainstem nuclei in the rat, *Brain Res.*, 619:1,1993.

28. Highstein, SM, Goldberg, JM, Moschovakis AK, and Fernandez, C, Inputs from regularly and irregularly discharging vestibular nerve afferents to secondary neurons in the vestibular nuclei of the squirrel monkey. II. Correlation with output pathways of secondary neurons, *J. Neurophysiol.*, 58:719,1987.

29. Hille, B, G-protein coupled mechanisms and nervous signaling, *Neuron.*, 9:187,1992.

30. Horii, A., Takeda, N., Matsunaga, T, Yamatodani, A, Mochizuki, T, Okakura-Mochizuki, K., and Wada, H, Effect of unilateral vestibular stimulation on histamine release from the hypothalamus of rats in vivo, *J. Neurophysiol.*, 70:1822,1993.

31. Hoyer, D, Clarke, DE, Fozard, JR, Hartig, PR, Martin, GR, Mylecharane, EJ, Saxena, PR, and Humphrey, PPA, International Union of Pharmacology classification of receptors for 5-hydroxytryptamine (serotonin), *Pharmacol. Rev.*, 46:157,1994.

32. Iasnetsov, VV, Pravdivtsev, VA, Drozd, IV, and Shashkov, VS, Effects of several neuromediators and neuromodulators on the electrical activity of the neurons of the medial vestibular nucleus, *Kosm. Biol. Aviakosm. Med.*, 24:59,1990.

33. Ito, M, Highstein, SM, and Fukuda, J, Cerebellar inhibition of the vestibulo-ocular reflex in rabbit and cat and its blockade by picrotoxine, *Brain Res.*, 17:524,1970.

34. Iwamoto, Y, Kitama, T, and Yoshida, K, Vertical eye movement-related secondary vestibular neurons ascending in medial longitudinal fasciculus in cat. I. Firing properties and projection pathways, *J. Neurophysiol.*, 63:902,1990.

35. Johnston, AR, Murnion, B, McQueen, DS, and Dutia, MB, Excitation and inhibition of rat medial vestibular nucleus neurones by 5-hydroxytryptamine, *Exp. Brain Res.*, 93:293,1993.

36. Katz, E, and Vianney, de Jong, JMB, Büttner-Ennever, J, and Cohen, B, Effects of midline medullary lesions on velocity storage and the vestibulo-ocular reflex, *Exp. Brain Res.*, 87:505,1991.

37. Kebabian, JW, and Calne, DB, Multiple receptors for dopamine, *Nature*, 277:93,1979.

38. Kirsten, EB, and Sharma, JN, Characteristics and response differences to iontophoretically applied norepinephrine, D-amphetamine and acetylcholine on neurons in the medial and lateral vestibular nuclei of the cat,. *Brain Res.*, 113:77,1976.

39. Kitahama, K, Buda, C, Sastre JP, Nagatsu, I, Raynaud, B, Jouvet, M, and Geffard, M, Dopaminergic neurons in the cat dorsal nucleus of the vagus, demonstrated by dopamine, AADC and TH immunohistochemistry, *Neurosci. Lett.*, 146:5,1992.

40. Kohl, KL, Calkins, DS, and Mandell, AJ, Arousal and stability: the effects of five new sympathicomimetic drugs suggest a new principle for the prevention of space motion sickness, *Aviat. Space Environ. Med.*, 57:137,1986.

41. Kohl, RL, and Lewis, MR, Mechanisms underlying the antimotion-sickness effects of psychostimulants, *Aviat. Space Environ. Med.*, 58:1215,1987.

42. Kumoi, K, Saito, N, Tanaka, C, Immunohistochemical localization of gamma-aminobutyric acid- and aspartate-containing neurons in the guinea pig vestibular nuclei, *Brain Res.*, 416:22,1987.

43. Licata, F, Li, Volsi, G, Maugeri, G, and Santangelo, F, Effects of 5-hydroxytryptamine on the firing rates of neurons of the lateral vestibular nucleus in the rat, *Exp. Brain Res.,* 79:293,1990.

44. Licata, F, Li, Volsi, G, Maugeri, G, and Santangelo, F, Excitatory and inhibitory effects of 5-hydroxytryptamine on the firing rate of medial vestibular nucleus neurons in the rat, *Neurosci. Lett.,* 154: 195,1993.

45. Licata, F, Li, Volsi, G, Maugeri, G, Ciranna, L, and Santangelo, F, Serotonin-evoked modifications of the neuronal firing rate in the superior vestibular nucleus: a microiontophoretic study in the rat, *Neuroscience,* 52:941,1993.

46. Licata, F, Li, Volsi, G, Maugeri, G, Ciranna, L, and Santangelo, F, Effects of noradrenaline on the firing rate of vestibular neurons, *Neuroscience,* 53:149,1993.

47. Lin, JS, Sakai, K, Vanni-Mercier, G, Arrang, JM, Garbarg, M, Schwartz, JC, and Jouvet, M, Involvement of histaminergic neurons in arousal mechanisms demonstrated with H_3 receptor ligands in the cat, *Brain Res.,* 523:325,1990.

48. Lisberger, SG, Miles, FA, and Optican, LM, Frequency-selective adaptation: evidence for channels in the vestibulo-ocular reflex?, *J. Neurosci.,* 3:1234,1983.

49. Maeda, M, Shimazu, H, and Shinoda, Y, Nature of synaptic events in cat abducens motoneurons at slow and quick phases of vestibular nystagmus, *J. Neurophysiol.,* 35:279,1972.

50. Matsuoka, I, Domino, EF, and Morimoto, H, Adrenergic and cholinergic mechanisms of single vestibular neurons in the cat, *Adv. Otorhinolaryngol.,* 19:163,1973.

51. Matsuoka, I, Ito, J, Sasa, M, and Takaori, S, Possible neurotransmitters involved in excitatory and inhibitory effects from inferior olive to contralateral lateral vestibular nucleus, *Adv. Oto-Rhino-Laryngol.,* 30: 58,1983.

52. McCormick, DA, and Wang, Z, Serotonin and noradrenaline excite GABAergic neurons of the guinea pig and cat nucleus reticularis thalami, *J. Physiol., (London)* 442:235,1991.

53. McElligott, JG, Freedman, W, Vestibulo-ocular reflex adaptation in cats before and after depletion of norepinephrine, *Exp. Brain Res.,* 69:509,1988.

54. Moore, RY, Bloom, FE, Central catecholamine neuron systems: anatomy and physiology of the dopamine systems, *Annu. Rev. Neurosci.,* 1:129,1978.

55. Mühlethaler, M, de Curtis, M, Walton, K, and Llinás, R, The isolated and perfused brain of the guinea pig in vitro, *Eur. J. Neurosci.,* 5:915,1993.

56. Nardo. L, Genazzani, AA, Lauria, N, and Grassi, M, Dihydroergocristine reduces experimental nystagmus in guinea pigs, *Acta. Ther.,* 16:323,1990.

57. Nicholas, AP, Hökfelt, T, and Pieribone, V, The distribution and significance of CNS adrenoceptors examined with in situ hybridization, *Trends Pharmacol. Sci.,* 17: 245,1996.

58. Nicholas, AP, Pieribone, V, and Hökfelt, T, Distribution of mRNAs for a_2 adrenergic receptor subtypes in rat brain, *J, Comp, Neurol.,* 328:575,1993.

59. Nicholls, DG, The glutamatergic nerve terminal, *Eur. J. Biochem.,* 212:613,1993.

60. North, RA, and Uchimura, N, 5-hydroxytryptamine acts at $5-HT_2$ receptors to decrease potassium conductance in rat nucleus accumbens neurones, *J. Physiol.,* (London) 417:1,1989.

61. Onodera, K, Yamatodani, A, Watanabe, T, and Wada, H, Neuropharmacology of the histaminergic neuron system in the brain and its relationship with behavioral disorders, *Prog. Neurobiol.,* 42:685,1994.

62. Palacios, JM, and Kuhar, MJ, a-adrenergic receptor localization in rat brain by light microscopic autoradiography, *Neurochem. Int.,* 4:473,1983.

63. Pazos, A, Cortes, R, and Palacios, JM, Quantitative autoradiographic mapping of serotonin receptors in the rat brain II. Serotonin-2 receptors, *Brain Res.,* 346:231,1985.

64. Pazos, A, Palacios, JM, Quantitative autoradiographic mapping of serotonin receptors in the rat brain. I. Serotonin-1 receptors, *Brain Res.,* 346:205,1985.

65. Petrosini, L, Dell'Anna, ME, Vestibular compensation is affected by treatment with dopamine active agents, *Arch. Ital. Biol.,* 131:159,1993.

66. Phelan, KD, Nakamura, J, and Gallagher, JP, Histamine depolarizes rat medial vestibular nucleus neurons recorded intracellularly in vitro, *Neurosci. Lett.,* 109:287,1990.

67. Pieribone, VA, Nicholas, AP, Dagerlind, Å, and Hökfelt, T, Distribution of a_1 adrenoceptors in rat brain revealed by *in situ* hybridization experiments utilizing subtype-specific probes, *J, Neurosci.,* 14: 4252,1994.

68. Pompeiano, O, Relationship of noradrenergic locus coeruleus neurones to vestibulo-spinal reflexes, *Prog. Brain Res.,* 80:329,1989.

69. Pompeiano, O, Manzoni, D, Barnes, CD, Stampacchia, G, and d'Ascanio, P, Responses of locus coeruleus and subcoeruleus neurons to sinusoidal stimulation of labyrinthine receptors, *Neuroscience,* 35: 227,1990.

70. Pompeiano O, Manzoni D, d'Ascanio P, and Andre P Noradrenergic agents in the cerebellar vermis affect adaptation of the vestibulospinal reflex gain, *Brain Res. Bull.,* 35:433,1994.

71. Precht, W, Schwindt, PC, Baker, R, Removal of vestibular commissural inhibition by antagonists of GABA and glycine, *Brain Res.,* 62:222,1973.

72. Rascol, O, Hain, TC, Brefel, C, Bénazet, M, Clanet, M, and Montastruc, JL, Antivertigo medications and drug-induced vertigo, *A pharmacological review. Drugs,* 50:777,1995.

73. Raymond, J, Demêmes, D, and Nieoullon, A, Neurotransmitters in vestibular pathways, *Prog. Brain Res.,* 76:29,1988.

74. Rosin, DL, Talley, EM, Lee, A, Stornetta, RL, Gaylinn BD, Guyenet, PG, and Lynch, KR, Distribution of a_{2C}-adrenergic receptor-like immunoreactivity in the rat central nervous system, *J. Comp. Neurol.,* 372:135,1996.

75. Ruat, M, Traiffort, E, Arrang, JM, Leurs, R, and Schwartz JC, Cloning and tissue expression of a rat histamine H_2 receptor gene, *Biochem. Biophys. Res. Commun.,* 179:1470,1991.

76. Satayavidad, J, and Kirsten, EB, Iontophoretic studies of histamine and histamine antagonists in the feline vestibular nuclei, *Eur. J. Pharmacol.,* 41:17,1977.

77. Scheinin, M, Lomasney, JW, Hayden-Hixson, DM, Schambra, UB, Caron MG, Lefkowitz, RJ, and Fremeau, RT, Jr, Distribution of alpha 2-adrenergic receptor subtype gene expression in rat brain, *Mol. Brain Res.,* 21:133,1994.

78. Schuerger, RJ, and Balaban, CD, Immunohistochemical demonstration of regionally selective projections from locus coeruleus to vestibular nuclei in rats, *Exp. Brain Res.,* 92:351,1993

79. Serafin, M, de Waele, C, Khateb, A, Vidal, PP, and Mühlethaler, M, Medial vestibular nucleus in the guinea pig. I. Intrinsic membrane properties in brainstem slices, *Exp. Brain Res.,* 84:417,1991.

80. Serafin, M, de Waele, C, Khateb, A, Vidal,PP, and Mühlethaler, M, Medial vestibular nucleus in the guinea pig. II. Ionic basis of the intrinsic membrane properties in brainstem slices, *Exp. Brain Res.,* 84: 426,1991.

81. Serafin, M, Khateb, A, Vibert, N, Vidal, PP, and Mühlethaler, M, Medial vestibular nucleus in the guinea pig. Histaminergic receptors. I. An in vitro study, *Exp. Brain Res.,* 93:242,1993.

82. Shimazu, H, and Precht, W, Tonic and kinetic responses of cat's vestibular neurons to horizontal angular acceleration, *J. Neurophysiol.,* 28:991,1965.

83. Shinoda, Y, and Yoshida, K, Dynamic characteristics of responses to horizontal head angular acceleration in vestibuloocular pathway in the cat, *J. Neurophysiol.,* 37:653,1974.

84. Smith, PF, Darlington, CL, and Hubbard, JI, Evidence for inhibitory amino acid receptors on guinea pig medial vestibular nucleus neurons *in vitro, Neurosci. Lett.,* 121:244-246.

85. Smith, PF, and Darlington, CL, Can vestibular compensation be enhanced by drug treatment? *J. Vest. Res.,* 4:169,1994.

86. Starke, K, Göthert, M, and Kilbinger, H, Modulation of neurotransmitter release by presynaptic autoreceptors, *Physiol. Rev.,* 69:864,1987.

87. Steinbusch, HWM, Distribution of serotonin immunoreactivity in the central nervous system of the rat: Cell bodies and terminals, *Neuroscience,* 6:557,1981.

88. Steinbusch, HWM, Distribution of histaminergic neurons and fibers in the rat brain. Comparison with noradrenergic and serotonergic innervation of the vestibular system, *Acta. Otolaryngol.,* (Suppl) 479: 12,1991.

89. Takada, M, Widespread dopaminergic projections of the subparafascicular thalamic nucleus in the rat, *Brain Res, Bull.,* 32:301,1993.

90. Takatani, T, Ito, J, Matsuoka, I, Sasa, M, and Takaori, S, Effects of diphenhydramine iontophoretically applied onto neurons in the medial and lateral vestibular nuclei, Japan *J. Pharmacol.,* 33:557,1983.

91. Takeda, N, Morita, M, Kubo, T, Yamatodani, A, Watanabe, T, Tohyama, M, Wada, H, and Matsunaga, T, Histaminergic projection from the posterior hypothalamus to the medial vestibular nucleus of rats and its relation to motion sickness. in: *The Vestibular System: Neurophysiologic and Clinical Research* (Graham MD, and Kemink JL, Eds.), pp 601-617, New York: Raven Press,1987.

92. Takeda, N, Morita, M, Hasegawa, S, Horii, A, Kubo, T, and Matsunaga, T, Neuropharmacology of motion sickness and emesis: a review, *Acta. Otolaryngol. (Suppl.),* 501:10,1993.

93. Talley, EM, Rosin, DL, Lee, A, Guyenet, PG, and Lynch, KR, Distribution of a_{2A}-adrenergic receptor-like immunoreactivity in the rat central nervous system, *J. Comp. Neurol.,* 372:111,1996.

94. Tan, HS, van Neerven, J, Collewijn, H, and Pompeiano, O, Effects of a-noradrenergic substances on the optokinetic and vestibulo-ocular responses in the rabbit: A study with systemic and intrafloccular injections, *Brain Res.,* 562:207,1991.

95. Ternaux, JP, and Gambarelli, F, Modulation of the vestibulo-ocular reflex by serotonin in the rat, *Pflügers Arch.,* 409:507,1987.

96. Tighilet, B, and Lacour, M, Distribution of histaminergic axonal fibres in the vestibular nuclei of the cat, *NeuroReport,* 7:873,1996.

97. Vallar, L, and Meldolesi, J, Mechanisms of signal transduction at the dopamine D_2 receptors, *Trends Pharmacol. Sci.,* 10:74,1989.

98. Vibert, N, Serafin, M, Vidal, PP, and Mühlethaler, M, Pharmacological properties of medial vestibular neurones in the guinea pig *in vitro.* in : *Information Processing Underlying Gaze Control* (Delgado-Garcia, JM, Godaux, E, Vidal, PP, Eds.), pp 159-173. New York: Pergamon Press,1994.

99. Vibert, N, Serafin, M, Crambes, O, Vidal, PP, and Mühlethaler, M, Dopaminergic agonists have both presynaptic and postsynaptic effects on the guinea pig's medial vestibular nucleus neurons, *Eur. J. Neurosci.,* 7:555,1995.

100. Wada, H, Inagaki, N, Yamatodani, A, and Watanabe, T, Is the histaminergic neuron system a regulatory center for whole brain activity?, *Trends Neurosci.,*14:415,1991.

101. de Waele, C, Abitbol, M, Chat M, Menini, C, Mallet, J, and Vidal, PP, Distribution of glutamatergic receptors and GAD mRNA-containing neurons in the vestibular nuclei of normal and hemilabyrinth-ectomized rats, *Eur. J. Neurosci.,* 6:565,1994.

102. de Waele, C, Mühlethaler, M, and Vidal, PP, Neurochemistry of the central vestibular pathways, *Brain Res. Rev.,* 20:24,1995.

103. Wanaka, A, Kiyama, H, Murakami, T, Matsumoto, M, Kamada, T, Malbon, CC, and Tohyama, M, Immunocytochemical localization of a-adrenergic receptors in rat brain, *Brain Res.,* 485:125,1989.

104. Wang, JJ, and Dutia, MB, Effects of histamine and betahistine on rat medial vestibular nucleus neurones: possible mechanism of action of anti-histaminergic drugs in vertigo and motion sickness, *Exp. Brain Res.,* 105:18,1995.

105. Wood CD, and Graybiel, A, A theory of motion sickness based on pharmacological reactions, *Cli. Pharmacol. Ther.,* n11:621,1970.

106. Woodward, DJ, Moises, HC, Waterhouse, BD, Yeh, HH, and Cheun, JE, Modulatory actions of norepinephrine on neural circuits, in: *Neuroreceptor Mechanisms in Brain* (Kito S, Ed.), pp 193-208. New York: Plenum Press,1991.

107. Wright, DE, Seroogy, KB, Lundgren, KH, Davis, BM, and Jennes, L, Comparative localization of serotonin-1A, 1C, and 2 receptor subtype mRNAs in rat brain, *J. Comp. Neurol.,* 351:357,1995.

108. Yabe, T, de Waele, C, Serafin, M, Vibert, N, Arrang, JM, Mühlethaler, M, and Vidal, PP, Medial vestibular nucleus in the guinea pig. Histaminergic receptors. II. An *in vivo* study, *Exp. Brain Res.,* 93:249,1993.

109. Yamamoto, C Pharmacologic studies of norepinephrine, acetylcholine, and related compounds in Deiters' nucleus and the cerebellum, *J. Pharmacol. Exp. Therap.,* 156: 39,1967.

110. Yokoyama, C, Okamura, H, Nakajima, T, Taguchi, JI, and Ibata, Y, Autoradiographic distribution of [^{3}H]YM-09151-2, a high affinity and selective antagonist ligand for the dopamine D_2 receptor group, in the rat brain and spinal cord, *J. Comp. Neurol.,* 344:121,1994.

111. Zifa, E, and Filion, G, 5-Hydroxytryptamine receptors, *Pharmacol, Rev.,* 44:401,1992.

9

Neuropeptides Associated with the Vestibular Nuclei

Dale W. Saxon and Alvin J. Beitz

CONTENTS

9.1 INTRODUCTION

Peptides are defined as compounds containing two or more amino acids in which the carboxyl group of one acid is linked to the amino group of the other. These compounds are widely distributed in biological systems and in many instances highly conserved across many phyla. A particularly rich assortment of peptides can be found in the central and peripheral nervous systems of mammals. Neurons which contain and release peptides are often referred to as peptidergic and the peptides which they release as neuropeptides. Peptides (e.g., vasopressin, oxytocin, gastrin, hypothalamic releasing hormones, etc.) have been known for many years to serve specialized endocrinological functions both within and outside the nervous system. An entirely new group of peptides (e.g., opioids, tachykinins, etc.) have proved to be strong candidates for a neurotransmitter/neuromodulator role throughout the nervous system. The concept that peptides might serve a neurotransmitter function stems in part from the original identification of the peptide, substance P, in neurons of peripheral ganglion.[46] A further push in the field of peptide research came with the discovery that opioid peptides had a broad range of physiological effects in the central nervous system, including the attenuation of nociception. Finally, with the widespread use of immunocytochemical techniques as well as more modern biochemical and molecular methods, it became possible to map the distribution of neuropeptides within the brain and spinal cord as well as in ganglia and fibers associated with the peripheral nervous system. To date, more than 40 different neuropeptides have been mapped in the central nervous system (CNS). While the majority of mapping studies have examined neuropeptide distribution throughout the brain, they have often failed to provide precise cartographic information regarding the vestibular nuclei. Despite this shortcoming anatomical, pharmacological, behavioral, and physiological information is beginning to accrue, implicating several neuropeptides in vestibular function. The following sections are an overview of the current state of knowledge as it pertains to peptides in the vestibular system.

9.2 PITFALLS ASSOCIATED WITH THE MAPPING OF BRAIN NEUROPEPTIDES

It should be noted at the outset that over the past two decades, a large number of reports on the distribution of neuropeptides have emerged, and with considerable frequency these reports are in part or in whole contradictory. This almost certainly stems from a combination of two or more of the following factors: technical aspects of the procedures used, methodological differences among laboratories, different sensitivities of various procedures to detect neuropeptides, species differences or the short half-life of some neuropeptides. In particular, discrepancies have arisen from studies using microdissection radioimmunoassay methodologies, and those employing immunocytochemical techniques to map the distributions of peptidergic neurons. This may be attributable in part to the fact that radioimmunoassays are unable to resolve peptide localization at a cellular level; despite this limitation, radioimmunoassays provide a useful and sensitive technique for measuring relative levels of peptides in particular regions or nuclei of the brain. A second crucial variable that affects radioimmunoassay results is the accuracy with which the investigator delineates a specific brain region or nucleus and the ability of the investigator to microdissect this region for radioimmunoassay. This latter point has significant consequences, since even a small amount of contamination by material from adjacent regions can result in false positives. This is a particularly important limitation when small nuclei or nuclear subdivisions are being sampled. It is likely that the above limitations may have contributed to some of the discrepancies observed in the literature between radioimmunoassay data and immunohistochemical data for particular peptides in the vestibular nuclei.

In contrast to radioimmunoassay procedures, immunocytochemical techniques have the advantage of being able to localize peptides to specific neurons or to components of the neuropil. Despite this advantage, immunocytochemical methods are still associated with technical problems that can potentially lead to contradictory results. These problems relate in part to antibody specificity, the sensitivity of different antibody visualization procedures, and to the actual compartmentalization of neuropeptides with the neuron. With respect to the latter issue, the richest supply of a releasable neuropeptide is often found in neuronal appendages (dendrites and axons). Such neuropeptides are produced in the cell body, but are typically transported rapidly to axon terminals where they are stored until their release. This concentration of neuropeptide in axonal processes with minimal levels in the perikarya where synthesis occurred probably also contributes to some of the contradictory reports (usually false negatives) regarding the presence or absence of selected neuropeptides in specific populations of neurons. To circumvent this problem, drugs are often administered that inhibit axonal transport (e.g., colchicine) and cause the neuropeptide to accumulate in the perikaryon, so that it can be visualized with immunocytochemical techniques. However, the administration of axonal inhibitors can also indirectly affect protein synthesis and thus alter the normal levels of certain peptides within neurons. The immunohistochemical maps of brain neuropeptide distribution can be confirmed to a large extent using *in situ* hybridization techniques which localize messenger RNAs associated with the neuropeptides of interest. However, because mRNA transcription can occur within a given neuron without accompanying neuropeptide translation, such correlations also have limitations. Neuropeptide receptors have also been localized using radiolabeled binding procedures, *in situ* hybridization, and immunocytochemical techniques. All of these procedures have limitations and the strongest evidence for the presence of neuropeptide receptors in a particular region stems from studies in which two or more of the above techniques are utilized to localize the receptor in question, in combination with physiological and pharmacological data. While the following discussion attempts to provide a balanced overview of neuropeptides and their receptors in the vestibular system, the immunohistochemical, radioimmunoassay, *in situ* hybridization, and receptor binding data must be interpreted with the above limitations in mind.

9.3 NEUROPEPTIDES IMPLICATED IN VESTIBULAR FUNCTION

Physiological and pharmacological studies have provided important evidence to suggest that several neuropeptides are involved in modulating vestibular system function.[2,19,70] A number of neuropeptides have now been investigated by direct injection of these neuroactive substances into the CNS of experimental animals. For example, intracerebroventricle (icv.) administration of somatostatin or arginine-vasopressin have been shown to elicit behavioral symptoms which resemble those associated with vestibular dysfunction. In particular, following administration of these peptides rats exhibit motor disturbances that resemble the barrel role behavior described in this species following labyrinthectomy.[2] A number of other hypothalamic peptides, including oxytocin, vasotocin, oxypressin, and leucine-vasopressin have also been shown to produce barrel role behavior following icv. administration[1,2,43] The sulfonated form of cholecystokinin[53] or large doses of opioid peptides administered intrathecally or intramedullarly can also elicit barrel role behavior,[2,26,60] as does the intraventricular or intracisternal administration of substance P[34,52]. The similarity of the responses to these peptides initially led to the postulation that they were directly involved in vestibular function. However, as the anatomical mapping studies reviewed below suggest, many of these peptides appear to affect the vestibular system indirectly, rather than acting at the level of the vestibular nuclei.

9.4 DISTRIBUTION OF NEUROPEPTIDES IN THE VESTIBULAR
COMPLEX

Direct anatomical data (obtained from *in situ* hybridization, radioimmunoassay, immunocytochemistry, or receptor binding studies) demonstrating the presence of a particular neuropeptide or its receptor in the vestibular system, in combination with direct pharmacological/physiological data showing that the neuropeptide has a direct effect on neurons in the vestibular system, provides the most compelling evidence for a direct role of a particular neuropeptide in vestibular function. Somatostatin is a good example of a peptide that has been shown to have direct effects on vestibular function. Immunocytochemical studies of the distribution of somatostatin in the rodent brain have shown small numbers of somatostatin-labeled perikarya as well as axons in the vestibular complex.[19,36,62,80] In addition, high-affinity somatostatin binding sites have been shown to be present in all four principal vestibular nuclei, with the highest concentration being in the medial vestibular nucleus (MVN).[2,12,19,78] Finally, microinjection of somatostatin directly into the vestibular nuclei produces a postural syndrome similar to that observed following hemilabyrinthectomy (including unilateral extension of the limbs and barrel role behavior).[2,10] The underlying mechanism(s) of somatostatin's action in the vestibular complex is not entirely understood, but behavioral/pharmacological studies suggest that somatostatin's action is dependent on cholinergic activity; somatostatin modulates or is modulated by muscarinic cholinergic activity.[2]

In contrast to the direct action of certain peptides, like somatostatin, in the vestibular system, the available evidence for several other neuropeptides suggest that their effect on the vestibular system is primarily indirect. Behavioral studies have clearly implicated two peptides, Arginine-vasopressin and oxytocin, in vestibular function; however, immunocytochemical and receptor-binding data fail to support the presence of these peptides or their receptors in the vestibular nuclei. At present, there is no evidence for vasopressin binding sites (see Tribollet et al.[75]) or for immunoreactive elements for vasopressin or oxytocin[62,71,72] in the vestibular nuclei. However, it is possible that these peptides or their receptors are present in the vestibular nuclei in very low levels beyond the detection sensitivity of the methods used. Use of reverse transcription-polymerase chain reaction on mRNA isolated from the vestibular nuclei is one approach that could be used to determine if the mRNA for these peptides or their receptors are present in the vestibular nuclei. If these peptides or their receptors are in fact absent from the vestibular nuclei and the inner ear, as the anatomical data predict, then it is likely that vasopressin and oxytocin affect the vestibular system indirectly.

9.4.1 OPIOID PEPTIDES IN THE VESTIBULAR COMPLEX

The classical opioid neuropeptides (enkephalins, β-endorphins, and dynorphins) have been extensively studied and mapped in the brains of various species, while many of the newer opioid-like neuropeptides, such as endomorphin 1,[85] endomorphin 2[85] and orphanin FQ[33] have not yet been carefully mapped, especially in the region of the vestibular nuclei. The presence of classical opioid neuropeptides in the vestibular complex has been consistently reported, but their distribution among the individual vestibular nuclei appears to be somewhat variable. Physiological evidence indicates that exogenous administration of the pentapeptides, met- and leu-enkephalin, depresses the spontaneous firing rate of neurons in the lateral vestibular nucleus (LVN) in a dose-dependent manner.[14,19] When GABA is applied concomitantly with met-enkephalin, the inhibitory effects are additive. Since cerebellar Purkinje cells exhibit immunoreactivity for enkephalin, and are known to inhibit LVN neurons via GABA$_A$ receptors, Chan-Palay and colleagues[14] suggested that this peptide may be a co-transmitter or neuromodulator within the LVN. Since enkephalin containing perikarya do not appear to be present in the LVN, the hypothesis that enkephalinergic input arises from an extrinsic source, such as Purkinje cells, remains the most tenable.

The action of enkephalins in the vestibular complex are believed to be mediated primarily through μ- and δ-opioid receptors via a G-protein effector molecule.[19] A radioimmunoassay study of enkephalin in the rhombencephalon showed the presence of low concentrations of both met- and leu- enkephalin in the MVN and LVN.[62] Preproenkephalin mRNA has been reported to be localized in neurons of the MVN and LVN[65] and the proenkephalin peptide, which serves as the precursor molecule to met- and leu-enkephalin, is found in fiber plexuses of the MVN and inferior vestibular nucleus (IVN).[2,55] Enkephalin-like immunoreactive terminals and perikarya are also found in the vestibular complex.[5,25,32,39,55,83] In particular, the MVN contains neurons expressing both the enkephalin mRNA and peptide.[87] This nucleus and the LVN are also rich in opiate receptors.[87] From a physiological viewpoint, both morphine and met-enkephalin have been shown *in vitro* to induce an acceleration of the resting discharge of neurons in the medial vestibular nucleus.[19] In addition, it has been shown that an intraperitoneal injection of the opioid receptor antagonist, naloxone, thirty minurtes pre-unilateral labyrinthectomy (UL) and five hours post-UL, significantly reduces the frequency of spontaneous nystagmus in the guinea pig, and thus serves to enhance ocular motor compensation.[20] In light of the above anatomical and physiological evidence, it is likely that enkephalin plays a significant role in the normal function of the vestibular system.

Endorphins (alpha and beta) act as endogenous ligands primarily at μ-opioid receptors. Unlike enkephalins, there is no immunocytochemical evidence of neurons or fibers containing either alpha- or beta-endorphin in the vestibular complex. There are, however, at least two reports based on radioimmunoassay data which indicate that low concentrations of these peptides are present in the MVN and LVN.[62] While these studies suggest that endorphins may be present in the vestibular complex, they should be interpreted with caution because there are several structures in the immediate vicinity of the vestibular complex (in particular, the nucleus of the tractus solitarius and the cuneate nucleus) which contain high to moderate levels of endorphin and could easily contaminate vestibular samples. The presence and possible function of endorphins in the vestibular nuclei is inconclusive.

Dynorphins (A and B) comprise the third classical opioid neuropeptide family and are potent opiate agonists which act at the κ-opioid receptor subtype. Using radioimmunoassay, several studies[62,86] identify low levels of dynorphin A and B in the MVN and LVN. These radioimmunoassay data are complemented by immunohistochemical data indicating that there are fibers in the MVN and IVN which stain positive for dynorphin.[58] While the immunohistochemical data do not support the presence of dynorphin in the LVN, it is possible that the region designated as the LVN by Palkovits[62] actually included part or all of the IVN. The physiological role of endorphins and dynorphins in the vestibular complex has not been rigorously tested, but it would seem reasonable to suggest based on their paucity relative to enkephalin that they play only a minor role in the function of the vestibular system.

9.4.2 TACHYKININS IN THE VESTIBULAR COMPLEX

The tachykinins are a family of undecapeptides defined by their similar pharmacological activities and conserved carboxy-terminal sequence -Phe-X-Gly-Leu-Met-NH$_2$, where X is an aromatic or branched aliphatic amino acid. The amino termini of the tachykinins shows a much wider variation, and most of the biological activities of this peptide family reside in their similar carboxyl sequences. Substance P (X = Phe) is the most widely known member of the tachykinin family, and was for many years generally believed to be the only mammalian tachykinin. In 1983, however, that belief was put to rest with the discovery of two novel mammalian tachykinins named substance K (now called neurokinin A) and neuromedin K (neurokinin B) to reflect their homology with the amphibian tachykinin, kassinin.[29,38,51] These tachykinins differ pharmacologically from substance P (Sub P) in their in both peripheral and central systems through the mediation of different physiological functions via different tachykinin receptors.[4,16,40,50] These tachykinin receptors are divided into three types, designated NK1 (Sub P-preferring), NK2 (neurokinin A-preferring), and NK3 (neurokinin B-preferring).[50] Natural tachykinins can act as full agonists at each one of the three receptors, albeit at different concentrations; synthetic agonists however, are available to selectively stimulate only one receptor[50]. Tachykinin receptors have been cloned and contain seven segments spanning the cell membrane, indicating their inclusion in the G-protein-linked receptor family.[40]

Sub P was the first neuropeptide to be discovered[23] and, during the six decades since its discovery, it has been the most extensively studied. It is found throughout the body, including the central and peripheral nervous systems, where it has been implicated as both a neurotransmitter and neuromodulator.[4,31] With respect to Sub P localization in the vestibular system, there are small numbers of neurons containing the signal for preprotachykinin mRNA (precursor to Sub P and neurokinin A) found in the caudal half of the MVN and also in the IVN.[31] Sakanaka and coworkers[66] reported the presence of a few positively stained Sub P-like perikarya in the MVN and IVN, while Ljungdahl et al.[47] found moderate numbers of stained perikarya in the IVN, but none in the MVN. An immunocytochemical study of the distribution of substance P containing fibers in the rodent CNS and PNS also revealed that positively stained fibers can be found in the caudal half of the MVN and to a lesser degree the IVN.[47] In another immunocytochemical study Nomura et al.[57] found a distribution of Sub P perikarya and fibers comparable to that described by Ljundahl and colleagues,[47] but also identified a sparse fiber network in the LVN and SVN. In the absence of any evidence for Sub P containing perikarya in the LVN and SVN, it is likely that the fibers in these nuclei take their origin from neurons located in the reticular formation in the MVN and IVN, or from primary vestibular sensory neurons located in the Scarpa's ganglion.[15,73,81]

With respect to tachykinin receptors, binding studies have demonstrated Sub P binding sites in the vestibular complex[87] and the mRNA for the NK-1 receptor has been identified in the MVN using *in situ* hybridization techniques.[49] Since it has been shown electrophysiologically that Sub P causes depolarization of a majority of neurons in the MVN,[19,79] it appears that this peptide mediates an excitatory effect in the vestibular complex. However, Vibert et al.[79] have shown that this excitatory effect of Sub P could not be reproduced with any one of the specific agonists of the three tachykinin receptor subtypes, nor was it blocked by the specific NK1 receptor antagonists GR 82664 and CP 99994. These data suggest that this depolarizing effect might be due to the activation of a new, pharmacologically distinct, "NK1-like" receptor. In addition, Vibert and colleagues[79] have demonstrated that approximately ten percent of MVN neurons are hyperpolarized and inhibited by Sub P, and that the membrane resistance of these cells is also decreased. Interestingly, only these hyperpolarizing effects could be mimicked by the specific NK1 receptor agonists GR 73632 and [Sar9, Met (O2)11]-SP, suggesting that this hyperpolarizing effect is mediated by the few typical NK1 receptors which have been demonstrated in the medial vestibular nucleus. To date, other mammalian tachykinins, such as neurokinin A and neurokinin B, have not been reported in the vestibular system.

9.4.3 Hypophysiotopic Peptides in the Vestibular Complex

The hypophysiotopic family of peptides include such members as corticotropin-releasing factor (CRF), adrenocorticotropin hormone (ACTH), lutenizing hormon-releasing hormone (LHRH), thyrotropin-releasing hormone (TRH), and somatostatin (see above). As indicated by their names, most of these peptides were originally believed to have only hormonal functions and in particular to affect anterior pituitary secretion. However, with the introduction of more sensitive radioimmunoassays and improved immunocytochemical techniques, it became evident that several of these peptides were also localized in cell groups or axons outside the hypothalamic/pituitary axis where they were originally isolated. In this regard, Skofitsch and Jacobowitz[69] used four different antibodies to CRF and radioimmunoassays to identify the distribution of CRF throughout the brain. With respect to the vestibular system, these authors reported the presence of CRF in the MVN. This was confirmed with immunocytochemical techniques that demonstrated the presence of perikarya immunoreactive for CRF in the MVN.[17,54,58,62,66] With regard to other vestibular nuclei, several studies have indicated that the IVN contains small numbers of CRF immunoreactive perikarya,[17,37,54,66] while only one study has indicated the presence of CRF perikarya in the SVN.[66] Unfortunately, the majority of these immunocytochemical studies were aimed at mapping the distribution of CRF throughout the brain; thus, details regarding the precise distribution of CRF neurons among the individual vestibular nuclei are lacking. However, Cummings et al.[17] indicate in their paper that CRF perikarya in the MVN tend to localize to the periphery of the nucleus. Staining of CRF immunoreactive fibers, presumably axons, is also present within the vestibular complex. With the exception of the study by Sakanaka et al.,[66] which reported that all four vestibular nuclei contain CRF immunoreactive fibers, the majority of reports indicate that CRF containing fibers are only present in the MVN and IVN, which coincidentally also contain the CRF immunoreactive perikarya. While the function of CRF in the vestibular complex has not been studied, the presence of CRF in numerous brainstem sensory nuclei has led to the suggestion that CRF may play a role in modulating sensory information prior to its relay to higher levels of the neuraxis[54]. This is also a likely function for CRF in the vestibular complex.

Adrenocorticotropic hormone (ACTH), also known as corticotropin, is principally found in neurons of the forebrain, although a few neurons in the hindbrain also contain the peptide.[58] A study by Pilcher and Joseph[63] suggested that ACTH fibers typically innervate regions of the brain containing CRF immunoreactive perikarya. If this hypothesis is true, one would predict that the MVN and IVN should contain fibers that stain positive for ACTH. However, to date, radioimmunoassay and immunocytochemical studies have failed to validate this hypothesis (see Palkovits;[62] and Nieuwenhuys.[58] Despite the lack of direct immunocytochemical evidence for ACTH fibers in the vestibular complex, electrophysiological studies of MVN neurons *in vitro* have indicated that these neurons respond to the 4-10 and 4-9 fragments of ACTH.[27] Moreover, Gilchrist et al.[26] have shown that administration of the synthetic ACTH-(4-9) analog, Org 2766, directly into the ipsilateral vestibular nucleus complex enhances vestibular compensation and produces a significant decrease in spontaneous ocular nystagmus. Thus, ACTH appears to have a significant effect on the vestibular system, despite the fact that it has not been detected in the vestibular complex by standard immunocytochemical and radioimmunoassay procedures. One explanation for this discrepancy is that ACTH may be acting allosterically on non-melanocortin receptors, such as the glutamate NMDA receptor.[18] Alternatively, ACTH may represent an example of a neuropeptide that may reach the vestibular nuclei indirectly via the cerebrospinal fluid (CSF) rather than being synthesized and secreted within the vestibular complex. Several hypothalamic/pituitary neuropeptides including LHRH, vasopressin, oxytocin, TRH, and ACTH, have been shown to be secreted into the CSF either directly or indirectly via the choroid plexus.[6,9,22] Thus, these peptides could reach the vestibular nuclei via the CSF of the fourth ventricle. In addition, it is known that the hypothalamus sends direct projections to the nucleus of the tractus solitarius (NTS), which lies adjacent to the vestibular complex. Thus, hypothalamic neuropeptides could be released into the NTS from these descending axons and diffuse into the vestibular complex. The above mechanisms could also explain the discrepancy between radioimmunoassay data, indicating that thyroid releasing hormone (TRH) is present in both the MVN and LVN[62] and immunocytochemical studies which fail to detect TRH in the vestibular complex.[58,62] Since TRH receptors are present in the MVN and LVN,[87] it is probable that this peptide affects vestibular function, but whether it is released locally in the vestibular complex, or diffuses to this area from the CSF or NTS, remains an open question. Further study is required to better characterize the mode of action of ACTH, TRH, somatostatin, CRF, and other hypothalamic peptides on central vestibular neurons.

9.4.4 OTHER NEUROPEPTIDES IN THE VESTIBULAR COMPLEX

9.4.4.1 Cholecystokinin

A number of other peptides have been identified in the vestibular complex. The octapeptide cholecystokinin (CCK), which was originally identified in the duodenum, and is responsible for secretion of pancreatic juice and the ejection of bile, is one of the most abundant neuropeptides in the brain and has been localized immunocytochemically in numerous cell groups along the neuraxis.[58] Studies in the early postnatal rat[42] indicate that small numbers of neurons containing CCK-like immunoreactivity can be found as early as day 6 in the vicinity of the MVN and prepositus hypoglossal nucleus. This same study shows scattered CCK neurons in the IVN and possibly the caudal LVN in the adult. In addition, a sparse plexus of fibers stained for CCK-like immunoreactivity has been described in the MVN and IVN.[44,62] Carpenter et al.[11] have also described CCK-like immunoreactive fibers and terminals surrounding the cells of the LVN in the squirrel monkey. Finally, King and co-workers[41] have described CCK immunoreactive neurons in the MVN of the opossum that project as mossy fibers to the cerebellar cortex. In sum, these results suggest that CCK immunoreactive perikarya are localized primarily in the MVN and IVN, that some CCK cells in the MVN project to the cerebellum, and that CCK immunoreactive fibers are concentrated in the MVN, IVN, and LVN. In addition, an autoradiographic study of CCK binding sites has indicated that the MVN possesses the highest density of CCK binding sites in the lower brainstem,[56] suggesting that CCK should have a potent effect on MVN neurons.

From a functional standpoint, CCK plays an important role in both the alimentary tract and the central nervous system. The overlapping distribution of opioid and cholecystokinin (CCK) peptides and their receptors (μ and δ-opioid receptors; CCK-A and CCK-B receptors) in the central nervous system have led to a large number of studies aimed at clarifying the functional relationships between these two neuropeptides. Pharmacological studies have suggested that CCK is an anti-opioid peptide and it appears to exert its anti-opioid actions mainly through the activation of CCK-B receptors.[13] However, CCK also exhibits opioid-like effects that seem to result from the stimulation of CCK-A receptors.[13] Most studies devoted to the interactions of CCK and opioids have focused on the control of pain[64] and have not examined possible relationships with respect to vestibular function. From a functional perspective, it is worth noting that CCK has been implicated in a wide variety of brain functions, including pain, feeding, control of the release of hypothalamic hormones, and the neurobiology of anxiety and panic disorder.[8] With respect to the vestibular system, it has been shown that the sulfonated form of cholecystokinin, when administered intramedullarly, can elicit barrel role behavior,[53] suggesting a possible role of CCK in vestibular function. Clearly, additional work is needed to clarify the effect of CCK on central vestibular neurons and the role that this peptide may play in vestibular function. It is worth mentioning that gastrin, a neuropeptide closely related to CCK, has not been localized to the vestibular complex.[48,58]

9.4.4.2 Vasoactive Intestinal Polypeptide

Vasoactive intestinal polypeptide (VIP), like CCK, is a gut hormone that can also be found in neurons of the CNS. At least one study utilizing radioimmunoassay indicates that VIP exists in low concentrations in the LVN.[21] However, immunocytochemical mapping studies of this peptide have failed to support the evidence that VIP is localized in the LVN (although it should be noted that VIP mRNA is found in the nucleus vestibularis descendens of the chicken; see Kuenzel et al.[45]). A likely explanation for this discrepancy is that the nucleus of the tractus solitarius (NTS) is the only lower brainstem structure that contains a significant number of VIP-like neurons, fibers, and receptors.[58] In light of this, it is probable that the radioimmunoassay detection of VIP in the LVN represents false positive data which is due to contamination of the vestibular sample with tissue containing the NTS. While VIP receptors are present in relatively high levels in the cerebellar cortex, they are generally present in low levels in the brainstem, with the exception of the NTS. Further study is needed to determine if VIP receptors are actually present in the vestibular complex and whether this peptide has any effect on central vestibular neurons.

9.4.4.3. Neuropeptide Y

Neuropeptide Y (NPY) is the most abundant peptide present in the mammalian central and peripheral nervous system,[3] and has been demonstrated in neurons of both the PNS and CNS (see Ref. 62). NPY exhibits a variety of central and peripheral effects, including those on feeding, blood pressure, cardiac contractility, and intestinal secretion.[3] Classical pharmacological studies and newer molecular studies have shown that NPY effects are mediated by five different receptor subtypes (Y1 - Y5).[3] With regard to the vestibular system, Harfstrand et al.[30] mapped the distribution of NPY in the medulla oblongata of the rat and found a distinct concentration of neurons containing the peptide in the caudal half of the MVN. From their diagrams and photomicrographs, it appears that NPY neurons are concentrated in the dorsomedial aspect of the MVN. In addition to NPY containing perikarya, Harfstand et al.[30] describe a sparse network of immunoreactive fibers that extend throughout the MVN and spread into the IVN. Unfortunately, because their study was restricted to the caudal brainstem, it is not clear if NPY perikarya and fibers are found at more rostral levels of the MVN and IVN or if NPY is found in the LVN and SVN. In contrast to the description of Harfstand and co-workers in the rat vestibular complex, Blessing et al.[7] failed to observe the presence of NPY neurons or fibers in the vestibular complex of the rabbit.   With respect to NPY immunoreactive fibers in the vestibular complex, it is interesting to note that a subpopulation of olivocerebellar fibers originating from the caudal inferior olive has been shown to express NPY.[76] Since the dorsal accessory olivary and beta subnuclei project to the vestibular complex, it is tempting to speculate that some of the neuropeptide Y fibers found in the vestibular complex may arise from the inferior olivary nucleus. Finally, it is worth noting that a recent study by Williams[84] has shown that NPY secretion is reduced in the feline vestibular nuclei during isometric contractions. The exact role of NPY in the vestibular system requires further investigation.

9.4.4.4 Pancreatic Polypeptide and Delta-Sleep-Inducing Peptide

With respect to other peptides in the vestibular nuclei, fibers staining positive for bovine pancreatic polypeptide (BPP) have been reported in the MVN, but details are lacking[61] and the significance of this innervation is unknown.   In an immunocytochemical study, Feldman and Kastin[24] have localized a peptide, known as delta-sleep-inducing peptide (DSIP), to perikarya in all four principal vestibular nuclei.   Details concerning the distribution and relative numbers of DSIP neurons in each of the vestibular nuclei were not provided, although a photomicrograph of the caudal half of the MVN indicates that these neurons are concentrated in the center of that particular nucleus. From a functional standpoint, DSIP has been shown to increase sleep in several species, but it also has some non-sleep-related effects including effects on locomotion.[24]   To our knowledge, the effects of BBP, DSIP, or NPY on vestibular neurons have not been reported.

9.4.4.5 Calcitonin Gene-Related Peptide

Both immunocytochemical and *in situ* hybridization studies have shown that calcitonin gene-related peptide (CGRP) is widely distributed in the CNS.[59,68,74,82]   The combined results of these studies indicate that CGRP is not found in the sensory portion of the vestibular system, but rather is associated with the efferent limb of the vestibular system (see Chapter 4 by Goldberg et al). Neurons comprising the efferent vestibular system are found in three groups: the largest group of vestibular efferent cell bodies lie between the vestibular and abducens nuclei; one group is located near the boarder of the genu of the facial nerve; and the third group is located in the caudal pontine reticular formation.[82]   All three groups of neurons are cholinergic and about 55 percent of these neurons colocalize with CGRP.[59]   The absence of CGRP-containing neurons in Scarpa's ganglion, combined with the finding that CGRP-containing fibers and terminals located in the vestibular sensory epithelium (i.e., cristae and maculae) disappear following sectioning of the vestibular efferent tract in the brainstem,[74] provide strong evidence that CGRP is an important component of the efferent vestibular pathway.   CGRP coexists with acetylcholine in efferent axonal endings that are presynaptic to type II hair cells and to afferent fiber.[67] The reader is referred to Chapter 4 for further details.

9.5 CONCLUSION

This chapter has summarized what is known regarding neuropeptides in the vestibular nuclei. While there are a myriad of neuropeptides that have been described in the brain, most anatomical mapping studies have neglected to carefully map these peptides among the vestibular nuclei; thus, the distribution of a small percentage only of neuropeptides has been described within the vestibular complex with any degree of accuracy. Of these, not enough have been tested with appropriate physiological and/or pharmacologcal approaches to justify their inclusion as bonified neurotransmitters/neuromodulators in the vestibular system. As reviewed above, there is ample evidence to suggest that somatostatin, enkephalin, substance P, and perhaps CCK play important roles in vestibular function. While several other neuropeptides are localized within the vestibular complex or have been shown physiologically or behaviorally to affect vestibular function, more investigation is required to verify that they are indeed neurotransmitters or neuromodulators in the vestibular system. Such is the case for several hypothalamic peptides which do not appear to be present in the vestibular nuclei, but which have been shown to affect vestibular function. There are many more neuropeptides in the brain for which anatomical distribution maps and functional data are as of yet incomplete and for which future studies may implicate a role in central vestibular function. In addition to the neuropeptides listed above, there are a number of other "non-neural" peptides that affect the nervous system, and these peptides are yet to be examined for their possible effects on vestibular function.   These include cytokines, certain chemokines, heat shock proteins, and a multitude of growth factors[88] (see Chapter 19, the summary to the book, for a list of some of these peptides).

Research in the areas of vestibular pharmacology and neurochemistry over the past decade has had a major impact on our understanding of the vestibular system. As reviewed in this and other chapters, multiple transmitter systems appear to converge on the vestibular nuclei, however, it is becoming apparent that glutamate is the major transmitter used by the vestibular nerve. It is also becoming clear that the vestibular nerve engages in complex transmitter interactions, possibly using neuropeptides as cotransmitters, in order to modify the sensitivity of postsynaptic receptors.   In addition, there appear to be neuropeptides that are intrinsic to the vestibular nuclei, such as CCK; those that are secreted from axons that arise primarily from extrinsic sources, such as thyroid-releasing hormone; and those that arise from both intrinsic and extrinsic sources, such as substance P. While the effects of several neuropeptides have been tested *in vitro* using brain slices, many neuropeptides must still be tested and there is an urgent need for more *in vivo* studies to confirm the results of *in vitro* studies and to verify that these neuropeptides do in fact play a role in the vestibular system of intact animals.

ACKNOWLEDGMENTS

We would like to thank Dr. David Brown for his helpful comments on neuropeptides in the brain. This work was supported by NIH grants NS31318 and DC01086.

REFERENCES

1. Abood, LG, Knapp, R, Mitchell, T, Booth, H, and Schwab, L, Chemical requirements of vasopressins for barrel rotation convulsions and reversal by oxytocin, *J. Neurosci. Res.,* 5: 191,1980.
2. Balaban, CD, Starcevic, VP, and Severs, WB, Neuropeptide modulation of central vestibular circuits, *Pharmacol. Rev.,* 41: 53,1989.
3. Balasubramaniam, AA, Neuropeptide Y family of hormones: receptor subtypes and antagonists, *Peptides,* 18:445,1997.
4. Barker, R, Tachykinins, neurotrophism and neurodegenerative diseases: a critical review of the possible role of tachykinins in the etiology of CNS diseases, *Rev. Neurosci.,* 7:187,1996.
5. Beitz, AJ, Clements, JR, Ecklund, L, and Mullett, MM, The nuclei of origin of brainstem enkephalin and cholecystokinin projections to the spinal trigeminal nucleus of the rat, *Neuroscience,* 20: 409,1987.
6. Bennett-Clarke, C, and Joseph, SA, Immunocytochemical distribution of LHRH neurons and processes in the rat: hypothalamic and extrahypothalamic locations, *Cell Tiss. Res.,* 221:493,1982.
7. Blessing, WW, Howe, PCR, Joh, TH, Oliver, JR, and Willoughby, JO, Distribution of tyrosine hydroxylase and neuropeptide Y-like immunoreactive neurons in rabbit medulla oblongata, with attention to colo-calization studies, presumptive adrenaline-synthesizing perikarya, and vagal preganglionic cells, *J. Comp. Neurol.,* 248: 285,1986.
8. Bourin, M, Malinge, M, Vasar, E, and Bradwejn, J, Two faces of cholecystokinin: anxiety and schizophrenia, *Fundam. Clin. Pharmacol.,* 10:116,1996.
9. Brownfield, MS, and Kozlowski, GP, The hypothalamo-choroidal tract. 1. Immunohistochemical demon-stration of neurophysin pathways to telencephalic choroidal plexuses and cerebrospinal fluid, *Cell Tiss. Res.,* 178:111,1977.
10. Burke, RE, and Fahn, S, Studies of somatostatin-induced barrel rotation in rats, *Regul. Peptides,* 7: 207,1983.
11. Carpenter, MB, Periera, AB, and Guha, N, Immunocytochemistry of oculomotor afferents in the squirrel monkey (*Saimiri sciureus*), *J. Hirnforsch.,* 33:151,1992.
12. Carpentier, V, Vaudry, H, Laquerrier, A, Tayot, J, and Leroux, P, Distribution of somatostatin receptors in the adult human brainstem, *Brain Res.,* 734:135,1996.
13. Cesselin, F, Opioid and anti-opioid peptides, *Fundam. Clin. Pharmacol.,* 9:409,1995.

14 Chan-Palay, V, Ito, M, Tongroach, P, Sakurai, M, and Palay, S, Inhibitory effects of motilin, somatostatin, leu-enkephalin, met-enkephalin and taurine on neurons of the lateral vestibular nucleus: interactions with GABA, *Proc. Natl. Acad. Sci.*, 79: 3355,1982.

15. Cuello, AC, and Kanazawa, I, The distribution of substance P. immunoreactive fibers in the rat central nervous system, *J. Comp. Neurol.*, 178: 129,1978.

16. Culman, J, and Unger, T, Central tachykinins: mediators of defence reaction and stress reactions, *Can. J. Physiol. Pharmacol.*, 73(7):885,1995.

17. Cummings, SL, Young, WS, Bishop, GA, DeSouza, EB, and King, JS, Distribution of corticotropin-releasing factor in the cerebellum and precerebellar nuclei of the opossum: A study utilizing immunohistochemistry, *in situ* hybridization histochemistry, and receptor autoradiography, *J. Comp. Neurol.*, 280: 501,1989.

18. Darlington, CL, Gilchrist, DP, and Smith, PF, Melanocortins and lesion-induced plasticity in the CNS: a review, *Brain Res. Rev.*, 22:245,1996.

19. deWaele, C, Muhlethaler, M, and Vidal, PP, Neurochemistry of the central vestibular pathways, *Brain Res. Rev.*, 20: 24,1995.

20. Dutia, MB, Gilchrist, DP, Sansom, AJ, Smith, PF, and Darlington, CL, The opioid receptor antagonist, naloxone, enhances ocular motor compensation in guinea pig following peripheral vestibular deafferentation, *Exp Neurol.*, 141:141,1996.

21. Eiden, LE, Nilaver, G, and Palkovits, M, Distribution of vasoactive intestinal polypeptide (VIP) in the rat brain, *Brain Res.*, 231: 472,1982.

22. Eiden, LE, and Brownstein, MJ, Extrahypothalamic distributions and functions of hypothalamic peptide hormones, *Federation Proc.*, 40:2553,1981.

23. Euler, US, and Gaddum, JH, An unidentified depressor substance in certain tissue extracts, *J. Physiol. (London)*, 72:74,1931.

24. Feldman, SC, and Kastin, AJ, Localization of neurons containing immunoreactive delta-sleep-inducing peptide in the rat brain, *Neuroscience*, 11: 303,1984.

25. Finley, JCW, Maderdrut, JL, and Petrusz, P, The immunocytochemical localization of enkephalin in the central nervous system of the rat, *J. Comp. Neurol.*, 33: 28,1981.

26. Frenk, H, Walkins, LR, and Meyer, DJ, Differential behavioral effects induced by intrathecal microinjection of opiates: comparison of convulsive and cataleptic effects produced by morphine, methadone and D-Ala2-methionine-enkephalinamide, *Brain Res.*, 299: 31,1984.

27. Gilchrist, DP, Darlington, CL, and Smith, PF, An *in vitro* investigation of the effects of the ACTH/MSH(4-9) analogue, Org 2766, on guinea pig medial vestibular nucleus neurons, *Peptides*, 17:681,1996.

28. Gilchrist, DP, Darlington, CL, and Smith, PF, Evidence that short ACTH fragments enhance vestibular compensation via direct action on the ipsilateral vestibular nucleus, *NeuroReport*, 7:1489,1996.

29. Harmar, AJ, Three tachykinins in mammalian brain, *Trends Neurosci.*, 7:57,1983.

30. Harfstrand, A, Fuxe, K, Terenuis, L, and Kalia, M, Neuropeptide Y-immunoreactive perikarya and nerve terminals in the rat medulla oblongata: Relationship to cytoarchitecture and catecholaminergic cell groups, *J. Comp. Neurol.*, 260: 20,1987.

31. Harlan, RE, Garcia, MM, and Krause, JE, Cellular localization of substance P and neurokinin A encoding preprotachykinin mRNA in the female rat brain, *J. Comp. Neurol.*, 287: 179,1989.

32. Hokfelt, T, Elde, R, Johansson, O, Terenius, L, and Stein, L, The distribution of enkephalin-immunoreactive cell bodies in the rat central nervous system. *Neurosci. Lett.*,5:25,1977.

33. Houtani, T, Nishi, M, Takeshima, H, Nukada, T, and Sugimoto, T, Structure and regional distribution of nociceptin/orphanin FQ precursor, *Biochem. Biophys. Res. Commun.*, 219:714,1996.

34. James, TA, and Starr, MS, Effects of Substance P injected into the substantia nigra, *Br. J. Pharmacol.*, 65: 423,1979.

35. Jennes, L, Stumpf, WE, and Kalivas, PW, Neurotensin: Topographical distribution in the rat brain by immunohistochemistry, *J. Comp. Neurol.*, 210: 211,1982.

36. Johansson, O, Hokfelt, T, and Elde, RP, Immunohistochemical distribution of somatostatin-like immunoreactivity in the central nervous system of the adult rat, *Neuroscience*, 13: 265,1984.

37. Joseph, SA, Pilcher, WH, and Knigge, KM, Anatomy of the corticotropin-releasing factor and opiomelanocortin systems of the brain, *Federation Proc.*, 44: 100,1985.

38. Kangawa, K, Minamino, N, Fukuda, A, and Matsuo, H, Neuromedin K: A novel mammalian tachykinin identified in porcine spinal cord, *Biochem. Biophys. Res. Commun.*, 114:533,1983.

39. Khachaturian, H, Lewis, ME, and Watson, SJ, Enkephalin systems in diencephalon and brainstem of the rat, *J. Comp. Neurol.,* 220: 310,1983.

40. Khawaja AM, and Rogers, DF, Tachykinins: receptor to effector, *Int. J. Biochem. Cell. Biol.,* 28:721,1996.

41. King, JS, and Bishop, GA, Distribution and brainstem origin of cholecystokinin-like immunoreactivity in the opossum cerebellum, *J. Comp. Neurol.,* 298:373,1990.

42. Kiyama, H, Shiosaka, S, Kubota, Y, Cho, HJ, Takagi, H, Tateishi K, Hashimura, E, Hamaoka T, and Tohyama, M, Ontogeny of cholecystokinin-8-containing neuron system of the rat: An immunohistochemical analysis-II. Lower brain stem, *Neuroscience,* 10: 1341,1983.

43. Kruse, H, Wimersma-Greidanus, TB, and deWied, D, Barrel rotation induced by vasopressin and related peptides in rats, *Pharmacol. Biochem. Behav.,* 7: 311,1977.

44. Kubota, Y, Inagaki, S, Shiosaka, S, Cho, HJ, Tateishi, K, Hashimura, E, Hamaoka, T, and Tohyama, M, The distribution of cholecystokinin octapeptide-like structures in the lower brain stem of the rat: An immunohistochemical analysis, *Neuroscience,* 9: 587,1983.

45. Kuenzel, WJ, Mccune, SK, Talbot, RT, Sharp, PJ, and Hill, JM, Sites of gene expression for vasoactive intestinal polypeptide throughout the brain of the chick (Gallus domesticus), *J. Comp. Neurol.,* 381:101,1997.

46. Leeman, S, and Mroz, EA, Substance P, *Life Sci.,* 15: 2033,1974.

47. Ljungdahl, A. Hokfelt, T, and Nilsson, G, Distribution of substance P immunoreactivity in the central nervous system of the rat. I. Cell bodies and nerve terminals, *Neuroscience,* 3: 861,1978.

48. Loren, I, Alumets, J, Hakanson, R, and Sundler F, Distribution of Gastrin and CCK-like peptides in rat brain, *Histochemistry,* 59: 249,1979.

49. Maeno, H, Kiyama, H, and Tohyama, M, Distribution of the substance P receptor (NK-1 receptor) in the central nervous system, *Mol. Brain Res.,* 18: 43,1993.

50. Maggi, CA, The mammalian tachykinin receptors, *Gen. Pharmacol.,* 26:911,1995.

51. Maggio, JE, Sandberg, BEB, Bradley, CV, Iversen, LL, Santikarn, S, Williams, DH, Hunte,r JC, and Hanley, MR, Substance K: A novel tachykinin in mammalian spinal cord. in: *Substance P* (Skrabanek P, Powell D, Eds.), pp. 20-21, Dublin, Boole Press,1983.

52. Magnusson, T, Carlsson, A, Fisher, GH, Chang, D, and Folkers, K, Effect of synthetic substance P on monoaminergic mechanisms in brain, *J. Neural Transm.,* 38: 89,1976.

53. Mann, JFE, Boucher, R, and Schiller PW, Rotational syndrome after central injection of C-terminal 7-peptide of cholecystokinin, *Pharmacol. Biochem. Behav.,* 13: 125,1980.

54. Merchenthaler, I, Corticotropin releasing factor (CRF)-like immunoreactivity in the rat central nervous system. Extrahypothalamic distribution, *Peptides,* 5: 53,1984.

55. Merchenthaler, I, Maderdrut, J, Altschuler, RA, and Pertrusz, P, Immunocytochemical localization of proenkephalin-derived peptides in the central nervous system of the rat, *Neuroscience,* 17: 325,1986.

56. Miceli, MO, and Steiner, M, Novel localizations of central- and peripheral-type cholecystokinin binding sites in Syrian hamster brain as determined by autoradiography, *Eur. J. Pharmacol.,* 169:215,1989.

57. Nomura, I, Semba, E, Kubo, T, Shirasishi T, Matsunaga T, Tohyama, M, Shiotani, Y, and Wu, JY, Neuropeptides and gamma-aminobutyric acid in the vestibular nuclei of the rat: An immunohistochemical analysis. I. Distribution, *Brain Res.,* 311: 109,1984.

58. Nieuwenhuys, R, (Ed.), *Chemoarchitecture of the Brain.* New York, Springer-Verlag, pp 1-113,1985.

59. Ohno, K, Takeda, N, Yamano, M, Matsunaga, T, and Tohyama, M, Coexistence of acetylcholine and calcitonin gene-related peptide in the vestibular efferent neurons in the rat, *Brain Res.,* 566: 103,1991.

60. Oley, N, Cordova, C, Kelly, ML, and Bronzino, JD, Morphine administration to the region of the solitary nucleus produces analgesia in rats. *Brain Res.,* 236:511,1982.

61. Olschowka, JA, O'Donohue, TL, and Jacobowitz, DM, The distribution of bovine pancreatic polypeptide-like immunoreactive neurons in the rat brain, *Peptides,* 2:309,1981.

62. Palkovits, M, Distribution of neuropeptides in the central nervous system: A review of biochemical mapping studies, *Prog. Neurobiology,* 23: 151,1984.

63. Pilcher, WH, and Joseph, SA, Colocalization of CRF-ir perikarya and ACTH-ir fibers in rat brain, *Brain Res.,* 299: 91,1984.

64. Roques, BP, and Noble, F, Association of enkephalin catabolism inhibitors and CCK-B antagonists: a potential use in the management of pain and opioid addiction, *Neurochem. Res.,* 21:1397,1996.

65. Saika, T, Takeda, N, Kiyama, H, Kudo, T, Tohyama, M, and Matsunaga, T, Changes in gene expression of neuropeptides after unilateral labyrinthectomy in the rat brain. In H. Krejocava and J. Jerabek (Eds.), *Proceeding of the XVII th Barany Society Meeting*, pp. 196-197,1992.

66. Sakanaka, M, Shibasaki, T, and Lederis, K, Corticotropin releasing factor-like immunoreactivity in the rat brain as revealed by modified cobalt-glucose oxidase-diaminobenzidine method, *J. Comp. Neurol.*, 260: 256,1987.

67. Scarfone, E, Ulfendahl, M, and Lundeberg, T, The cellular localization of the neuropeptides substance P, neurokinin A, calcitonin gene-related peptide and neuropeptideY in guinea pig vestibular sensory organs: a high-resolution confocal microscopy study, *Neuroscience*, 75:587,1996.

68. Skofitsch, K, and Jacobowitz, DM, Calcitonin gene-related peptide: detailed immunohistochemical distribution in the central nervous system, *Peptide*, 6: 721,1985a.

69. Skofitsch, K, and Jacobowitz, DM, Distribution of corticotropin releasing factor-like immunoreactivity in the rat brain by immunohistochemistry and radioimmunoassay: Comparison and characterization of ovine and rat/human CRF antisera, *J. Comp. Neurol.*, 6: 319,1985b.

70. Smith, PF, and Darlington, CL, Recent advances in the pharmacology of the vestibulo-ocular reflex, *Trends Pharmacol. Sci.*, 17:421,1996.

71. Sofroneiw, MV, Morphology of vasopressin and oxytocin neurons and their central processes, *Prog. Brain Res.*, 60: 101,1983a.

72. Sofroneiw, MV, Vasopressin and oxytocin in the mammalian brain and spinal cord, *Trends Neurologic Sci.*, 6: 467,1983b.

73. Standaert, DG, Saper, CB, Rye, DB, and Wainer, BH, Colocalization of atriopeptin-like immunoreactivity with choline acetyltransferase and substance P-like immunoreactivity in the pedunculopontine and laterodorsal tegmental nuclei in the rat, *Brain Res.*, 382: 163,1986.

74. Tanaka, M, Takeda, N, Senba, E, Tohyama, M, Kubo, T, and Matsunaga, T, Localization, origin and fine structure of calcitonin gene-related peptide-containing fibers in the vestibular end-organs of the rat, *Brain Res.*, 504: 31,1989.

75. Tribollet, E, Barberis, C, Jard, S, Dubois-Dauphin, M, and Dreifuss, JJ, Localization and pharmacological characterization of high affinity binding sites for vasopressin and oxytocin in the rat brain by light microscopic autoradiography, *Brain Res.*, 442: 105,1988.

76. Ueyama, T, Houtani, T, Nakagawa, H, Baba, K, Ikeda, M, Yamashita, T, and Sugimoto, T, A subpopulation of olivocerebellar projection neurons express neuropeptide Y, *Brain Res.*, 634:353,1994.

77. Uhl, GR, Kuhar, MJ, and Snyder, SH, Neurotensin: Immunohistochemical localization in the rat central nervous system, *Proc. Natl. Acad. Sci.*, 74: 4059,1977.

78. Uhl, GR, Tran, V, Snyder, SH, and Martin, JS, Distribution of somatostatin receptors in rat central nervous system and human frontal cortex, *J. Comp. Neurol.*, 240: 288,1985.

79. Vibert, N, Serafin, M, Vidal, PP, and Muhlethaler, M, Effects of substance P on medial vestibular nucleus neurons in guinea pig brainstem slices, *Eur. J. Neurosci.*, 8:1030,1996.

80. Vincent, SR, McIntosh, CHS, Buchan, AMJ, and Brown, CJ, Central somatostatin systems revealed with monoclonal antibodies, *J. Comp. Neurol.*, 238: 169,1985.

81. Vincent, SR, Satoh, K, Armstrong DM, Fibiger HC Substance P in the ascending reticular system, *Nature*, 306: 688,1983.

82. Wackym, PA, Popper, P, and Micevych, PE, Distribution of calcitonin gene-related peptide mRNA and immunoreactivity in the rat central and peripheral vestibular system, *Acta Otolaryngol.*, 113: 601,1993.

83. William, RG, and Dockray, GJ, Distribution of enkephalin-related peptides in rat brain: Immunohistochemical studies using antisera to met-enkephalin and met-enkephalin Arg6Phe7, *Neuroscience, 9*: 563,1983.

84. Williams, CA, Neuropeptide Y-like substances are released from the rostral brainstem of cats during the muscle pressor response, *J. Physiol. (London)*, 495:267,1996.

85. Zadina, JE, Hackler, L, Ge, LJ, and Kastin, AJ, A potent and selective endogenous agonist for the mu-opiate receptor, *Nature*, 386:499,1997.

86. Zamir, N, Palkovits, M, and Brownstein, MJ, Distribution of immunoreactive dynorphin in the central nervous system of the rat, *Brain Res.*, 280: 81,1983.

87. Zanni, M, Giardino, L, Toschi, L, Galetti, G, and Calza, L, Distribution of neurotransmitters, neuropeptides, and receptors in the vestibular nuclei complex of the rat: an immunocytochemical, *in situ* hybridization and quantitative receptor autoradiographic study, *Brain Res. Bull.*, 36:443,1995.

88. Zheng, JL, Stewart, RR, and Gao, WQ, Neurotrophin-4/5 brain-derived neurotrophic factor, and neutor-ophin-3 promote survival of cultured vestibular ganglion neurons and protect them against neurotox-icity of ototoxins, *J. Neurobiol.,* 28: 330,1995.

Section IV

Neurochemical Mechanisms Underlying Vestibular Control of Movement and Posture

10

Neuropharmacological Aspects of the Vestibulo-Ocular Reflex

James G. McElligott and Robert F. Spencer

CONTENTS

10.1 INTRODUCTION

The vestibulo-ocular reflex stabilizes targets of visual interest on the retina in order to maintain clear vision during head motion. Eye movements elicited by motion of the whole visual field are often not adequate to compensate for these head movements, because the visual related associated neuronal processes involve a slower control system.[78] Head rotation stimulates the peripheral vestibular endorgan that sends afferent signals related to head position and velocity into the central nervous system. Ultimately, these signals are integrated and transmitted to the extra-ocular muscles to produce compensatory eye movements. In order to clearly view a stationary, earth-fixed visual target, the resulting eye movement must be equal in amplitude and velocity, but in the opposite direction of the head (Figure 10.1A: NORMAL).

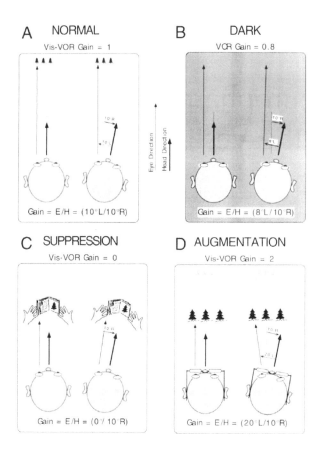

FIGURE 10.1 The vestibulo-ocular reflex is schematically represented in four different head movement associated conditions; (A: NORMAL) operation of the reflex in the light while viewing an earth-fixed target (trees); (C: SUPPRESSION) in the light while viewing a target (handheld book) that moves with the head; (D: AUGMENTATION) in the light while viewing an earth-fixed target (trees) through magnifying glasses (magnifying factor = 2X); and (B: DARK) in the dark. Light and heavy arrows, respectively, represent eye (E) and head (H) directions. For the pictoral representations (A-D) that describe operation of the reflex, the term Vis-VOR (visuo-vestibulo-ocular reflex) and VOR are used to describe operation of the reflex respectively in the light (A,C,D) and in the dark (B).

In many of our daily activities, it is not always appropriate to produce eye movements that are equal in amplitude and velocity, but opposite to that of the head. For example, while looking at a book in a moving vehicle, if there is a sudden alteration of the vehicle's motion, momentum causes the head to undergo acceleration, resulting in stimulation of the peripheral vestibular end-organs. In this situation, it would not be proper for the eyes to move since the book is held stationary with respect to the head. Thus, the reflex must be suppressed and canceled to minimize motion of the eyes with respect to the head and book (Figure 10.1C: SUPPRESSION).

There are also situations where the eyes should move in the opposite direction but with a greater degree of rotation than the head, for example, when wearing eyeglasses. To correct for hyperopia, or farsightedness, the lenses of the eyeglasses in combination with those of the eye result in a telescope with a positive magnification. In this situation, the resulting eye movement must be of greater magnitude than that of the head in order to avoid image blurring (Figure 10.1D: AUG-MENTATION). In many cases, especially for people who have never worn glasses, proper adjustments or adaptations do not occur immediately and take place over a period of several days. This lack of instantaneous adaptation can cause visual and vertigo problems and, frequently, is the reason for complaints to the ophthalmologist. Often, these problems are corrected without any intervention other than having the patient wear the glasses at increasing time intervals for a period of several days while gradual adaptation of the vestibulo-ocular reflex takes place.

During sinusoidal stimulation, the vestibulo-ocular reflex is usually described in terms of gain and phase. Gain refers to the magnitude (position) or the rate (velocity) of eye movement compared to that of the head. Phase refers to the timing of eye and head movements with respect to each other. In the example where the eye is centered on an earth-fixed target during head movement (Figure 10.1A: NORMAL), the resulting equal and opposite eye movement has a gain equal to one and a phase equal to 180 degrees.

The phrase vestibulo-ocular reflex will be used in this chapter to describe the reflex in a general or generic way. However, a distinction should be made with respect to the two prominent modes in which the reflex is studied. The vestibulo-ocular reflex did not evolve to operate in isolation, but was meant to work in cooperation with the visual system. Visual following during head movement using the visual-oculomotor system alone is often not fast enough and must be "supplemented" by the vestibular system. Since both vestibular and visual inputs are contributing here, the reflex is more appropriately called the visuo-vestibulo-ocular reflex, or Vis-VOR. Experimental studies of the vestibulo-ocular reflex also are carried out in the dark when there is no visual input. In this case, the reflex is operating only with vestibular information and here the specific term VOR will be used. Typically, in normal operation, the gain of the Vis-VOR in the light is one (Figure 10.1A: NORMAL) and the gain of the VOR in the dark is approximately 0.8 to 0.9 (Figure 10.1B: DARK).

In the dark or in the light during normal operation, the vestibulo-ocular reflex functions unaltered. While operation in this mode is of interest in understanding the basic or normal operation of the reflex, the reflex can also be studied while being adapted to change gain over periods ranging from hours to days. It is this gradual adaptation that has served as a model to investigate neuroplastic phenomena in the brain and to study mechanisms of learning and memory in the visuo-vestibular sensorimotor system. These adaptive changes are necessary for this reflex to function properly during developmental as well as during age-related changes. For example, after birth, the size of a baby's head increases; thus the afferent signal from the peripheral vestibular end-organ for the same head rotation will increase because the peripheral sensors are farther from the center of rotation. Central nervous system mechanisms must compensate and adjust for this alteration in the afferent input signal. During aging, extra-ocular muscles sometimes weaken and/or eyeglasses are required. Both of these conditions also necessitate a readjustment or adaptation for the vestibulo-ocular reflex to function properly.

The purpose of this chapter, after providing a background summary for the basic anatomy and physiology of this reflex, is to describe the neuropharmacology involved with the Vis-VOR and the VOR with respect to their normal operation and adaptive processes.

10.2 ANATOMICAL AND PHYSIOLOGICAL SUBSTRATES OF THE VESTIBULO-OCULAR REFLEX

Correlated morphological, physiological, pharmacological, and behavioral studies in the mammalian oculomotor system have identified the neurons, nuclei, and pathways that are related to the many different types of eye movements we perform (e.g., smooth pursuit, optokinetic, saccadic, and vestibulo-ocular). This involves a complex interrelated circuitry of brainstem and cerebellar structures related to the vestibular and the visual systems (Figure 10.2).

In general, the premotor circuitry involves the reciprocal excitatory regulation of one set of extra-ocular motoneurons and the antagonistic inhibition of a second set of oculomotor neurons. It is now well established that different types of neurotransmitters are involved in the synaptic connections of pre-oculomotor neurons, and these can be related to a differential role in horizontal and vertical eye movements.[149-151]

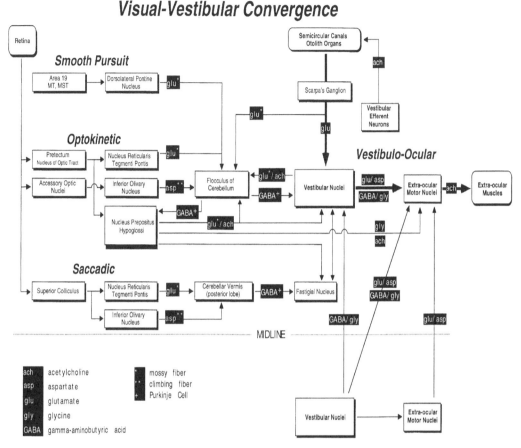

FIGURE 10.2 Different aspects of the visual-oculomotor system (smooth pursuit, optokinetic, and saccadic) converge on the vestibular system by a variety of pathways in the nervous system. The traditional direct 'three neuron arc' vestibulo-ocular pathway is represented by the thick arrows. The principal pathways and structures of the reflex are presented, along with the most likely associated excitatory and inhibitory neurotransmitters involved in the circuitry for the vestibulo-ocular reflex.

10.2.1 PERIPHERAL VESTIBULAR COMPONENTS

The peripheral vestibular endorgan consists of a membranous labyrinth comprising the three semicircular canals (horizontal, anterior, and posterior) and the otolith organs (utricle and saccule). The receptors for the vestibular system are hair cells, which are located in specialized regions of these structures: the crista ampullares of the semicircular canals and the maculae of the utricle and the saccule. The hair cells in the cristae of the semicircular canals are stimulated by an appropriate angular acceleration of the head that causes movement of the endolymph with a resulting deflection of the cilia. The hair cells in the maculae of the utricle and the saccule are stimulated by linear acceleration of the head, also producing deflection of the cilia by the overlying otoliths. Vestibular hair cells are excited (depolarized) or inhibited (hyperpolarized) by deflection of the cilia toward or in the direction opposite to the polarization of the stereocilia, respectively. In the crista ampullaris, the cilia are displaced by movement of the cupula caused by flow of the endolymph through the canals when the head moves. In the maculae, changes in the position of the head produce a shearing effect of the otoliths on the cilia.

Neural innervation of hair cells in the semicircular canals and the otolith organs is provided by bipolar neurons in the vestibular (Scarpa's) ganglion. The central processes of the bipolar neurons form the superior and inferior branches of the vestibular portion of the VIIIth (vestibulo-cochlear) cranial nerve. The frequency of discharges of the first-order vestibular axons encode head velocity and head position. There are also efferent projections from the CNS that synapse onto the receptor cells in the peripheral endorgan.

10.2.2 Central Vestibular Components

The central vestibular components are responsible for behaviors involving compensatory eye movements and the maintenance of postural balance. This is implemented by connections of the vestibular nuclei with motoneurons, respectively, in the extra-ocular motor nuclei and the spinal cord. Motoneurons in the extra-ocular motor nuclei are the final common elements receiving convergent afferents from brainstem premotor areas. The basic three-neuron chain of the vestibulo-ocular reflex—that is, —the peripheral vestibular afferents (first-order vestibular neurons), vestibular nucleus neurons (second-order vestibular neurons) and the extra-ocular motoneurons - comprise the traditional direct pathway of the reflex (Figure 10. 2; thick arrows). It is clear, however, that the nucleus prepositus hypoglossi and the cerebellum, as well as vestibular commissural connections which comprise part of the indirect pathway, have an important role in normal reflex operation (Figure 10.2).

Most first-order vestibular fibers terminate in one or more of the vestibular nuclei, which are arranged in two longitudinal columns along the dorsolateral extent of the medulla. The lateral column consists of the inferior (spinal or descending) vestibular nucleus, the lateral vestibular nucleus, and the superior vestibular nucleus. The medial vestibular nucleus constitutes the medial cell column. Upon entering the vestibular complex, most of these vestibular fibers bifurcate into ascending and descending rami. In general, the vestibular fibers that innervate the cristae ampullares of the semicircular canals project primarily to rostral portions of the vestibular complex, including the superior vestibular nucleus, ventral portions of the lateral vestibular nucleus, and rostral portions of the medial vestibular nucleus.[15] By contrast, primary vestibular fibers that innervate the maculae of the saccule and the utricle terminate in caudal portions of the medial and inferior vestibular nuclei.

Some first-order vestibular fibers bypass the vestibular nuclei and ascend directly to the cerebellum via the juxtarestiform body. Cells in all parts of the vestibular ganglion send mossy fiber projections to the ipsilateral nodulus, uvula, and flocculus. The cerebellum plays a major role in the integration of vestibular and visual information (Figure 10.2). In addition to first-order vestibular ganglion neurons, the second-order neurons in the superior, medial, and inferior vestibular nuclei project their axons through the juxtarestiform body bilaterally to the cortex (nodulus, uvula, and flocculus) and deep nuclei (fastigial) of the cerebellum. Reciprocal connections with the cerebellum are achieved by cerebello-vestibular fibers from the fastigial nuclei of both sides, and from Purkinje cells in the ipsilateral flocculus that course through the juxtarestiform body to terminate in all four vestibular nuclei. In this manner, the cerebellum exerts an influence on the entire vestibular complex.

Canal-specific, differential efferent projections of second-order vestibular neurons with motoneurons in the extra-ocular motor nuclei provide the basis for the vestibulo-ocular reflex. Each semicircular canal is related to two pairs of muscles in each eye through reciprocal excitatory and inhibitory synaptic connections with the extra-ocular motoneurons. The basis of this interaction is due to the relationship of the spatial orientation of the semicircular canals with the pulling actions of the extra-ocular muscles. However, pathways between the vertical semicircular canals and horizontal eye muscles, as well as vice versa, also exist as demonstrated by behavioral[69] and electrophysiological[67] cross-axis adaptation studies.

10.2.2.1 Vertical Vestibulo-Ocular Pathways

The vertical VOR (Figure 10.3A) involves reciprocal excitatory and inhibitory synaptic connections of second-order vestibular neurons with motoneurons in the oculomotor (III, Figure 10.3a) and trochlear (IV, Figure 10.3a) nuclei[151,152].

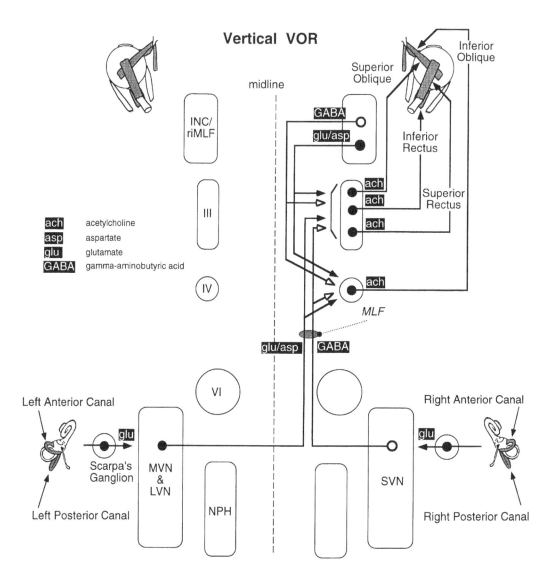

FIGURE 10.3 The principal pathways for the vertical (A) and horizontal (B) VOR identifying major excitatory and inhibitory neurotransmitters. (A) The structures involved in the vertical VOR include the MVN (medial), LVN (lateral), SVN (superior) vestibular nuclei, the III (oculomotor) and IV (trochlear) nuclei, as well as the irMLF (interstitial rostral nuclei of the medial longitudinal fasciculus) and the ICN (interstitial nucleus of Cajal). (B) The structures involved in the horizontal VOR include, the MVN, LVN, III, IV, PPRF (paramedian zone of the pontine reticular formation), IBN (inhibitory burst neurons), and NPH (nucleus prepositus hypoglossi). The associated semicircular canals and extra-ocular muscles that participate in the vertical and horizontal eye movements are darkened in each representation. Excitatory and inhibitory paths are, respectively, represented by filled and open circles and arrowheads.

Most, if not all, of the inhibitory vestibular neurons that are associated with the anterior and posterior vertical semicircular canals are located in the superior vestibular nucleus. The axons of these inhibitory neurons ascend in the ipsilateral medial longitudinal fasciculus (MLF) and establish synaptic connections predominantly on the somata and proximal dendrites of motoneurons in the oculomotor and trochlear nuclei. Excitatory second-order vestibular neurons that are related to the posterior vertical canal are located in the medial and ventral lateral vestibular nuclei, while those that are related to anterior vertical canal are located in the superior vestibular nucleus. The axons of these excitatory second-order vestibular neurons ascend in the contralateral MLF or course through the ventral tegmentum and terminate predominantly on the distal dendrites of oculomotor motoneurons. The connections of excitatory and inhibitory second-order vestibular neurons with motoneurons in the oculomotor and trochlear nuclei are, to a certain extent, specific to the semicircular canal from which they receive synaptic inputs. Excitatory second-order vestibular neurons that are related to the posterior vertical semicircular canal establish synaptic connections with superior oblique and inferior rectus motoneurons, while those that are related to the anterior vertical semicircular canal establish connections with inferior oblique and superior rectus motoneurons.

10.2.2.2 Horizontal Vestibulo-Ocular Pathways

The horizontal VOR (Figure 10.3B) is mediated predominantly by reciprocal excitatory and inhibitory synaptic connections of second-order vestibular neurons with lateral rectus motoneurons and internuclear neurons in the abducens nucleus.[151,152] Both the inhibitory and excitatory second-order vestibular neurons that project to the abducens nucleus are located in the medial and ventral lateral vestibular nuclei. A smaller direct excitatory second-order vestibular input to medial rectus motoneurons originates from neurons in the ventral portion of the lateral vestibular nucleus, where axons course ipsilaterally via the ascending tract of Deiters.[59,99,125]

10.2.2.3 Nucleus Prepositus Hypoglossi

The nucleus prepositus hypoglossi (NPH, Figure 10.2 and Figure 10.3b) represents one site of interaction between the visual and vestibular systems. When a compensatory eye movement is made in response to head rotation, the head velocity signals of vestibular neurons per se are incapable of maintaining gaze in the new position. A fundamentally important role in gaze holding is played by the nucleus prepositus hypoglossi.[66] This nucleus is regarded as the neural integrator that is responsible for converting the head velocity signals of vestibular neurons to eye position signals that are carried by extra-ocular motoneurons.[24,56] Neurons in this nucleus have visual receptive fields and exhibit eye movement-related activity.[37] Lesions of the prepositus disrupt horizontal optokinetic, vestibulo-ocular, and saccadic integration processing.[12,23,25] The nucleus prepositus hypoglossi has extensive interconnections with the vestibular nuclei, as well as the pontomedullary reticular formation and cerebellum, and efferent connections with the extraocular motor nuclei.[4,5,98] Like the vestibular and reticular inputs to abducens neurons, those from the nucleus prepositus hypoglossi also have excitatory and inhibitory components.[49] A similar neural integrator function in the control of vertical eye movements has been posulated for neurons in the interstitial nucleus of Cajal (INC, Figure 10.3b).[56,57]

10.2.2.4 Vestibular Commissural Connections

Most of the vestibular nuclei and cell group y on one side of the brainstem have homotopic and, to a lesser extent, heterotopic commissural connections with their contralateral counterparts.[14] Many commissural connections are mediated by second-order vestibular neurons in the medial vestibular nucleus that simultaneously provide axon collateral projections to the contralateral medial vestibular nucleus and to the extra-ocular motor nuclei.[99] In addition to the direct commissural pathway that courses through the dorsal tegmentum of the brainstem, indirect connections utilize vestibulocerebellar pathways, as well as the nucleus prepositus hypoglossi, which is intimately related to the vestibular nuclei by extensive reciprocal connections. The vestibular commissural pathway is regarded as largely inhibitory in function.[11,135,144] Two mechanisms have been proposed for commissural inhibition. Shimazu and Precht[135] postulated a disynaptic inhibitory commissural pathway involving tonic excitatory Type I medial vestibular nucleus neurons that synapse on contralateral inhibitory Type II neurons, which in turn synapse on ipsilateral Type I neurons. Kasahara et al.[81] described the presence of inhibitory commissural projections with latencies, suggesting a mono-synaptic connection originating from phasic Type I neurons and projecting onto contralateral Type I vestibular neurons. Stimulation of the semicircular canals inhibited the majority of contralateral neurons in the excitatory medial vestibular nucleus, while inhibiting only a few contralateral inhibitory neurons.[161]

10.2.3 Central Visual Components

Direct projections from retinal ganglion cells, as well as from primary and associational cerebral-cortical visual areas, to the midbrain superior colliculus, pretectum, and accessory optic nuclei, are involved in the visual control of eye movements (Figure 10.2). Regions of the occipito-temporal and parieto-occipital cortex appear to be important for the generation of optokinetic and smooth pursuit eye movements.[160] Optokinetic eye movements are tracking movements elicited by move-ment of the whole visual field. Optokinetic nystagmus (OKN) is characterized by a slow-phase eye movement in the direction of a moving stimulus and a quick-phase return eye movement in the opposite direction when the excursion limit of the oculomotor range has been reached. The opto-kinetic system works synergistically with the vestibular system to stabilize images in the visual field on the retina during movements of the head. As shown in Figure 10.2, the vermal and floccular regions of the cerebellum play an essential role in the convergence of different aspects of the visual-oculomotor (smooth pursuit, optokinetic, and saccadic) and vestibular systems.

The vestibulo-cerebellum (i.e., flocculus, nodulus, and uvula) functions in the control of eye movements that stabilize images on the retina. There are extensive reciprocal connections between the vestibular nuclei and all areas of the vestibulo-cerebellum.[9,17,48,129] Cerebellectomy produces persistent deficits in the optokinetic reflex and in holding eccentric positions of gaze. The flocculus of the cerebellum can be considered as the principal site of convergence of vestibular and visual information. Signals from these two sensory systems derive, respectively. from the first- and second-order vestibular fibers and from brainstem precerebellar nuclei (e.g., via the mossy fiber input from the nucleus reticularis tegmenti pontis and dorsolateral pontine nucleus and the climbing fiber input from the inferior olivary nucleus) that are synaptically related to the pretectal nucleus of the optic tract, the accessory optic nuclei, and the occipitotemporal cortex. This integrated information is then transmitted to the oculomotor system via cerebello-vestibular projections from the flocculus.

10.3 NEUROTRANSMITTERS UTILIZED IN THE VESTIBULO-OCULAR REFLEX

10.3.1 Inhibitory Neurotransmitters

Physiological, anatomical, and neurochemical studies have provided compelling evidence for a role of gamma-aminobutyric acid (GABA) as an inhibitory neurotransmitter in the vestibulo-ocular reflex pathways (Figure 10.2). The blockade of vestibular-evoked disynaptic IPSPs elicited in oculomotor and trochlear motoneurons by bicuculline or picrotoxin suggests that GABA is the neurotransmitter utilized in vertical vestibulo-ocular reflex inhibitory synaptic connections[79,116,123] (Figure 10.3a). Picrotoxin also blocks the slow muscle potential recorded from the extra-ocular muscles in a manner similar to removal of the second-order vestibular input to oculomotor moto-neurons, following lesions of the dorsolateral brainstem that interrupt the inhibitory vestibular pathway.[80] These physiological findings are correlated with the uptake of [3H]GABA by synaptic endings in the oculomotor nucleus[90,146] and a high density of GABA-immunoreactive synaptic endings in the oculomotor and trochlear nuclei[33,146,149,150,152]. Lesions of the superior vestibular nucleus or MLF reduce the levels of GABA in the trochlear nucleus.[122,126] Consistent with these findings, neurons containing GABA or its synthesizing enzyme, glutamate decarboxylase (GAD), have been identified in the lateral, medial, inferior, and, to a lesser extent, in the superior vestibular nuclei.[18,34,74,89,114,149,170] Further evidence for a role of GABA in vestibular function is derived from *in vitro* physiological studies, which have demonstrated that the application of GABA or muscimol, a $GABA_A$ agonist, in slice preparations inhibits the tonic activity of neurons in the vestibular nuclei.[47,144,167] Microinjections of bicuculline, a non-specific GABA antagonist, into oculomotor-related regions of the vestibular nuclei *in vivo* induce a spontaneous nystagmus, whereas muscimol injections impair horizontal gaze holding.[153] The sources of GABAergic inputs to the vestibular nuclei include vestibular commissural projections, the cerebellum, (Figure 10.2) and intrinsic interneurons. Unilateral vestibular nerve lesions produce an ipsilateral increase and a contralateral decrease in GABA-immunoreactive staining in the lateral vestibular nucleus.[158] On the other hand, ablation of the anterior cerebellar vermis results in a 73 percent reduction in GAD-immunoreactive terminals in the lateral vestibular nucleus.[74] GABA also has been implicated as the neurotransmitter associated with inhibitory input to the vestibular nuclei from the inferior olive.[97] Consistent with the role of GABA as a putative neurotransmitter, both $GABA_A$[47,70,75,144] and $GABA_B$[47,72,144,168] receptors have been localized in the vestibular nuclei. Differential roles of GABA acting on $GABA_A$ and $GABA_B$ receptors have been demonstrated for the neural integrator and velocity storage mechanisms, respectively, in the vestibular nuclei through vestibular commissural and cerebellar pathways[28,104,105,153,169] (Figure 10.4).

Both GABA and GAD,[95,162] as well as $GABA_A$ receptors,[54] have been localized in hair cells and afferent synaptic endings in the vestibular end-organs. GABA has also been suggested as a neurotransmitter associated with the vestibular efferent system in the cristae and maculae in the monkey.[163] However, the role of GABA as a neurotransmitter in the periphery remains unclear.

Correlated morphological and physiological studies have provided conclusive evidence that glycine is the inhibitory neurotransmitter utilized by second-order neurons in horizontal vestibulo-ocular reflex connections (Figure 10.3b), and probably also by inhibitory burst neurons that are related to horizontal gaze and neurons in the nucleus prepositus hypoglossi.[149,151,152] The high density of glycine-immunoreactive synaptic endings with a widespread soma-dendritic distribution in relation to motoneurons and internuclear neurons in the abducens nucleus is correlated with glycine immunoreactivity in these brainstem nuclei. Furthermore, the disynaptic IPSPs recorded from physiologically identified abducens motoneurons following electrical stimulation of the ipsilateral vestibular nerve are blocked by strychnine, but are unaffected by bicuculline or picrotoxin. These differences between vertical and horizontal canal-related inhibitory vestibular neurons may be related not only to the differential roles of GABA and glycine in vestibular commissural inhibition,[60,91,105,123,144] but also to the axonal branching patterns of second-order vestibular neurons. These project to both the ipsilateral abducens nucleus and the spinal cord,[76,77,100] where the vestibular-evoked disynaptic IPSPs in neck motoneurons are also blocked by strychnine.[50] While the possibility also exists that neurons may exhibit co-localization of GABA and glycine,[25,170] in most instances only one or the other appears to have a synaptic effect, as indicated by the specificity of pharmacological antagonism,[144] presumably dictated by the type and presence of the postsynaptic receptor with which the input is associated.

10.3.2 EXCITATORY NEUROTRANSMITTERS

Substantial evidence has accumulated regarding the roles of glutamate and aspartate as excitatory neurotransmitters in the vestibulo-ocular reflex pathways (reviewed in Ref. 35,142, and 145). The synaptic connections between vestibular hair cells and vestibular nerve afferent endings utilize glutamate and/or aspartate acting through kainate/AMPA (a-amino-3-hydroxy-5-methyl-4-isoxazole-propionic acid) receptors.[38,42,68,164] Recent findings indicate that twelve individual subunits that potentially form functional kainate, AMPA and NMDA (N-methyl-D-aspartate) receptors are expressed in vestibular ganglion neurons, and that all three receptor types may be expressed at the same hair cell/vestibular nerve synapse.[112] Consistent with the localization of glutamate in the

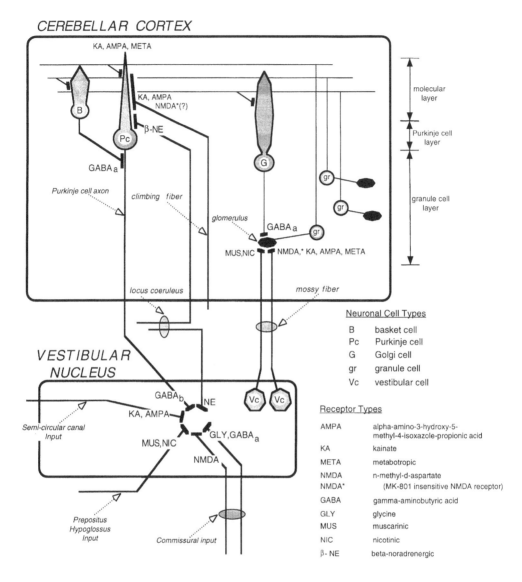

FIGURE 10.4 Principal receptor types and sub-types within the cerebellar cortex and the vestibular nucleus that have been associated with the vestibulo-ocular reflex.

vestibular end-organs,[68,164] vestibular ganglion neurons are labelled autoradiographically or immunocytochemically with glutamate or aspartate.[18,41,42,124] Synaptic endings in the vestibular nuclei are labelled by high-affinity uptake of [3H]-glutamate.[159] The synaptic terminations of the primary vestibular afferent axons are associated with kainate or AMPA receptors.[27,43] On the other hand, the localization of NMDA receptors in the vestibular nuclei[34] is associated predominantly with vestibular commissural inputs[43] (Figure 10.4). The application of the selective NMDA receptor/channel antagonists MK-801 and CPP to brainstem slices *in vitro* produces a decrease in the resting tonic activity of vestibular neurons.[141,143] Neurons in the vestibular nuclei also possess NMDA receptors that regulate the resting discharge properties of tonic and phasic types of neurons and produce oscillatory patterns that might be associated with nystagmic eye movements.[36,134,143] The application of NMDA causes a depolarization and increased spontaneous firing rate of vestibular neurons.[134] The *in vitro* application of 2-amino-5-phosphovalerate (AP5), another selective

NMDA antagonist, suppresses the monosynaptic commissural activation of vestibular neurons.[43] Chronic infusion of AP5 into the vestibular nuclei *in vivo* induces postural and oculomotor deficits similar to those resulting from hemilabyrinthectomy.[36] Within the nucleus prepositus hypoglossi, antagonism of NMDA receptors by microinjections of ketamine or AP5 causes a failure of the horizontal gaze-holding system, consistent with its role as the neural integrator for horizontal eye movements.[26,105]

Both glutamate- and aspartate-immunoreactive neurons have been localized within all four of the vestibular nuclei.[18,19,170] Individual vestibular neurons may colocalize glutamate and aspartate, in addition to GABA and/or glycine.[170,176] Aspartate-immunoreactive neurons predominate in the vestibular[89] and prepositus hypoglossi[176] nuclei. Vestibulospinal neurons in the superior, medial, descending (inferior), and particularly lateral vestibular nuclei are immunoreactive toward aspartate.[84] Vestibulo-ocular neurons that project to the oculomotor and trochlear nuclei (Figure 10.3a), however, are immunoreactive toward glutamate and/or aspartate.[19,85] Synaptic endings in the oculomotor nucleus are labeled by high-affinity uptake of [3H]-glutamate.[39] Both glutamate and NMDA produce depolarization of abducens motoneurons.[45] NMDA responses are voltage-dependent and are characterized by bursts of action potentials followed by stable repetitive, rhythmic firing.[44] NMDA receptors appear to have a predominantly dendritic location, but are not associated with the excitatory second-order vestibular input to oculomotor motoneurons.[46]

Acetylcholine is the likely excitatory neurotransmitter of vestibular efferent neurons.[16,40,87] Neurons in the nucleus prepositus hypoglossi that are immunoreactive to choline acetyltransferase (ChAT), the synthesizing enzyme of acetylcholine, are the source of a cholinergic input to the cat abducens nucleus.[148] ChAT-immunoreactive neurons also have been described in the prepositus hypoglossi and medial/inferior vestibular nuclei in the monkey,[18] as well as in the rabbit and rat,[2] and may provide a cholinergic mossy fiber input to the uvula, nodulus, and flocculus of the cerebellum. Microinjections of the cholinergic agonist carbachol into the flocculus increase the gain of both optokinetic responses (OKR) and of the VOR in the dark, whereas injections of cholinergic antagonists acting on muscarinic or nicotinic receptors reduce the gain of the OKR.[154]

The excitatory neurotransmitters also play a major role in the cerebellum. The main afferents to this structure are mossy fibers (MF) conveying vestibular information via the granule cell-parallel fibers to the Purkinje cells. Excitation from the mossy fiber to the granule cell is by means of glutamate, which acts on the NMDA as well as the metabotropic and the ionotropic AMPA receptors.[127,171]

Climbing fibers are a second major source of inputs to the cerebellum and arise from a single source, the inferior olivary nucleus in the brainstem. An individual climbing fiber makes extensive synaptic contact with the dendritic tree of a Purkinje cell, bringing visual information from the accessory optic system. Aspartate or glutamate is thought to be the neurotransmitter acting at the climbing fiber/Purkinje cell synapse.[127] Although experiments in the guinea pig have identified NMDA as one of the possible receptors present at this synapse,[86,133] other work in the rat has identified the non NMDA receptor at this synapse[62,88] (Figure 10.4). NMDA is particularly abundant in the developing cerebellum,[127,171] but may have a less prominent role in the adult nervous system. NMDA stimulation produces a voltage-dependent response that is enhanced when accompanied by concomitant excitatory stimulation.[171] NMDA receptors have also been identified on cerebellar granule cells.[127]

A third afferent source into the cerebellum is a noradrenergic input that arises from the locus coeruleus, a small brain stem nucleus that has a widespread projection throughout the cerebellum[110] and also to the vestibular nuclei[130] (Figure 10.4). Considered to be a neuromodulator, it has been shown to affect afferent input into the cerebellum and appears to be related to cerebellar adaptive processes.[83,102,103,166]

The climbing and mossy fibers provide excitatory input into the cortex. Interneurons, namely the basket and stellate cells, are inhibitory to Purkinje neurons. The other interneuron, the Golgi cell, produces inhibitory feedback at the mossy fiber-granule cell synapse in the glomeruli. The Purkinje cell axon is the only output of the cerebellar cortex and synapses in the deep cerebellar and the vestibular nuclei. GABA is the inhibitory transmitter that has been identified with the Purkinje cell and the other inhibitory interneurons in the cerebellum.

10.4 BEHAVIORAL NEUROPHARMACOLOGY

In an effort to understand the vestibulo-ocular reflex operation, a number of investigations have used the traditional "elimination" or "destruction" approach employed in brain research. Thus, an area of the brain involved with the vestibulo-ocular reflex circuitry is removed permanently, either by a surgical or an electrolytic lesion. The main disadvantage or drawback with this approach is that this technique can be non-specific.

As has been described earlier in this chapter, the circuitry responsible for the vestibulo-ocular reflex involves many interrelated pathways. Individual characteristics of the reflex may be a distributed process of the network. Thus, removal or inactivation of a pathway or site may not necessarily isolate or pinpoint a specific area or region as being the one responsible for some unique property of the reflex. Efforts to overcome this limitation have involved experiments that consider the latency or timing characteristics of the reflex in attempts to understand the anatomical and physiological bases for the vestibulo-ocular reflex.[94]

A more selective "lesion" can be produced either permanently or reversibly using a pharmacological approach. Pharmacological inactivation possesses a degree of specificity that depends on the particular drug used. For example, different neurotransmitters are involved and these transmitter systems are in turn associated with several receptor subtypes. Thus, the degree of specificity achieved is related to the selectivity of a particular ligand that blocks or activates a particular receptor. There have been a number of investigations that have attempted to use behavioral pharmacological approaches in the investigations on the vestibulo-ocular reflex. Both the basic normal operation and the adaptive processes of the reflex have been investigated in this manner.

10.4.1 GLUTAMATE

Lesions of the VIIIth cranial nerve, either experimentally in animals or surgically in humans as treatment for Meniere's disease or acoustic neuroma, result in characteristic oculomotor and postural deficits. Static symptoms occur in the absence of head movement and include deviation of the eyes toward the lesioned side, spontaneous nystagmus with the quick phase directed toward the intact side, and head tilt toward the lesioned side.[10,51,136,138,177] Dynamic symptoms occur in response to head movement and include a reduced amplitude and abnormal timing of the vestibulo-ocular and vestibulospinal reflexes.[51,136,138,140,178] Neurons in the medial vestibular nucleus ipsilateral to the nerve lesion are decreased in number and their activity is reduced,[137] presumably due to persistent commissural inhibition and an increase in neuronal activity contralaterally.[136] Behavioral recovery from unilateral labyrinthectomy occurs through the poorly-understood clinical phenomenon of vestibular compensation. Compensation is commonly attributed to the recovery of resting activity ipsilateral to the labyrinthectomy.[31,32,36,52,138,140] Studies involving the pharmacological manipulation of the compensated state in experimental animals have demonstrated that administration of NMDA receptor antagonists such as AP5[36] or MK-801[53] and CPP[139] interfere with vestibular compensation in hemilabyrinthectomized animals, implicating a role for NMDA receptors in recovery. Administration of calmidazolium chloride into the ipsilateral medial vestibular nucleus or fourth ventricle reduces spontaneous nystagmus,[128] suggesting that Ca^{2+} channels directly participate in compensation. Disruption of the vestibular commissural pathway also results in spontaneous partial decompensation, involving a sudden reappearance of precompensation dynamic deficits,[138] indicating that commissural pathways are involved in recovery from such dynamic deficits. Destruction of the commissures in the frog prior to unilateral labyrinthectomy completely prevents compensation from taking place.[61]

10.4.2 GABA

Studies described previously in this chapter have detailed the action of GABA on neuronal activity in the structures that participate in the vestibulo-ocular reflex. There have also been a number of *in vivo* studies directly involving GABA's action in the normal vestibulo-ocular reflex only and not in the adaptive aspects of this reflex.

In the cerebellum, GABA is the prime transmitter for the inhibitory interneurons such as the basket and the Golgi cells which synapse, respectively, onto the Purkinje and the granule cells. $GABA_A$ and not $GABA_B$ receptors are considered to be responsible for basket-cell inhibition of the Purkinje cell[127] (Figure 10.4). Both muscimol ($GABA_A$ agonist) and baclofen ($GABA_B$ agonist) when injected directly into the flocculus of rabbits, leads to a significant reduction of the Vis-VOR and the VOR (40 to 60 percent reduction) as well as the OKR gain.[165] A complementary study in cats has also shown that a floccular injection of muscimol also produces a similar impairment in the vestibulo-ocular reflex, in addition to a gaze-holding deficit reflecting an effect on the neural integrator.[55] Although these agonists are probably acting at different sites on both the $GABA_A$ and the $GABA_B$ receptors within the cerebellar cortex, most notably in the granule layer or at the level of the Purkinje cell, their effect appears to be similar; that is, a reduction or elimination of cerebellar output via the Purkinje cell. Van Neerven et al.[165] interpret this finding to indicate that the effect of normal Purkinje cell activity is to enhance the vestibulo-ocular reflex, rather than to inhibit its response as is generally supposed.

This appears to be in contradiction to electrophysiological studies which demonstrate that P-cell activity inhibits activity in the vestibular nucleus and should be correlated with an increase in the gain of the vestibulo-ocular reflex. However, it should be noted that often, apparent contradictions arise because results from two different types of studies are being compared. In this case, the comparison is between behavioral pharmacological[55,165] and cellular pharmacological studies.[22,120] Studies that have attempted to remove the influence of the cerebellum by cerebellectomy or by lidocaine injection may shed some light on this. These studies report that removal of the cerebellar influence can have a minor and somewhat variable effect on vestibulo-ocular reflex gain.[107] The one point generally conceded by present thought is that removal of the cerebellar influence affects the adaptive capability of the vestibulo-ocular reflex. Presently, no studies have been carried out with respect to the effect of cerebellar GABAergic activity on reflex adaptation.

GABA agonists and antagonists have also been injected into other CNS structures related to the operation of the vestibulo-ocular reflex. Muscimol, injected unilaterally into the vestibular nucleus, reduces the VOR gain to approximately zero and produces a gaze-holding deficit in the monkey.[153] Thus, the increased inhibition via the feedback pathways is thought to produce a neural integrator deficit. In contrast to this, a similar injection of bicuculline, the non-specific GABA antagonist, did not alter the overall Vis-VOR or VOR gain.[153] However, velocity bias and nystagmus did develop. Thus, the loss of inhibition in the vestibular nucleus had no effect on neural integration and reflex gain. Another study in the cat has demonstrated that muscimol injected into the central medial vestibular nucleus or the nucleus prepositus hypoglossi produces the same deficit for horizontal vestibulo-ocular reflex.[106] Injection in the rostral medial vestibular nucleus did not produce this deficit for the horizontal vestibulo-ocular reflex. Increasing the degree of inhibition through the use of GABA agonists by a direct localized injection into these discrete brainstem nuclei provides us with no information about the particular pathways involved, since there are many sources of GABA inhibition. However, it is interesting to note that an even less discrete and more generalized systemic application of a $GABA_B$ agonist, baclofen, produces the same effect in the rat[113] and the monkey.[28] However, since these ligands may be acting at several places, it is difficult to interpret the data from these experiments.

10.4.3 ACETYLCHOLINE

The effect of manipulations on acetylcholine has also been investigated with respect to the vestibulo-ocular reflex. The suggestion that acetylcholine may affect this reflex centrally is derived from several sources. First, anticholinergic drugs are used to treat motion sickness.[119] Second, acetylcholine afferents project from the vestibular and nucleus prepositus hypoglossi to the vestibulo-cerebellum.[2] Third, non specific cholinergic agonists (carbachol) have been shown to increase the gain of the vestibulo-spinal reflex when injected into the cerebellum.[121] There also appears to be an intimate and possibly synergistic relationship between norepinephrine and acetylcholine associated with developmental cortical plasticity in the visual system.[3] This could be an important consideration, especially since norepinephrine has been shown to affect vestibulo-ocular reflex adaptation.[83,102,103,166]

Application of cholinergic agonists and antagonists into the cerebellar flocculus affects the vestibulo-ocular reflex. The initial studies conducted in the pigmented Dutch belted rabbit indicated that bilateral injection of carbachol, a non specific acetylcholine agonist, significantly elevated normal VOR but not Vis-VOR gain during sinusoidal stimulation. In addition, a more pronounced increase in the gain of the OKR was observed.[154] Other work has shown that this effect of the OKR is probably mediated by the muscarinic receptor within the cerebellum.[156] Subsequent work by Tan and Collewijn[155] was not able to confirm any affect on Vis-VOR or VOR gain following carbachol floccular injection. However, the previously observed elevation of OKR gain by carbachol[154] was enhanced by a concomitant injection of a noradrenergic agonist.

These experiments indicate that the acetylcholine system acts within the cerebellar flocculus, but this action is primarily on the OKR. There appears to be a synergistic relationship between the noradrenergic and the cholinergic systems here. These effects on the OKR are similar to that reported on the vestibulo-spinal reflex.[121] From the two studies published by Tan and Collewijn[154,155] the action of acetylcholine in the flocculus on the vestibulo-ocular reflex is minimal and perhaps nonexistent. Presently, no studies have been carried out using cholinergic antagonists, and no work has been carried out in areas other than that in the cerebellum. In addition, attempts to alter the vestibulo-ocular adaptive process using an intravenous injection of scopolamine, a non-specific muscarinic antagonist, have also been unsuccessful.[131]

10.4.4 BIOGENIC AMINES

Norepinephrine is one of the biogenic amines that affects the operation of the vestibulo-ocular reflex. Noradrenergic fibers emanate from the locus coeruleus, a small bilateral brain stem nucleus that has a diffuse and widespread projection throughout the central nervous system. This nucleus supplies noradrenergic input to two prime areas involved with the vestibulo-ocular reflex: namely the vestibular nuclei[130] and the cerebellum[110] (Figure 10.4). Norepinephrine is thought to act as a modulator by regulating gating mechanisms in individual cells located in selective areas of the brain, including the cerebellum.[175]

Norepinephrine is considered to be involved with neuroplastic phenomena, most notably with hippocampal long-term potentiation (LTP), an *in vitro* model for learning and memory,[73] and with developmental plasticity in the visual cortex.[82] There have also been a number of experimental studies that have investigated the relationship of this noradrenergic system to adaptation of the vestibulo-ocular reflex.

Studies in the cat have shown that, following 6-hydroxy-dopamine (6-OHDA) ventricle injections, the ability to produce adaptive vestibulo-ocular reflex gain decreases[83] and increases[103] are hindered by permanent systemic central nervous system depletion of norepinephrine. A more selective noradrenergic lesion in the cerebellum, after 6-OHDA application directly into the brachium conjunctivum, also resulted in a deficit in the ability to produce VOR adaptive gain increases.[102]

Studies conducted by Miyashita and Watanabe[108] are also relevant as they investigated the role of serotonin as well as norepinephrine (NE) depletion on adaptive vestibulo-ocular reflex gain changes. These investigators reported that depletion of serotonin and norepinephrine by intraventricular injection of 5,7-dihydroxytryptamine (5,7-DHT) results in the loss of vestibulo-ocular reflex adaptation in pigmented rabbits. Since there was no loss in vestibulo-ocular reflex adaptation in the rabbits after injection of 6-OHDA, which depleted cerebellar NE but not 5-HT, it was concluded that only 5-HT plays a role in adaptive modifiability of the vestibulo-ocular reflex. This result does not agree with the previous three studies that had been carried out in the cat.[83,102,103] The possibile lack of sufficient NE depletion or a species difference in the work of Miyashita and Watanabe could account for this difference.

Other studies using more specific noradrenergic ligands for a and b noradrenergic receptors have also been carried out. Tan et al.,[157] injecting a variety of a-noradrenergic compounds directly into the rabbit flocculus, found that the a_1 and a_2 receptors do not participate in the normal performance or adaptation of the vestibulo-ocular reflex, even though these receptors are prominent in the cerebellum. In contrast, van Neerven et al.[166] demonstrated that injection of a b-agonist, isoproterenol, directly into the flocculus of rabbits enhanced adaptive gain increases. Moreover, a b-antagonist, sotalol, reduced the adaptation of the VOR gain (Figure 10.4). No changes in normal VOR operation were detected after injection of these compounds. Confirmatory studies using another non-specific b-antagonist, propranolol, injected directly into the vestibulo-cerebellum of the goldfish, has also demonstrated impairment of adaptive increases but not decreases in gain.[172]

It has also been reported that the gain of the vestibulo-ocular reflex varies in a manner that is related to arousal level.[58] For example, increased alertness in human subjects was associated with a constant and elevated vestibulo-ocular reflex gain.[29] Studies have shown that amphetamine,[58,102] which increases arousal, also ensures high reflex gain. Norepinephrine could be involved here since amphetamine acts in part by a release of this neurotransmitter. In spite of this implied relationship between norepinephrine and normal vestibulo-ocular reflex gain, data is lacking that specifically and directly relates the gain of the normal reflex to norepinephrine.

10.4.5 NITRIC OXIDE

Besides investigations on the traditional neurotransmitters and neuromodulators, recent work indicates that other less-conventional molecules such as nitric oxide (NO) are involved with CNS neuropharmacological mechanisms (see Chapter 6). NO is a small gaseous intercellular messenger that has been implicated in several different models of learning[6,71,111] and in neuronal developmental processes.[96] Although nitric oxide is produced in individual neurons, its sphere of influence as a gas extends beyond these cells and may act via a presynaptic[115] and/or a heterosynaptic[132] effect.

Recent work has shown that experimentally induced reductions in cerebellar NO availability inhibits the adaptive processes of the vestibulo-ocular reflex. Investigations carried out in cerebelli of rabbit and monkey[111] suggested that floccular injection of hemoglobin, which binds NO non-specifically, inhibited the acquisition phase of both adaptive VOR gain increases and decreases. Other studies in the goldfish conducted by Li et al.[93] demonstrated that a specific blocker of NO production (l-N_G-monomethyl-arginine; l-NMMA) inhibited the acquisition but not the retention phases of VOR adaptation. In addition, this effect was reversed by restoration of NO production following co-administration of l-NMMA and l-arginine.

This study carried out in the goldfish reported that only increases and not decreases in VOR gain were affected. The study by Nagao and Ito,[111] resulting in the inhibition of adaptive VOR gain decreases after floccular hemoglobin injection, was an observation from an individual monkey. Besides the obvious species difference between the two studies, this discrepancy may also have arisen because hemoglobin acts non-specifically, producing other effects unrelated to NO inhibition.

These investigations suggest that cerebellar production of NO is essential for acquisition of adaptive changes in VOR gain. NO has also been shown to mediate other neuroplastic changes in the cerebellum, such as long-term depression.[30] NO is produced from l-arginine by the enzymatic action of nitric oxide synthase (NOS).[64] This enzyme is localized in cerebellar neuronal and glial cells in the molecular and granule cell layers.[7] Cerebellar NO levels increase after (1) activation of glutamate receptors (ionotropic[8,63,65] or metabotropic[117]), (2) depolarization of climbing fiber afferents,[147] or (3) increases in intracellular calcium.[117]

Nitric oxide directly affects blood flow by its action on the cerebral vasculature. Because l-NMMA impeded only adaptive gain increases but not decreases, at least in one study, and did not

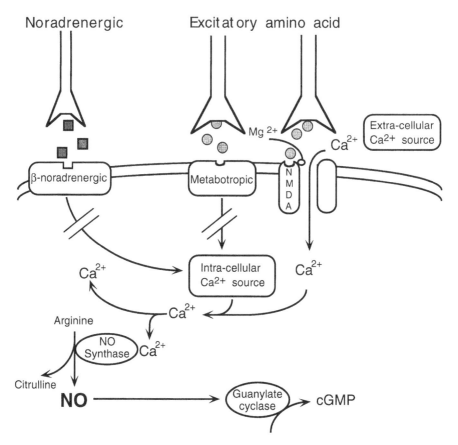

FIGURE 10.5 Illustrated are three possible biochemical pathways that depict how Ca^{2+} is made available from extracellular (NMDA receptor activation) or intracellular (via b-noradrenergic or metabotropic glutamate G-protein-linked receptor activation) sources. Nitric oxide (NO) production is produced from l-arginine by a calcium-dependent nitric oxide synthase (NOS) reaction. Increased NO production is associated with the elevation of cGMP levels. In turn, this leads to protein phosphorylation and/or an increase in neurotransmitter release, two events that are associated with adaptive and neuroplastic changes. A number of pharmacological studies (on b-noradrenergic, metabotropic, or NMDA antagonism, as well as NO inhibition) that have been carried out in the cerebellum present data that are consistent with this representation.

alter post-adaptive recovery, an indirect effect on adaptation of the vestibulo-ocular reflex due to blood flow reduction levels is unlikely. Although NOS inhibitors can produce cerebral vasoconstriction, recent work has shown that they do not interfere with cerebellar neuronal activity.[92] Since NO is prevalent throughout the cerebellar cortex, it may be acting at non-Purkinje cell sites. Alternatively, NO could reach and act in Purkinje cells via diffusion from other sources, although these cells do not contain NOS, the enzyme necessary for the production of NO (see also Chapter 6).

Retinal slip during adaptive vestibulo-ocular reflex training activates climbing fiber responses that consequently release NO at Purkinje cell terminals.[147] Thus, common pharmacological and physiological mechanisms are associated with both reflex adaptation and NO production.

A recent study indicated that antagonism of the cerebellar metabotropic receptors also inhibited the acquisition of adaptive increases but not decreases in vestibulo-ocular reflex gain.[20] Other related work has also shown that cerebellar NMDA antagonism by AP5[21] but not by MK-801[101] also inhibits adaptive gain increases. This would indicate that the cerebellum possesses a subtype of the NMDA receptor that is MK-801-insensitive and participates in the vestibulo-ocular reflex adaptive process

(Figure 10.4).

The NMDA and the metabotropic receptors that have been identified in the cerebellum,[127] are, respectively, a voltage-dependent and enzyme-linked second messengers receptors; both have been associated with modulatory and neuroplastic processes.[109] These receptors are also present in the vestibular nucleus[1,13,145] and could play a role in vestibulo-ocular reflex adaptation in this structure as well.

All these events may be taking part in a common mechanism within the cerebellum since metabotropic and NMDA receptor excitation increases intracellular levels of Ca^{2+}, which is an essential component for NO production. The NMDA receptor is associated with an ion channel that produces an EPSP whose amplitude is dependent upon concomitant excitatory inputs or on the state of depolarization of the neuron,[171] allowing calcium to enter the cell. One form of the metabotropic receptor acts via the phosphoinositide (PI) system and releases stored intracellular calcium.[127] It has been demonstrated that metabotropic[118] and NMDA[173] receptor stimulation increases cGMP in cerebellar cells by a nitric oxide-dependent process. Elevation of cGMP levels leads to protein phosphorylation and/or an increase in neurotransmitter release, two events that are associated with adaptive and neuroplastic changes. Thus, it is possible that cerebellar metabotropic or NMDA antagonism and inhibition of NO production are blocking different parts of a common cerebellar mechanism, responsible for the acquisition phase of adaptive VOR gain increase. Figure 10.5 graphically illustrates the biochemical processes that may be associated with vestibulo-ocular reflex adaptation in the cerebellum.

Norepinephrine could also be playing a role in this process since NMDA-dependent increases in cGMP have been shown to be dependent upon monoaminergic innervation of the cerebellum.[174]

10.5 SUMMARY

This chapter has described the anatomical, physiological, and pharmacological aspects that underlie the operation of the vestibulo-ocular reflex. In addition, the functional characteristics of the normal and adaptive facets of the vestibulo-ocular reflex were treated with regard to the behavioral neuropharmacology of the reflex. Over the past several decades, our understanding of this reflex has made gradual, consistent progress due to these anatomical-physiological, behavioral, and modeling studies. We now know considerably more about the reflex and its operation than we did in the past. Some might argue that the word "slow" would not be an inappropriate adjective to describe this progress since there are still many unresolved arguments, disagreements, and questions that have been around for a considerable period of time. It is our hope and anticipation that in addition to the the classical anatomical and physiological studies, the more recent work on the pharmacological aspects will provide the extra knowledge and insight that will lead us to a fuller understanding of the vestibulo-ocular reflex operation and in turn to a fuller and more complete understanding of brain function.

ACKNOWLEDGMENTS

This work was supported by NIH grants DC01094 to JMcE and EY02191 to RFS.

REFERENCES

1. Alford, S, and Dubuc, R, Glutamate metabotropic receptor mediated depression of synaptic inputs to lamprey reticulospinal neurons, *Brain Res.,* 605,175,1993.
2. Barmack, NH, Baughman, RW, Eckenstein, FP, and Shojaku, H, Secondary vestibular cholinergic projection to the cerebellum of rabbit and rat as revealed by choline acetyltransferase immunohistochemistry, retrograde and orthograde tracers, *J. Comp. Neurol.,* 317,250,1992.

3. Bear, MF, and Singer, W, Modulation of visual cortical plasticity by acetylcholine and noradrenaline, *Nature,* 320,172,1986.

4. Belknap, DB, and McCrea, RA, Anatomical connections of the prepositus and abducens nuclei in the squirrel monkey, *J. Comp. Neurol.,* 268,13,1988.

5. Blanks, RH, Cerebellum, [Review]. *Rev. Oculomot. Res.,* 2,225,1988.

6. Bohme, GA, Bon, C, Lemaire, M, Reibaud, M, Piot, O, Stutzmann, JM, Doble, A, and Blanchard, JC, Altered synaptic plasticity and memory formation in nitric oxide synthase inhibitor-treated rats, *Proc. Natl. Acad. Sci. USA,* 90,9191,1993.

7. Bredt, DS, Hwang, PM, and Snyder, SH, Localization of nitric oxide synthase indicating a neural role for nitric oxide, *Nature,* 347,768,1990.

8. Bredt, DS, and Snyder, SH, Nitric oxide mediates glutamate-linked enhancement of cGMP levels in the cerebellum, *Proc. Natl. Acad. Sci. USA,* 86,9030,1989.

9. Brodal, A, and Brodal, P, Observations on the secondary vestibulocerebellar projections in the macaque monkey, *Exp. Brain Res.,* 58,62,1985.

10. Bruning, G, Wiese, S, and Mayer, B, Nitric oxide synthase in the brain of the turtle *Pseudemys scripta elegans, J. Comp. Neurol.,* 348,183,1994.

11. Cabrera, B, Portillo, F, Pasaro, R, and Delgado-Garcia, JM, Location of motoneurons and internuclear neurons within the rat abducens nucleus by means of horseradish peroxidase and fluorescent double labeling, *Neurosci. Lett.,* 87,1,1988.

12. Cannon, SC, and Robinson, DA, Loss of the neural integrator of the oculomotor system from brain stem lesions in monkey, *J. Neurophysiol.,* 57,1383,1987.

13. Capocchi, G, Della Torre, G, Grassi, S, Pettorossi, VE, and Zampolini, M, NMDA receptor-mediated long term modulation of electrically evoked field potentials in the rat medial vestibular nuclei, *Exp. Brain Res.,* 90,546,1992.

14. Carleton, SC, and Carpenter, MB, Afferent and efferent connections of the medial, inferior and lateral vestibular nuclei in the cat and monkey, *Brain Res.,* 278,29,1983.

15. Carpenter, MB, Vestibular nuclei, afferent and efferent projections. [Review], *Prog. Brain Res.,* 76,5,1988.

16. Carpenter, MB, Chang, L, Pereira, AB, Hersh, LB, Bruce, G, and Wu, JY, Vestibular and cochlear efferent neurons in the monkey identified by immunocytochemical methods, *Brain Res.,* 408,275,1987.

17. Carpenter, MB, and Cowie, RJ, Connections and oculomotor projections of the superior vestibular nucleus and cell group "y," *Brain Res.,* 336,265,1985.

18. Carpenter, MB, Huang, Y, Pereira, AB, and Hersh, LB, Immunocytochemical features of the vestibular nuclei in the monkey and cat, *J. Hirnforsch.,* 31,585,1990.

19. Carpenter, MB, Periera, AB, and Guha, N, Immunocytochemistry of oculomotor afferents in the squirrel monkey (*Saimiri sciureus*), *J. Hirnforsch.,* 33,151,1992.

20. Carter, TL, and McElligott, JG, Metabotropic glutamate receptor antagonist (L-AP3) inhibits vestibulo-ocular reflex adaptation when administered into the vestibulo-cerebellum of the goldfish,*Soc. Neurosci. Abstr.,* 20,#17.10,1994.

21. Carter, TL, and McElligott, JG, Cerebellar NMDA receptor antagonism by AP5 blocks vestibulo-ocular reflex adaptation in the goldfish, *Soc. Neurosci. Abstr.,* 21,#115.5.

22. Chan-Palay, V, Palay, SL, Wu, JY, Gamma-aminobutyric acid pathways in the cerebellum studied by retrograde and anterograde transport of glutamic acid decarboxylase antibody after in vivo injections, *Anat. Embryol., (Berl),* 157,1,1979.

23. Cheron, G, Gillis, P, and Godaux, E, Lesions in the cat prepositus complex: effects on the optokinetic system, *J. Physiol. (London),* 372,95,1986.

24. Cheron, G, and Godaux, E, Disabling of the oculomotor neural integrator by kainic acid injections in the prepositus-vestibular complex of the cat, *J. Physiol., (London),* 394,267,1987.

25. Cheron, G, Godaux, E, Laune, JM, and Vanderkelen, B, Lesions in the cat prepositus complex: effects on the vestibulo-ocular reflex and saccades, *J. Physiol., (London),* 372,75,1986.

26. Cheron, G, Mettens, P, and Godaux, E, Gaze-holding defect induced by injections of ketamine in the cat brainstem, *NeuroReport,* 3,97,1992.

27. Cochran, SL, Kasik, P, and Precht, W, Pharmacological aspects of excitatory synaptic transmission to second-order vestibular neurons in the frog, *Synapse,* 1,102,1987.

28. Cohen, B, Helwig, D, and Raphan, T, Baclofen and velocity storage: a model of the effects of the drug on the vestibulo-ocular reflex in the rhesus monkey, *J. Physiol., (London),* 393,703,1987.

29. Collins, WE, and Guedry, FE, Arousal effects and nystagmus during prolonged constant angular acceleration, *Acta. oto-Laryng.,* 54,349,1962.

30. Daniel, H, Hemart, N, Jaillard, D, and Crepel, F, Long-term depression requires nitric oxide and guanosine 3':5' cyclic monophosphate production in rat cerebellar Purkinje cells, *Eur. J. Neurosci.,* 5,1079,1993.

31. Darlington, CL, Flohr, H, and Smith, PF, Molecular mechanisms of brainstem plasticity. The vestibular compensation model, [Review], *Mol. Neurobiol.,* 5,355,1991.

32. Darlington, CL, Smith, PF, and Hubbard, JI, Neuronal activity in the guinea pig medial vestibular nucleus *in vitro* following chronic unilateral labyrinthectomy, *Neurosci. Lett.,* 105,143,1989.

33. De la Cruz, RR, Pastor, AM, Martinez-Guijarro, FJ, Lopez-Garcia, C, and Delgado-Garcia, JM, Role of GABA in the extraocular motor nuclei of the cat: A postembedding immunocytochemical study, *Neuroscience,* 51,911,1992.

34. de Waele, C, Abitbol, M, Chat, M, Menini, C, Mallet, J, and Vidal, PP, Distribution of glutamatergic receptors and GAD mRNA-containing neurons in the vestibular nuclei of normal and hemilabyrinthectomized rats, *Eur. J. Neurosci.,* 6,565,1994.

35. de Waele, C, Muhlethaler, M, and Vidal, PP, Neurochemistry of the central vestibular pathways, [Review]. *Brain Res. Brain Res. Rev.,* 20,24,1995.

36. de Waele, C, Vibert, N, Baudrimont, M, and Vidal, PP, NMDA receptors contribute to the resting discharge of vestibular neurons in the normal and hemilabyrinthectomized guinea pig, *Exp. Brain Res.,* 81,125,1990.

37. Delgado-Garcia, JM, Vidal, PP, Gomez, C, and Berthoz, A, A neurophysiological study of prepositus hypoglossi neurons projecting to oculomotor and preoculomotor nuclei in the alert cat, *Neuroscience,* 29,291,1989.

38. Dememes, D, Lleixa, A, and Dechesne, CJ, Cellular and subcellular localization of AMPA-selective glutamate receptors in the mammalian peripheral vestibular system, *Brain Res.,* 671,83,1995.

39. Dememes, D, and Raymond, J, Radioautographic identification of [³H]glutamic acid labeled-nerve endings in the cat oculomotor nucleus, *Brain Res.,* 231,433,1982.

40. Dememes, D, Raymond, J, and Sans, A, Selective retrograde labelling of vestibular efferent neurons with [³H]choline, *Neuroscience,* 8,285,1983.

41. Dememes, D, Raymond, J, and Sans, A, Selective retrograde labeling of neurons of the cat vestibular ganglion with [³H]D-aspartate, *Brain Res.,* 304,188,1984.

42. Dememes, D, Wenthold, RJ, Moniot, B, and Sans, A, Glutamate-like immunoreactivity in the peripheral vestibular system of mammals, *Hearing Res.,* 46,261,1990.

43. Doi, K, Tsumoto, T, and Matsunaga, T, Actions of excitatory amino acid antagonists on synaptic inputs to the rat medial vestibular nucleus: an electrophysiological study *in vitro, Exp. Brain Res.,* 82,254,1990.

44. Durand, J, NMDA actions on rat abducens motoneurons, *Eur. J. Neurosci.,* 3,621,1991.

45. Durand, J, Engberg, I, and Tyc-Dumont, S, L-glutamate and N-methyl-D-asparatate actions on membrane potential and conductance of cat abducens motoneurons, *Neurosci. Lett.,* 79,295,1987.

46. Durand, J, and Gueritaud, JP, Excitatory amino acid actions on membrane potential and conductance of brainstem motoneurons, in: *Amino Acids: Chemistry, Biology and Medicine* (Lubec G, and Rosenthal GA Eds.), pp. 255-262. Amsterdam: Escom.,1990.

47. Dutia, MB, Johnston, AR, and McQueen, DS, Tonic activity of rat medial vestibular nucleus neurons in vitro and its inhibition by GABA, *Exp. Brain Res.,* 88,466,1992.

48. Epema, AH, Gerrits, NM, and Voogd, J, Secondary vestibulo-cerebellar projections to the flocculus and uvulo-nodular lobule of the rabbit: A study using HRP and double fluorescent tracer techniques, *Exp. Brain Res.,*1990.

49. Escudero, M, and Delgado-Garcia, JM, Behavior of reticular, vestibular and prepositus neurons terminating in the abducens nucleus of the alert cat, *Exp. Brain Res.,* 71,218,1988.

50. Felpel, LP, Effects of strychnine, bicuculline and picrotoxin on labyrinthine-evoked inhibition in neck motoneurons of the cat, *Exp. Brain Res.,* 14,494,1972.

51. Fetter, M, and Zee, DS, Recovery from unilateral labyrinthectomy in rhesus monkey, *J. Neurophysiol.,* 59,370,1988.

52. Flohr, H, Abeln, W, and Luneburg, U, Neurotransmitter and neuromodulator systems involved in vestibular compensation, [Review]. *Rev. Oculomotor. Res.,* 1,269,1985.

53. Flohr, H, and Luneburg, U, Role of NMDA receptors in lesion-induced plasticity, *Arch. Ital. Biol.,* 131,173,1993.

54. Foster, JD, Drescher, MJ, and Drescher, DG, Immunohistochemical localization of GABA$_A$ receptors in the mammalian crista ampullaris, *Hear Res.*, 83,203,1995.

55. Fukushima, K, Buharin, EV, and Fukushima, J, Responses of floccular Purkinje cells to sinusoidal vertical rotation and effects of muscimol infusion into the flocculus in alert cats, *Neurosci. Res.*, 17,297,1993.

56. Fukushima, K, Kaneko, CR, and Fuchs, AF, The neuronal substrate of integration in the oculomotor system, [Review]. *Prog. Neurobiol.*, 39,609,1992.

57. Fukushima, K, and Kaneko, CRS, Vestibular integrators in the oculomotor system, *Neurosci. Res.*, 22,249,1995.

58. Furman, JM, O'Leary, DP, and Wolfe, JW, Changes in the horizontal vestibulo-ocular reflex of the rhesus monkey with behavioral and pharmacological alerting, *Brain Res.*, 206,490,1981.

59. Furuya, N, and Markham, CH, Arborization of axons in oculomotor nucleus identified by vestibular stimulation and intra-axonal injection of horseradish peroxidase, *Exp. Brain Res.*, 43,289,1981.

60. Furuya, N, Yabe, T, and Koizumi, T, Neurotransmitters regulating vestibular commissural inhibition in the cat, *Acta. Otolaryngol. Suppl. (Stockh)*, 481,205,1991.

61. Galiana, HL, Flohr, H, and Jones, GM, A re-evaluation of intervestibular nuclear coupling: its role in vestibular compensation, *J. Neurophysiol.*, 51,242,1984.

62. Garthwaite, J, and Beaumont, PS, Excitatory amino acid receptors in the parallel fibre pathway in rat cerebellar slices, *Neurosci. Lett.*, 107,151,1989.

63. Garthwaite, J, Charles, SL, and Chess-Williams, R, Endothelium-derived relaxing-factor release on activation of NMDA receptors suggests role as intercellular messenger in the brain, *Nature*, 336,385,1988.

64. Garthwaite, J, Garthwaite, G, Palmer, RM, and Moncada, S, NMDA receptor activation induces nitric oxide synthesis from arginine in rat brain slices, *Eur. J. Pharmacol.*, 172,413,1989.

65. Garthwaite, J, Southam, E, and Anderton, M, A kainate receptor linked to nitric oxide synthesis from arginine, *J. Neurochem.*, 53,1952,1989.

66. Godaux, E, Mettens, P, and Cheron, G, Differential effect of injections of kainic acid into the prepositus and the vestibular nuclei of the cat, *J. Physiol. (London)*, 472,459,1993.

67. Graf, W, Baker, J, and Peterson, BW, Sensorimotor transformation in the cat's vestibulo-ocular reflex system. I. Neuronal signals coding spatial coordination of compensatory eye movements, *J. Neurophysiol.*, 70,2425,1993.

68. Harper, A, Blythe, WR, Grossman, G, Petrusz, P, Prazma, J, and Pillsbury HC, Immunocytochemical localization of aspartate and glutamate in the peripheral vestibular system, *Hear. Res.*, 86,171,1995.

69. Harrison, RE, Baker JF, Isu, N, Wickland, CR, and Peterson, BW, Dynamics of adaptive change in vestibulo-ocular reflex direction. I. Rotations in the horizontal plane, *Brain Res.*, 371,162,1986.

70. Hironaka, T, Morita, Y, Hagihira S, Tateno, E, Kita, H, and Tohyama, M, Localization of GABA$_A$-receptor alpha 1 subunit mRNA-containing neurons in the lower brainstem of the rat, *Brain Res. Mol. Brain Res.*, 7,335,1990.

71. Holscher, C, and Rose, SP, Inhibiting synthesis of the putative retrograde messenger nitric oxide results in amnesia in a passive avoidance task in the chick, *Brain Res.*, 619,189,1993.

72. Holstein, GR, Martinelli, GP, and Cohen, B, L-baclofen-sensitive GABA$_B$ binding sites in the medial vestibular nucleus localized by immunocytochemistry, *Brain Res.*, 581,175,1992.

73. Hopkins, WF, and Johnston, D, Noradrenergic enhancement of long-term potentiation at mossy fiber synapses in the hippocampus, *J. Neurophysiol.*, 59,667,1988.

74. Houser, CR, Barber, RP, and Vaughn, JE, Immunocytochemical localization of glutamic acid decarboxylase in the dorsal lateral vestibular nucleus: evidence for an intrinsic and extrinsic GABAergic innervation, *Neurosci. Lett.*, 47,213,1984.

75. Hutchinson, M, Smith, PF, and Darlington, CL, Further evidence on the contribution of GABA$_A$ receptors to the GABA-mediated inhibition of medial vestibular nucleus neurons *in vitro*, *Neuroreport.*, 6,1649,1995.

76. Isu, N, Sakuma, A, Hiranuma, K, Uchino, H, Sasaki, S, Imagawa, M, and Uchino, Y, The neuronal organization of horizontal semicircular canal activated inhibitory vestibulocollic neurons in the cat, *Exp. Brain Res.*, 86,9,1991.

77. Isu, N, and Yokota, J, Morphophysiological study on the divergent projection of axon collaterals of medial vestibular nucleus neurons in the cat, *Exp. Brain Res.*, 53,151,1983.

78. Ito, M, *The Cerebellum and Neural Control*, New York: Raven Press,1984

79. Ito, M, Highstein, SM, and Tsuchiya T, The postsynaptic inhibition of rabbit oculomotor neurons by secondary vestibular impulses and its blockage by picrotoxin, *Brain Res.,* 17,520,1970.

80. Ito, M, Nisimaru, N, and Yamamoto, M, Postsynaptic inhibition of oculomotor neurons involved in vestibulo-ocular reflexes arising from semicircular canals of rabbits, *Exp. Brain Res.,* 24,273,1976.

81. Kasahara, M, Mano, N, Oshima, T, Ozawa, S, and Shimazu, H, Contralateral short latency inhibition of central vestibular neurons in the horizontal canal system, *Brain Res.,* 8,376,1968.

82. Kasamatsu, T, Norepinephrine hypothesis for visual cortical plasticity: thesis, antithesis, and recent development, [Review], *Curr. Top. Dev. Biol.,* 21,367,1987.

83. Keller, EL, and Smith, MJ, Suppressed visual adaptation of the vestibulo-ocular reflex in catecholamine-depleted cats, *Brain Res.,* 258,323,1983.

84. Kevetter, GA, and Coffey, AR, Aspartate-like immunoreactivity in vestibulospinal neurons in gerbils, *Neurosci. Lett.,* 123,273,1991.

85. Kevetter, GA, and Hoffman, RD, Excitatory amino acid immunoreactivity in vestibulo-ocular neurons in gerbils, *Brain Res.,*19;554,348,1991.

86. Kimura, H, Okamoto, K, and Sakai, Y, Pharmacological evidence for L-aspartate as the neurotransmitter of cerebellar climbing fibres in the guinea pig, *J. Physiol. (London),* 365,103,1985.

87. Kong, WJ, Egg, G, Hussl, B, Spoendlin, H, and Schrott-Fischer, A, Localization of ChAT-like immuno-reactivity in the vestibular endorgans of the rat, *Hearing Res.,* 75,191,1994.

88. Konnerth, A, Llano, I, and Armstrong, CM, Synaptic currents in cerebellar Purkinje cells, *Proc. Natl. Acad Sci. USA,* 87,2662,1990.

89. Kumoi, K, Saito, N, and Tanaka, C Immunohistochemical localization of gamma-aminobutyric acid- and aspartate-containing neurons in the guinea pig vestibular nuclei, *Brain Res.,* 416,22,1987.

90. Lanoir, J, Soghomonian, JJ, and Cadenel, G, Radioautographic study of 3H-GABA uptake in the oculo-motor nucleus of the cat, *Exp. Brain Res.,* 48,137,1982.

91. Lapeyre, PN, and de Waele, C, Glycinergic inhibition of spontaneously active guinea pig medial vestibular nucleus neurons *in vitro, Neurosci. Lett.,* 188,155,1995.

92. Li, J, and Iadecola, C, Nitric oxide and adenosine mediate vasodilation during functional activation in cerebellar cortex, *Neuropharmocology,* 33,1453,1994.

93. Li, J, Smith, SS, and McElligott, JG, Cerebellar nitric oxide is necessary for vestibulo-ocular reflex adaptation, a sensorimotor model of learning, *J. Neurophysiol.,* 74,489,1995.

94. Lisberger, SG, The latency of pathways containing the site of motor learning in the monkey vestibulo-ocular reflex, *Science,* 225,74,1984.

95. Lopez, I, Wu, JY, and Meza, G, Immunocytochemical evidence for an afferent GABAergic neurotrans-mission in the guinea pig vestibular system, *Brain Res.,* 589,341,1992.

96. Matsumoto, T, Pollock, JS, Nakane, M, and Forstermann, U, Developmental changes of cytosolic and particulate nitric oxide synthase in rat brain, *Brain Res. Dev. Brain Res.,*1994.

97. Matsuoka, I, Ito, J, Sasa, M, and Takaori, S, Possible neurotransmitters involved in excitatory and inhibitory effects from inferior olive to contralateral vestibular nucleus, *Adv. Otorhinolaryngol.,*1983.

98. McCrea, RA, and Baker, R, Anatomical connections of the nucleus prepositus of the cat, *J. Comp. Neurol.,*1985.

99. McCrea, RA, Strassman, A, May, E, Highstein, SM, Anatomical and physiological characteristics of vestibular neurons mediating the horizontal vestibulo-ocular reflex of the squirrel monkey, *J. Comp. Neurol.,*1987.

100. McCrea, RA, Yoshida, K, Berthoz, A, and Baker, R, Eye movement-related activity and morphology of second-order vestibular neurons terminating in the cat abducens nucleus, *Exp. Brain Res.,* 40,468,1980.

101. McElligott, JG, Carter, TL, and Baker, R, Effects of MK-801 on adaptive vestibulo-ocular reflex modi-fications in the goldfish, *Soc. Neurosci. Abstr.,* 17,#127.9.

102. McElligott, JG, and Freedman, W, Central and cerebellar norepinephrine depletion and vestibulo-ocular reflex (VOR) adaptation. in: *Post-lesion Neuronal Plasticity* (Flohr H, ed), pp 661-674. Berlin: Springer Verlag, 1988.

103. McElligott, JG, and Freedman, W, Vestibulo-ocular reflex adaptation in cats before and after depletion of norepinephrine, *Exp. Brain Res.,*1988.

104. Mettens, P, Cheron, G, and Godaux, E, NMDA receptors are involved in temporal integration in the oculomotor system of the cat, *NeuroReport.,*1994.

105. Mettens, P, Cheron, G, and Godaux, E, Role of the vestibular commissure in gaze-holding in the cat: a pharmacological evaluation, *NeuroReport.*,1994.
106. Mettens, P, Godaux, E, Cheron, G, and Galiana, HL, Effect of muscimol microinjections into the prepositus hypoglossi and the medial vestibular nuclei on cat eye movements, *J. Neurophysiol.*,1994.
107. Miles, FA, The cerebellum. in: *Eye Movements* (R.H.S.Carpenter Ed.), pp. 224-243. Boca Raton, FL: CRC Press, Inc.,1991.
108. Miyashita, Y, and Watanabe, E, Loss of vision-guided adaptation of the vestibulo-ocular reflex after depletion of brain serotonin in the rabbit, *Neurosci. Lett.*,1984.
109. Monaghan, DT, Bridges, RJ, and Cotman, CW, The excitatory amino acid receptors: their classes, pharmacology, and distinct properties in the function of the central nervous system, [Review]. *Annu. Rev. Pharmacol. Toxicol.*,1989.
110. Moore, RY, and Bloom, FE, Central catecholamine neuron systems: anatomy and physiology of the norepinephrine and epinephrine systems, [Review]. *Annu. Rev. Neurosci.*,1979.
111. Nagao, S, and Ito, M, Subdural application of hemoglobin to the cerebellum blocks vestibuloocular reflex adaptation, *Neuroreport.*,1991.
112. Niedzielski, AS, and Wenthold, RJ, Expression of AMPA, kainate, and NMDA receptor subunits in cochlear and vestibular ganglia, *J. Neurosci.*,1995.
113. Niklasson, M, Tham, R, Larsby, B, and Eriksson, B, Effects of GABA$_B$ activation and inhibition on vestibulo-ocular and optokinetic responses in the pigmented rat, *Brain Res.*,1994.
114. Nomura, I, Senba, E, Kubo, T, Shiraishi, T, Matsunaga, T, Tohyama, M, Shiotani, Y, and Wu, JY, Neuropeptides and gamma-aminobutyric acid in the vestibular nuclei of the rat: An immunohistochemical analysis, I. Distribution. *Brain Res.,* 311,109,1984.
115. O'Dell, TJ, Hawking, RD, Kandel, ER, and Arancio, O, Tests of the role of two diffusible substances in long term potentiation: Evidence for nitric oxide as a possible early retrograde messenger, *Proc. Natl. Acad. Sci., USA,*1994.
116. Obata, K, and Highstein, SM, Blocking by picrotoxin of both vestibular inhibition and GABA action on rabbit oculomotor neurons, *Brain Res.*,1970.
117. Okada, D, Two pathways of cyclic GMP production through glutamate receptor-mediated nitric oxide synthesis, *J. Neurochem.*,1992.
118. Okada, D, Protein kinase C modulates calcium sensitivity of nitric oxide synthase in cerebellar slices, *J. Neurochem.,* 64,1298,1995.
119. Parrott, AC, Transdermal scopolamine: Review of its effects upon motion sickness, psychological performance, and physiological functioning, *Aviat. Space. Environ. Med.,* 60,1,1989.
120. Pettorossi, VE, Troiani, D, and Petrosini, L, Diazepam enhances cerebellar inhibition on vestibular neurons, *Acta. Otolaryngol. (Stockholm),* 93,363,1982.
121. Pompeiano, O, Noradrenergic and cholinergic modulations of corticocerebellar activity modify the gain of vestibulospinal reflexes, *Ann. NY Acad. Sci.,* 656,519,1992.
122. Precht, W, Baker, R, and Okada, Y, Evidence for GABA as the synaptic transmitter of the inhibitory vestibulo-ocular pathway, *Exp. Brain Res.,* 18,415,1973.
123. Precht, W, Schwindt, PC, and Baker, R, Removal of vestibular commissural inhibition by antagonists of GABA and glycine, *Brain Res.,* 62,222,1973.
124. Raymond, J, Nieoullon A, Dememes, D, and Sans, A, Evidence for glutamate as a neurotransmitter in the cat vestibular nerve: Radioautographic and biochemical studies, *Exp. Brain Res.,*56,523,1984.
125. Reisine, H, Strassman, A, and Highstein, SM, Eye position and head velocity signals are conveyed to medial rectus motoneurons in the alert cat by the ascending tract of Deiters', *Brain Res.,* 211,153,1981.
126. Roffler-Tarlov, S, and Tarlov, E, Reduction of GABA synthesis following lesions of inhibitory vestibulo-trochlear pathway, *Brain Res.,* 91,326,1975.
127. Ross, CA, Bredt, D, and Snyder, SH, Messenger molecules in the cerebellum. [Review], *Trends Neurosci.,* 13,216,1990.
128. Sansom, AJ, Darlington, CL, Smith, PF, Gilchrist, DP, Keenan, CJ, and Kenyon, R, Injections of calmidazolium chloride into the ipsilateral medial vestibular nucleus or fourth ventricle reduce spontaneous ocular nystagmus following unilateral labyrinthectomy in guinea pigs, *Exp. Brain Res.,* 93,271,1993.
129. Sato, Y, Kanda, K, Ikarashi, K, and Kawasaki, T, Differential mossy fiber projections to the dorsal and ventral uvula in the cat, *J. Comp. Neurol.,* 279,149,1989.

130. Schuerger, RJ, and Balaban, CD, Immunohistochemical demonstration of regionally selective projections from locus coeruleus to the vestibular nuclei in rats, *Exp. Brain Res.*, 92,351,1993.
131. Schultheis, LW, and Robinson, DA, The effect of scopolamine on the vestibulo-ocular reflex, gain adaptation, and the optokinetic response, *Ann. NY Acad. Sci.*, 656,880,1992.
132. Schuman, EM, and Madison, DV, Locally distributed synaptic potentiation in the hippocampus, *Science*, 263,532,1994.
133. Sekiguchi, M, Okamoto, K, and Sakai, Y, NMDA-receptors on Purkinje cell dendrites in guinea pig cerebellar slices, *Brain Res.*, 437,402,1987.
134. Serafin, M, Khateb, A, de Waele, C, Vidal, PP, and Muhlethaler, M, Medial vestibular nucleus in the guinea pig: NMDA-induced oscillations, *Exp. Brain Res.*, 88,187,1992.
135. Shimazu, H, andPrecht, W, Inhibition of central vestibular neurons from the contralateral labyrinth and its mediating pathway, *J. Neurophysiol.*, 29,467,1966.
136. Smith, PF, Curthoys, IS, Neuronal activity in the contralateral medial vestibular nucleus of the guinea pig following unilateral labyrinthectomy, *Brain Res.*, 444,295,1988.
137. Smith, PF, and Curthoys, IS, Neuronal activity in the ipsilateral medial vestibular nucleus of the guinea pig following unilateral labyrinthectomy, *Brain Res.*, 444,308,1988.
138. Smith, PF, and Curthoys, IS, Mechanisms of recovery following unilateral labyrinthectomy: A review, [Review], *Brain Res. Brain Res. Rev.*, 14,155,1989.
139. Smith, PF, and Darlington, CL, The NMDA antagonists MK801 and CPP disrupt compensation for unilateral labyrinthectomy in the guinea pig, *Neurosci. Lett.*, 94,309,1988.
140. Smith, PF, and Darlington, CL, Neurochemical mechanisms of recovery from peripheral vestibular lesions (vestibular compensation) [published erratum appears in *Brain Res. Brain Res. Rev.*, 17(2),183,1992][Review]. *Brain Res. Brain Res. Rev.*, 16,117,1991.
141. Smith, PF, and Darlington, CL, Comparison of the effects of NMDA antagonists on medial vestibular nucleus neurons in brainstem slices from labyrinthine-intact and chronically labyrinthectomized guinea pigs, *Brain Res.*, 590,345,1992.
142. Smith, PF, and Darlington, CL, Pharmacology of the vestibular system. [Review], *Baillieres Clin. Neurol.*, 3,467,1994.
143. Smith, PF, Darlington, CL, and Hubbard, JI, Evidence that NMDA receptors contribute to synaptic function in the guinea pig medial vestibular nucleus, *Brain Res.*, 513,149,1990.
144. Smith, PF, Darlington, CL, and Hubbard, JI, Evidence for inhibitory amino acid receptors on guinea pig medial vestibular nucleus neurons *in vitro*, *Neurosci. Lett.*, 121,244,1991.
145. Smith, PF, de Waele, C, Vidal, PP, and Darlington, CL, Excitatory amino acid receptors in normal and abnormal vestibular function. [Review], *Mol. Neurobiol.*, 5,369,1991.
146. Soghomonian, JJ, Pinard, R, and Lanoir, J, GABA innervation in adult rat oculomotor nucleus: A radioautographic and immunocytochemical study, *J. Neurocytol.*, 18,319,1989.
147. Southam, E, and Garthwaite, J, Climbing fibers as a source of nitric oxide in the cerebellum, *Eur. J. Neurosci.*, 3,379,1994.
148. Spencer, RF, and Baker, R, A cholinergic input to the cat abducens nucleus, *Soc. Neurosci. Abstr.*, 15, 99.12, 1989.
149. Spencer, RF, Baker, R, GABA and glycine as inhibitory neurotransmitters in the vestibulo-ocular reflex. [Review], *Ann. NY Acad. Sci.*, 656,602,1992.
150. Spencer, RF, Wang, SF, Immunohistochemical localization of neurotransmitters utilized by neurons in the rostral interstitial nucleus of the medial longitudinal fascicularis (riMLF) that project to the oculomotor and trochlear nuclei in the cat, *J. Comp. Neurol.*, in Press,1995.
151. Spencer, RF, Wang, SF, and Baker, R, The pathways and functions of GABA in the oculomotor system. [Review]. *Prog. Brain Res.*, 90,307,1992.
152. Spencer, RF, Wenthold, RJ, and Baker, R, Evidence for glycine as an inhibitory neurotransmitter of vestibular, reticular, and prepositus hypoglossi neurons that project to the cat abducens nucleus. *J. Neurosci.*, 9,2718,1989.
153. Straube, A, Kurzan, R, and Buttner, U, Differential effects of bicuculline and muscimol microinjections into the vestibular nuclei on simian eye movements, *Exp. Brain Res.*, 86,347,1991.
154. Tan, HS, and Collewijn, H, Cholinergic modulation of optokinetic and vestibulo-ocular responses: A study with microinjections in the flocculus of the rabbit, *Exp. Brain Res.*, 85,475,1991.

155. Tan, HS, and Collewijn, H, Cholinergic and noradrenergic stimulation in the rabbit flocculus have synergistic facilitatory effects on optokinetic responses, *Brain Res.*, 586,130,1992.
156. Tan, HS, and Collewijn, H, Muscarinic nature of cholinergic receptors in the cerebellar flocculus involved in the enhancement of the rabbit's optokinetic response, *Brain Res.*, 591,337,1992.
157. Tan, HS, van Neerven, J, Collewijn, H, and Pompeiano, O, Effects of alpha-noradrenergic substances on the optokinetic and vestibulo-ocular responses in the rabbit: a study with systemic and intrafloccular injections, *Brain Res.*, 562,207,1991.
158. Thompson, GC, Igarashi, M, and Cortez, AM, GABA imbalance in squirrel monkey after unilateral vestibular end-organ ablation, *Brain Res.*, 370,182,1986.
159. Touati, J, Raymond, J, and Dememes, D, Quantitative autoradiographic characterization of L-[^{3}H] glutamate binding sites in rat vestibular nuclei, *Exp. Brain Res.*, 76,646,1989.
160. Tusa, RJ, and Zee, DS, Cerebral control of smooth pursuit and optokinetic nystagmus, *Curr. Neuro-Opthalmol.*, 2,115,1989.
161. Uchino, Y, Ichikawa, T, Isu, N, Nakashima, H, and Watanabe, S, The commissural inhibition on secondary vestibulo-ocular neurons in the vertical semicircular canal systems in the cat, *Neurosci. Lett.*, 70,210,1986.
162. Usami, S, Hozawa, J, Tazawa, M, Igarashi, M, Thompson, GC, Wu, JY, and Wenthold, RJ, Immunocytochemical study of the GABA system in chicken vestibular endorgans and the vestibular ganglion, *Brain Res.*, 503,214,1989.
163. Usami, S, Igarashi, M, and Thompson, GC, GABA-like immunoreactivity in the squirrel monkey vestibular endorgans, *Brain Res.*, 417,367,1987.
164. Usami, S, and Ottersen, OP, Differential cellular distribution of glutamate and glutamine in the rat vestibular endorgans: An immunocytochemical study, *Brain Res.*, 676,285,1995.
165. van Neerven, J, Pompeiano, O, and Collewijn, H, Effects of GABAergic and noradrenergic injections into the cerebellar flocculus on vestibulo-ocular reflexes in the rabbit, *Prog. Brain Res.*, 88,485,1991.
166. van Neerven, J, Pompeiano, O, Collewijn, H, and van der Steen, J, Injections of beta-noradrenergic substances in the flocculus of rabbits affect adaptation of the VOR gain, *Exp. Brain Res.*, 79,249,1990.
167. Vibert, N, Serafin, M, Vidal, PP, and Muehlethaler, M, Direct and indirect effects of muscimol on medial vestibular nucleus neurons in guinea pig brainstem slices, *Exp. Brain Res.*, 104,351,1995.
168. Vibert, N, Serafin, M, Vidal, PP, and Muhlethaler, M, Effects of baclofen on medial vestibular nucleus neurons in guinea pig brainstem slices, *Neurosci. Lett.*, 183,193,1995.
169. Waespe, W, Cohen, B, and Raphan, T, Dynamic modification of the vestibulo-ocular reflex by the nodulus and uvula, *Science*, 228,199,1985.
170. Walberg, F, Ottersen, OP, and Rinvik, E, GABA, glycine, aspartate, glutamate and taurine in the vestibular nuclei: An immunocytochemical investigation in the cat, *Exp. Brain Res.*, 79,547,1990.
171. Watkins, JC, and Collingridge, GL, *The NMDA Receptor*, Oxford: IRL Press,1989.
172. Williams, L, and McElligott, JG, Cerebellar b-noradrenergic receptor antagonism by propranolol inhibits the vestibulo-ocular reflex adaptation in goldfish, *Soc. Neurosci. Abstr.*, 20, #488.14.
173. Wood, PL, and Rao, TS, A review of *in vivo* modulation of cerebellar cGMP levels by excitatory amino acid receptors: Role of NMDA, quisqualate and kainate subtypes, [Review]. *Prog. Neuropsychopharmacol. Biol. Psychiatry*, 15,229,1991.
174. Wood, PL, Ryan, R, and Li ,M, NMDA-, but not kainate- or quisqualate-dependent increases in cerebellar cGMP are dependent upon monoaminergic innervation, *Life Sci.*, 51,PL267,1992.
175. Woodward, DJ, Moises, HC, Waterhouse, BD, Yeh, HH, and Cheun, JE, Modulatory actions of norepinephrine on neural circuits, [Review], *Adv. Exp. Med. Biol.*, 287,193,1991.
176. Yingcharoen, K, Rinvik, E, Storm-Mathisen, J, and Ottersen, OP, GABA, glycine, glutamate, aspartate and taurine in the perihypoglossal nuclei: An immunocytochemical investigation in the cat with particular reference to the issue of amino acid colocalization, *Exp. Brain Res.*, 78,345,1989.
177. Zennou-Azogui, Y, Borel, L, Lacour, M, Ez-Zaher, L, and Ouaknine, M, Recovery of head postural control following unilateral vestibular neurectomy in the cat. Neck muscle activity and neuronal correlates in Deiters' nuclei, *Acta. Otolaryngol. Suppl. (Stockholm)*, 509,1,1993.
178. Zennou-Azogui, Y, Xerri, C, and Harlay, F, Visual sensory substitution in vestibular compensation: neuronal substrates in the alert cat, *Exp. Brain Res.*, 98,457,1994.

11

Control of Vestibulospinal and Vestibulo-Ocular Reflexes by Noradrenergic and ACh Systems

O. Pompeiano

CONTENTS

0-8493-7679-3/00/$0.00+$.50

11.1. SYNOPSIS

The dorsal pontine tegmentum in mammals contains noradrenergic (NAergic) and NA-sensitive neurons, located in the locus coeruleus (LC)-complex as well as cholinergic (ACh) and ACh-sensitive neurons located in the pedunculo-pontine (PPT), laterodorsal tegmental (LDT) nuclei, and the neighboring dorsal pontine reticular formation. These two populations of neurons, which show reciprocal changes in their firing rate in relationship to the awake/sleep state of the animal, exert a regulatory influence on vestibular function. They influence vestibular function by utilizing either direct projections to the spinal cord, the vestibular nuclei, and the cerebellar cortex, or via indirect projections passing through precerebellar structures, such as the caudal part of the vestibular nuclear complex, the lateral reticular nucleus, or the inferior olive. Some of these structures also contain choline-acetyltransferase (ChAT)-immunoreactive neurons, which send ACh afferents to the cerebellar cortex. Radioligand binding studies, as well as *in situ* hybridization studies, have revealed the regional distribution of the different subtypes of NAergic and ACh receptors. In addition, electrophysiological experiments employing microiontophoretic application of NAergic and ACh agents have demonstrated their influences on single-unit activity recorded either extracellularly or intracellularly from different target structures. Evidence is presented indicating that in some of these structures both NA and ACh act as neuromodulators, by producing long-lasting increases in responsiveness of their target neurons to conventional excitatory (glutamate) and inhibitory (GABA) responses. A large number of units, recorded from both the NAergic LC-complex and the ACh pontine tegmental region, responded in decerebrate cats to sinusoidal roll tilt of the animal, leading to stimulation of labyrinth receptors. Moreover, experiments involving the local microinjection of NAergic and ACh agonists and antagonists either into the LC-complex and the related pontine tegmental region, or into different regions of the cerebellar cortex, have shown that these structures exert a prominent role in the gain regulation of both the vestibulo-spinal (VSR) and the vestibulo-ocular (VOR) reflexes, as recorded in decerebrate cats or in intact unanesthetized rabbits, respectively. Additional experiments performed in these preparations demonstrated that microinjection into the cerebellar anterior vermis or in the flocculus of b-adrenergic agonists or antagonists increased or decreased, respectively, the adaptive changes in gain of the VSR or the VOR elicited during a sustained neck-vestibular or visuo-vestibular stimulation. The molecular and/or cellular mechanisms responsible for these effects are discussed.

11.2 INTRODUCTION

The central noradrenergic (NAergic) system is comprised of seven discrete subgroups of neurons (A1-A7), which have been delineated in the rat brainstem. The A6 subgroup, which correlates to the locus coeruleus (LC), is the largest noradrenergic group.[63] The A6 neurons show immunoreactivity for both tyrosine hydroxylase (TH) and dopamine-b-hydroxylase (Db H), two enzymes that contribute to the synthesis of NA (cf. Ref. 27). The LC-complex, composed of a dorsal part (LCd), a ventral part (LCa), and the subcoeruleus (SC), is located in the dorsal pontine tegmentum and has extensive projections to broad regions of the central nervous system, including vestibular and reticular structures, the cerebellum, and the spinal cord (cf. Refs. 27, and 63). This finding led us to postulate that the NAergic system could play a role in the control of vestibular functions.

In addition to the NAergic system, there is a central cholinergic (ACh) system that originates mainly from eight major groups of neurons, designated in the rat Ch1-Ch8[121,122,130] and characterized by their immunoreactivity for choline-acetyltransferase (ChAT), the enzyme that provides the synthesis of ACh.[87,93,190,205] These neuronal groups have widespread projections to the brain, with the pedunculo-pontine (PPT) nucleus, the laterodorsal tegmental (LDT) nucleus and the parabrachial nuclei sending ACh afferents to several brainstem structures, such as the vestibular nuclei (VN), the pontine (PRF), and medullary (MRF) reticular formation and the cerebellum. This finding suggests that the ACh system could also be relevant in the regulation of vestibular functions. Additional groups of ACh neurons have also been found in precerebellar structures, including the vestibular nuclei (see Section 5).

The complexity of the effects that the NAergic and the ACh systems exert on the target structures involved in the vestibular functions depends in part on the topographical localization of the corresponding afferents, in part on the various subtypes of receptors used. Radioligand binding studies and *in situ* hybridization studies have revealed the regional distribution of the different subtypes of adrenergic receptors (a_1-, a_2-, b_1- and b_2-adrenoceptors[222]), as well as of the cell bodies that synthetize different sequences of adrenoceptor mRNA.[134,135,141a] This high number of receptors increases the number of possibilities for signal transduction in adrenergic pathways.[187] In particular, a_1-adrenoceptors are G_q-proteins coupled to phosphoinositide-specific phospolipase C, an enzyme which initiates the hydrolysis of a membrane phospholipid to produce a series of second messengers, such as diacylglycerol (DAG) and inositol 1,4,5-triphosphate (IP3). Alpha1-receptors are also coupled directly to Ca^{2+} influx, but in some cases the resulting excitatory responses are largely due to decrease in resting K^+ conductance. Alpha2-adrenoceptors are Gi-proteins coupled to adenylate cyclase which is inhibited, thus reducing intracellular levels of cyclic adenosine monophosphate (cAMP), or linked to ion channels such as K^+ channels. Finally, b-adrenoceptors are Gs-proteins coupled to adenylate cyclase, which is stimulated to increase intracellular levels of cAMP. Adrenoceptors can be found not only postsynaptically, but also presynaptically on NAergic neurons, as shown for the a_2-receptors which occur either on NAergic terminals or on the somatodendritic membrane of LC neurons.[184]

With regard to ACh receptors, they have been subdivided into two classes, referred to as nicotinic and muscarinic receptors. While nicotinic receptors of the ganglionic type or the neuromuscular type belong to the family of receptor-gated ion channels which act by directly regulating the opening of cations channels, muscarinic receptors belong to the family of G-protein coupled receptors, thus acting through second messengers.[72,169] Neuronal nicotinic receptors are different from skeletal muscle nicotinic receptors in subunit composition, pharmacological characteristics, and functional role. In particular, while the latter receptors are composed of four types of subunits, (a, b, g or e, and d), the former appear to contain two types of subunits, a and b (cf. Ref. 206). Four subtypes of muscarinic receptors (M_1 - M_4) have been identified on the basis of their differential sensitivity to antagonists (see Ref. 74), but five subtypes (m_1-m_5) have been recognized using *in situ* hybridization techniques,[31] the pharmacologically M_1 to M_4 subtypes corresponding to the cloned m^1 to m^4 receptors, respectively (cf. Ref. 205). Generally m_1, m_3, and m_5 receptors stimulate phosphoinositide hydrolysis, while m_2 and m_4 subtypes inhibit adenylate cyclase (cf. Ref. 72). ACh receptors are found not only postsynaptically, but also on presynaptic terminals, including those of cholinergic neurons, where they can either inhibit ACh release, as shown for the muscarinic m_2 receptors (Ref. 102; cf. Ref. 205), or enhance ACh release, as shown for nicotinic receptors.[42]

11.3 NORADRENERGIC LOCUS COERULEUS (LC) SYSTEM AND ITS RELATION WITH THE ACH SYSTEM IN THE DORSAL PONTINE TEGMENTUM.

The study of the physiological properties of both the NAergic LC neurons and the ACh neurons located in the dorsal pontine tegmentum and their anatomo-functional relationships are essential in order to understand the role that these structures exert in the regulation of vestibular functions. Radioligand binding studies have demonstrated not only of a_2-, but also a_1- and b-adrenoceptors in the LC-complex (cf. Ref. 43,53,152). However, *in situ* hybridization studies have shown that LC neurons may express only a_2 (i.e. the a_{2A}) mRNA, but not subtypes of a_1- and b-mRNA.[134,135,141a] Possible explanations for these discrepancies have been discussed in the latter studies. From a functional point of view, LC neurons can be inhibited by microiontophoretic application, either of NA or of the a_2-adrenergic agonist clonidine (cf. Ref. 152), and similar results are obtained after systemic injection of a_1-agonists.[167] LC neurons can also be inhibited by local application of the b-adrenergic agonist isoproterenol,[53] but they are excited under given conditions through a_1-adrenoceptors (cf. Ref. 43 for ref.). As a result of these findings, it appears that the LC neurons are not only NAergic, but also NA-sensitive, particularly due to self-inhibitory synapses which act on a_2-adrenoceptors by utilizing either recurrent collaterals from LC axons or dendro-dendritic contacts between neighboring LC neurons.[63,153] It also appears that activity in LC neurons produces a Ca^{2+}-activated increase in K+ conductance, which hyperpolarizes the cells and inhibits impulse activity.[2,215] These findings explain the very slow and regular discharge of LC neurons (1 to 2 imp./s) in the intact animal at rest, as well as their prolonged inhibition following a transient excitation in response to nociceptive stimulation (cf. Ref. 63).

In addition to NAergic LC neurons, the dorsal pontine tegmentum contains ACh (ChAT-positive) neurons, which were particularly located in the PPT/LDT (see Section 11.2), but they were also found to be intermingled with NAergic neurons in the LC-complex and the peri-LC.[86,171] These neurons send ACh afferents to pontine tegmental structures, such as the PRF,[124,179] and either directly or through a descending tegmento-reticular projection, to the medial part of the MRF,[88,173] where the inhibitory reticulospinal (RS) neurons are located (see Section 11.4.2). These ACh afferents are considered to be excitatory.[171,172] Within these pontine tegmental structures there are not only ACh, but also cholinosensitive neurons (cf. Ref. 198). In particular, ACh tends to excite noncholinergic neurons in the PRF while it inhibits, presumably through autoreceptors, ACh neurons in the PPT/LDT.[198]

An important feature in the organization of the pontine tegmental structures is that both the NAergic and ACh systems are reciprocally interconnected. In fact, neurons of the LC-complex send NAergic afferents to the cholinergic PPT/LDT, the peri-Lca, and the dorsal PRF, on which they exert an inhibitory influence.[153] This effect was apparently mediated either directly through b-adrenoceptors (cf. Ref. 54), or indirectly through a_1-activation of GABAergic inhibitory interneurons (cf. Ref. 43), which have been particularly observed in this tegmental region of cats.[86, 156] On the other hand, the PPT/LDT and dorsal PRF send ACh fibers and terminals to the LC-complex, on which they exert an excitatory influence.[153] This effect is apparently mediated through muscarinic (M_2) receptors, as shown in physiological[59] and *in situ* hybridization studies.[205]

It is of interest that the structures indicated above play a prominent role in the regulation of the sleep-waking cycle. Several lines of evidence have, in fact, shown that in intact animals during waking, there is an increased discharge of the NAergic LC neurons,[63,172,185,198] which inhibit the activity of the dorsal PRF neurons and the related medullary inhibitory RS neurons. This finding may account for the postural activity observed in the awake animal. However, as soon as the discharge of the LC neurons disappear during rapid eye movement (REM) sleep, the activity of the presumably ACh and cholinosensitive PRF neurons and the related inhibitory RS neurons increases, thus accounting for the postural atonia that occurs during this phase of sleep.[63,172,185,198] The following sections summarize data showing that these NAergic and ACh systems, located in the dorsal pontine tegmentum, play an important role in the control of vestibular function. They do so by acting at the level of different target structures, including the spinal cord (Section 11.4), the vestibular nuclei (Section 11.5), and the cerebellum (Section 11.6).

11.4 NORADRENERGIC AND ACH SYSTEMS PROJECTING TO THE SPINAL CORD

11.4.1 DIRECT PROJECTION FROM THE LC-COMPLEX TO THE SPINAL CORD: THE NORADRENERGIC COERULEO-SPINAL (CS) SYSTEM

Figure 11.1 illustrates schematically the organization of the NAergic and the ACh systems projecting to the spinal cord.

The NAergic CS projection, which originates from the LC and the SC,[151b,166] sends thin terminal fibers into the ventral horn, where they surround both large- and small-sized motoneurons (Figure 11.1), presumably a- and g-motoneurons, and also interneurons.[23,47] Ventral horn motoneurons express different sequences of a_1-receptor mRNA,[141a] while a_2- (i.e. a_{2A}) and b_1-positive cells, but not b_2-cells, can be found in the dorsal part of the ventral horn.[134,135]

Stimulation experiments performed in decerebrate cats have shown that the CS projection facilitates the monosynaptic reflexes involving the ipsilateral hindlimb extensor and flexor motoneurons.[23] The demonstration that NA does not act as a neurotransmitter, but rather as a neuromodulator, is shown by the fact that microiontophoretic application of this agent produces in rats and cats a slowly developing small-amplitude depolarization in intracellularly recorded spinal motoneurons, which outlasts the ejection period.[214] This effect was associated with an increase in glutamate-evoked motoneuronal firing[214] and was apparently mediated by a- (presumably a_1-) adrenoceptors.[23,48,195, 101,106] Stimulation of the NAergic CS pathway may also inhibit the activity of Renshaw (R)-cells[65,66] (cf. Ref. 149; Figure 11.1) and spinal interneurons interposed in the inhibitory pathway from group II muscle afferents to extensor motoneurons.[85] The possibility that these effects were mediated by a_2- and/or b-adrenoceptors remains to be investigated. In conclusion, it appears that in addition to a direct facilitatory influence on spinal motoneurons, the CS pathway suppresses the activity of inhibitory interneurons, thus leading to disinhibition of tonic a- as well as static g-motoneurons (cf. Ref. 147).

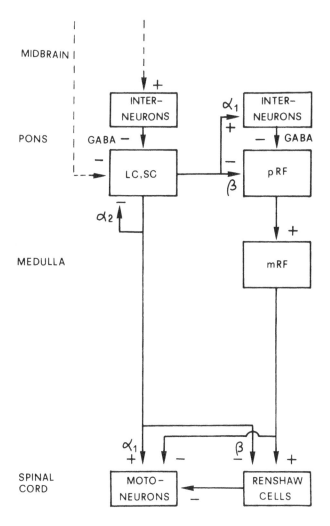

FIGURE 11.1 Descending projections from the adrenergic LC-SC nuclei to the spinal cord. They Include a direct coeruleospinal projection, as well as an indirect projection passing through the dorsal pontine reticular formation (pRF) and the related medullary reticular formation (mRF), from which inhibitory reticulospinal systems originate. a_1, a_2, and b refer to adrenoceptors; + and - indicate excitatory and inhibitory postsynaptic effects on various targets. Dashed lines refer to supramesencephalic descending systems which presumably inhibit the LC-SC nuclei either directly or through interposed GABAergic neurons. (Reprinted from Pompeiano (Ref. 151b) with permission from Elsevier Science, NL)

11.4.2   INDIRECT PROJECTION FROM THE LC TO THE SPINAL CORD THROUGH PONTINE TEGMENTAL STRUCTURES: THE ACh SYSTEM

The NAergic LC neurons may contribute to the postural activity not only by utilizing the direct CS projection, but also by suppressing the activity of the presumably ACh and/or cholinosensitive PRF neurons. These PRF neurons utilize a descending tegmento-reticular projection to activate two distinct regions of the medial medulla: the nucleus magnocellularis of the rostral medulla and the nucleus paramedianus of the caudal medulla.[151b,153] These medullary areas give rise to inhibitory RS projections, which may suppress the activity of spinal motoneurons by activating R-cells[147,151b] (Figure 11.1). However, only the nucleus paramedianus is made up of cholinosensitive neurons and receives an excitatory input from ACh pontine neurons.[99,100]

Electrophysiological experiments using single unit recording have shown that the discharge of NAergic LC and CS neurons, which is very low (1 - 2 imp./s) in the intact animal at rest (cf. Ref. 63,185), increased to about 10/s following decerebration[154,155,158] (cf. Ref. 146) due to interruption of a supramesencephalic pathway exerting an inhibitory influence on the LC neurons. This effect could be mediated either by a direct GABAergic inhibitory projection from the substantia nigra, pars reticulata to the pontine tegmentum, or by an indirect projection acting through GABAergic, inhibitory neurons, some of which appear to be located close to or within the LC-complex.[86] The increased discharge of the CS neurons after decerebration actually contributes to the postural rigidity because their axons exert not only a direct facilitatory influence on a-extensor (and flexor) moto-neurons, but also an inhibitory influence on R-cells and other inhibitory interneurons (see Section 11.4.1). On the other hand, the increased activity of the LC neurons may enhance posture by suppressing the discharge of the ACh and/or cholinosensitive pontine tegmental neurons and the related medullary inhibitory RS neurons. Indeed, there is evidence that in decerebrate cats, the high resting discharge rate of the LC-complex neurons is associated with a low discharge rate of the PRF neurons.[73,154,146] However, as soon as the LC neurons cease firing, as after systemic injection of an anticholinesterase,[146,154,155] the discharge rate of presumably ACh and/or cholinosensitive PRF neurons[73,146,154] and inhibitory medullary RS neurons[182] increased, suppressing posture.[145] Similar results may also account for the postural atonia that occurs in intact animals during REM sleep (see Section 11.3).

In addition to the findings reported above, experiments performed in decerebrate cats have shown that a decrease or a suppression of postural activity in the ipsilateral limbs occurred after: (1) unilateral microinjection into the LC complex of the a_2-adrenergic agonist clonidine, which inhibits the discharge of the NA LC neurons;[152] (2) microinjection into the dorsal PRF either of the b-adrenergic antagonist propranolol[54] or the a_1-adrenergic antagonist prazosin,[43] which blocks the inhibitory influence exerted by the NAergic LC neurons on dorsal pontine structures either directly (through b-adrenoceptors) or through a_1 activation of local GABAergic interneurons; and (3) microinjection into the dorsal pontine reticular area of ACh agonists (carbachol, bethanechol),[22,55,99] which activate the cholinosensitive PRF neurons and the related medullary inhibitory RS neurons, thus inhibiting the extensor motoneurons postsynaptically.[128] On the other hand, lesion of the pontine reticular area suppresses the episodes of postural atonia that normally follow systemic injection of an acetylcholinesterase (AChE) inhibitor.[52,148]

11.5 NORADRENERGIC AND ACH SYSTEMS PROJECTING TO THE VESTIBULAR NUCLEI (VN)

11.5.1 NORADRENERGIC AFFERENTS TO THE VN

11.5.1.1 Anatomical Distribution

Original fluorescence studies were unable to detect catecholamine containing neurons or nerve terminals in the vestibular nuclei (cf. Ref. 220). Using an antibody against bovine dopamine b-hydroxylase (DbH), Swanson and Hartman[188] were able to find only very scattered DbH-positive fibers in the VN of rats, suggesting that these structures did not receive a significant NAergic input. However, using an antiserum against rat DbH, Nicholas et al.[134] revealed that the VN receive some DbH-immunoreactive fibers, providing a possible source for NA release. Moreover, TH-positive fibers were found in the VN of guinea pigs.[73b]   Schuerger and Balaban[176] have recently investigated direct projections from the LC and SC to the VN in rats. These investigators found that TH- and DbH-immunoreactive fibers segregated into two separate pathways: (1) a lateral descending bundle which projects to the superior vestibular nucleus (SVN), the cochlear nuclei, and the cerebellar cortex; and (2) a medial descending bundle which sends afferents to the lateral (LVN), medial (MVN), and descending (DVN) vestibular nuclei before continuing on to the cochlear and cerebellar nuclei. Exposure to DSP-4 (a neurotoxin selective for LC axons) abolished DbH-immunoreactivity in the VN, cochlear nuclei, and cerebellum.[64] The NAergic innervation of the VN-complex was particularly dense in the dorsal LVN (an area which receives a heavy inhibitory projection from the cerebellar vermis[49,80]), was less dense in the SVN and ventral LVN, and relatively sparse in MVN and DVN. Variable levels of NA in the VN were also found in biochemical studies (cf. Ref. 176).

Different types of adrenoceptors control information processing in the VN. In particular, radioligand binding studies have shown the existence of a_2-adrenoceptors in the rostral part of the MVN and DVN of rats,[32,69,202] a_1- and b-adrenoceptors in the VN-complex,[139] and b- (particularly b_2) receptors in the LVN.[139,210] Moreover, *in situ* hybridization studies have shown a_1 A-D mRNA,[141a] a_2 A mRNA, and b_1 mRNA[134,135] labeling in the VN-complex, particularly in the MVN.

In addition to the NAergic innervation in the VN-complex which originates from the LC and SC, there is also a NAergic innervation in the vestibular labyrinth which is supplied by sympathetic fibers originating from the superior cervical ganglion.[73a,b] These fibers, which show TH-like immunoreactivity, are unmyelinated and of the varicose-type. They course along the labyrinthine artery before reaching Scarpa's ganglion, the cristae ampullaris of the semicircular canals, and the maculae of the saccule and utricule. These fibers terminate beneath the sensory epithelia either near capillaries or among the vestibular nerve fibers, particularly at the level of the nodes of Ranvier. Sympathetic fibers were not observed in sensory epithelia. It has been postulated that these fibers control not only the permeability of capillaries through perivascular varicosities, but also the excitability of the vestibular afferent fibers either directly or indirectly through inhibition or excitation of the efferent fibers.

11.5.1.2   Physiological Findings

Experiments performed in decerebrate, partially cerebellectomized cats[96,220] have shown that microiontophoretic application of NA increased the firing rate of a large proportion of LVN neurons (52 and 63 %, respectively). This increased firing occurred with a latency of 6 to 20 seconds from the beginning of the stimulus, and exhibited a long duration (up to 4 minutes) before returning to baseline values. In Yamamoto's[220] experiments, the excitatory effects of NA on LVN neurons were not blocked by the a-receptor antagonist phentolamine, but it was reduced in two of eight units by the b-receptor blocker dichloroisoproterenol. On the other hand, the b-adrenergic agonist isopropylarterenol did not have any effect on LVN neurons. In contrast, in the experiments by Kirsten and Sharma,[96] the NA-induced excitation of LVN neurons was antagonized by the a-receptor antagonist phentolamine, but not by the b-receptor blockers propranolol or MJ-1999. However, compared to the LVN, Kirsten and Sharma found that 76% of the MVN neurons tested were inhibited by iontophoresis of NA, an effect which was not modified by iontophoresis of the a-adrenergic (phentolamine) or b-adrenergic (propranolol or MJ-1999) blockers.

A reinvestigation of the problem was performed by Licata et al.[103] in urethane-anesthetized rats with the cerebellum intact. Of the 120 units tested in all the VN, 85% modified their resting discharge upon microiontophoretic application of NA, with inhibition being observed in 86% of the responsive units, while excitation was observed only in 14% of the units. The inhibitory effects appeared 5 to 20 seconds after the beginning of NA application and lasted up to 60 sec after the end of the injection. The responses were dose dependent and affected equally well VN neurons antidromically activated by stimulation either of the ascending medial longitudinal fascicle (MLF), or of the descending lateral vestibulo-spinal (VS) tract. The majority of the inhibitory responses of VN neurons to NA are apparently mediated through a_2-adrenoceptors, although a small contribution of a_1- and b-receptors cannot be excluded. The coexistence of various adrenoceptors inducing opposite effects on VN neurons is considered to be responsible for the small number of excitatory or biphasic effects (inhibition followed by excitatory rebound) recorded in some VN neurons. The presence of various adrenoceptors is also thought to underlie the increase of the inhibitory responses to NA induced in some cases by b-antagonists.

The inhibitory response of LVN neurons obtained by Licata et al.[103] after microiontophoretic application of NA in rats with the cerebellum intact contrasts with the excitatory response obtained by Yamamoto[220] and Kirsten and Sharma[96] in cerebellectomized cats. Since the dorsal LVN receives a heavy GABAergic projection from the cerebellar vermis, we may postulate that the NA-evoked inhibition reported by Licata et al.[103] depends on potentiation of GABA activity. With regard to MVN neurons, which are particularly inhibited by microiontophoretic NA (either in the presence[103] or after partial removal of the cerebellum[96]), this effect can be attributed to NA-induced potentiation of GABA inhibition. This GABAergic inhibition may originate either from the flocculonodular lobe or from the commissural inhibitory system interconnecting the MVN of both sides[80]. Observations made by Gallagher et al.,[68] using *in vitro* brain stem slice preparations where most of these GABAergic inhibitory pathways had been interrupted, have shown that NA consistently depolarizes MVN neurons and usually induces action potentials or increases their firing rates. No attempt has been made thus far to investigate whether NA acts only on the resting discharge of VN neurons, or whether it can also act to enhance the response of VN neurons to excitatory (glutamate) or inhibitory (GABA) neurotransmitters, as has been shown for cerebellar Purkinje (P-) cells (see Section 11.5.1.2). Since the responses of VN neurons to peripheral signals represent the result of integration of both labyrinthine excitation and corticocerebellar and/or commissural inhibition, we may postulate that one of the main roles of the NAergic system is to improve the information processing within the VN-complex.

11.5.2 CHOLINERGIC AFFERENTS TO THE VN

11.5.2.1 Anatomical Distribution

The VN-complex is rich in ACh[36] as well as in ChAT activity, as shown by a micropunch technique.[34] Moreover, immunohistochemical studies have shown the existence of ChAT-containing terminals in all of the four main V.[12,40,93] These ACh afferents to the VN do not correspond to primary vestibular afferents, which utilize an excitatory aminoacid as a neurotransmitter (cf. Ref. 208 and Chapters 2, 4, and 5), but originate either from the vestibular complex (intrinsic source) or from extravestibular structures (extrinsic source).

11.5.2.1.1 Intrinsic Source

The hypothesis that ACh afferents to the VN originated from ACh neurons in the vestibular complex of the rat was excluded by Burke and Fahn,[35] who showed that kainic acid lesion of the VN did not affect ChAT activity in this region, as assayed by a micropunch technique. However, ChAT-positive cells have been convincingly demonstrated in the caudal part of the MVN, the dorsal part of the DVN,[19,40,136,190] as well as in the nucleus prepositus hypoglossi[40,93,136,190] of several animal species. These ChAT-positive neurons appear to project predominantly to the flocculonodular lobe and the dorsal cap of the inferior olive, IO.[19,20] However, it is likely that these ACh vestibulo-cerebellar neurons send collaterals to the VN-complex.

11.5.2.1.2 Extrinsic Source

ACh afferents to the VN may also originate from ChAT-immunoreactive neurons of PPT/LDT nuclei, which provide a descending input to the VN in rats.[217] ACh afferents to the VN could also come from the contralateral IO.[117] Receptor binding and immunohistochemical studies have shown that both muscarinic[37,38,170,178,209] and nicotinic[45,178,189] receptors can be found in the VN-complex, particularly in the MVN. *In situ* hybridization histochemistry has also been used to identify cells containing mRNA coding for the m_2 subtype of muscarinic receptors in the VN of rats,[205] as well as different types of nicotinic receptor subunits.[207]

11.5.2.2 Physiological Findings

It is generally agreed that not only the LVN but also the MVN are cholinosensitive and that ACh agents (like ACh, the nonselective agonist carbachol, and the anticholinesterase physostigmine) predominantly increase the spontaneous firing rates of these cells (cf. Ref. 208). These effects are mediated by muscarinic receptors.[94-96,116] The results of these *in vivo* experiments were confirmed and extended by experiments in which extracellular recording of VN neurons was performed *in vitro*.[39,58,200,201] MVN neurons recorded in rat brainstem slices exhibited an endogenous regular discharge, which was associated with membrane potential changes independent of individual transmitter systems, thus behaving like pacemaker neurons[39]. Using this preparation, Ujihara et al.[200,201] have shown that 66.7% of extracellularly recorded MVN neurons exhibit a dose-dependent increase in spontaneous firing after carbachol application to the perfusion medium, 22.2% showed a decrease, with no changes in the remaining units (11.1%). Similar results have been obtained by other investigators.[39,58] The excitatory or inhibitory effects of carbachol were mimicked by muscarine, but not by nicotine, and were antagonized by atropine.[58,200]

Intracellular recording experiments performed in slice preparations of the rat VN-complex, however, have shown that the ACh system acts on MVN neurons not only through muscarinic, but also through nicotinic receptors.[68,141] MVN neurons responded to the mixed cholinergic agonists, ACh or carbachol, with membrane depolarization, and similar results were also obtained after application of the selective muscarinic agonist muscarine or the selective nicotinic agonists nicotine or 1,1-dimethyl-4-phenylpiperazinium (DMPP). Moreover, the muscarinic-induced depolarization (which was reversibly blocked by the selective muscarinic antagonist, atropine) was attributed to a decreased conductance due to a closure of K+ channels. Conversely, the nicotine- and DMPP-induced depolarization (which was suppressed by the selective ganglionic nicotinic antagonist, mecamylamine) was attributed to an increase in conductance due to the opening of cationic channels.

In addition to these direct excitatory influences on MVN neurons, Phelan and Gallagher[141] found that carbachol or muscarine produced a pure membrane hyperpolarization (in 17% of MVN neurons) or transient hyperpolarizing responses during membrane depolarization (in about 40% of the tested units). These effects were reversibly blocked by pretreatment of the slice with the selective $GABA_A$ receptor antagonist, bicuculline. These data indicate that the effects were due to muscarinic receptor activation exerted either postsynaptically on intrinsic GABAergic interneurons or presynaptically on GABAergic terminals which surround VN neurons, thus leading to release of the inhibitory transmitter. Selective nicotinic agonists failed to elicit membrane hyperpolarizations. To date, there have been no investigations that have examined whether application of a muscarinic (or nicotinic) agonist (independent of the cholinergic influences on the resting discharge of VN neurons) could enhance the responses of VN neurons to excitatory (glutamate), or to inhibitory (GABA) neurotransmitters, as was shown for cerebellar Purkinje cells (see above). If this were the case, one might expect that the ACh system would improve information processing within the VN-complex, as has been proposed for the NAergic system (see Section 11.4.2).

11.5.3 NORADRENERGIC AND ACh INFLUENCES ON POSTURE AND OCULOMOTOR ACTIVITY

No attempt has been made thus far to investigate the effects on posture and motor activities of unilateral microinjection of NAergic agents into the VN-complex. On the other hand, unilateral injection of a direct-acting ACh agent (oxotremorine) or of an anticholineserase (eserine) into the rat VN-complex produced contraversive roll and twist about the longitudinal axis of the animal. This effect can be attributed to ACh activation of VN neurons exerting an excitatory influence on ipsilateral limb extensor motoneurons.[33] As expected, unilateral injection of chlorpromazine methiodide (CPZ MI), which behaves as an antimuscarinic agent, into the VN produces ipsiversive roll and twist movement, an effect similar to that induced after unilateral labyrinthectomy (UL).

There are few data relating to the function of the NAergic system in vestibular compensation. This would seem worthy of investigation given that the VN neurons receive NAergic afferents. Abeln and Flohr (cited in Ref. 181) have reported that intracisternal injection of NAergic agonists (e.g., clonidine) in compensated frogs causes decompensation of roll head tilt, whereas NAergic antagonists (e.g., phentolamine) causes overcompensation. Moreover, unpublished observations made in our laboratory have shown that unilateral or bilateral microinjection into the LC-complex of small doses of clonidine, leading to inactivation of the NAergic LC neurons, produces the reappearance of the postural deficits in the compensated cat. Other aspects of the problems relating the NAergic LC system to vestibular compensation have been discussed in a recent review.[151a]

More recently, the effects of UL on the concentration of NA and dopamine and their respective metabolites (3-methoxy-4-hydroxyphenylglycol [MHPG] and 3,4-dihydroxy-phenylacetic acid, DOPAC) in the LC and the MVN have been investigated.[49a] This neurochemical study was conducted 6 hours after UL in both albino and pigmented rats. In UL-albino rats the main changes were observed in the contralateral LC, which exhibited a large increase of NA and DOPAC, indicating enhanced NA synthesis. On the other hand, in the MVN there was a bilateral increase of MHPG and an ipsilateral increase of dopamine, suggesting activation of NA synthesis and metabolism. One possible explanation for these findings is that the labyrinthine input of the intact side activates the ipsilateral LC and this, through uncrossed and crossed pathways, activates the NAergic projection to both MVNs. This is in agreement with previous electrophysiological findings.[157,158] In contrast, in UL-pigmented rats the only change observed was a decrease in NA content in the ipsilateral LC. This was interpreted as a consequence of an earlier strong increase of NA release. These findings confirm the relevant role of the NAergic LC innervation in vestibular compensation and also indicate the involvment of NA in the MVN during the early stage of this process. The different strain-related NAergic responses observed 6 hours after UL suggest that the involvement of basal NA, particularly from the LC innervation, may be more crucial and sustained in albino than in pigmented rats.

The role of the vestibular ACh system in the compensation of the vestibular syndrome induced by UL has also been investigated. The lack of modification of ACh content in the deafferented LVN[218] does not support the hypothesis of a cholinergic mechanism involved in vestibular compensation. However, ACh-modulating drugs (cf. Ref. 181) modify the time course of vestibular compensation. Observations made in compensated animals have in fact shown that systemic or intracranial injection of ACh agonists (arecoline, carbachol, nicotine, muscarine, oxotremorine, methacholine, or bethanechol) or AChE inhibitors produces decompensation of the postural and motor deficits, while ACh antagonists (atropine and scopolamine) generate an overcompensation.[1,29,78]

Experiments performed in rats have recently shown that there is a strong reduction of AChE activity in the MVN and the prepositus hypoglossi on the deafferented side.[199] This finding, which was particularly strong 6 hours after UL, would potentiate the discharge of the deafferented VN neurons, thus contributing to recovery of the vestibular syndrome. The suggestion that a depletion of AChE is the basis for a reactivation of the deafferented VN is congruent with the proposition by Flohr et al.[61] that vestibular compensation results from a hypersensitivity of the neurons on the lesioned side. By contrast, a permanent influence of AChE activity on the intact VC persists as in normal animals.[199] Flohr's hypothesis is reinforced by the recent immunohistochemical demonstration of an increase in the number of cholinergic neurons in the deafferented VN-complex of lesioned cats,[197] but his hypothesis is contradicted by the finding of an increase in muscarinic receptor density in the intact VN-complex in rats.[38]

In addition to changes in posture and movements, cholinergic agents may also affect the activity of the oculomotor system. For example, experiments performed in precollicular decerebrate cats have shown that IV injections of small doses of eserine sulfate (0.025 mg/kg) produced, after 5 minutes, both rotatory and postrotatory nystagmus, which were always absent prior to the injection and were suppressed by IV administration of the muscarinic antagonist atropine sulfate.[17] Intravenous injection of higher doses of eserine sulfate (0.05 to 0.1 mg/kg) in the same type of preparation (which produced episodes of postural atonia; see Section 11.4.2) elicited the regular occurrence of bursts of horizontal rapid eye movements (REM) in stationary cats. This is similar to those which occur in intact, unanesthetized animals during REM sleep (cf. Ref. 146). Moreover, single unit recording performed in decerebrate cats[120] has shown that second-order MVN neurons monosynaptically activated by stimulation of the ipsilateral labyrinth and antidromically driven by stimulation of the contralateral abducens nucleus showed rhythmic discharges related to the cholinergically induced ocular jerks. Similar results are obtained in intact, unanesthetized cats during spontaneously occurring REM sleep (cf. Ref. 142). Lesion experiments indicate that the discharge of these vestibulo-oculomotor neurons originate from presumably ACh and/or cholinosensitive neurons located in the dorsal pontine tegmentum.[148]

11.6 NORADRENERGIC AND ACH SYSTEMS PROJECTING TO THE CEREBELLAR CORTEX

11.6.1 NORADRENERGIC AFFERENTS TO THE CEREBELLAR CORTEX

11.6.1.1 Anatomical Distribution

In addition to the classical mossy fibers (MF) and climbing fibers (CF) which utilize excitatory amino acids as neurotransmitters, the cerebellar cortex also receives an NAergic input that originates largely from the LC-complex.[27,63] This projection is bilateral but with an ipsilateral preponderance, and reaches the entire cerebellar cortex including the cerebellar vermis and the flocculus.[27,63] The NAergic afferents terminate within the molecular and the granular layers of the cerebellar cortex, where they make synaptic contacts primarily on Purkinje cell dendrites in the molecular layer and, to a lesser extent, on the Purkinje cell body and superficial granular cell layers.[92,221]

The NAergic system may act on the cerebellar cortex of mammals through a_1- and a_2-receptors (cf. Ref. 3) as well as b_1- and b_2- receptors,[4] with the b_1 receptors being more numerous than the b_2 receptors. The selective distribution of these adrenoceptors in the different layers of the cerebellar cortex, as demonstrated in *in vitro* autoradiographic studies, has been reviewed recently (cf. Ref. 8,151). An important question in these studies is whether the binding sites relate to cell bodies or fiber terminals of target neurons. Using *in situ* hybridization with probes for genes coding for different types of adrenoceptors, signals were found for a_1B receptor mRNA[141a] and a_2C receptor mRNA[135] in the granule cell and the Purkinje cell layers, while neurons expressing b_2 mRNA corresponded primarily to the granule cells.[34,138] The mismatch found between the localization of the b_2 receptors in the molecular layer, as shown in *in vitro* autoradiographic studies[8] and the localization of the corresponding mRNA in the granule cell layer, indicate that these receptors are synthesized in the granule cell layer, transported to their axon terminals, and are ultimately located presynaptically on parallel fiber terminals. It is of interest that b-adrenoceptors, particularly of the

b_1-subtype, have been found not only on neurons, but also on glial cells, where they control metabolic and trophic actions.[186]

11.6.1.2 Physiological Findings

In vivo experiments have shown that microiontophoretic applications of NA,[127] as well as LC stimulation,[126] decreased the resting discharge of Purkinje cells.[8] This effect was associated with hyperpolarization of the cell membrane[211] which was coupled with an increase in their input resistance, probably due to NAergic activation of adenylate cyclase, leading to an increased synthesis of cAMP.[213] These effects were attributed to a-adrenoceptors.[127,211] These findings differ from those obtained in cerebellar slices[28] as well as in intraocular cerebellar grafts,[70] showing that NA could induce not only inhibition of Purkinje cell activities through a-receptors, but also excitation through b receptors (cf. Ref. also 50). Finally, observations made in anesthetized rats have shown that pressure microinjection of NA elicited a dose-dependent inhibition of Purkinje cell firing rate that was blocked by a-adrenergic antagonists (particularly of the a_2-subtype), but was rarely affected by b- antagonists.[140] The fact that the NA-induced excitation is rarely observed *in vivo* but is observed *in vitro* and in oculo preparations, where many mossy afferents to Purkinje cells are interrupted or inactive, suggests that these afferents play an important role in the NA-induced inhibition.

Independent of the NAergic influence on the resting discharge of the Purkinje cells, observations made on *in vivo* preparations have shown that local applications of NA as well as LC stimulation enhance the response of these cells to both excitatory (MF and CF) and inhibitory (basket and stellate cells) inputs. Similarly, iontophoretically applied NE or LC stimulation enhances the response of Purkinje cells to excitatory (glutamate, aspartate) and inhibitory (GABA) neurotransmitters associated with afferents to the cerebellar cortex.[8,213] These effects were mediated by b-adrenergic receptors, probably acting through second messengers.[211,213] These findings indicated that one of the main functions of the NAergic input in the normal operation of the cerebellum is to augment target neuron responsiveness to conventional afferent systems which are directly concerned with detailed information transfer. This may occur by increasing the signal-to-noise ratio of the evoked vs. spontaneous activity, a finding which justifies the implication of NA as a neuromodulator of the cerebellar cortex.[216] The same input could also act to gauge the efficacy of subliminal synaptic inputs conveyed by classical afferent systems.[212]

11.6.2 CHOLINERGIC AFFERENTS TO THE CEREBELLAR CORTEX

11.6.2.1 Anatomical Distribution

Immunohistochemical studies have shown the presence of ChAT-like immunoreactivity in the cerebellar cortex, which occurs in a subpopulation of MF-terminals in the granular layer as well as in the corresponding rosettes.[18,75,77,89,90,137] These fibers terminate particularly in the archicerebellum, that is, in the vermal lobules IX and X and the flocculus they are moderately accumulated in the paleocerebellum, including the vermal lobules I and VI, and, to a lesser extent, in the neocerebellum.[13,75] ChAT-immunoreactivity was also found in a distinct net of thin, varicose fibers which spread within the granular layer and the lower part of the molecular layer to contact the Purkinje cell bodies and their initial dendritic shafts.[18,75,77,137]

Although some of the ChAT-positive fibers may originate within the cerebellar cortex and nuclei,[10,136,217] the majority of the ACh cerebellar afferents originate from precerebellar structures and course through the inferior and, to a lesser extent, the middle cerebellar peduncle.[13,137] Double-labeling experiments in several animal species have shown the extensive ACh MF projection to the flocculo-nodular lobe (which receives vestibulo-cerebellar afferents and controls the vestibulo-ocular reflex [VOR] and the optokinetic reflex [OKR]) comes primarily from the MVN and DVN.

In addition, some ACh fibers arise from the nucleus prepositus hypoglossi, where ChAT-immunoreactive neurons are also located.[19,76] With regard to the ACh afferents terminating in the cerebellar vermis (which does not receive direct,[80] but only indirect vestibulo-cerebellar afferents[144] and controls the vestibulospinal reflex [VSR]), these fibers apparently originate from the lateral reticular nucleus,[16,19,93,190,196] the medullary tegmentum,[137,180] and the Group X of Brodal and Pompeiano. The neurons of Group X in the monkey are all immunoreactive for ChAT.[40] It should be noted that all of these structures convey proprioceptive inputs to the cerebellar cortex. A further candidate for ACh afferents to the cerebellar vermis is represented by the dorsal pontine tegmental area, namely the PPT/LDT and the parabrachial nucleus, which contain ChAT-immunoreactive cells (see Section 11.3). In particular, double-labeling experiments in kittens in our laboratory have allowed us to identify tegmento-cerebellar projecting neurons following injection of the retrograde tracer rhodamine-labeled latex microspheres in lobules V to VII of the cerebellar vermis. The cholinergic nature of this projection was subsequently confirmed using immunohistochemical staining for ChAT.[44] A small proportion of ChAT-positive tegmental neurons project to the cerebellar vermis. However, among the whole population of retrogradely-labeled tegmental neurons, one-third were ChAT-positive. These ACh neurons were located not only in the PPT/LDT but, to a lesser extent, also in the LC-complex and the peri-LCa, where they were intermingled with the more numerous NAergic coeruleo-cerebellar neurons. It is worth noticing that in contrast to the unique source of origin of the NAergic afferents from the LC to the whole cerebellar cortex, there are apparently multiple sources of origin of the cholinergic afferents, each of them terminating in more limited areas of the cerebellar cortex.

Both muscarinic (cf. Ref. 10,151) and nicotinic receptors (cf. Ref. 5,151) have been identified in the cerebellar cortex. In particular, muscarinic receptors showed in rodents a higher density in the granular and P-cell layers than in the molecular layer,[133] these receptors being of the m_2 subtype.[205] The nicotinic receptors, however, were located in the granular layer, where weak signals were detected for a_4 and b_2 -subunits of the nicotinic receptors.[207] Moreover, most if not all P-cells displayed dense hybridization for b_2 subunits.[207]

11.6.2.2 Physiological Findings

The action of ACh on individual P-cells as well as the identification of the receptor types involved in the postsynaptic ACh effects are rather controversial.[151] These findings question the hypothesis that detailed information transfer occurs through cholinergic fibers. The hypothesis that ACh acts in the cerebellar cortex as a neuromodulator rather than a neurotransmitter has been tested by our laboratory. In these experiments we examined the effects of microiontophoretic application of the cholinergic muscarinic agonist bethanecol (10 to 60 nA, 5 minutes) on the Purkinje cell responses in anesthetized rats to 10- to 20-second pulses of glutamate (the putative neurotransmitter of the excitatory parallel fiber input to the P-cells[11]). In addition, we have examined effects of bethanecol application to 15-second pulses of GABA (the putative neurotransmitter of inhibitory basket and stellate interneurons acting on Purkinje cells[9]). The results obtained are illustrated in Figures 11.2A and B, respectively.

Bethanecol significantly increased the glutamate-evoked excitatory responses in 22/33 P-cells, as well as the GABA-evoked inhibitory responses in 22/25 P-cells, regardless of whether this agent altered the basal discharge rate of the cell. The increase of the glutamate or GABA-evoked responses developed slowly after the onset of the bethanecol application and greatly outlasted (up to 30 to 40 minutes) the offset of the injection current. Similar results were also obtained by ACh application (20 nA, 5 minutes), which produced a small increase of basal firing rate during the injection and a slowly developing, long-lasting increase in the glutamate response, lasting up to 50 minutes after the injection offset. The specificity of the results was shown by the fact that the muscarinic antagonist scopolamine prevented the bethanecol-induced increase in the glutamate or GABA responses (Figures 11.2A and B). In conclusion, it appears that the ACh input to the cerebellum may facilitate the action of conventional afferent systems by increasing the signal-to-noise ratio of the Purkinje cell responses to the corresponding excitatory and inhibitory neurotransmitters, thus providing a mechanism for improving information processing within the cerebellar circuits. In particular, the ACh system may increase the amplitude of the Purkinje cell responses to a given signal without modifying the spatio-temporal characteristics which result from appropriate inter-action between excitatory and inhibitory volleys. Although the long-lasting modulatory effect of

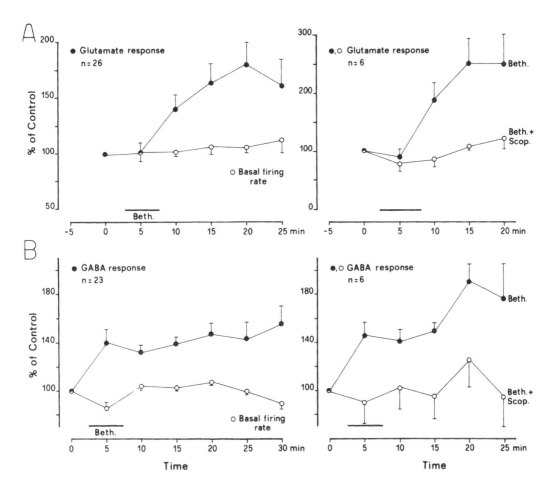

FIGURE 11.2 Effect of microiontophoretic application of the muscarinic agonist bethanecol (Beth.) on Purkinje cell responses to glutamate (upper diagrams) or GABA (lower diagrams) and scopolamine (Scop.) antagonism (right diagrams). (A): Left diagram: Beth. effect on basal firing rate (open circles) and responses (filled circles) of 26 Purkinje cells to glutamate. Right diagrams, suppression of the bethanecol-induced increase in the glutamate responses of six Purkinje cells following combined Scop. and Beth. ejection. (B): Left diagram: Beth. effect on basal firing rate (open circles) and responses (filled circles) of 23 Purkinje cells to GABA. Right diagram, suppression of the bethanecol-induced increase in the GABA responses of six Purkinje cells following combined Scop. and Beth. injection. Means ± S.E.M. are plotted. Beth. application is indicated by horizontal bars. (Reprinted from Andre et al. (9 and 11) with permission from Elsevier Scientific, NL.)

the ACh muscarinic system on the Purkinje cells resembles that exerted by the NAergic system, possible differences in the relative effectiveness of the two inputs on the same Purkinje cell, as well as in the molecular mechanisms responsible for these actions, remain to be investigated.

11.6.3 NORAGRENERGIC AND ACh AFFERENTS TO PRECEREBELLAR STRUCTURES

The NAergic and ACh systems may influence the activity of the cerebellar cortex by not only acting directly on Purkinje cells, but also by acting indirectly on the precerebellar structures projecting to them. These target structures, including the inferior olive (IO) vestibular complex and LRN, are of great relevance since they may convey labyrinthine signals either to the flocculonodular lobe which controls the VOR, or to the cerebellar anterior vermis which controls the VSR.[80] The NAergic and ACh innervation of the VN-complex (i.e., of the MVN, DVN, and the nucleus prepositus hypoglossi) has been reported previously (see Sections 11.5.1 and 11.5.2). We will, therefore, limit this section to the IO and the LRN.

The principal source of the dense plexus of NAergic afferents which surround all subdivisions of the IO[188] originates bilaterally from the LC, SC, and medial parabrachial nucleus[97,213a]. Different types of adrenoceptors were found in the IO of several animal species by using *in vitro* autoradiography.[62,164,202] However, *in situ* hybridization methods have shown a_1 A - D receptor mRNA[141a] in the rat IO, while a_2 A, b_1, and b_2 receptor mRNA expression was lacking.[134,135] It is known that the IO neurons, which fire at a frequency of 1 to 10/s in *in vivo* experiments, may spontaneously exhibit oscillatory firing. Observations made by Llinas and Yarom[105] in guinea pig brainstem slices *in vitro* indicate that in 10% of these experiments, subthreshold oscillations of the membrane potential at 4 to 6/s occurred spontaneously in all cells tested. These oscillations reflect the properties of neuronal ensembles made by a large number of electrically coupled elements. This oscillatory rhythm was attributed to an inward Ca^{2+} current which activates a Ca^{2+}-dependent K^+ conductance that, in turn, leads to hyperpolarization followed by a Ca^{2+}-dependent rebound conductance change. Addition of NA to the bath reduced or blocked this ensemble oscillation of IO neurons, probably due to block of Ca^{2+}-conductance. The nature of the adrenoceptors involved in this effect was not investigated. We believe that the suppression of synchronous firing of IO neurons following activation of the NAergic system may improve the processing of sensory information passing through the IO during labyrinth stimulation.

In addition to NAergic afferents, the entire IO-complex contains, in the cat, small-sized cells covered by ChAT-positive terminals.[93] Observations made by Barmack et al.[20] have shown that in rats and monkeys this ACh pathway terminates only in the dorsal cap of the medial accessory olive, a structure that receives horizontal optokinetic information from the nucleus of the optic tract and projects to the flocculus. In the rabbit, however, the cholinergic innervation affects the dorsal cap as well as the adjacent nucleus b, a structure that receives vertical vestibular information. Experimental anatomical methods have shown that in rats and rabbits these ACh afferents originate ipsilaterally and/or contralaterally from the MVN, DVN, and nucleus prepositus hypoglossi. It also appears that the IO contains the M_2 subclass of muscarinic receptors (cf. Ref. 20). Since in other systems activation of M_2 receptors may hyperpolarize the neurons by increasing the K+ conductance (Ref. 20), it has been postulated that this mechanism could reduce the sensitivity of the dorsal cap neurons to the optokinetic stimulus.[20] According to this hypothesis, the ACh projection would be functionally inhibitory. An alternative possibility, however, would be that the ACh-induced hyperpolarization suppresses the tendency of the IO neurons to fire spontaneously, thus improving the processing of sensory information through the IO. In addition to muscarinic receptors, the IO shows distinct labeling for nicotinic receptors a_4 and b_2 subunits.[207] Direct experiments are required to investigate the effects of activation of muscarinic and nicotinic receptors on the neuronal activity in the IO.

NAergic[188,213a] and ACh afferents[93] also project to the LRN; in particular, a dense terminal field of ACh fibers surrounds noncholinergic and cholinergic ChAT-positive neurons located in this structure.[93] The LRN of rats contains cells labeled for a_1B-receptor mRNA[141a] as well as scattered cells labeled for $_2$A-, but not for the b_1- and b_2- receptor mRNA.[134,135] LRN neurons also contain cells with mRNA coding for the M_2 subtype of muscarinic receptors.[205] The functional role of these afferent systems to the LRN remains to be investigated.

11.7 VESTIBULAR INFLUENCES ON NORADRENERGIC AND ACH SYSTEMS

As early as 1971, Barnes and Pompeiano[25,26] performed experiments in decerebrate cats and found that brainstem NAergic and ACh systems were activated by vestibular electrical stimulation influencing the spinal cord. Since then, attempts have been made to (1) provide the anatomical evidence for the existence of direct or indirect projections from the VN-complex to the NAergic LC-complex, as well as to the ACh and/or cholinoceptive structures located in the dorsal pontine tegmentum and the related medullary RF; and (2) to investigate whether the LC neurons and possibly the CS neurons respond to natural stimulation of labyrinth receptors, thus contributing to the regulation of vestibular reflexes (cf. Ref. 150).

11.7.1 VESTIBULAR AFFERENTS TO THE DORSAL PONTINE TEGMENTUM

Using the technique of retrograde HRP transport, bilateral ascending projections from all four VN to the LC-complex were found in rats, with the contralateral projections being considered more extensive than the ipsilateral ones[41] or vice versa.[46] The possible sources of afferents from the VN to the LC are difficult to examine with tracing methods because some of the labeled cells in the VN may result from uptake of HRP by damaged vestibulo-cerebellar fibers or vestibular axons coursing through the ascending MLF. Moreover, a faint "halo" of the injection site centered in LC may spread to the neighboring peri-LCa and the dorsal PRF, so that the retrogradely labeled VN cells could project to the dorsal tegmental structures rather than to the LC.[15,174] More recently, observations made by Fung et al.[67] have shown that after unilateral injection of HRP into the cat and the rat LC, bilateral retrograde labeling was observed in all four VN, but with an ipsilateral predominance. The frequency distribution was in the order of LVN > MVN > SVN > DVN. However, even in these experiments, the spread of HRP at the injection site usually ranged from 300 to 1000 mm in diameter. In addition to these direct projections, the vestibular nuclei may influence the activity of the LC-complex by utilizing indirect projections passing through the RF.[98] In addition to the LC neurons, vestibular projections could influence the ACh and/or cholinosensitive neurons located in the dorsal pontine tegmentum and the related medullary inhibitory RF.

11.7.2 VESTIBULAR INFLUENCES ON THE LC AND THE RELATED PONTINE AND MEDULLARY RETICULAR STRUCTURES

Slow roll tilt of the animal leading to sinusoidal stimulation of macular, utricular receptors produces a VSR characterized by contraction of limb estensors during ipsilateral side-down tilt and relaxation during contralateral side-up lift, while just the opposite response pattern affects the dorsal neck muscles for the same directions of animal orientation.[111,175,104] These postural changes were related to animal position and could in part be attributed to the activity of VS neurons originating from the LVN.[109] In fact, most of these antidromically-activated neurons projecting to the lumbosacral segments of the spinal cord (cf. Ref. 115) and exerting a monosynaptic and polysynaptic excitatory influence on ipsilateral limb extensor motoneurons (cf. Ref. 143) increased their discharge rates during side-down tilt and decreased their discharges during side-up tilt (a-responses). Only a small proportion of the VS neurons showed the opposite response pattern (b-responses; Figure 3A).

11.7.2.1.   Response of LC Neurons to Animal Tilt

Pompeiano et al.[157,15] (cf. Ref. also 149,150) have recorded the activity of 141 neurons located in the LC-complex of decerebrate cats.   Most of these units had the characteristics usually attributed to NAergic neurons as recorded in the rat[63], that is, (1) a typical positive-negative extracellular spike of long duration (>1.5 ms); (2) a slow and regular resting discharge; and (3) a response to a pinch stimulus applied to the limbs characterized by a transient excitation followed by a prolonged inhibition (see Section 11.3).   Moreover, 16 of the recorded units were CS neurons antidromically activated from the ipsilateral spinal cord at T12-L1, thus projecting to the lumbosacral segments of the spinal cord.

Among the whole population of 141 LC-complex neurons tested during roll tilt of the animal at 0.15 Hz, ±10°, 80 (57%) showed a periodic modulation of their firing rates in response to the sinusoidal input. In particular, the majority of the units (45/80, i.e., 56%) increased their discharge rate during side-up tilt of the animal (b-responses), while a smaller number of units (22/80, i.e., 25%) were activated during side-down tilt (a-responses, Figure 11.3B).   Moreover, the former group of units showed more than a twofold larger gain (expressed as imp./s/deg.) than the latter group.   Similar response properties were also found in 11/16 antidromically identified CS neurons which responded to animal tilt.   The responses of LC-complex neurons to increasing frequencies of tilt (from 0.008 to 0.32 Hz and at the fixed amplitude of ±10°) did not show changes in gain and phase angle which remained related to the extreme animal position and was thus attributed to macular stimulation.   In other units, however, there was an increase in gain and phase lead of the responses to increasing frequencies of tilt, thus becoming related to the velocity signal. This effect was attributed to vertical canal stimulation.

More recently, the effects of caloric stimulation on the neuronal activity of the LC have been investigated in urethane-anesthetized rats and the results compared with those obtained from vestibular nuclei neurons.[135a]   Surprisingly, the predominant effect of caloric stimulation with hot and cold water on LC neuronal activity was inhibitory, an effect "which occurred" only 1 minute after the cessation of the caloric stimulation and persisted for 3 to 5 minutes. Transient excitation, however, "was observed" in some neurons during cold (ice-water) irrigation.   These findings contrasted with the reciprocal pattern of response of vestibular nuclei neurons, which showed excitation by caloric irrigation with hot water and inhibition by cold (ice-water) irrigation; moreover, the responses occurred during caloric stimulation and disappeared immediately after the end of the stimulation. The reasons for these discrepancies (high sensitivity of LC neurons to nociceptive stimuli, stress, or anesthesia) have not been investigated.

11.7.2.2 Responses of Pontine Reticular and Related Medullary Inhibitory Reticulospinal (RS) Neurons to Animal Tilt

Experiments performed in decerebrate cats have shown that, similar to the NAergic LC-complex neurons, even the presumably ACh and/or cholinosensitive neurons located in the dorsal pontine tegmentum responded to roll tilt of the animal at 0.15Hz, ±10 (unpublished). Their predominant response patterns closely corresponded to those obtained from presumably inhibitory RS neurons which are, in part at least, under the excitatory influence of the cholinergic pontine system. In particular, Manzoni et al[114] recorded in decerebrate cats the activity of 168 neurons from the inhibitory area of the medullary RF, 93 of which were antidromically activated by electrical stimulation of the spinal cord at T12-L1.   The cholinosensitivity of these presumably inhibitory RS neurons was shown by the fact that they underwent an increased discharge after systemic injection of an anticholinesterase, a finding which was associated with a suppression of posture.[182] Among the entire population of 168 recorded units, 113 (67%) responded to roll tilt of the animal at 0.026 Hz, ±10°. Most of the responsive units (71/113, i.e., 63%) increased their discharge rate during side-up tilt (b-responses).   On the other hand, a smaller group of units (24/113, i.e., 21%)

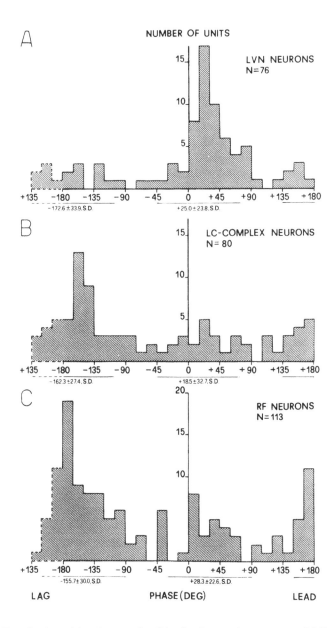

FIGURE 11.3 - Contribution of the phase angle of the first harmonic responses of different populations tested during roll tilt of the animal at 0.026 Hz (A,C) and 0.15 Hz (B) ±10° All the experiments were performed in precollicular decerebrate cats. (A): distribution of the phase angle of responses to animal tilt of 76 LVN neurons, antidromically activated by stimulation of the spinal cord at T12-L1. Thus, these neurons project to the lumbosacral segments of the spinal cord. (B): Responses of 80 LC-complex neurons, 11 of which project to the lumbosacral segments of the spinal cord. (C): Responses of 113 medullary RF neurons, 64 of which project to the lumbosacral segments of the spinal cord. Positive numbers in the abscissa indicate the phase lead in degrees, whereas negative numbers indicate the phase lag of responses with respect to the extreme side-down position of the animal, as indicated by 0.° Responses of the neurons to tilt, underlined by horizontal bars, were used to evaluate the average phase angle of units excited during or near the side-down (0°) or side-up displacement of the animal (180°). Most of the LVN neurons (51/76) were excited during the side-down tilt of the animal, while most of the LC-complex (45/80) and the medullary RF (71/113) neurons were excited during the side-up tilt. (Adapted from Marchand et al. (Ref. 115), Pompeiano et al. (Ref. 158), and Manzoni et al. (Ref. 114). Reprinted from Pompeiano (Ref. 150) with permission from Springer Verlag, FRG.)

was activated during side-down tilt of the animal (a-responses, Figure 11.3C). Even in this case, the former group of units showed almost a twofold larger gain than the latter group. Similar response properties were also found among the antidromically-activated RS neurons. By increasing frequencies of tilt from 0.008 to 0.32 Hz at the fixed amplitude of ±10°, some of the RS neurons showed a stable gain and phase angle of the responses which were related to animal position. Other units, however, showed an increase in gain and phase lead of the responses that became related to the velocity signal due to macular-canal convergence.

In conclusion, it appears that both the LC-complex neurons[158] and the medullary RS neurons[114] projecting to the lower segments of the spinal cord showed a predominant response pattern to tilt (b-responses). This response was opposite in sign to that of the VS neurons[115] projecting to the same segments of the spinal cord (a-responses). These findings can be attributed either to labyrinthine activation of uncrossed projections from b-responsive VN neurons to the ipsilateral pontine and medullary structures,[98] or to activation of crossed projections from a-responsive VN neurons to the ventral part of the contralateral medullary RF.[144] In this region there are, in fact, not only neurons contributing to the inhibitory RS pathway,[114] but also neurons which project to the LC-complex and related structures[15] where they exert a prominent excitatory influence.[60]

11.8 NORADRENERGIC AND ACH INFLUENCES ON THE VESTIBULO-SPINAL REFLEX (VSR) MEDIATED BY DESCENDING SYSTEMS

Experiments performed in decerebrate cats have shown that electrolytic lesion of the LC-complex or of the reticular structures located in the dorsal pontine tegmentum greatly modified the response characteristics of limb extensors to sinusoidal stimulation of labyrinth receptors.[51,52,56] The selective contribution of the NAergic and ACh systems to these lesion-induced effects has been evaluated in the following studies.[153]

11.8.1 NORADRENERGIC INFLUENCE ON THE VSR

Pompeiano et al.[53,152] performed the initial series of experiments in decerebrate cats and showed that neuronal inactivation of the LC-complex elicited by unilateral microinjection of small doses of the a_2-adrenergic agonist clonidine (0.25 ml at 0.012 to 0.15 mg/ml) or the nonselective b-adrenergic agonist isoproterenol (0.25 ml at 4.5 to 9.0 mg/ml) into this structure decreased the postural activity in the ipsilateral and, to a lesser extent, in the contralateral limbs, while the gain of the multiunit EMG responses of the triceps brachii on both sides to roll tilt of the animal (at 0.15 Hz, ±10°) increased significantly (Figure 11.4, A-C).

Only slight changes in the phase angle of the EMG responses were observed. These findings occurred even when the limb position was adjusted to produce the same background discharge as in the control experiments. The results appeared 10 minutes after the injection, fully developed within 30 to 60 minutes, and persisted for more than 2 hours before disappearing. The effects of clonidine and isoproterenol were site-specific and dose-dependent. They were attributed to inactivation of both the NAergic CS neurons, which exert a facilitatory influence on ipsilateral limb extensor motoneurons, and the LC neurons, which inhibit the dorsal pontine tegmental neurons and the related medullary inhibitory RS neurons of both sides, thus releasing their discharge from inhibition. The latter hypothesis is supported by the results of a second series of experiments[43,54] showing that in decerebrate cats, unilateral microinjection of either the selective a_1-antagonist prazosin (0.25 ml at 0.1-1.0 mg/ml) or the nonselective b-antagonist propranolol (0.25 ml at 4.5-9.0 mg/ml) into the peri-LCa and the dorsal PRF, as well as in the PPT nucleus, decreased the postural activity in the ipsilateral limb, while the response gain of the ipsilateral triceps brachii to animal tilt increased significantly. On the other hand, opposite postural and reflex changes usually affected the contralateral limbs. In these experiments, administration of a_1 and b-adrenergic antagonists would block the inhibitory influence that the NAergic LC neurons exert on the ACh and/or cholinosensitive neurons located in the ipsilateral dorsal pontine tegmentum either directly, through b-adrenoceptors, or indirectly, through a_1-activation of local GABAergic inhibitory interneurons (see Section 11.3). The resulting increase in discharge of these pontine neurons and the related medullary inhibitory RS neurons would then serve to decrease the postural activity in the ipsilateral limbs. Moreover, due to the reciprocal changes in discharge of the inhibitory RS neurons (b-responses[114]) with respect to the excitatory VS neurons (a-responses[115]) during animal tilt, we postulate that for the same labyrinth signal giving rise to excitatory VS volleys ending on ipsilateral limb extensor motoneurons, the higher the firing rate of the medullary RS neurons in the animal at rest, the greater would be the disinhibition which affects the ipsilateral limb estensor motoneurons during side-down tilt. These motoneurons would then respond more efficiently to the same excitatory VS volleys elicited by given parameters of stimulation, thus leading to an increased gain of the EMG responses of the forelimb extensor triceps brachii to labyrinth stimulation. In addition to ipsilateral effects, the increased discharge of the peri-LCa and the dorsal PRF could also activate, through crossed projections,[43,88] either the contralateral PRF and the related medullary inhibitory RS neurons,[88] or the contralateral LC-complex and the related CS neurons.[54] This would account for the increase or decrease in the response gain of the contralateral triceps brachii to labyrinth stimulation that occurs after prazosin or propranolol injection.

11.8.2 CHOLINERGIC INFLUENCES ON THE VSR

Systemic (i.v.) injection in decerebrate cats of the anticholinesterase eserine sulfate (at a dose of 0.05 to 0.075 mg/kg, which reduces postural activity) increased the gain[161] or revealed (if absent)[113,163] the EMG responses of limb extensors to roll tilt of the animal without modifying the pattern and phase angle of the responses. This effect occurred 5 to 10 minutes after injection, reached its highest value in about 1 hour, and then slowly declined. The central origin of these changes was shown by the fact that injection of neostigmine methylsulfate (0.05 to 0.30 mg/Kg, i.v., which does not cross the blood-brain barrier) was ineffective. Moreover, administration of the muscarinic blocker atropine sulfate (0.1 to 0.5 mg/kg, i.v., which crosses the blood-brain barrier) but not of atropine metylnitrate (0.1-2.0 mg/kg, i.v.) suppressed the action of eserine.[161] Electrolytic lesions of the dorsal PRF, while increasing the postural activity in the ipsilateral limbs, decreased the response gain of the triceps brachii to labyrinth stimulation, and also prevented the gain from increasing after systemic injection of small doses of the eserine sulfate.[52,148] These findings indicated that cholinoceptive neurons activated by the anticholinesterase were located in the dorsal PRF.

After these experiments had been performed, Barnes et al.[22] demonstrated that microinjection of small doses of the nonselective cholinergic agonist carbachol (0.1 to 0.2 ml at 0.01 to 0.2 mg/ml, which acts on both muscarinic and nicotinic receptors) or of an equal dose of bethanecol (which is a selective muscarinic agonist) into the peri-LCa and the dorsal PRF of decerebrate cats caused decreased postural activity in the ipsilateral limbs, but increased the amplitude of modulation and thus the response gain of the ipsilateral triceps brachii to roll tilt of the animal at 0.15 Hz, ±10° (Figure 11.4, A-C). However, no changes in the dynamic characteristics of the responses were observed in the contralateral triceps brachii. The effects described above, which were site-specific and dose-dependent, appeared soon after the injection and persisted up to 3 hours before disappearing. They were suppressed by local injection into the PRF of the muscarinic antagonist atropine sulfate (0.25 ml at 6 mg/ml). The postural and reflex changes described above were attributed to the same mechanisms that occur when the discharge of the dorsal PRF and the related medullary inhibitory RS neurons are released from NAergic inhibition[43,54] (see Section 11.8.1) rather than being directly activated by ACh agonists. Evidence was also presented indicating that dorsal PRF neurons contribute cholinergic excitatory input to the LC-complex[71], the response of which to

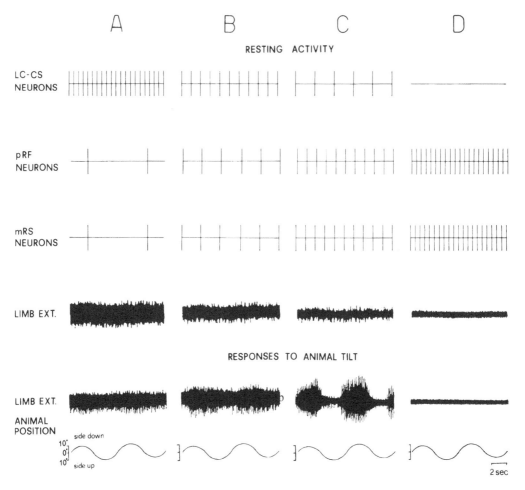

FIGURE 11.4 - Schematic diagram illustrating the modalities of operation of the LC-neurons and the adjacent dorsal pontine tegmental neurons leading to "reciprocal" or "parallel" changes in posture and gain of the VS reflexes. LC-CS neurons: resting discharge of neurons projecting to the spinal cord either through the pRF (LC) or directly through the coeruleo-spinal projection (CS). pRF neurons: discharge of neurons located in the peri-LC a and the adjacent dorsal pontine reticular formation. mRS neurons: discharge of medullary inhibitory reticulospinal neurons. Limb ext.: EMG activity of the triceps brachii in the animal at rest (upper traces) or during roll tilt of the animal at 0.15 Hz, ±10° (lower traces). (A): In decerebrate cats, the high resting discharge of the LC-CS neurons is associated with a reduced activity of the pRF and the related mRS neurons; this leads to a prominent extensor rigidity and a small amplitude EMG modulation of the limb extensor during tilt. (B,C): A reduced discharge of the LC-CS neurons leading to an increased activity of the pRF-mRS neurons reduces the postural activity while the amplitude of the EMG modulation of the limb extensor during tilt increases. (D): A further decrease or suppression of the discharge of the LC-CS neurons leading to a prominent activity of the pRF-mRS neurons produces a postural atonia and a suppression of the VS reflexes. (Adapted from Pompeiano et al. (Ref. 161) and Manzoni et al. (Ref. 112). Reprinted from Pompeiano et al. (Ref. 153) with permission from Elsevier Science, NL.)

unilateral local microinjection of carbachol (0.25 ml at 0.02 to 0.1 mg/ml) produced changes in posture and VSR opposite to those induced by clonidine application.[183]

11.8.3 NORADRENERGIC AND ACH INFLUENCES SUPPRESSING THE VSR

When high doses of NAergic and ACh agents are injected into the dorsal pontine tegmentum leading either to a complete suppression of LC neuronal activity or to a prominent increase in discharge of ACh and/or cholinosensitive PRF neurons, the decrease in postural activity (which is typically observed in the ipsilateral limbs after small doses) is followed by transient episodes of postural atonia. These episodes last several minutes and affect the ipsilateral and, occasionally, also the contralateral limbs.[22,54-56] In these instances, the EMG modulation of the corresponding triceps brachii to animal tilt at 0.15 Hz, ±10° disappears (Figure 11.4D). Moreover, the discharge of the PRF neurons and the related medullary inhibitory RS neurons increases to such a high level that it produces a postsynaptic inhibition of the extensor a-motoneurons.[128] These episodes closely resembled those obtained either in intact animals during REM sleep,[145] or in decerebrate cats after systemic injections of 0.10 to 0.15 mg/Kg, i.v., of eserine sulfate,[146] in which case the postural atonia is also associated with a suppression of the VSR recorded from triceps brachii during tilt.[112]

11.8.4 ROLE OF RENSHAW-CELLS IN THE GAIN REGULATION OF THE VSR

Experiments reported in Section 11.7.2 have shown that the VSR elicited by slow roll tilt of the animal and acting on limb extensors depends on the discharge of lateral VS neurons, which increases during side-down tilt and decreases during side-up tilt (a-responses[115]). The demonstration that most of the CS neurons,[157,158] as well as the medullary RS neurons,[114] display the same response pattern to tilt (characterized by an increased discharge during side-up and a decreased discharge during side-down tilt; b-responses, see Section 11.7.2) while producing opposite influences on Renshaw (R-) cells (Ref. 147, cf. Section 11.4) can be understood only if we consider that, from time to time, the activity of the LC neurons predominates over that of the PRF neurons and the related medullary inhibitory RS neurons, and vice versa (see Section II). In the decerebrate cats, in which the activity of LC-complex neurons, and thus of the CS neurons, predominates over that of the dorsal PRF and the related medullary inhibitory RS neurons,[154,155] the gain of the VSR was very low or absent in limb estensors.[113,162] In these preparations, the R-cells linked with the gastrocnemius-soleus (GS) motoneurons fired at a low rate in the animal at rest, due to the tonic inhibitory influence exerted by the discharge of the CS neurons. Moreover, the same R-cells showed only a-reponses to tilt,[162] due in part to an increased discharge during side-down tilt of VS neurons exerting an excitatory influence on GS motoneurons,[115] in part due to a reduced discharge of CS neurons,[157,158] leading to disinhibition of these R-cells. This would enhance the recruitment of these inhibitory interneurons during the motoneuronal discharge induced by the VS volleys, thus limiting the response gain of limb estensors to labyrinth stimulation.

An opposite situation, however, occurred in the same preparation after local inactivation of the NAergic LC neurons (see Section 11.9.1) or direct activation of the cholinoceptive PRF neurons and the related medullary inhibitory RS neurons (see Section 11.8.2). Similar results were obtained after systemic injection of eserine sulfate,[161] which also activated PRF neurons and related medullary inhibitory RS neurons.[52,148] Under these circumstances, the firing rate of the R-cells linked with the GS motoneurons increased after injection of the anticholinesterase due to ACh activation of the RS neurons, which reduced the postural activity in the animal at rest.[163] The same R-cells, however, showed b-responses to tilt.[163] In this instance, the decreased discharge of the RS neurons during side-down tilt[114] could lead to the derecruitment of the R-cells. The limb extensor motoneurons would then be decoupled from their own R-cells during side-down tilt, a finding which could increase the response gain of the corresponding muscles to the same parameters of labyrinth stimulation.[149]

The results described above only occurred when a partial inactivation of LC-complex neurons (associated with a moderate increase in the discharge rate of the PRF neurons and the related medullary inhibitory RS neurons) led to a slight decrease in postural activity (see Sections 11.8.1 and 11.8.2, and Figures 11.4, A-C). We defined this as the primary state of operation of the system.[153] However, when complete inactivation of the LC-complex neurons (associated with a prominent increase in firing rate of the pontine and medullary reticular neurons) greatly inhibited the extensor motoneurons, we observed a suppression of both the postural activity in the animal at rest and the VSR[22,54-56] (Figure. 11.4D). This was defined as the secondary state of operation of the system. We conclude, therefore, that the dorsal pontine tegmentum, with the related CS and RS pathways, operates as a variable gain regulator that acts during the VSR by modifying the functional coupling of the limb extensor motoneurons with the corresponding R-cells. Since the LC-complex neurons and the related pontine and medullary reticular neurons show reciprocal changes in their firing rate[63,172,185,198] leading to changes in posture[145] during the sleep-waking cycle, they could intervene in order to adapt the amplitude of the VSR to the animal state.

11.9   NORADRENERGIC AND ACH INFLUENCES ON THE VSR MEDIATED BY THE CEREBELLAR CORTEX

11.9.1   INFLUENCES OF THE CEREBELLAR VERMIS ON THE VSR

The LVN, which gives rise to the VS projection exerting an excitatory influence on ipsilateral limb extensor motoneurons, is under the inhibitory control of the cerebellar cortex,[80] particularly of the lateral Zone B of the cerebellar anterior vermis which projects to the ipsilateral LVN.[49] We have shown in decerebrate cats that a proportion of the Purkinje (P) cells located in Zone B of the cerebellar anterior vermis responds to sinusoidal roll tilt of the animal (at 0.026 Hz, ±10 ° or 15°) with a modulation of their simple spike discharges.[57] This was opposite in phase to that of the VS neurons.[115] The conclusion of these experiments (i.e., that the cerebellar vermis exerts a positive influence on the VSR gain) was supported by the finding that, in decerebrate cats, unilateral microinjection of a $GABA_A$ (muscimol) or a $GABA_B$ (baclofen) agonist (0.25 ml at 2 to 16 mg/ml) into Zone B of the culmen leads to inactivation of the P-cells and significantly decreases the response gain of the ipsilateral (but not of the contralateral) triceps brachii to animal tilt without modifying the phase angle of the reflex.[7] These effects were site-specific and dose-dependent, and persisted for at least 1 to 2 hours before disappearing. The opposite changes in gain of the VSR were obtained after local microinjection of a $GABA_A$ (bicuculline) or a $GABA_B$ (saclofen) antagonist (0.25 ml at 4 to 8 mg/ml). These findings provide information that is useful in understanding the changes in posture and VSR induced by intravermal administration of NAergic and ACh agents.[8,151,151a]

11.9.2   EFFECTS OF INTRAVERMAL INJECTION OF NORADRENERGIC AGONISTS AND ANTAGONISTS

Experiments performed by Andre et al.[3,4] in decerebrate cats have shown that unilateral microinjections of selective a_1- or a_2-adrenergic agonists (clonidine and metoxamine, respectively) into Zone B of the cerebellar anterior vermis (culmen), as well as the nonselective b-adrenergic agonist isoproterenol (0.25 ml at 4 to 16 mg/ml), reduced postural activity in the ipsilateral limbs. In addition, these agonists significantly increased the response gain of the ipsilateral and, to some extent, also of the contralateral triceps brachii to roll tilt of the animal (at 0.15 Hz, $\pm 10°$). These effects, which were associated only with slight changes in the phase angle of the responses, were more prominent after activation of the a_2- than the a_1- or b-adrenoceptors. The effects began 5 to 10 minutes after the injection, reached a peak after 20 to 30 minutes, and were followed for about 2 hours before disappearing. The results described above were obtained in the absence of any change in base frequency, which was kept constant due to appropriate changes in limb position.   These results were also site-specific and dose-dependent.   The selectivity of the results was demonstrated by the finding that opposite changes in posture and gain of the VSR were obtained following unilateral microinjection of adrenergic antagonists into the cerebellar vermis.   In particular, adrenergic antagonists either prevented the gain from increasing after a successive injection of the corresponding agonist, or reduced the response gain of the ipsilateral triceps brachii to animal tilt previously enhanced by the agonist injection. Considering the distribution of the different types of adrenoceptors among the three layers of the cerebellar cortex (see Section 11.6.1.1), we attribute the ipsilateral effects to NAergic agents acting directly on the Purkinje cell dendrites and somata or indirectly through postsynaptic (local interneurons) and presynaptic (terminals of parallel fibers) mechanisms.   On the other hand, we attribute the contralateral effects, which were most obvious after activation of a_2-adrenoceptors, to the activity of granule cells which interconnect the hemivermal zones of both sides via their parallel fiber axons.

11.9.3   EFFECTS OF INTRAVERMAL INJECTION OF ACH AGONISTS AND ANTAGONISTS.

Experimental studies performed in decerebrate cats[5,10] have shown that unilateral microinjection of the nonselective ACh agonist carbachol (0.25 ml at 0.5 mg/ml) or the anticholinesterase inhibitor eserine sulfate (0.25 ml at 0.2 to1.0 mg/ml, which increases the naturally present amount of ACh) into Zone B of the cerebellar anterior vermis decreased the postural activity in the ipsilateral limbs, but significantly increased the response gain of the triceps brachii of both sides to labyrinth stimulation (Figure 11.5A). Only slight changes in the phase angle of the responses were observed. Similar results were obtained after unilateral administration of the muscarinic agonist bethanecol or nicotine (0.25 ml at 0.1 mg/ml, Figures 11.5, B and C).

Moreover, the effects of bethanecol injection were more prominent on the ipsilateral side, while those induced by nicotine were more prominent on the contralateral side, probably due to differential distribution of muscarinic and nicotinic receptors on Purkinje cells and granule cells, respectively. These effects followed the same time course as that produced by injection of adrenergic agonists and were site-specific and dose-dependent. Unilateral intravermal microinjection of the muscarinic antagonist scopolamine (0.25 ml at 4 to 8 mg/ml) significantly decreased the response gain of both the ipsilateral and the contralateral triceps brachii to animal tilt. A smaller depression, however, was obtained after local administration of the same doses of nicotinic antagonists of the ganglionic type (hexametonium) or the neuromuscolar type (D-tubocurarine).   The depression of the VSR induced by muscarinic and nicotinic blockers indicates that the ACh system impinging on the cerebellar vermis is tonically active in the decerebrate cats.

11.9.4   FUNCTIONAL SIGNIFICANCE

Although the effects of NA and ACh on the resting discharge of Purkinje cells and the related GABAergic interneurons are rather controversial (see Sections 11.6.1.2 and 11.6.2.2), we postulate that the reduced postural activity in the ipsilateral limbs (which occurs in decerebrate cats after unilateral intravermal injection of NAergic and ACh agonists) depends on the overall increase in the discharge of the Purkinje cells. This in turn leads to inhibition of the corresponding LVN. In addition to the changes in posture, intravermal injection of NAergic and ACh agonists increases the gain of the EMG

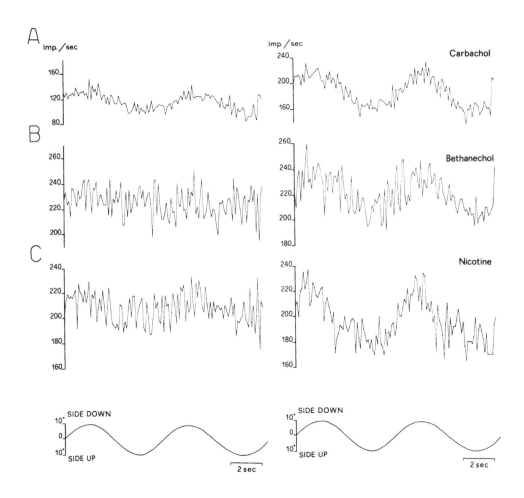

FIGURE 11.5 - Increase of the response gain of the triceps brachii to animal tilt after microinjection of the cholinergic agonists carbachol (A), bethanecol (B), and nicotine (C) into the ipsilateral vermal cortex of the cerebellar anterior lobe (culmen) in precollicular, decerebrate cats. Sequential pulse density histograms show the averaged multiunit responses of the triceps brachii of one side to animal tilt at 0.15 Hz, ±10.° Each record is the average of six sweeps (128 bins, 0.1 s bin width). The lower traces indicate the animal displacement. The traces on the left side were taken before, while those of the right side after individual injections of 0.25 μl of carbachol (0.5 μg/μl of saline), bethanechol (0.1 μg/μl), and nicotine (0.1 μg/μl) solutions. The response gain increased on the average from 1.46 to 2.87 imp./s/deg in A, from 0 to 1.48 imp./s/deg in B, and from 0.68 to 1.83 imp./s/deg in C. (Reprinted from Andre et al. (Ref. 5, 10) with permission of Pisa University, Italy.)

response of the triceps brachii of both sides to labyrinth stimulation. We have observed that microiontophoretic application of either NA (see Section 11.6.1.2), ACh, or a muscarinic agonist (see Section 11.6.2.2) increases the response of Purkinje cells to both the major excitatory (glutamate) and inhibitory (GABA) neurotransmitters. It appears, therefore, that one of the main functions of the NAergic and ACh systems is to facilitate the response of the target neurons to conventional excitatory (MF and CF) and inhibitory (basket and stellate cells) inputs and thereby improve information transfer through the cerebellar cortex.[212,213,216] Since a proportion of Purkinje cells of the cerebellar vermis fire out of phase with respect to the VS neurons during roll tilt of the animal, we postulate that both the NAergic and ACh systems act on these Purkinje cells by enhancing the amplitude of modulation to a given labyrinthine signal. This increases the gain of the ipsilateral VSR. The same system could also increase the gain of the contralateral VSR by acting on granule cells, which may transmit the signal to the Purkinje cells of the opposite side.

Since both the NAergic LC neurons and the presumably ACh and/or cholinosensitive PRF neurons undergo reciprocal changes in firing rate during the sleep-waking cycle (see Section 11.3), we postulate that their relative contribution to the cerebellar control of the VSR may vary in relationship to the different animal states.

11.10 NORADRENERGIC AND ACH INFLUENCES ON VESTIBULO-OCULAR REFLEX (VOR) AND OPTOKINETIC REFLEX (OKR) MEDIATED BY THE CEREBELLAR CORTEX

11.10.1 Influences of the Cerebellar Flocculus on the VOR and OKR

The cerebellar flocculus has been implicated in the gain regulation of the VOR and OKR.[79,80] In fact, ablation of the flocculus in rabbits results in a decrease in gain of the VOR and the OKR,[131] or of the OKR alone.[21] The most convincing results in this respect were obtained by van Neerven et al.[203] who showed the effects of functional and reversible inactivation of floccular Purkinje cells in unanesthetized, young adult, Dutch belted rabbits. Bilateral intrafloccular microinjections of the $GABA_A$ agonist muscimol (1 ml at 16 mg/ml) or the $GABA_B$ agonist baclofen (1 ml at 5 mg/ml) strongly reduced the gain of the VOR in both light and darkness (by 50%). In addition, these agonists reduced the gain of the OKR (by 65%) without modifying the phase angle of the eye movement responses at the two tested frequencies of 0.10 Hz, 2.5° and 0.25 Hz, 5.° Recovery from this depression was slow. If the assumption that GABA agonists inhibit floccular Purkinje cells is correct, we may conclude that these cells contribute positively to the gain of the VOR and OKR in the rabbit. This conclusion is supported by the fact that, although Purkinje cells of the cerebellar flocculus inhibit the VN cells,[80] the simple spike discharge of the floccular Purkinje cells in the rabbit is usually out of phase with respect to that of the vestibulo-ocular neurons.[83; cf. 79,80] Thus, deepening the ongoing modulation of Purkinje cell activity will magnify the ongoing eye movements. In contrast, inactivation of these Purkinje cells results in a decrease in gain of the VOR and the OKR.

11.10.2 Effects of Intrafloccular Injection of Noradrenergic Agonists and Antagonists

Van Neerven et al.[204] studied the effects of bilateral microinjection in the flocculus of b-agonists and antagonists in unanesthetized, young adult, Dutch belted rabbits. Both the VOR (in light and in darkness) and the OKR, were elicited at the frequency of 0.15 Hz, 5° and recorded every 15 minutes for 2.5 hours. After intrafloccular injection of the nonselective b-adrenergic agonist isoproterenol (1 ml at 16 mg/ml), the VOR in light showed no difference in the average gain change; the VOR in the dark showed only a slight increase in gain. The gain of the OKR either did not change[204] or significantly increased at 30 minutes after injection.[192] On the other hand, injection of the nonselective b-adrenergic antagonist sotalol (1 ml at 4 mg/ml) did not modify the gain of the VOR in the light, while the VOR in darkness and the OKR showed a slight decrease in gain during the first hour after the injection. When interpreting these findings, it should be noted that in darkness, the gain of the VOR, which occurs under feed-forward conditions, is never fully compensatory for the head movement. This condition is the most sensitive indicator of subtle changes in the parameters of oculomotor control. During rotation of the animal in the light, additional visual input will provide feedback and will lead to compensation for the residual retinal slip. This feedback operation will tend to conceal the effects of the b-adrenergic substances on the basic VOR gain. By using the same type of preparation, Tan et al.[194] found that bilateral intrafloccular injections of a_1- and a_2-adrenergic agonists (phenilephrine and clonidine) or antagonists (prazosin and idazoxan) at variable concentrations (1 ml at 4 to 16 mg/ml) did not significantly affect the basic gain of the VOR in light and in darkness, or the OKR. These negative results, obtained in intact, unanesthetized rabbits, differ from the positive effects of intravermal injections of these agents on the VSR gain in decerebrate cats (see Section 11.9.2). Differences in animal species and in the type of preparations, which result in different levels of resting discharge of the NAergic LC neurons, as well as regional distribution of different types of adrenoceptors, may account for these findings.

11.10.3 EFFECTS OF INTRAFLOCCULAR INJECTIONS OF ACh AGONISTS AND ANTAGONISTS

Tan and Collewijn[191,193] studied the effects of intrafloccular microinjection of ACh agents on the VOR and the OKR tested in the same type of preparation and with the same parameters of stimulation indicated in the previous section. In particular, a bilateral intrafloccular injection of the nonselective ACh agonist carbachol (1 ml at 1 mg/ml) significantly increased the gain of the OKR and, to a lesser extent, that of the VOR in darkness; these effects persisted for several hours. On the other hand, the gain of the VOR in the light was weakly affected by carbachol due to the high gain level measured for this reflex under baseline conditions (around unity gain) and the nature of the control of compensatory eye movements (see Section 11.10.2). Effects similar to those induced by carbachol but smaller in amplitude were also obtained after injection of eserine sulfate (1 ml at 15 mg/ml), which locally increases the amount of ACh.

The enhancing effect of carbachol on the OKR gain was mediated in part through muscarinic receptors and in part through nicotinic receptors. Thus, the gain of the OKR increased after bilateral intrafloccular injection of a muscarinic (bethanecol) or nicotinic (DMPP) agonist (1 ml at 10 mg/ml), but decreased after administration of a muscarinic (atropine sulfate) or a nicotinic (mecamylamine) antagonist, which acts on hexametonium-sensitive receptors. A comparison between these findings and those obtained on the VSR (see Section 11.9.3) indicates that intravermal injections of both muscarinic and nicotinic agents are quite effective on the VSR gain, while intrafloccular injections of similar agents are particularly effective on the OKR, but not on the VOR in light or in darkness. It is of interest that a floccular injection of a mix of ACh agonist (carbachol) and a NAergic agonist (isoproterenol) exerts a synergistic effect on the gain of the OKR.[192]

11.10.4 EFFECTS OF SYSTEMIC INJECTIONS OF NORADRENERGIC AND ACh AGENTS

Since the discharge of the NAergic LC neurons can be inhibited by both local and systemic application of the a_2-adrenergic agonist clonidine (see Section 11.3), efforts were performed in unanesthetized, young adult, Dutch belted rabbits to investigate the effects of systemic injection of a-agonists on the VOR and OKR gain.[194] Intravenous injection of clonidine, which passes the blood-brain barrier, produced a dose-dependent immediate reduction in the gains of the OKR and VOR in darkness, followed by a recovery of the gains of the two reflexes in less than half an hour. The gain of the VOR in light, however, was not affected by this agent, probably because the combined effects of VOR and OKR were strong enough to overcome the depressive effects on these reflexes acting alone. Systemic i.v. injections of the a_1-agonist phenylephrine did not affect the VOR and OKR, but this agent does not easily pass the blood-brain barrier. The effects of clonidine on the VOR and OKR did not depend on drowsiness or sedation, which may occur after systemic injection of that agent,[107] since no obvious change in alertness and behavior occurred in rabbits after injection of clonidine at the dosage used. The decreased discharge of the LC neurons after systemic injection of clonidine may reduce the NAergic input to several structures involved in the VOR and OKR. These include not only the flocculus (Section IX, B), but also the MVN (Section IV, A) and the inferior olive (Section V, C), where specific labeling of a_2-receptors has been shown.

Systemic injection of ACh agents was also performed in intact preparations. In particular, administration of the muscarinic blocker scopolamine, a potent anti-motion sickness drug that crosses the blood-brain barrier, lowers the gain of the VOR in darkness and the OKR in humans[165] and cats.[177] This finding could be partially attributed to an action on the cerebellar flocculus.[191]

11.11 NORADRENERGIC AND ACH INFLUENCES ON ADAPTIVE CHANGES IN VESTIBULAR REFLEXES

11.11.1 NORADRENERGIC INFLUENCES

11.11.1.1 Adaptation of the VSR

Rotation of the whole animal on one side, leading to stimulation of labyrinth receptors, induces a contraction not only of the ipsilateral limb extensors, but also of the contralateral dorsal neck extensors.[175] This righting of the head would lead to stimulation of neck receptors, which contributes synergistically to the labyrinth-induced contraction of the limb extensors to maintain the support of the body over the limbs during animal tilt.[104] In the decerebrate animal, in which the proprioceptive neck input does not occur due to fixation of the head to the stereotaxic equipment, the VSR acting on limb extensors is barely compensatory, but its amplitude can be enhanced by out-of-phase body-to-head rotation, leading to synergistic stimulation of neck receptors.[cf. 104] This enhanced response has been called neck-vestibulospinal reflex (N-VSR)[6]. In addition to these rapid postural adjustments, the neck input may also induce a slow adaptive change in gain of the VSR.[6] In particular, a 3-hour period of sustained sinusoidal roll tilt of the head (at 0.15 Hz, ±10°, leading to stimulation of labyrinth receptors associated with a 2.5° out-of-phase body-to-head rotation, leading to stimulation of neck receptors; see Figure 11. 6A) always produced in decerebrate cats an adaptive increase in gain of the VSR (Figure 11. 6B, triangles). The amplitude of this reflex, which was recorded every 10 to 15 minutes from the triceps brachii during roll tilt of the whole animal (at 0.15 Hz, ±10°) increased on the average to 240% of the control value at the end of the third hour of stimulation, and lasted during the postadaptation period. Similar results also affected the N-VSR (Figure 11.6C, triangles).

On the other hand, an inconsistent increase in gain of the VSR was obtained during a 3-hour period of sustained roll tilt of the whole animal not associated with neck rotation, while no change in gain occurred when the VSR was tested intermittently in the nonadaptation experiments.

The Zone B of the cerebellar anterior vermis, which receives labyrinthine signals from macular receptors[57] and supplies inhibitory synapses to the VS neurons, [cf. 80] has been implicated in the adaptation of the VSR. We have, in fact, shown that in decerebrate cats, microinjection of the GABA$_A$ (muscimol) or GABA$_B$ agonist (baclofen) at concentrations and/or sites which produced only a slight decrease in gain of intermittently recorded VSR (see Section 11.9.1) either prevented the occurrence of adaptive changes in gain of the VSR during a sustained out-of-phase neck-vestibular stimulation, or decreased the already adapted VSR gain.[108] The role of Zone B of the cerebellar anterior vermis in VSR adaptation can be understood if we assume that under normal conditions, some of the corresponding Purkinje cells projecting to the LVN[57] fire out of phase with respect to the VS neurons,[115] thus contributing positively to the VSR gain. The same Purkinje cells also responded to neck rotation by utilizing not only MF, but also CF.[57] We postulate, therefore, that during a sustained out-of-phase neck-vestibular stimulation, this neck input could not only increase the proportion of Purkinje cells showing an out-of-phase modulatory response to labyrinth stimulation, but also give rise to an efficient process of adaptation, whatever this process might be.

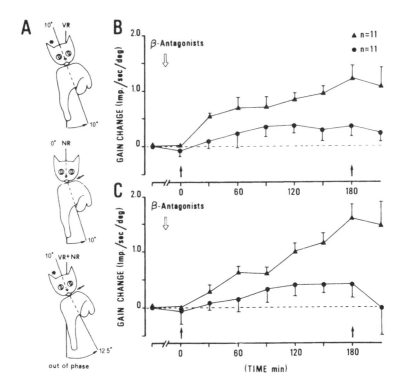

FIGURE 11.6 - A: Schematic representation of different head and/or body rotations. In particular, the upper lower diagrams illustrate, respectively, the postural changes induced by selective vestibular stimulation (10° rotation about the longitudinal axis of the whole animal to the right side, leading to a vestibulo-spinal reflex, VSR), neck stimulation (10° rotation about the longitudinal axis of the body to the left side over a fixed head, leading to a cevico-spinal reflex), or combined neck-vestibular stimulation (10° tilt of the head to the right side, asterisk, associated with 12.5° rotation of the body in the opposite direction). This produced a 2.5° body-to-head rotation to the left (arrow) which was thus out of phase with respect to the head displacement. The resulting neck input increased the postural asymmetry elicited by the pure labyrinth signal, leading to a neck-vestibulo-spinal reflex (N-VSR). (B and C): Averaged curves of the adaptive changes in the gain of the VSR (B) and N-VSR (C) recorded from the triceps brachii of one or both sides. Recordings were made during and after a 3-hour period of sustained out-of-phase neck-vestibular stimulation (between the arrows). Averaged curves obtained in normal control experiments are marked by triangles, those after intravermal microinjection of propranolol or sotalol are marked by dots. Means ± S.E.M. are plotted. (Reprinted from Pompeiano et al. (Ref. 159) with permission of Pisa University, Italy.)

11.11.1.2 Effects of Intravermal Injection of Noradrenergic Agents

Since NA plays a facilitatory role in synaptic plasticity,[cf. 27] attempts were made to investigate whether, in decerebrate cats, the NAergic system acting on the cerebellar vermis could modify not only the steady-state input-output relations of the VSR (see Section 11.9.2), but also the adaptation of the VSR.[159,160, cf. 151a] Under normal conditions, the VSR recorded from the triceps brachii at regular intervals of 10 to 15 minutes during a 3-hour period of sustained neck-vestibular stimulation consistently showed an adaptive increase in gain at the end of the third hour of stimulation with respect to the baseline value (**D**G), which corresponded to 1.20 ± 0.76, SD imp./s/deg (average of 11 curves of adaptation recorded in 8 experiments; $p < 0.001$, paired t-test). In other experiments, however, unilateral intravermal injection of the nonselective **b**-adrenergic antagonists propranolol or sotalol (0.25 ml at 8 mg/ml) prevented the occurrence of an adaptive increase in gain of the

VSR during a sustained out-of-phase neck-vestibular stimulation (Figure 11.6B, dots). In these instances, the gain of the VSR increased insignificantly to 0.35 ± 0.59, SD imp./s/deg after 3 hours of adaptation (average of 11 curves of adaptation recorded in 6 experiments). Moreover, the **D**G value evaluated for the VSR was severely depressed after intravermal injection of the **b**-adrenergic blockers with respect to the experiments performed under normal conditions ($p < 0.0001$, MANOVA). Similar results affected also the N-VSR (Figure 11.6C, dots). The effects described above were site-specific since they were observed only when the **b**-adrenergic agents were injected within Zone B of the cerebellar anterior vermis. In conclusion, local administration of **b**-adrenergic antagonists decreases or suppresses the adaptive capacity of the VSR that always occurs in normal experiments during a sustained out-of-phase neck-vestibular stimulation. No attempts were made to investigate whether the vermal **a**$_1$- and **a**$_2$-adrenoceptors are involved in adaptation of the VSR.

11.11.1.3 Adaptation of the VOR

Changes in head position elicit the VOR, characterized by compensatory eye movements that occur in the opposite direction of the head movement.[80] Although these eye movements are not fully compensatory in darkness, the effectiveness of these movements increases in the light and is particularly potentiated during an out-of-phase moving visual field, leading to a synergistic visual-vestibular interaction. In this case, an adaptive increase in gain occurs in a few hours and persists some time after the end of the sustained stimulation.[79-80] The cerebellar cortex of the flocculus, which receives primary vestibular afferents and supplies inhibitory synapses to the vestibulo-oculomotor neurons, has been implicated in the adaptation of the VOR.[79,80,84; cf. however 123] In particular, it was found that chemical (by kainic acid) or surgical flocculectomy not only decreased the gain of the VOR in several mammals,[cf. 203] but also abolished the adaptive changes of this reflex elicited during combined visual-vestibular stimulation.[82,131,168; cf. 79]

The role of the cerebellar flocculus in VOR adaptation can be understood only if we consider that, under normal conditions, some of the floccular Purkinje cells fired out of phase with respect to the vestibulo-ocular neurons during animal rotation, thus contributing positively to the gain of the VOR.[79,80] In addition to the vestibular input, the flocculus also receives retinal motion information, which is considered as the error signal subserving the recalibration of the VOR.[79,80,123] It has been postulated that the latter input determines the shift of dominance from the in-phase to the out-of-phase modulatory responses of floccular Purkinje cells during the VOR adaptation,[132] and also contributes to the plastic changes underlying this phenomenon.[81] In particular, it appears that the visual input, conveyed to the flocculus through climbing fibers,[79,80] produces a long-term depression (LTD) of the MF responses of the Purkinje cells to labyrinthine stimulation due to desensitization of the excitatory synapses made by parallel fibers on Purkinje cell dendrites,[80,81] thus leading to an adaptive increase in the VOR gain.

11.11.1.4 Effects of Intrafloccular Injection of Noradrenergic Agents

A generalized depletion of the central stores of NA by intracisternal injection of 6-hydroxydopamine (6-OHDA) in cats reduces or abolishes the ability to produce adaptive changes in the VOR gain;[91,118; cf. however 125] a similar result is obtained after injection of 6-OHDA in the coeruleo-cerebellar pathways.[119] Van Neerven et al.[204] have recently studied whether the NAergic system acts through the cerebellar flocculus to modify the adaptation of the VOR. An adaptive increase of the VOR gain was elicited in young adult, Dutch belted rabbits by a sustained 3-hour period of sinusoidal oscillation of the platform at 0.15 Hz, 5,° while the optokinetic drum moved out-of-phase at 0.15 Hz, 2.5.° Baseline measurements of the VOR were taken during oscillation of the platform (at 0.15 Hz, 5°), in light as well as in darkness, while the drum remained stationary (Figure 11.7, dots).

After bilateral intrafloccular injection of a **b**-adrenergic agonist (isoproterenol, 1 ml at 16 mg/ml) or antagonist (sotalol, 1 ml at 4 mg/ml), motion of both the platform and the drum in

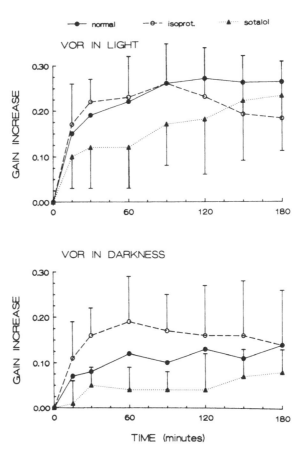

FIGURE 11.7 - Averaged curves of the adaptive changes in gain of the VOR during a 3-hour period of sustained out-of-phase visual-vestibular stimulation obtained in control (normal) experiments or after intra-floccular microinjection of isoproterenol or sotalol. The upper graph represents the effects in light; the lower graph shows the effects in darkness. Mean values are 10 rabbits, with bars representing 1 SD for the isoproterenol and sotalol experiments. The zero values represent the average initial values obtained in four baseline measurements which were taken during the first 15-minute interval. (Reprinted from van Neerven et al. (Ref. 204) with permission of Springer Verlag, FRG.)

counterphase was started, but was interrupted every 15 minutes to record the VOR in light as well as in darkness. In most rabbits, administration of isoproterenol did not result in a clear effect on the VOR measured in the light as compared to the normal adaptation experiments. However, the values of the VOR measured in darkness were always higher after floccular injection of isoproterenol than during normal adaptation, particularly in the first 1.5 hours (Figure 11.7, circles). On the other hand, injection of sotalol produced a clear decrease in the adaptive gain change that was statistically significant, both in light ($p<0.009$, MANOVA) and in darkness ($p<0.003$, MANOVA), and recovered particularly during the third hour of adaptation (Figure 11.7, triangles). Since floccular injection of a **b**-adrenergic agonist or antagonist did not modify the steady-state gain values but increased or decreased, respectively, the adaptation of the VOR gain, we conclude that the NAergic system exerts a prominent influence on the plasticity of the oculomotor system. Since isoproterenol and sotalol bind aselectively to both b_1- and b_2-receptors, experiments are needed to find out which **b**-receptor subtype is responsible for the changes in adaptation of the VOR gain.

In contrast to b-receptors, the floccular a_1- and a_2-receptors were not involved in the adaptation of the VOR, as shown by intrafloccular injection of appropriate agonists or antagonists.[194] The effects of NA on the VOR adaptation should then be considered as mediated solely through b-adrenergic receptors.

11.11.1.5 Modalities of Action of the Noradrenergic System in Adaptation of the Vestibular Reflexes

The noradrenergic LC neurons respond to stimulation not only of labyrinth receptors[110,157,158] (see Section 11.7.2.1), but also of neck[24,110,157] and visual receptors,[14] that is, to the same inputs that act on the cerebellar anterior vermis and the flocculus, respectively, to induce adaptation of the VSR and the VOR. If some of these responsive neurons could be identified as projecting to the cerebellum, one might assume that both the cerebellar vermis and the flocculus receive an increased NAergic input during sustained out-of-phase neck-vestibular or visuo-vestibular stimulation.

The molecular and cellular mechanisms by which the NAergic system acts on the cerebellar cortex during adaptation remain to be investigated. It is known that the LTD, which is at the basis of the VOR (and possibly also of the VSR) adaptation, depends on a complex chain of events involving transmitter receptors and second messengers at the Purkinje cell level.[79-81] There is now evidence that second-messenger systems cause the induction of immediate early genes, such as c-fos and Jun-B, the protein products of which may function as third messengers to regulate the transcription of a number of target genes.[129] Since these latter genes control cellular functions, the c-fos and Jun-B induction may represent a mechanism by which relatively short-term signals at the cell membrane are transduced into long-term biochemical changes that are required for vestibular adaptation (see Chapter 15). Recent experiments have indeed shown that stimuli that cause LTD can trigger c-fos and Jun-B gene transcription in cerebellar Purkinje cells, possibly contributing to cerebellar long-term plasticity.[219] There is now evidence that the release of NA in the rodent cerebral cortex causes a local increase in c-fos expression by acting through b-adrenoceptors.[30] Based on this finding, it is tempting to postulate that NA release in the cerebellar cortex resulting from labyrinth, neck, and/or visual stimulation may act through b-adrenoceptors to increase c-fos and possibly also Jun-B expression during vestibular adaptation. A similar effect could also occur in other structures, such as the VN and/or the IO, which receive a b-noradrenergic innervation and also play a role in the adaptive process.

11.11.2 ACh INFLUENCES

Although muscarinic antagonists (applied locally into the flocculus) may interfere with specific processing of vestibular information during the normal development of the VOR,[191] systemic administration of scopolamine 3 mg/kg/h does not prevent the VOR gain adaptation in the cat[177]. Adaptive changes in VOR gain are thus retained after scopolamine administration.

11.12 CONCLUSION

In the dorsal pontine tegmentum, there are NAergic and NA-sensitive neurons located in the LC-complex, as well as ACh and cholino-sensitive neurons located in the PPT, LDT, and dorsal PRF. These structures are interconnected by reciprocal connections, which explains why, periodically, the increased discharge of the LC neurons is associated with a reduced discharge of the pontine tegmental neurons and vice versa. We have demonstrated that these two populations of neurons exert a regulatory influence on the vestibular reflexes by acting on different target structures such as the spinal cord, the vestibular nuclei (VN), and the cerebellar cortex. At this level, however, the NAergic and the ACh systems could act either directly or indirectly through precerebellar structures, which represent the relay stations of labyrinthine, proprioceptive (neck), and visual signals to the cerebellar cortex. Some of these structures, particularly those located in the caudal part of the VN-complex and the lateral reticular nucleus, contain ChAT-immunoreactive neurons which send ACh afferents to different areas of the cerebellar cortex.

The effects of microiontophoretic application of NA on spinal cord motoneurons and of NA and ACh on single neurons located in the VN and the cerebellar cortex have been investigated. NA and/or ACh were found to exert excitatory or inhibitory changes on the resting activity of the tested units. However, one of the most prominent changes induced either by NA on spinal moto-neurons or by NA and ACh on Purkinje cells of the cerebellar cortex was to increase the respon-siveness of their target neurons to conventional excitatory (glutamate) and inhibitory (GABA) neurotransmitters. This serves to increase the signal-to-noise ratio of the evoked versus spontaneous activity. Such data clearly strengthen the implication that NA and ACh act as neuromodulators of motoneuronal and corticocerebellar activity. The effects of these agents were long-lasting, sug-gesting that they act through second messengers, and that their effects can be mediated by different subtypes of NAergic and ACh receptors.

The NAergic and ACh systems located in the dorsal pontine tegmentum are under the reflex control of labyrinthine signals. The results of single unit recording experiments, performed in decerebrate cats, have in fact shown that a large proportion of NAergic LC neurons, and of ACh and/or cholinosensitive neurons located in the dorsal pontine tegmentum, respond to slow sinusoidal roll tilt of the animal, leading to selective stimulation of labyrinth receptors. This finding suggests that the neuromodulatory influence exerted by the NAergic and the ACh systems at spinal, vestibular, and corticocerebellar levels could dynamically change during labyrinthine stimulation.

The method of local microinjection of NAergic and ACh agonists and antagonists into the LC-complex or in the neighboring pontine tegmental structures has proven to be a powerful tool to clarify the role that these regions exert in the gain regulation of vestibular reflexes. Experiments have also been performed in decerebrate cats or in intact, unanesthetized rabbits to investigate whether microinjection of NAergic and ACh agents into the cerebellar anterior vermis or in the flocculus modify the gain of the VSR and/or the VOR, respectively. Preliminary experiments have shown that inactivation of these corticocerebellar areas after local microinjections of GABA agonists decreased the gain of the VSR or the VOR. These findings were attributed to the fact that, although the Purkinje cells are inhibitory in function, their units discharge out of phase with respect to the vestibulo-spinal and vestibulo-ocular neurons during sinusoidal animal rotation, thus exerting a facilitatory influence on the amplitude of the VSR and the VOR. Intravermal microinjections of a_1-, a_2-, and b-NAergic agents or ACh (muscarinic and nicotinic) agonists in decerebrate cats increase the gain of the VSR, while the opposite result is obtained after injection of the correspond-ing antagonists. These findings led us to postulate that the NAergic and ACh systems acted on Purkinje cells of the cerebellar vermis by enhancing their amplitude of modulation to a given labyrinthine signal, thus increasing the gain of the VSR. Similarly, experiments made in intact, unanesthetized rabbits demonstrated that intrafloccular injection of a b-NAergic agonist facilitate both the OKR and the VOR in dark (but not in light), while a depression of these reflexes occurs after injection of a b-antagonist. Finally, intrafloccular injections of ACh agonists increase the gain of the OKR and VOR in dark, an effect which is particularly mediated through muscarinic receptors. It appears, therefore, that even at the level of the flocculus, the NAergic and the ACh systems act on the corresponding Purkinje cells by enhancing their amplitude of response to a given optokinetic and vestibular stimulation, thus increasing the gain of the corresponding OKR and VOR. The possibility that the NAergic and the ACh systems intervene in the compensation of the postural and motor deficits following UL is also discussed.

The Purkinje cells of the cerebellar anterior vermis and the flocculus exert a prominent role not only in the gain regulation of the VSR and the VOR, but also in the adaptation of these reflexes. These adaptive changes were elicited either in decerebrate cats or in intact, unanesthetized rabbits, respectively. We have pointed out that microinjection of b-adrenergic agonists or antagonists into the cerebellar anterior vermis or in the flocculus increase or decrease, respectively, the adaptive changes in gain of the VSR or the VOR during a sustained neck-vestibular or visuo-vestibular stimulation.   Thus, we postulate that the b-noradrenergic system could activate both second-messenger systems and the expression of immediate early genes in the cerebellar cortex . These latter genes then act as third messengers to regulate the transcription of target genes controlling cellular functions. Induction of these genes may thus represent a mechanism by which relatively short-term signals at the cell membrane are transduced into long-term biochemical changes that are required for vestibular adaptation.

ACKNOWLEDGMENTS

I thank Drs. P. d'Ascanio and P. Arrighi for kindly typing the manuscript. This work was supported by the National Institute of Neurological and Communicative Disorders and Stroke Research Grant NS 07685-26, and by Grants of the Ministero dell'Universita' e della Ricerca Scientifica e Tecnologica, and the Agenzia Spaziale Italiana (ASI 1994 RS-124 and 1995 RS-53), Rome, Italy.

REFERENCES

1.   Abein, W, Bienhold, H, and Flohr, H, Influence of cholinomimetics and cholinolytics on vestibular compensation, *Brain Res.,* 222:458,1981.
2.   Andarde, R, and Aghajanian, GK, Locus coeruleus activity in vitro: intrinsic regulation by a calcium-dependent potassium conductance but not by alpha2-adrenoceptors, *J. Neurosci.,* 4:161,1984.
3.   Andre, P, d'Ascanio, P, Gennari, A, Pirodda, A, and Pompeiano, O, Microinjections of a1- and a2-noradrenergic substances in the cerebellar vermis of decerebrate cats affect the gain of the vestibulospinal reflexes, *Arch. Ital. Biol.,* 129:113,1991.
4.   Andre, P, d'Ascanio, P, Manzoni, D, Pompeiano, O Microinjections of b-noradrenergic substances in the cerebellar vermis of decerebrate cats modify the gain of the vestibulospinal reflexes, *Arch. Ital. Biol.,* 129:161,1991.
5.   Andre, P, d'Ascanio, P, Manzoni, D, and Pompeiano, O, Nicotinic receptors in the cerebellar vermis modulate the gain of the vestibulospinal reflexes in decerebrate cats, *Arch. Ital.,* Biol 131:1,1993.
6.   Andre, P, d'Ascanio, P, Manzoni, D, and Pompeiano, O, Adaptive modifications of the cat's vestibulospinal reflex during sustained vestibular and neck stimulation, *Pflugers. Arch.,* 425:469,1993.
7.   Andre, P, d'Ascanio, P, Manzoni, D, and Pompeiano, O, Depression of the vestibulopsinal reflex by intravermal microinjection of GABAA and GABAB agonists in the decerebrate cat, *J. Vest. Res.,* 4:251,1994.
8.   Andre, P, d'Ascanio, P, and Pompeiano, O, Noradrenergic agents in the cerebellar anterior vermis modify the gain of the vestibulospinal reflexes in the cat. in: *Progress in Brain Research,* Vol. 88 (Barnes CD, Pompeiano O, Eds.), pp. 463-484, Amsterdam: Elsevier.
9.   Andre, P, Fascetti, F, Pompeiano, O, and White, SR, The muscarinic agonist bethanecol enhances GABA-induced inhibition of Purkinje cells in the cerebellar cortex, *Brain Res.,* 637:1,1994.
10.   Andre, P, Pompeiano, O, Manzoni, D, and d'Ascanio, P, Muscarinic receptors in the cerebellar vermis modulate the gain of the vestibulospinal reflexes in decerebrate cats, *Arch. Ital. Biol.,* 130:213,1992.
11.   Andre, P, Pompeiano, O, and White, S, Activation of muscarinic receptors induces a long-lasting enhancement of Purkinje cell responses to glutamate, *Brain Res.,* 617: 28,1993.
12.   Armstrong, DM, Saper, CB, Levey, AI, Wainer, BH, and Terry RD, Distribution of cholinergic neurons in rat brain: demonstrated by immunocytochemical localization of choline acethyltransferase,*J. Comp. Neurol.,* 216:53,1983.
13.   Asin, KE, Satoh, K, and Fibiger, HC, Regional cerebellar choline acetyltransferase activity following peduncular lesions, *Exp. Brain Res.,* 53:370,1984.
14.   Aston-Jones, G, Chiang, C, and Alexinsky, T, Discharge of noradrenergic locus coeruleus neurons in behaving rats and monkeys suggests a role in vigilance. in: *Progress in Brain Research,* Vol.88 (Barnes, CD and Pompeiano, O, Eds.), pp. 501-520. Amsterdam: Elsevier, 1991.
15.   Aston-Jones, G, Ennis, M, Pieribone, VA, Nickel, WT, and Shipley, MT, The brain nucleus locus coeruleus: restricted afferent control of a broad efferent network, *Science,* 234:734,1986.

16. Azizi, SA, Painchand, AJ, and Woodward, DJ, Mapping of choline acetyl transferase (ChAT) cells in the rat brain: possible evidence for cholinergic input to the cerebellum, *Soc. Neurosci. Abstr.*, 16:1057.

17. Barale, F, Ghelarducci, B, and Pompeiano, O, Nistagmo da rotazione prodotto da un anticolinesterasi nel gatto decrebrato, *Arch. Fisiol.*, 69:52,1972.

18. Barmack, NH, Baughman, RW, and Eckenstein, FP, Cholinergic innervation of the cerebellum of rat, rabbit, cat, and monkey as revealed by choline acetyltransferase activity and immunohistochemistry, *J. Comp. Neurol.*, 317:233,1992.

19. Barmack, NH, Baughman, RW, Eckenstein, FP, and Shojaku, H, Secondary vestibular cholinergic projection to the cerebellum of rabbit and rat as revealed by choline acetyltransferase immunohistochemistry, retrograde and orthograde tracers, *J. Comp. Neurol.*, 317:250,1992.

20. Barmack, NH, Fagerson, M, and Errico, P, Cholinergic projection to the dorsal cap of the inferior olive of the rat, rabbit and monkey, *J. Comp. Neurol.*, 328:263,1993.

21. Barmack, NH, and Pettorossi, VE, Effects of unilateral lesions of the flocculus on optokinetic and vestibuloocular reflexes of the rabbit, *J. Neurophysiol.*, 53:481,1985.

22. Barnes, CD, d'Ascanio, P, Pompeiano, O, and Stampacchia, G, Effects of microinjections of cholinergic agonists into the pontine reticular formation on the gain of the vestibulospinal reflexes in decerebrate cats, *Arch. Ital. Biol.*, 125:71,1987.

23. Barnes, CD, Fung, SJ, and Pompeiano, O, Descending catecholaminergic modulation of spinal cord reflexes in cat and rat, *Ann. NY Acad. Sci.*, 563:45,1989.

24. Barnes, CD, Manzoni, D, Pompeiano, O, Stampacchia, G, and d'Ascanio, P, Responses of locus coeruleus and subcoeruleus neurons to sinusoidal neck rotation in decerebrate cat, *Neuroscience*, 31:371,1989.

25. Barnes, CD, and Pompeiano, O, Vestibular nerve activation of a brain stem cholinergic system influencing the spinal cord, *Neuropharmacol.*, 10:425,1971.

26. Barnes, CD, and Pompeiano, O, The interaction of brain stem adrenergic systems and VIIIth nerve stimulation on the spinal cord, *Neuropharmacol.*, 10:437,1971.

27. Barnes, CD, and Pompeiano, O, Neurobiology of the locus coeruleus. *Progress in Brain Research*, Vol.88, pp. XIV-642. Amsterdam: Elsevier,1991.

28. Basile, AS, and Dunwiddie, TV, Norepinephrine elicits both excitatory and inhibitory reponses from Purkinje cells in the in vitro rat cerebellar slice, *Brain Res.*, 296:15,1984.

29. Bienhold, H, and Flohr, H, Role of cholinergic synapses in vestibular compensation, *Brain Res.*, 195:476,1980.

30. Bing, G, Stone, EA, Zhang, Y, and Filer, D, Immunohistochemical studies of noradrenergic-induced expression of c-fos in the rat CNS, *Brain Res.*, 592:57,1992.

31. Bonner, TI, Young, AC, Brann, MR, and Buckley, NJ, Cloning and expression of the human and rat m5 muscarinic acetylcholine receptor genes, *Neuron.*, 1:403,1988.

32. Boyajian, CL, Loughlin, SE, and Leslie, FM, Anatomical evidence for a2-adrenoceptor heterogeneity:differential autoradiographic distributions of [3H]rauwolscine and [3H]idazoxan in rat brain, *J. Pharmacol. Exp. Ther.*, 241:1079,1987.

33. Burke, RE, and Fahn, S, Chlorpromazine methiodide acts at the vestibular nuclear complex to induce barrel rotation in the rat, *Brain Res.*, 288:273,1983.

34. Burke, RE, and Fahn, S, Choline acetyltransferase activity of the principal vestibular nuclei of rat, studied by micropunch technique, *Brain Res.*, 328:196,1985.

35. Burke, RE, and Fahn, S, The effect of selective lesions on vestibular nuclear complex choline acetyltransferase activity in the rat, *Brain Res.*, 360:172,1985.

36. Butcher, LL, and Wolf, NJ, Histochemical distribution of acetylcholinesterase in the central nervous system: clues to the localization of cholinergic neurons. in: *Handbook of Chemical Neuroanatomy*, Vol. 3 (Bjorklund A, Hokfelt T,Kuhar MJ, Eds.), pp. 1-50. Amsterdam: Elsevier,1984.

37. Calza,' L, Giardino, L, Zanni, M, and Galetti, R, Muscarinic and gamma-aminobutyric acid-ergic receptor changes during vestibular compensation. A quantitative autoradiographic study of the vestibular nuclei complex in the rat, *Eur. Arch. Otorhinolaryngol.*, 249:34,1992.

38. Calza,' L, Giardino, L, Zanni, M, Galetti, R, Parchi, P, and Galetti, G, Involvement of cholinergic and GABA-ergic systems in vestibular compensation. In: *Vestibular Compensation, Facts, Theories and Clinical Perspectives* (Lacour, M, Toupet, M, Denise, P, and Christen, Y, Eds.), pp.189-199. Paris: Elsevier,1989.

39. Carpenter, DO, and Hori, N, Neurotransmitter and peptide receptors on the medial vestibular nucleus neurons, *Ann. NY Acad. Sci.,* 656:668,1992.

40. Carpenter, MB, Chang, L, Pereira, AB, and Hersh, LB, Comparisons of the immunocytochemical localization of choline acetyltransferase in the vestibular nuclei of the monkey and rat, *Brain Res.,* 418:403,1987.

41. Cedarbaum, JM, and Aghajanian, GK, Afferent projections to the rat locus coeruleus as determined by a retrograde tracing tecnique, *J. Comp. Neurol.,* 178:1,1978.

42. Chesselet, M-F, Presynaptic regulation of neurotransmitter relase in the brain : facts and hypothesis, *Neuroscience,* 12:347,1984.

43. Cirelli, C, d'Ascanio, P, Horn, E, Pompeiano, O, and Stampacchia, G, Modulation of vestibulospinal reflexes through microinjection of an a1-adrenergic antagonist in the dorsal pontine tegmentum of decerebrate cats, *Arch. Ital. Biol.,* 131:275,1993.

44. Cirelli, C, Liu, R, Fung, S, Barnes, CD, and Pompeiano, O, Cholinergic projection of the dorsal pontine tegmentum to the vermal cortex of the kitten, *Pflugers. Arch.,* 428:R19 n.59,1994.

45. Clarke, PBS, Schwartz, RD, Paul, SM, and Pert, A, Nicotinic binding in the rat brain: autoradiographic comparison of [^{3}H]acetylcholine, [^{3}H]nicotine, [125I]bungarotoxin, *J. Neurosci.,* 5:1307,1985.

46. Clavier, RM, Afferent projections to self-stimulation regions in the dorsal pons, including the locus coerulus in the rat as demostrated by the horseradish peroxidase technique, *Brain Res. Bull.,* 4:497,1979.

47. Commissiong, JW, Spinal monoaminergic systems: an aspect of somatic motor functions, *Fed. Proc.,* 40:2771,1981.

48. Connell, LA, Majiol, A, and Wallis, DI, Involvement of a1-adrenoceptors in the depolarizing but not the hyperpolarizing responses of motorneurones in the neonate rat to noradrenaline, *Neuropharmacology,* 28:1399,1989.

49. Corvaja, N, and Pompeiano, O, Identification of cerebellar corticovestibular neurons retrogradely labeled by horseradish peroxidase, *Neuroscience,* 4:507,1979.

49a. Cramsac, H, Peyrin, L, Farhat, F, Cottet-Emard, JM, Pequignot, JM, and Reber, A, Effect of hemilabyrinthectomy on monoamine metabolism in the medial vestibular nucleus, locus coeruleus and other brainstem nuclei of albino and pigmented rats, *J. Vest. Res.,* 6:243,1996.

50. Crepel, F, Debono, M, and Flores, R, a-Adrenergic inhibition of rat cerebellar Purkinje cells in vitro: a voltage-clamp study, *J. Physiol. (London),* 383:487,1987.

51. d'Ascanio, P, Bettini, E, and Pompeiano, O, Tonic inhibitory influences of locus coeruleus on the response gain on limb extensors to sinusoidal labyrinth and neck stimulations, *Arch. Ital. Biol.,* 123:69,1985.

52. d'Ascanio P, Bettini, E, and Pompeiano, O, Tonic facilitatory influences of dorsal pontine reticular structures on the response gain of limb extensors to sinusoidal labyrinth and neck stimulations, *Arch. Ital. Biol.,* 123:101,1985.

53. d'Ascanio, P, Horn, E, and Pompeiano, O, Stampacchia G Injections of b-adrenergic substances in the locus coeruleus affect the gain of vestibulospinal reflexes in decerebrate cats, *Arch. Ital. Biol.,* 127:187,1989.

54. d'Ascanio, P, Horn, E, Pompeiano, O, and Stampacchia, G, Injections of a b-adrenergic antagonist in pontine reticular structures modify the gain of vestibulospinal reflexes in decerebrate cats, *Arch. Ital. Biol.,* 127:275,1989.

55. d'Ascanio, P, Pompeiano, O, Stampacchia, G, and Tononi, G, Inhibition of vestibulospinal reflexes following cholinergic activation of the dorsal pontine reticular formation, *Arch. Ital. Biol.,* 126:291,1988.

56. d'Ascanio, P, Pompeiano, O, and Tononi, G, Inhibition of vestibulospinal reflex during episodes of postural atonia induced by unilateral lesion of the locus coeruleus in the decerebrate cat, *Arch. Ital. Biol.,* 127:81,1989.

57. Denoth, F, Magherini, PC, Pompeiano, O, and Stanoievic M Responses of Purkinje cells of the cerebellar vermis to neck and macular vestibular inputs, *Pflugers. Arch.,* 381:87,1979.

58. Dutia, MB, Neavy, P, and Mc Queen, DS, Effects of cholinergic agonists on spontaneously active rat medial vestibular neurons in vitro, *J. Physiol. (London),* 425:90P,1990.

59. Egan, TM, and North, RA, Acetylcholine acts on m2 muscarinic receptors to excite rat locus coeruleus neurons, *Br. J. Pharmacol.,* 85:733,1985.

60. Ennis M, and Aston-Jones, G A potent excitatory input to nucleus locus coeruleus from the ventrolateral medulla, *Neurosci. Lett.,* 71:299,1986.

61. Flohr, H, Bienhold, H, Abehn, W, and Macskovicz, I, Concepts of vestibular compensation. in: Lesion-induced neuronal plasticity in sensorimotor systems (Flohr H, Precht WE,eds), pp. 153-172, Berlin: Springer Verlag,1981.

62. Flugge, G, Iurdzinski, A, Brandt, S, and Fuchs, E, Alpha2-adrenergic binding sites in the medulla oblongata of three shrews demonstrated by in vitro autoradiography: species related differences in comparison to the rat, *J. Comp. Neurol.,* 297:253,1990.

63. Foote, SL, Bloom, FE, Aston-Jones, G, Nucleus locus coeruleus: New evidence of anatomical and physiological specificity, *Physiol. Rev.,* 63: 844,1983.

64. Fritschy, J-M, and, Grzanna, R, Immunohistochemical analysis of the neurotoxic effects of DSP-4 identifies two populations of noradrenergic axon terminals, *Neuroscience,* 30:181,1989.

65. Fung, SJ, Pompeiano, O, and Barnes, CD, Suppression of the recurrent inhibitory pathway in lumbar cord segments during locus coeruleus stimulation in cats, *Brain Res.,* 402:351,1987.

66. Fung, SJ, Pompeiano, O, and Barnes, CD, Coeruleospinal influence on recurrent inhibition of spinal motonuclei innervating antagonistic hindleg muscles, *Pflugers Arch.,* 412:346,1988.

67. Fung, SJ, Reddy, VK, Bowker, RM, and Barnes, CD, Differential labeling of the vestibular complex following unilateral injections of horseradish peroxidase into the cat and rat locus coeruleus, Brain Res., 401:347,1987.

68. Gallagher, JP, Phelan, KD, and Shinnick-Gallagher, P, Modulation of excitatory transmission at the rat medial vestibular nucleus synapse, *Ann. NY Acad. Sci.,* 656:630,1992.

69. Giardino, L, Calza,' L, Galetti, R, Parchi, P, and Galetti, G, Mappatura autoradiografica di recettori per neurotrasmettitori e farmaci nel complesso dei nuclei vestibolari, *Acta. Otorhinol. Ital.,* 7:581,1987.

70. Granholm, A-CE, and Palmer, MR, Electrophysiological effects of nerepinephrine on Purkinje neurons in intraocular grafts: a- versus b-specificity, *Brain Res.,* 459:256,1988.

71. Horn, E, d'Ascanio, P, Pompeiano, O, and Stampacchia, G, Pontine reticular origin of cholinergic excitatory afferents to the locus coeruleus controlling the gain of vestibulospinal and cervicospinal reflexes in the decerebrate cats, *Arch. Ital. Biol.,* 125:273,1987.

72. Hosey, MM, Diversity of structure, signaling and regulation within the family of muscarinic cholinergic receptors, *FASEB Journal,* 6:845,1992.

73. Hoshino, K, and Pompeiano, O, Selective discharge of pontine neurons during the postural atonia produced by an anticholinesterase in the decerebrate cat, *Arch. Ital. Biol.,* 114:244,1976.

73a. Hozawa, K, and Kimura, RS, Vestibular sympathetic nervous system in guinea-pig, *Acta. Otolaryngol, Stockholm,* 107:171,1988.

73b. Hozawa, K, and Takasaka, T, Catecholaminergic innervation in the vestibular labyrinth and vestibular nucleus of guinea-pigs, *Acta. Otolaryngol., Stockholm, Suppl.,* 503:111,1993.

74. Hulme, EC, Birdsall, NJM, and Buckley, NJ Muscarininc receptor subtypes, *Ann. Rev. Pharmacol. Toxicol.,* 30: 633,1990.

75. Ikeda, M, Houtani, T, Ueyama, T, and Sugimoto, T, Choline acetyltransferase immunoreactivity in the cat cerebellum, *Neuroscience,* 45:671,1991.

76. Ikeda, M, Houtani, T, Ueyama, T, and Sugimoto, T, Cholinergic and CRF-containing neurons in the vestibular nuclei and related cell groups provide cerebellar mossy fiber projections, *Soc. Neurosci. Abstr.,* 17(2):1574,1991.

77. Illing, RB, A subtype of cerebellar Golgi cells may be cholinergic, *Brain Res.,* 522:267,1990.

78. Ishikawa, I, and Igarashi, M, Effects of atropine and carbachol on vestibular compensation on squirrel monkey, *Am. J. Otolaryngol.,* 6:290,1985.

79. Ito, M, Cerebellar control of the vestibulo-ocular reflex: around the flocculus hypothesis, *Annu. Rev. Neurosci.,* 5:275,1982.

80. Ito, M, The cerebellum and neural control, pp XVII-580. New York: Raven Press,1984.

81. Ito, M, Long-term depression in the cerebellum. Semin Neurosci 2:381,1990.

82. Ito, M, Jastreboff PJ, and Miyashita, Y, Specific effects of unilateral lesions in the flocculus upon eye movements in albino rabbits, *Exp. Brain Res.,* 45:233,1982.

83. Ito, M, Nisimaru, N, and Yamamoto, M, Specific patterns of neuronal connexions involved in the control of the rabbit's vestibulo-ocular reflexes by the cerebellar flocculus, J. Physiol. (London) 265:833,1977.

84. Ito, M, Shiida, T, Yagi, N, and Yamamoto, M, Visual influence on rabbit horizontal vestibulo-ocular reflex presumably effected via the cerebellar flocculus, *Brain Res.,* 65:170,1974.

85. Jankowska, E, Riddel, JS, Skoog, B, and Noga, BR, Gating of transmission to motoneurons by stimuli applied in the locus coeruleus and raphe nuclei of the cat, *J. Physiol.*, (London) 461:705,1993.

86. Jones, BE, Noradrenergic locus coeruleus neurons: their distant connections and their relationship to neighboring (including cholinergic and GABAergic) neurons of the central gray and reticular formation, in: Progress in Brain Research, Vol.88, (Barnes CD, Pompeiano O, eds), pp 15-30, Amsterdam: Elsevier,1991.

87. Jones, BE, and Beaudet, A, Distribution of acetylcholine and catecholamine neurons in the cat brain stem: a choline acetyltransferase and tyrosine hydroxylase immunohistochemical study, *J. Comp. Neurol.*, 261:15,1987.

88. Jones, BE, and Yang, T-Z, The efferent projections from the reticular formation and the locus coeruleus studied by anterograde and retrograde axonal transport in the rat, *J. Comp. Neurol.*, 242:56,1985.

89. Kan, K-SK, Chao, L-P, and Eng, LF, Immunohistochemical localization of choline acetyl-transferase in rabbit spinal cord and cerebellum, *Brain Res.*, 146:221,1978.

90. Kan, K-SK, Chao, L-P, and Forno, IS, Immunohistochemical localization of choline acetyltransferase in the human cerebellum, *Brain Res.*, 193:165,1980.

91. Keller, EL, and Smith, MJ, Suppressed visual adaptation of the vestibulo-ocular reflex in catecholamine depleted cats, *Brain Res.*, 258:323,1983.

92. Kimoto, Y, Tohyama, M, Satch, K, Sakumoto, T, Takahashi, Y, and Shimizu, N, Fine structure of rat cerebellar noradrenaline terminals as visualized by potassium permanganate in situ perfusion fixation method, *Neuroscience*, 6:47,1981.

93. Kimura,. H, McGeer, PL, Peng, JH, and McGeer, EG, The central cholinergic system studied by choline acetyltransferase immunohistochemistry in the cat, *J. Comp. Neurol.*, 200:151,1981.

94. Kirsten, EB, and Schoener, EP, Action of anticholinergic and related agents on single vestibular neurons, Neuropharmacology, 12:1167,1973.

95. Kirsten, EB, and Sharma, JN, Microiontophoresis of acetylcholine, histamine, and their antagonists on neurones in the medial and lateral vestibular nuclei of the cat, *Neuropharmacology*, 15:743,1976.

96. Kirsten, EB, and Sharma, JN, Characteristics and response differences to iontophoretically applied norepinephrine, D-amphetamine and acetylcholine on neurons in the medial and lateral vestibular nuclei of the cat, *Brain Res.*, 112:77,1976.

97. Kobayashi, RM, Parkovits, M, Kopin, IJ, and Jacobowitz, DM, Biochemical mapping of noradrenergic nerves arising from the rat locus coeruleus, *Brain Res.*, 77:269,1974.

98. Ladpli, R, and Brodal, A, Experimental studies of commissural and reticular formation projections from the vestibular nuclei of the cat, *Brain Res.*, 8:65,1968.

99. Lai, YY, and Siegel, JM, Medullary regions mediating atonia, *J. Neurosci.*, 8:4790,1988.

100. Lai, YY, and Siegel, JM, Ponto-medullary glutamate receptors mediating locomotion and muscle tone suppression, *J. Neurosci.*, 11:2931,1991.

101. Lai, Y-Y, Strahlendorf, HK, Fung, SJ, and Barnes, CD, The actions of two monoamines on spinal motoneurons from stimulation of the locus coeruleus in the cat, *Brain Res.*, 484:268,1989.

102. Lapchak, PA, Aranyo, DM, Quirion, R, and Collier, B, Binding sites for [³H] AF-DX116 and effect of AF-DX116 on endogenous acetylcholine release from rat brain slices, *Brain Res.*, 496:285,1989.

103. Licata, F, Li, Volsi, G, Maugeri, G, Ciranna, L, and Santangelo, F, Effects of noradrenaline on the firing rate of vestibular neurons, *Neuroscience*, 53:149,1993.

104. Lindsay, KW, Roberts, TDM, and Rosenberg, JR, Asymmetric tonic labyrinth reflexes and their interaction with neck reflexes in the decerebrate cat, *J. Physiol. (London)*, 261:583,1976.

105. Llinas, R, and Yarom, Y, Oscillatory properties of guinea-pig inferior olivary neurons and their pharmachological modulation: an in vitro study, *J. Physiol. (London)*, 376:163,1986.

106. Lui, PW, Lee, TY, and Chan, SH, Involvement of locus coeruleus and noradrenergic neurotransmission in fentanyl-induced muscular rigidity in the rat, *Neurosci. Lett.*, 96:114,1989.

107. Makela, JP, and Hilakivi, IT, Evidence for the involvement of alpha-2 adrenoceptors in the sedation but not REM sleep inhibition by clonidine in the rat, *Med. Biol.*, 64:355,1986.

108. Manzoni, D, Andre, P, d'Ascanio, P, and Pompeiano, O, Depression of the vestibulospinal reflex adaptation by intravermal microinjection of GABA-A and GABA-B agonists in the cat, *Arch. Ital. Biol.*, 132:243,1994.

109. Manzoni, D, Andre, P, Pompeiano, O, and Sarkisian, VH Postspike facilitation of muscle activity in the triceps brachii by Deiters' neurons responsive to animal tilt, *Pflugers. Arch.*, 421:R32 (n.106),1992.

110. Manzoni, D, Pompeiano, O, Barnes, CD, Stampacchia, G, and d'Ascanio, P, Convergence and interaction of neck and macular vestibular inputs on locus coeruleus and subcoeruleus neurons, *Pflugers. Arch.,* 413:580,1989.

111. Manzoni, D, Pompeiano, O, Srivastava, UC, and Stampacchia, G, Responses of forelimb extensors to sinusoidal stimulation of macular labyrinth and neck receptors, *Arch. Ital. Biol.,* 121:205,1983.

112. Manzoni, D, Pompeiano, O, Srivastava, UC, and Stampacchia, G, Inhibition of vestibular and neck reflexes in forelimb extensor muscles during the episodes of postural atonia induced by an anticholinesterase in decerebrate cat, *Arch. Ital. Biol.,* 121:267,1983.

113. Manzoni, D, Pompeiano, O, Srivastava, UC, and Stampacchia, G, Gain regulation of vestibular reflexes in fore- and hindlimb muscles evoked by roll tilt, *Boll. Soc. Ital. Biol. Sper.,* 60 (Suppl 3):9,1984.

114. Manzoni, D, Pompeiano, O, Stampacchia, G, and Srivastava, UC, Responses of medullary reticulospinal neurons to sinusoidal stimulation of labyrinth receptors in decerebrate cat, *J. Neurophysiol.,* 50:1059,1983.

115. Marchand, AR, Manzoni, D, Pompeiano, O, and Stampacchia, G, Effects of stimulation of vestibular and neck receptors on Deiters' neurons projecting to the lumbosacral cord, *Pflugers. Arch.,* 409:13,1987.

116. Matsuoka, I, Domino, EF, and Morimoto, H, Effects of cholinergic agonists and antagonists on nucleus vestibularis lateralis unit discharge to vestibular nerve stimulation in the cat, *Acta. Otolaryngol.,* 80:422,1975.

117. Matsuoka, I, Ito, J, Sasa, M, and Takaori, S, Possible neurotransmitter involved in excitatory and inhibitory effects from inferior olive to controlateral lateral vestibular nucleus, *Adv. Otorhinolaryngol.,* 30:58,1983.

118. Mc Elligott, JG, and Freedman, W, Vestibulo-ocular reflex adaptation in cats before and after depletion of norepinephrine, *Exp. Brain Res.,* 69:509,1988.

119. Mc Elligott, JG, and Freedman, W, Central and cerebellar norepinephrine depletion and vestibulo-ocular reflex (VOR) adaptation. in: Post-lesion neuronal plasticity (Flohr H,ed), pp. 661-674. Berlin, Heidelberg, New York, Tokyo: Springer,1988.

120. Mergner, T, and Pompeiano, O, Single unit firing patterns in the vestibular nuclei related to saccadic eye movement, in the decerebrate cat, *Arch. Ital. Biol.,* 116:91,1978.

121. Mesulam, M-M, Levey, AI, and Wainer, BH, Atlas of cholinergic neurons in the forebrain and upper brainstem of the macaque based on monoclonal choline acetyltransferase immunoistochemistry and acetylcholinesterase histochemistry, *Neuroscience,* 12:669,1984.

122. Mesulam, M-M, Mufson, EJ, Wainer, BH, and Levey, AI, Central cholinergic pathways in the rat: an overview based on an alternative nomenclature (Ch1-Ch6), *Neuroscience,* 10:1185,1983.

123. Miles, FA, and Lisberger, SG, Plasticity in the vestibulo-ocular reflex: a new hypothesis, *Annu. Rev. Neurosci.,* 4:273,1981.

124. Mitani, A, Ito, K, Hallanger, AE, Wainer, BH, Kataoka, K, and Mc Carley, RW, Cholinergic projections from the laterodorsal and pedunculopontine tegmental nuclei to the pontine gigantocellular tegmental field in the cat, *Brain Res.,* 451:397,1988.

125. Miyashita, Y, and Watanabe, E, Loss of vision-guided adaptation of the vestibulo-ocular reflex after depletion of brain serotonin in the rabbit, *Neurosci. Lett.,* 51:177,1984.

126. Moises, HC, and Woodward, DJ, Potentiation of GABA inhibitory action in cerebellum by locus coeruleus stimulation, *Brain Res.,* 182:327,1980.

127. Moises, HC, Woodward, DJ, Hoffer, BJ, and Freedman, R, Interactions of norepinephrine with Purkinje cell responses to putative amino acid neurotransmitters applied by microiontophoresis, *Exp. Neurol.,* 64:493,1979.

128. Morales, FR, Engelhardt, JK, Soja, PJ, Pereda, AE, and Chase, MH, Motoneurons properties during motor inhibition produced by microinjection of carbachol into the pontine reticular formation of the decerebrate cat, *J. Neurophysiol.,* 57:1118,1987.

129. Morgan, JI, and Curran, T, Stimulus-transcription coupling in the nervous system: involvement of inducible proto-oncogenes fos and jun, *Annu. Rev. Neurosci.,* 14:421,1991.

130. Mufson, EJ, Martin, TL, Mash, DC, Wainer, BH, and Mesulam, M-M, Cholinergic projections from the parabigeminal nucleus (Ch8) to the superior colliculus in the mouse: a combined analysis of horseradish peroxidase transport and choline acetyltransferase immunohistochemistry, *Brain Res.,* 370:144,1986.

131. Nagao, S, Effect of vestibulocerebellar lesions upon dynamic characteristics and adaptation of vestibulo-ocular and optokinetic responses in pigmented rabbits, *Exp. Brain Res.,* 53:36,1983.

132. Nagao, S, Role of cerebellar flocculus in adaptive gain control of the vestibulo-ocular reflex. In: Vestibular and brain stem control of eye, head and body movements (Shimazu H, Shinoda Y, Eds.), pp. 439-449, Tokyo: Japan Sci Soc Press; Basel:S Karger,1992.

133. Neustadt, A, Frostholm, A, and Rotter, A, Topographical distribution of muscarinic cholinergic receptors in the cerebellar cortex of the mouse, rat, guinea pig, and rabbit: a species comparison, *J. Comp. Neurol.,* 272:317,1988.

134. Nicholas, AP, Pieribone, VA, and Hokfelt, T, Cellular localization of messenger RNA for beta-1 and beta-2 adrenergic receptors in rat brain: an in situ hybridization study, *Neuroscience,* 56:1023,1993.

135. Nicholas, AP, Pieribone, VA, and Hokfelt, T, Distributions of mRNAs for alpha-2 adrenergic receptor subtypes in rat brain: an in situ hybridization study, *J. Comp. Neurol.,* 328:575,1993.

135a. Nishiike, S, Takeda, N, Nakamura, S, Arakawa, S, Kubo, T, Responses of locus coeruleus neurons to caloric stimulation in rats, *Acta. Otolaryngol, Stockholm Suppl.,* 520:105,1995

136. Oh, JD, Woolf, NJ, Rogani, A, Edwards, RH, and Butcher, LL, Cholinergic neurons in the rat central nervous system demonstrated by in situ hybridization of choline acetyltransferase mRNA, *Neuroscience,* 47:807,1992.

137. Ojima, H, Kawajiri, S-I, and Yamasaki, T, Cholinergic innervation of the rat cerebellum: qualitative and quantitative analyses of elements immunoreactive to a monoclonal antibody against choline acetyltransferase, *J. Comp. Neurol.,* 290:41,1989.

138. Palacios, JM, Mapping brain receptors by autoradiography, *ISI. Atlas. of Science, Pharmachology,* 2:71,1988.

139. Palacios, JM, and Kuhar, MJ, Beta-adrenergic receptor localization in rat brain by light microscopic autoradiography, *Neurochem. Int.,* 4:473,1983.

140. Parfitt, KD, Freedman, R, and Bickford-Wimer, PC, Electrophysiological effects of locally applied noradrenergic agents at cerebellar Purkinje neurons: receptor specificity, *Brain Res.,* 462:242,1988.

141. Phelan, KD, and Gallagher, JP, Direct muscarinic and nicotinic receptor-mediated excitation of rat medial vestibular nucleus neurones in vitro, *Synapse,* 10:349,1992.

141a. Pieribone, VA, Nicholas, AP, Dagerlind, A, and Hokfelt, T, Distribution of a1-adrenoceptors in rat brain revealed by in situ hybridization experiments utilizing subtype-specific probes, *J. Neurosci.,* 14:4252,1994.

142. Pompeiano, O Vestibular influences during sleep. in: Handbook of sensory physiology, Vol 6 (1), Vestibular system, Pt I (Kornhuber HH, ed), pp. 583-622. Berlin, Heidelberg, New York: Springer.

143. Pompeiano, O, Vestibulo-Spinal Relationships. in: *The Vestibular System* (Naunton, RF, Ed.), pp. 147-180, New York: Academic Press,1975.

144. Pompeiano, O, Macular input to neurons of the spinoreticulocerebellar pathway, *Brain Res.,* 95:351,1975.

145. Pompeiano, O, Mechanisms Responsible for Spinal Inhibition during Desynchronized Sleep: Experimental Study. in: *Narcolepsy, Advances in Sleep Research,* Vol 3 (Guilleminault, C, Dement, WC, and Passouant, P, Eds.), pp. 411-449. New York: Spectrum,1976.

146. Pompeiano, O, Cholinergic activation of reticular and vestibular mechanisms controlling posture and eye movements. in: The reticular formation revisited, IBRO monograph series, Vol 6 (Hobson JA, Brazier MAB, eds), pp. 473-512. New York: Raven Press.

147. Pompeiano, O, Recurrent inhibition. in: Handbook of the spinal cord, Vol 2-3 (Davidoff RA, ed), pp 461-557. New York: Marcel Dekker.

148. Pompeiano, O, Cholinergic mechanisms involved in the gain regulation of postural reflexes. in: Sleep: neurotransmitters and neuromodulators (Waquier A, Gaillard JM, Monti JM, Radulovacki M, eds), pp 165-184. New York: Raven Press.

149. Pompeiano, O, Influence of the noradrenergic coeruleospinal system on recurrent inhibition in the spinal cord and its role during postural reflexes. in: Post-lesion neural plasticity (Flohr H, ed), pp 259-278. Berlin. Heidelberg: Springer,1988.

150. Pompeiano, O, Excitatory and Inhibitory Mechanisms Involved in the Dynamic Control of Posture during the Vestibulospinal Reflexes. in: *From Neuron to Action* (Deecke, L, Eccles ,JC, and Mountcastle, VB,Eds.), pp 107-123. Berlin, Heidelberg: Springer,1990.

151. Pompeiano, O, Noradrenergic and cholinergic modulation of corticocerebellar activity modify the gain of vestibulospinal reflexes, *Ann. NY. Acad. Sci.,* 656:519,1992.

151a. Pompeiano, O, Noradrenergic Control of Cerebello-Vestibular Functions: Modulation, Adaptation, Compensation. In:*Progress in Brain Research,* Vol. 100 (Bloom,F, Ed.),pp.105-114,Amsterdam:Elsevier,1994.

151b. Pompeiano, O, Neural Mechanisms of Postural Control. In: *Vestibular and Neural Front,* (Taguchi, K, Igarashi, M, and Mori, S, Eds.), pp. 423-436, Amsterdam:Elsevier,1994.

152. Pompeiano, O, d'Ascanio, P, Horn E, and Stampacchia, G, Effects of local injection of the a2-adrenergic agonist clonidine into the locus coeruleus complex on the gain of vestibulospinal and cervicospinal reflexes in decerebrate cats, *Arch. Ital. Biol.,* 125:225,1987.

153. Pompeiano, O, Horn E, and d'Ascanio, P, Locus Coeruleus and Dorsal Pontine Reticular Influences on the Gain of Vestibulospinal Reflexes. in: *Progress in Brain Research,* Vol.88 (Barnes CD, Pompeiano O, Eds.), pp. 435-462, Amsterdam: Elsevier,1991.

154. Pompeiano, O, and Hoshino, K, Central control of posture: Reciprocal discharge by two pontine neuronal groups leading to suppression of decerebrate rigidity, *Brain Res.,* 116:131,1976.

155. Pompeiano, O, and Hoshino, K, Tonic inhibition of dorsal pontine neurons during the postural atonia produced by an anticholinesterase in the decerebrate cats, *Arch. Ital. Biol.,* 114:310,1976.

156. Pompeiano, O, Joffe, M, and Andre, P, Modulation of vestibulospinal reflexes through microinjection of GABAergic agents in the dorsal pontine tegmentum of decerebrate cats, *Arch Ital. Biol.,* 133:149,1995.

157. Pompeiano, O, Manzoni, D, and Barnes, CD, Responses of Locus Coeruleus Neurons to Labyrinth and Neck Stimulation. in: *Progress in Brain Research,* Vol. 88 (Barnes, CD and Pompeiano, O, Eds.), pp. 411-434, Amsterdam:Elsevier,1991.

158. Pompeiano, O, Manzoni, D, Barnes, CD, Stampacchia, G, and d'Ascanio, P, Responses of locus coeruleus and subcoeruleus neurons to sinusoidal stimulation of labyrinth receptors. Neuroscience 35:227-248.

159. Pompeiano, O, Manzoni, D, d'Ascanio, P, and Andre, P, Injections of b-noradrenergic substances in the cerebellar anterior vermis of cats affect adaptation of the vestibulospinal reflex gain, *Arch. Ital. Biol.,* 132:117,1994.

160. Pompeiano, O, Manzoni, D, d'Ascanio, P, and Andre, P Noradrenergic agents in the cerebellar vermis affect adaptation of the vestibulospinal reflex gain, *Brain Res. Bull.,* 35:433,1994

161. Pompeiano, O, Manzoni, D, Srivastava, UC, and Stampacchia, G, Cholinergic mechanisms controlling the response gain of forelimb extensor muscles to sinusoidal stimulation of macular labyrinth and neck receptors, *Arch. Ital. Biol.,* 121:285,1983.

162. Pompeiano, O, Wand, P, and Srivastava, UC, Responses of Renshaw cells coupled with hindblimb extensor motoneurons to sinusoidal stimulation of labyrinth receptors in the decerebrate cat, *Pflugers Arch.,* 403:245,1985.

163. Pompeiano, O, Wand, P, and Srivastava, UC, Influence of Renshaw cells on the gain of hindlimb extensor muscles to sinusoidal labyrinth stimulation, *Pflugers Arch.,* 404:107,1985.

164. Probst, A, Cortes, R, Palacios, JM, Distribution of a2-adrenergic receptors in the human brain stem: an autoradiographic study using [^{3}H]p-aminoclonidine, *Eur. J. Pharmacol.,* 106:477,1985.

165. Pyykko, I, Schalen, J, and Matsuoka, I, Transdermally administered scopolamine vs dimenhydrinate II. Effect on different types of nystagmus, *Acta. Otolaryngol.,* 99:597,1985.

166. Reddy, VK, Fung, SJ, Zhuo, H, and Barnes, CD, Spinally projecting noradrenergic neurons of the dorsolateral pontine tegmentum: A combined immunocytochemical and retrograde labeling study, *Brain Res.,* 491:144,1989.

167. Reiner, PB, Clonidine inhibits central noradrenergic neurons in unanesthetized cats, *Eur. J. Pharmacol.,* 111:249,1985.

168. Robinson, DA, Adaptive gain control of vestibulo-ocular reflex by the cerebellum, *J. Neurophysiol.,* 39:954,1976.

169. Role, LW, Diversity in primary structure and function of neuronal nicotinic acetylcholine receptor channels, *Curr. Opin. Neurobiol.,*2:254,1992.

170. Rotter, A, Birdsall, NJM, Field, PM, and Raisman, G, Muscarinic receptor in the central nervous system of the rat. I. Distribution of the binding of H3-propylbenzilylcholine mustard in the midbrain and hindbrain, *Brain Res. Rev.,* 1:167,1979.

171. Sakai, K, Physiological properties and afferent connections of the locus coeruleus and adjacent tegmental neurons involved in the generation of paradoxical sleep in the cat. in: Progress of Brain Research, Vol.88 (Barnes CD, Pompeiano O, eds), pp. 31-45, Amsterdam: Elsevier,1991.

172. Sakai, K, Sastre, J-P, Kanamori, N, and Jouvet, M, State-specific neurons in the ponto-medullary reticular formation with special references to the postural atonia during paradoxical sleep in the cat. In: Brain mechanisms of perceptual awareness and purposefull behaviour, IBRO Monograph Series, Vol 8 (Pompeiano O, Ajmone-Marsan C, eds), pp. 405-429, New York: Raven Press,1981.

173. Sakai, K, Sastre, J-P, Salvert, D, Touret, M, Tohyama, M, and Jouvet, M, Tegmentoreticular projections with special reference to the muscolar atonia during paradoxical sleep in the cat: An HRP study. *Brain Res.,* 176:233,1979.

174. Sakai, K, Touret M, Salvert, D, Leger, L, and Jouvet, M, Afferent projections to the cat locus coeruleus as visualized by horseradish peroxidase technique, *Brain Res.,* 119:21,1977.

175. Schor, RH, and Miller, AD, Vestibular reflexes in neck and forelimb muscles evoked by roll tilt, *J. Neurophysiol.,* 46:167,1981.

176. Schuerger, RJ, and Balaban, CD, Immunoistochemical demonstration of regionally selective projection from locus coeruleus to the vestibular nuclei in rats, *Exp. Brain Res.,* 92:351,1993.

177. Schultheis, LW, and Robinson, DA, The effect of scopolamine on the vestibuoocular reflex, gain adaptation, and the optokinetic response. *Ann. NY Acad. Sci.,* 656:880,1992.

178. Schwartz. RD. Autoradiographic distribution of high affinity muscarinic and nicotinic cholinergic receptor labeled with H3-acetylcholine in rat brain, *Life,* Sci 38: 2111,1986.

179. Shiromani, PJ, Armstrong, DM, Gillin, JC, Cholinergic neurons from the dorsolateral pons project to the medial pons: a WGA-HRP and choline acetyltransferase immunohistochemistry study, *Neurosci. Lett.,* 95:19,1988.

180. Shute, CCD, Lewis, PR, Cholinesterase-containing pathways of the hindbrain: afferent cerebellar and centrifugal cochlear fibres, *Nature.,* 205:242,1965.

181. Smith, PF, and Darlington, CL, Neurochemical mechanisms of recovery from peripheral vestibular lesions (vestibular compensation), *Brain Res. Rev.,* 16:117,1991.

182. Srivastava, UC, Manzoni, D, Pompeiano, O, and Stampacchia, G, State-dependent properties of medullary reticular neurons involved during the labyrinth and neck reflexes, *Neurosci. Lett.,* 10:S461.

183. Stampacchia, G, Barnes, CD, d'Ascanio, P, and Pompeiano, O, Effects of microinjection of cholinergic agonist into the locus coeruleus on the gain of vestibulospinal reflexes in decerebrate cats, Arch. Ital. Biol., 125:107,1987.

184. Starke, K, Presynaptic a-autoreceptors. Rev Physiol Biochem Pharmacol 107:73,1987.

185. Steriade, M, and McCarley, RW, *Brainstem Control of Wakefulness and Sleep*, pp. XV-499. New York: Plenum,1990.

186. Stone, EA, and Ariano, MA, Are glial cells targets of the central noradrenergic system? A review of the evidence, *Brain Res. Rev.,* 14:297,1989.

187. Summers, RJ, Mc and Martin, LR, Adrenoceptors and their second messenger systems, *J. Neurochem.,* 60:10,1993.

188. Swanson, LW, and Hartman, BK, The central adrenergic system. An immunofluorescence study of the localization of cell bodies and their efferent connections in the rat utilizing dopamine-b-hydroxylase, *J. Comp. Neurol.,* 163: 467,1975.

189. Swanson, LW, Simmons, DM, Whiting, PJ, and Lindstrom, J, Immunohistochemical localization of neuronal nicotinic receptors in the rodent central nervous system, J. Neurosci. 7:3334,1987.

190. Tago, H, Mc Geer, PL, Mc Geer, EG, Akiyama, H, and Hersch, LB, Distribution of choline acetyltransferase immunopositive structures in rat brainstem, *Brain Res.,* 495:271,1989.

191. Tan, HS, and Collewijn, H Cholinergic modulation of optokinetic and vestibulo-ocular responses: a study with microinjections in the flocculus of the rabbit, *Exp. Brain Res.,* 85:475,1991.

192. Tan, HS, and Collewijn, H, Cholinergic and noradrenergic stimulation in the rabbit flocculus have synergistic facilitatory effects on optokinetic responses, *Brain Res.,* 586:130,1992.

193. Tan, HS, and Collewijn, H, Muscarinic nature of cholinergic receptors in the cerebellar flocculus involved in the enhancement of the rabbit's optokinetic response, *Brain Res.,* 591:337,1992.

194. Tan, HS, van Neerven, J, Collewijn, H, and Pompeiano, O, Effects of a-noradrenergic substances on the optokinetic and vestibulo-ocular responses in the rabbit: a study with systemic and intrafloccular injections, *Brain Res.,* 562:207,1991.

195. Tanabe, M, Ono, H, and Fukuda, H, Spinal alpha1- and alpha2-adrenoceptors mediated facilitation and inhibition of spinal motor transmission, respectively, *Jap. J. Pharmacol.*, 54:69,1990.

196. Tatehata, T, Shiosaka, S, Wanaka, A, Rao, ZR, and Tohyama, M, Immunocytochemical localization of the choline acetyltransferase containing neuron system in the rat lower brain stem, *J. Hirnforsch.*, 28:707,1987.

197. Tighilet, B, and Lacour, M, Etude immunohistochimique des modifications du marquage cholinergique dans les noyaux vestibulaires apres neurectomie unilaterale chez le chat, *Coll. Soc. Neuro-sciences.*,Lyon B41:95,1995.

198. Tononi, G, and Pompeiano, O, Pharmacology of the cholinergic system. in: Handbook of experimental pharmacology, Vol 116, The pharmacology of sleep (Kales, A, Ed.), pp. 143-210, Berlin:Springer,1995.

199. Torte-Hoba, MP, Leroy, MH, Courjon, JH, Boyer, N, Dominey, P, and Reber, A, Acetylcholinesterase activity changes in medial vestibular complex after hemilabyrinthectomy in the rat. *J Vest .Res.*, 6:243,1996.

200. Ujihara, H, Akaike, A, Sasa, M, and Takaori, S, Electrophysiological evidence for cholinoceptive neurons in the medial vestibular nucleus studies on rat brainstem in vitro, *Neurosci. Lett.*, 93:231,1988.

201. Ujihara, H, Akaike, A, Sasa M, and Takaori, S, Muscarinic regulation of spontaneously active medial vestibular neurons in vitro, *Neurosci. Lett.*, 106:205,1989.

202. Unnerstall, JR, Kopajtic, TA, and Kuhar, MJ, Distribution of a_2 agonist binding sites in the rat and human central nervous system: analysis of some functional, anatomic correlates of the pharmacologic effects of clonidine and related adrenergic agents, *Brain Res. Rev.*, 7:69,1984.

203. Van Neerven, J, Pompeiano, O, and Collewijn, H, Depression of the vestibulo-ocular and optokinetic responses by intra-floccular microinjection of GABA-A and GABA-B agonists in the rabbit, *Arch. Ital. Biol.*, 127:243,1989.

204. Van Neerven, J, Pompeiano, O, Collewijn, H, and Van der Steen, J, Injections of b-noradrenergic substances in the flocculus of the rabbits affect adaptation of the VOR gain, *Exp. Brain Res.*, 79:249,1990.

205. Vilaro, MT, Wiederhold, K-H, Palacios, JM, and Mengod, G, Muscarinic m2 receptor mRNA expression and receptor binding in cholinergic and non-cholinergic cells in the rat brain: a correlative study using in situ hybridization histochemistry and receptor autoradiography, *Neuroscience,* 47:367,1992.

206. Wada, E, McKinnon, D, Heinemann, S, Patrick, J, Swanson, LW, The distribution of mRNA encoded by a new number of the neuronal nicotinic acetylcholine receptor gene family (a5) in the rat central nervous system, *Brain Res.*, 526:45,1990.

207. Wada, E, Wada, K, Boulter, J, Deneris, E, Heinemann, S, Patrick, J, and Swanson, LW, The distribution of alpha 2, alpha 3, alpha 4, and beta 2 neuronal nicotinic receptor subunit mRNAs in the central nervous system. A hybridization histochemical study in the rat, *J. Comp. Neurol.*, 284:314,1989.

208. Waele, C de, Muhlethaler, M, and Vidal, PP, Neurochemistry of the central vestibular pathways, *Brain Res. Rev.*, 20:24,1995.

209. Walmsley, JK, Lewis, MS, Young, WS, and Kuhar, MJ, Autoradiographic localization of muscarinic cholinergic receptors in rat brainstem, *J. Neurosci.*, 1:176,1981.

210. Wanaka, A, Kiyama, H, Murakami, T, Matsumoto, M, Kamada, T, Malbon, CC, and Tohyama, M, Immunocytochemical localization of b-adrenergic receptors in rat brain, *Brain Res.*, 485:125,1989.

211. Waterhouse, BD, Moises, HC, Yeh, HH, and Woodward, DJ, Norepinephrine enhancement of inhibitory synaptic mechanisms in cerebellum and cerebral cortex: Mediation by beta adrenergic receptors, *J. Pharmacol. Exp. Ther.*, 221:495,1982.

212. Waterhouse, DB, Sessler, FM, Cheng, JT, Woodward, JD, Azizi, SA, and Moises, HD, New evidence for a gating action of norepinephrine in central neuronal circuits of mammalian brain, *Brain Res. Bull.*, 21:425,1988.

213. Waterhouse, BD, Sessler, FM, Liu,W, and Lin, C-S, Second Messenger-Mediated Actions of Norepinephrine on Target Neurons in Central Circuits: A New Perspective on Intracellular Mechanisms and Functional Consequences. in: *Progress in Brain Research*, Vol.88 (Barnes, CD and Pompeiano, O, Eds.), pp 351-362. Amsterdam: Elsevier,1991.

213a. Westlund, KN, and Coulter, JD, Descending projections of the locus coeruleus and subcoeruleus/medial parabrachial nuclei in the monkey:axonal transport studies and dopamine-b-hydroxylase immunocytochemistry, *Brain Res. Rev.*, 2:235,1980.

214. White, SR, Fung, SJ, and Barnes, CD, Norepinephrine Effects on Spinal Motoneurons. In: *Progress in Brain Research*, Vol.88 (Barnes CD, and Pompeiano O, Eds.), pp 343-350. Amsterdam: Elsevier,1991.
215. Williams, JT, North, RA, Shefner, SA, Nishi, S, and Egan, TM, Membrane properties of rat locus coeruleus neurons, *Neuroscience*, 13:137,1984.
216. Woodward, JD, Moises, HC, Waterhouse, DB, Hoffer, BJ, and Freedman, R, Modulatory actions of norepinephrine in the central nervous system, *Fed. Proc.*, 38:2109,1979.
217. Woolf, NJ, and Butcher, LL, Cholinergic systems in the rat brain. IV. Descending projections of the pontomesencephalic tegmentum, *Brain Res. Bull.*, 23:519,1989.
218. Yamada, C, Tachibana, M, and Kuruyama, K, Neurochemical changes in the cholinergic system of the rat lateral vestibular nucleus following hemilabyrinthectomy, *Acta. Otorhinolaryngol.*, 245:197,1988.
219. Yamamori, T, Mikawa, S, and Kado, R, Jun-B expression in Purkinje cells by conjunctive stimulation of climbing fibre and AMPA, *NeuroReport*, 6:793,1995.
220. Yamamoto, C, Pharmacologic studies of norepinephrine, acetylcholine and related compunds in Deiters' nucleus and the cerebellum, *J. Pharmacol. Exp. Ther.*, 156:39,1967.
221. Yamamoto, T, Ishikawa, M, and Tanaka, C, Catecholaminergic terminals in the developing and adult rat cerebellum, *Brain Res.*, 132:355,1977.
222. Young, WS, and Kuhar, MJ, Noradrenergic ɑ1 and ɑ2 receptors: Autoradiographic visualization, *Eur. J Pharmacol.*, 59:317,1979.

12

Cholinergic Pathways and Functions Related To the Vestibulo-Cerebellum

Neal H. Barmack

CONTENTS

12.1 TRANSMITTER-SPECIFIC VESTIBULAR PATHWAYS

The vestibular system projects to the uvula-nodulus of the cerebellum via several transmitter-specific pathways, some of which we are beginning to understand in terms of their physiological activity, topography, and transmitter content. This chapter focuses on two of these transmitter-specific pathways: (1) a mossy fiber cholinergic pathway originating from secondary vestibular neurons and projecting to the uvula-nodulus, and (2) a descending cholinergic pathway that originates in the nucleus prepositus hypoglossi and terminates on the contralateral dorsal cap of the inferior olive. The assumption underlying this review is that knowledge of transmitter-specific pathways will lead to a deeper functional understanding of brain function.

More than 70% of *first-order vestibular afferents* project directly to the *ipsilateral* cerebellar nodulus where they terminate in the granule cell layer as mossy fiber terminals[8, 63] (Figure 12.1).

These mossy fiber first-order vestibular afferents are collaterals of first-order vestibular afferents projecting to the ipsilateral vestibular complex. The transmitter for first-order vestibular afferents has not been identified conclusively, but it is probably glutamate.[35, 92] Cerebellar granule cells[28,46,75] and secondary vestibular neurons[32,38,90,100,101] express ionotropic glutamate receptors.

Ascending secondary vestibular afferent pathways originate from the caudal medial and descending vestibular nuclei (MVN, DVN) and project *bilaterally* as mossy fibers onto granule

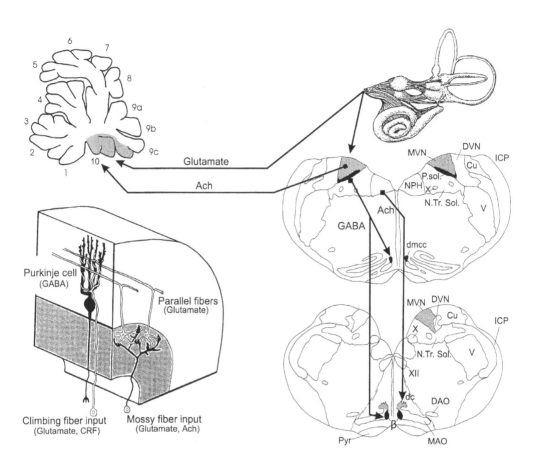

FIGURE 12.1 Illustration of some vestibularly-related transmitter-specific pathways to the uvula-nodulus of the cerebellum and the dorsal cap of the inferior olive. Vestibular primary afferents project to the ipsilateral uvula-nodulus and to the vestibular nuclei. The parasolitary nucleus (black) projects to the ipsilateral b-nucleus and dorsomedial cell column (dmcc). These olivary projections are GABAergic. The b-nucleus and dmcc project to the contralateral uvula-nodulus. The caudal medial vestibular nucleus (oblique lines) projects bilaterally to the nodulus and uvula. This projection is cholinergic. The dorsal cap also receives a cholinergic input from the contralateral nucleus prepositus hypoglossi. This projection is illustrated as cholinergic, but also includes GABAergic fibers. The dorsal cap, in turn, projects to the contralateral uvula-nodulus and flocculus. **Abbreviations:** b, b-nucleus; Cu, cuneate nucleus; dc, dorsal cap of Kooy; dmcc, dorsomedial cell column; DVN, and MVN, descending and medial vestibular nuclei, respectively; NPH, nucleus of the optic tract; Pyr, pyramidal tract; P. Sol., parasolitary nucleus; ICP, inferior cerebellar peduncle; DAO, and MAO, dorsal and medial accessory olive, respectively; N. Tr. Sol., solitary nucleus; Pyr, pyramidal tract; SpV, spinal tract of V; X, dorsal motor nucleus of the vagus; V, spinal trigeminal nucleus; XII, hypoglossal nucleus; CRF, corticotropin-releasing factor; GABA, gamma-aminobutyric acid; Ach, acetylcholine.

cells in the uvula-nodulus. At least one of the transmitters for this mossy fiber pathway is acetylcholine.[6, 7] This transmitter-specific secondary vestibular cholinergic afferent pathway has been demonstrated with double-label techniques in which horseradish peroxidase (HRP) and choline acetyltransferase (ChAT) have been colocalized to cells in the caudal MVN following injections of HRP into the uvula-nodulus.[6]

Descending secondary vestibular afferent pathways include ipsilateral projections from the parasolitary nucleus, MVN, and DVN onto neurons located in the b-nucleus and dorsal medial cell column (dmcc) of the inferior olive. These vestibular projections are GABAergic.[9,42,83,84,97,107]

Descending GABAergic projections from the lateral aspect of the MVN and from the DVN also convey vestibular information to neurons located in the nucleus reticularis gigantocellularis (NRGc).[41] Presently, the distribution of GABAergic receptor subtypes within the inferior olive is not known. Another *descending secondary vestibular afferent pathway* originates from the MVN and DVN and projects ipsilaterally onto the nucleus prepositus hypoglossi (NPH).[47,78] The transmitter for this pathway is unknown.

Tertiary vestibular afferent pathways originate from the b-nucleus and dmcc and synapse as climbing fibers on Purkinje cells in the contralateral uvula-nodulus.[1,76] The transmitter for this pathway is probably glutamate,[77,110,114] but it also contains the neuropeptide corticotropin releasing factor (CRF).[16,36,37,82,113] Another *tertiary vestibular pathway* includes the projection from the NPH onto the contralateral dorsal cap of the inferior olive.

12.2. ACETYLCHOLINESTERASE HISTOCHEMISTRY

Classical histochemical stains for either acetylcholinesterase (AChE) and pseudocholinesterase (pAChE) have shown that the cerebellum is rich in both these enzymes[18,23,27,48,58,60,74,94,99] and that they are distributed as sagittal bands throughout the cerebellar vermis.[48,73,74] The sagittal bands cover all of the cerebellar layers, including the white matter. AChE is also distributed differentially within separate divisions of the cerebellum and is found in greatest density in the uvula and nodulus.[23,27,58,99] These zonal patterns of AChE distribution might have a significance that is independent of classical hydrolysis of acetylcholine. The enzyme might be released in a calcium-dependent manner following activation of cerebellar afferents, and this release, at least *in vitro*, increases the excitability of Purkinje cells.[3,4] The expression of AChE may also influence the survival of Purkinje cells during the development of the cerebellum. Exposure of cerebellar cultures to the AChE inhibitor physostigmine causes increased survival of Purkinje cells.[81] Similarly, Purkinje cell survival is enhanced by stimulation with the cholinergic agonist carbachol if given simultaneously with nerve growth factor.[81]

12.3. CHOLINE ACETYLTRANSFERASE IMMUNOHISTOCHEMISTRY

In contrast to the results obtained with AChE staining, immunohistochemical staining for choline acetyltransferase (ChAT), the synthetic enzyme for acetylcholine, has revealed a different picture. In the cerebellum of rat,[6,60,87] rabbit,[6,59] cat,[6,53] guinea pig,[60] monkey,[6] and human,[31] ChAT activity is localized to large, "grape-like," mossy fiber rosettes distributed throughout the granule cell layer (Figure 12.2).

These ChAT-positive mossy fiber rosettes are concentrated in three separate regions of the cerebellum: (1) uvula-nodulus (lobules 9 and 10), (2) flocculus-ventral paraflocculus, and (3) anterior lobe vermis (lobules 1 and 2). The regional differences in ChAT-positive afferent terminations in the cerebellar cortex, measured immunohistochemically, agree with independent regional measurements of ChAT activity in rat, rabbit, and cat.[6] In ChAT-positive regions there is a decrease in immunostaining following surgical deafferentation, indicating that the major cholinergic inputs to this region are afferent fibers of extrinsic origin.[60]

12.4. CHOLINERGIC PROJECTIONS FROM THE VESTIBULAR NUCLEI TO THE CEREBELLUM

The cholinergic afferent projection sites in the cerebellum correspond to regions of the cerebellar cortex that receive vestibular primary afferent projections, suggesting that the ChAT-positive terminals could be primary afferents. However, in experiments directed at this possibility, Scarpa's ganglion was found to be devoid of ChAT-positive cell bodies. Furthermore, ChAT-positive mossy

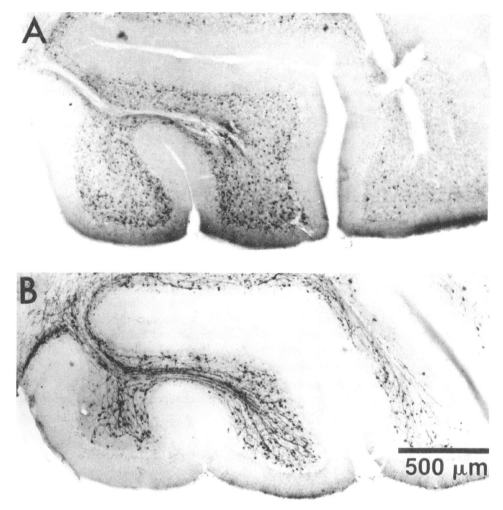

FIGURE 12.2 Cholinergic innervation of the uvula-nodulus of the rat.
(A) Cholinergic mossy fibers terminate in the granule cell layer of lobules 9d and 10. **(B)** The same pattern of mossy fiber projections can be seen following an iontophoretic injection of the orthograde tracer *phaseolus leucoagglutinin* into the ipsilateral medial vestibular nucleus.

fiber terminals in the uvula-nodulus remain intact following a unilateral labyrinthectomy that destroyed Scarpa's ganglion (Barmack, unpublished observations).

The uvula-nodulus also receives secondary afferent vestibular projections. A subset of these secondary afferents are cholinergic. This cholinergic projection was demonstrated by double-label experiments in which cells in the vestibular complex were examined for both ChAT and HRP following injections of HRP into the nodulus, flocculus, or ventral paraflocculus.[7] In the rat and rabbit, the caudal MVN and, to a lesser extent, the NPH contain ChAT-positive neurons. Neurons of the caudal MVN are double-labeled following HRP injections into the uvula-nodulus. Fewer ChAT-positive neurons in the MVN and some ChAT-positive neurons in the NPH are double-labeled following HRP injections into the flocculus. Few ChAT-positive neurons are double-labeled in the MVN or NPH following HRP injections into the ventral paraflocculus. Injections of *phaseolus leucoagglutinin* (PHA-L) into the caudal MVN of both the rat and rabbit demonstrate projection patterns to the uvula-nodulus and flocculus that are similar to those observed using ChAT immunohistochemistry (Figure 12.2B).

The cholinergic mossy fiber pathways to the cerebellum in general, and to the uvula-nodulus in particular, are likely to mediate secondary vestibular information related to postural adjustments. This secondary vestibular information combines with primary vestibular afferent information also conveyed by mossy fibers at the level of cerebellar granule cells.[8] Vestibular climbing fibers originating from the contralateral b-nucleus also synapse on Purkinje cells in the uvula-nodulus and convey information concerning activation of the vertical semicircular canals and utricular otoliths.[1,15,102]

12.5 INTRINSIC CEREBELLAR CHOLINERGIC NEURONS

Although the vestibular complex is the major source of the cholinergic input to the uvula-nodulus, flocculus, and ventral paraflocculus, there is some disagreement concerning the possibility of intrinsic cholinergic cerebellar cell types. In particular, immunohistochemical staining for ChAT in both the feline cerebellum[54] and human cerebellum[31] provide evidence for the labeling of a subtype of cholinergic Golgi cell. This particular subtype comprises less than 5% of the total Golgi cell population and does not seem to be concentrated in the regions of the cerebellum in which cholinergic mossy fiber terminals are the most dense. It is possible that these cholinergic Golgi cells were overlooked by previous immunohistochemical studies.[6] Alternatively, the immunohistochemical demonstration of ChAT-positive Golgi cells might be an artifact related to particular immunohistochemical protocols.

12.6 CHOLINERGIC PROJECTION FROM THE NUCLEUS PREPOSITUS HYPOGLOSSI TO THE INFERIOR OLIVE

Cells from the NPH of the rat and monkey not only project to the uvula-nodulus and the flocculus-ventral paraflocculus, but NPH neurons comprise one of two major projections to the dorsal cap of the inferior olive (Figures 12.1 and 12.3).

The first major projection to the dorsal cap is excitatory and originates from directionally selective (posteriorÆanterior) ganglion cells in the contralateral eye. This retinal ganglion cell projection is relayed through the contralateral nucleus of the optic tract (NOT) before descending to the ipsilateral dorsal cap.[1,61,71,72,104] Although the information conveyed by the optokinetic pathway to the dorsal cap is well understood, the transmitter for this pathway is unknown, but presumed to be glutamate.

The second major projection to the dorsal cap originates from the contralateral NPH. In the rat, both GABA and acetylcholine have been identified as transmitters for this pathway.[9] The origin of the cholinergic projection from the NPH in the rat is supported by both the orthograde transport of PHA-L, and by the reduction of ChAT staining in the contralateral dorsal cap following subtotal unilateral lesions placed in the NPH and MVN. In the rabbit, although both orthograde and retrograde transport experiments reveal a projection from the NPH to the contralateral dorsal cap, both the paucity of ChAT labeling of terminals within the dorsal cap and the lack of HRP retrograde labeling of ChAT-positive neurons within the NPH suggest that this projection is not primarily cholinergic; rather, the transmitter for this projection appears to be GABA.[33,84] These observations raise the interesting possibility that the same anatomical projection in the rat and rabbit is mediated by different neurotransmitters: acetylcholine in the rat and GABA in the rabbit. Both transmitters are likely coexpressed in the cells of the NPH of both species, but in different quantities. Such coexpression of acetylcholine and GABA has been described in amacrine cells of the retina.[86]

Although, the transmitter content of the NPH projection to the dorsal cap is known, the function subserved by this pathway remains obscure. Electrophysiological recordings from single neurons in the NPH suggest that these cells encode vestibular and eye movement-related information.[17,70,79] Consequently, the two pathways to the dorsal cap from the NOT and NPH could provide an

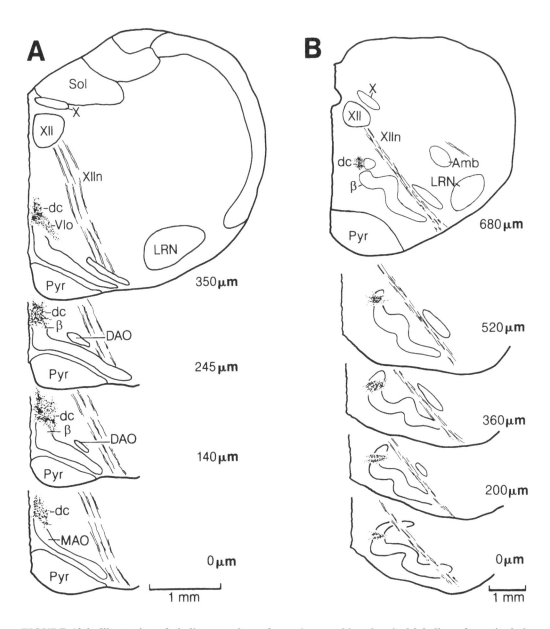

FIGURE 12.3 Illustration of choline acetyltransferase immunohistochemical labeling of terminals in the dorsal caps of the rat and monkey. (A) Cholinergic innervation of the rat dorsal cap. Depletion experiments show that this pattern of labeling is lost following destruction of the contralateral nucleus prepositus hypoglossi. (B) ChAT-labeled sparse innervation of the dorsal cap of the rhesus monkey. **Abbreviations**: Amb, nucleus ambiguus; b, beta nucleus; Vlo, ventrolateral outgrowth; DAO, dorsal accessory olive; dc, dorsal cap; LRN, lateral reticular nucleus; MAO, medial accessory olive; Pyr, pyramidal tract; Sol, nucleus and tractus solitarius; X, dorsal motor nucleus of the vagus; XII and XIIn, hypoglossal nucleus and nerve, respectively.

opportunity for visual, vestibular, neck proprioceptive, and eye movement-related information to interact at the level of the inferior olive. The climbing fibers from the dorsal cap project to topographically discrete sagittal strips in the nodulus and flocculus.[1,12,13,49,67,71,103]

12.7 MUSCARINIC RECEPTORS IN THE CEREBELLUM AND INFERIOR OLIVE

Although there is general agreement about regional distribution of a cholinergic projection to the cerebellum and the termination of most of this cholinergic input within the granule cell layer, there is wide-ranging disagreement about the type of cholinergic receptors that exist in the cerebellum and their cellular localization. Histochemical experiments that have used classic ligand binding techniques characterize muscarinic receptors within the cerebellar cortex. These experiments have shown that [^{3}H]quinuclidinyl benzilate (QNB) and [^{3}H]propylbenzilylcholine mustard (PrBCM) bind to different regions of the cerebellum, but particularly lobules 9 and 10.[85,96] The distribution of the QNB and PrBCM binding sites within the cerebellum is species dependent. In the mouse, rat, guinea pig, and rabbit, the granule cell layer is most heavily labeled by QNB in most of the cerebellum. However, in lobules 9 and 10, where the innervation by ChAT-labeled mossy fiber terminals is most dense, QNB and PrBCM preferentially label the molecular layer in the rat and guinea pig. These same ligands label the Purkinje cell layer in lobules 9 and 10 in the rabbit cerebellum.[57,85] In the rat, sodium-dependent [^{3}H]hemicholinium-3 binding sites appear to be restricted to the granular layer, the site of termination of cholinergic mossy fibers.[57] This pattern of labeling is not consistent with the pattern of labeling obtained with a monoclonal antibody specific for the m_2 muscarinic receptor. This particular antibody labels the molecular layer.[57]

Pharmacological experiments suggest that the cerebellar homogenates contain high-affinity binding sites for both pirenzepine (m_1), (11[[2](diethylamino)methyl][1-piperidinyl]acetyl-5,11-dihydro-6H-pyrido[2,3-b][1,4]-benzodiazepine-6-one) (AF-DX 116) (m_2),[40] and 4-diphenylacetoxy-N-methylpiperidine methiodide (4-DAMP) (m_3).[109] Displacement binding studies of cultured granule cells demonstrate that [^{3}H]N-methyl scopolamine is displaced with high affinity by both m_2 (methoctramine) and m_3 (4-DAMP) antagonists, but not with the m_1 antagonist pirenzepine.[2] Immunoprecipitation studies with cerebellar homogenates, in which a fusion protein-induced antibody to the m_2 receptor was used to immunoprecipitate m_2 muscarinic receptors, showed that m_2 receptors comprise 75% of the cerebellar muscarinic receptors.[68]

Cultured cerebellar granule cells have mRNA transcripts for both m_2 and m_3 receptors, but not m_1 receptors. These mRNAs and receptor proteins are down-regulated by bathing granule cells with carbachol, a cholinergic muscarinic agonist.[43,44,52]

We have used five oligonucleotide probes, m_{1-5}, to determine the levels of possible muscarinic receptor transcripts in noncultured cerebellar tissue taken from adult rats. The probes for the muscarinic receptors coded unique leader sequences. Of the five oligonucleotide probes, only the probe for the m_2 receptor hybridized to cerebellar poly A+ mRNA prepared from rat cerebellum. Each of the other four oligonucleotide probes was negative at the concentrations of poly A+ mRNA that were used (Figure 12.4).

A hybridization histochemical experiment, using an oligonucleotide probe for the m_2 receptor, showed weak labeling of the cerebellar granule cell and Purkinje cell layers.[106] Regionally, the labeling was more intense over lobules 9 and 10.[106]

Using an antibody developed against muscarinic m_2 receptors, we showed that the granule cell layer of the rat cerebellum is immunoreactive. This same antibody labels inferior olivary neurons (not illustrated). At the light microscopic level, it is not possible to determine whether the epitope recognized by the m_2 antibody is expressed on granule cell dendrites or on mossy fiber terminals (Figure 12.5).

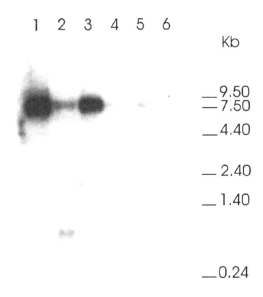

FIGURE 12.4 Northern hybridization for mRNA isolated from rat heart and cerebellum with an oligo-nucleotide probe for the m2 muscarinic receptor mRNA. Rat heart poly A+ mRNA was loaded into Lanes 1, 3, 5. Rat cerebellar poly A+ mRNA was loaded into Lanes 2, 4, 6. The concentration of poly A+ mRNA in each lane was: Lanes 1, 2 -6 bM; Lanes 2, 3 - 4 bM; and Lanes 5, 6 - 2 bM. Bands of 7.6 Kb were found in the mRNA isolated from heart at all concentrations (Lanes 1, 3, 5). Bands of 7.6 Kb were found in the mRNA isolated from the cerebellum at concentrations of 6 mM (Lane 2) and 4 mM (Lane 4). The bands in the lower parts of Lanes 2 and 4 indicate detection with a second oligonucleotide probe for cyclophilin mRNA as a positive control. Cyclophilin is expressed in neurons, but not in heart muscle.

However, an immuno-ultrastructural study of the rat uvula-nodulus provides clear evidence that cholinergic mossy fiber rosettes make presynaptic contact with both granule cells and unipolar brush cells.[56]

In other neural systems, m_2 receptors are coupled via a pertussis toxin-sensitive G protein to a potassium channel.[19,24,55,64,88,93] Activation of m_2 receptors causes an increase in conductance for potassium and hyperpolarizes neurons on which the m_2 receptors are located. Therefore, the release of acetylcholine by mossy fiber terminals could hyperpolarize cerebellar granule cells, thereby antagonizing the excitatory action of glutamatergic mossy fibers.

12.8 NICOTINIC CEREBELLAR RECEPTORS

Nicotinic as well as muscarinic receptors are expressed in the cerebellum. In immature rats, the binding of [³H]-nicotine and [¹²⁵I]-a-bungarotoxin are elevated relative to the binding of these ligands in adults by a factor of 6, suggesting a developmental modulation of cerebellar function linked to a nicotinic receptor.[69] An a7 subunit of the neuronal nicotinic receptor has been localized to cerebellar Purkinje cells in the rat using immunohistochemistry.[34] The expression of the a7

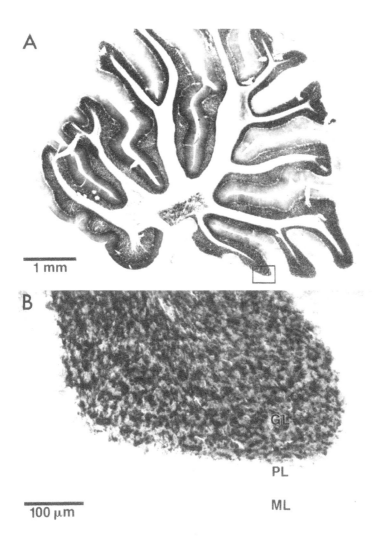

FIGURE 12.5 Immunohistochemical labeling of granule cell layer of the cerebellum of the rat with an antibody to the m$_2$ muscarinic receptor. (**A**) Low-power photomicrograph of sagittal section of the rat cerebellum, labeled with an antibody to the m$_2$ muscarinic receptor. The box indicates the region that is displayed at higher power in (**B**) Note the presence of immunolabeling in the granule cell layer (GL) and the absence of labeling in Purkinje cell (PL) and molecular layers (ML).

subunit is also developmentally regulated with a time course that appears to parallel the final differentiation of Purkinje cells. Nicotinic agonists, whether applied by pressure injections or by iontophoresis, appear to decrease the excitability of cerebellar Purkinje neurons.[29,30,65] These actions can be reversed by either the ganglionic blockers, K-bungarotoxin[29] or mecamylamine.[30]   In cerebellar slices, N-methylcarbamylcholine (MCC) binds to nicotinic receptor sites and evokes the release of acetylcholine.[66] This release is antagonized by d-tubocurarine and a-bungarotoxin, butnot by muscarinic receptor antagonists (atropine, AF-DX 116).[66] It appears that the MCC acts on nicotinic binding sites that are located on cholinergic afferent terminals, since there is no evidence for nicotinic receptors on either the somata or dendrites of Purkinje cells.

Intracellular recordings in cerebellar tissue slices obtained from the uvula-nodulus of the rat showed that both electrically evoked climbing fiber responses and electrically evoked mossy fiber-granule cell-evoked EPSPs in Purkinje neurons can be recorded even if the tissue slice is bathed with relatively high concentrations of the muscarinic receptor antagonist atropine.[26] These data prompted the conclusion that "massive" cholinergic projections to the uvula-nodulus were unlikely. Based on present evidence, one might suppose that more subtle tests of muscarinic receptor function are needed to clarify the role of these receptors in cerebellar function. For example, muscarinic receptors on granule cells or mossy fiber terminals could alter the thresholds for granule cells to other noncholinergic mossy fiber inputs. This could be achieved either by direct inhibition of granule cells or by an autocrine regulation of transmitter release from mossy fiber terminals. This possible autocrine function might be subserved either by muscarinic or nicotinic receptors.

12.9 FUNCTIONAL ROLE OF CHOLINERGIC MOSSY FIBERS IN THE UVULA-NODULUS

The cholinergic pathway to the uvula-nodulus might have particular importance for vestibular compensation since the nodulus is implicated in habituation to vestibular stimulation.[14,25] Removal of the uvula-nodulus in dogs decreases the likelihood of vestibularly-induced vomiting, suggesting the possibility that perturbations of the adaptive functions of the uvula-nodulus might play a role in the genesis of motion sickness.[80] Some effective anti-motion sickness agents include H_1 receptor blockers that have antimuscarinic actions.[39] The belladonna alkaloids, such as scopolamine, are also antimuscarinics.[62,108] In human subjects, administration of a centrally acting muscarinic receptor blocker, atropine, suppresses gastric dysrhythmias induced by brief exposure to circular vection, whereas the administration of a peripherally acting antimuscarinic, methscopolamine, does not.[50] Where do antimuscarinics act? Presently, there is no direct evidence that the ameliorative effects of scopolamine or atropine are achieved by the blockage of either muscarinic receptors in the cerebellum or in the inferior olive. Nonetheless, it is tempting to speculate that anti-motion sickness antimuscarinics act on the cholinergic pathway to cerebellum.

The uvula-nodulus receives sensory-specific climbing fiber signals that include visual, vestibular, and probably neck proprioceptive information. These signals are distributed in parasagittal zones in an orderly topographic map that comprises a sensory-motor space. Each parasagittal climbing fiber zone is activated by a subset of olivary neurons originating from the contralateral b-nucleus, dmcc, and dorsal cap.[10,11,15,47,98,111,112] The topography of these olivary projections has been previously characterized at the level of the inferior olive[5,10] and at the uvula-nodulus.[15] For example, in the left uvula-nodulus, a parasagittal zone related to stimulation in the plane of the ipsilateral posterior semicircular canal and contralateral anterior semicircular canal ($L_{PC} . R_{AC}$) is located near the midline. Further laterally, a parasagittal zone related to stimulation of the ipsilateral anterior semicircular canal and contralateral posterior semicircular canal ($L_{AC} . R_{PC}$) exists.[15,45] (Figure 12.6).

Stimulation (or adaptation) of any one of these sagittal zones could cause a selective bias on neurons that receive synaptic innervation from the Purkinje cells. If these target neurons were elements of an appropriate postural response, then the execution of these postural responses could be influenced by activity in the sagittal zones. For example, stimulation of the lateral parasagittal zone would be expected to differentially influence postural responses that were coupled to activation of the ipsilateral anterior and contralateral posterior semicircular canals, as well as otolith-related with polarization vectors aligned with these canals in the same sagittal strip ($L_{AC} . R_{PC}$).[45] Conversely, activation of the medial sagittal zone would influence postural responses aligned with stimulation of the ipsilateral posterior and contralateral anterior semicircular canals ($L_{PC} . R_{AC}$).

Autonomic adjustments such as respiration and blood pressure are evoked by electrical stimulation of the uvula-nodulus,[20-22,51,89,91,95,105] These autonomic functions might also be influenced by local activation of sagittal zones in the uvula-nodulus.

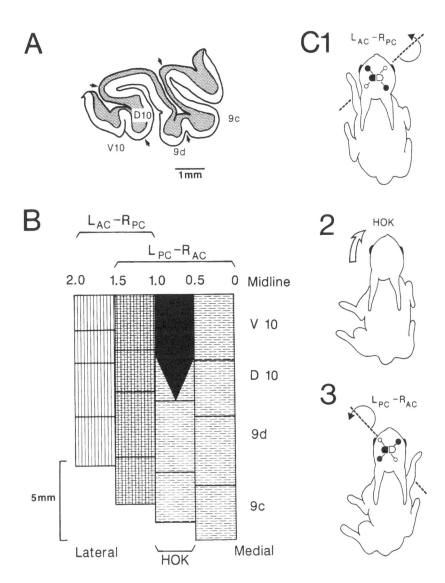

FIGURE 12.6 Topographic map of vestibularly and visually responsive regions in the uvula-nodulus.
(A) A sagittal view of the uvula-nodulus is illustrated. The arrows demarcate arbitrary boundaries for the ventral nodulus (V10), dorsal nodulus (D10), ventral uvula (9d) and dorsal uvula (9c). **(B)** The surface of the most medial 2.0 mm of the left uvula-nodulus is represented as an unfolded two-dimensional sheet. Note that the scales for the horizontal and vertical representations of this sheet are different. The most medial 1.5 mm of this sheet (horizontal dashed lines) contains the region in which CFRs evoked by vestibular stimulation in the plane of the left posterior - right anterior semicircular canals (L_{PC} - R_{AC}) are found. The black region (HOK) indicates the area in which CFRs were evoked by posterior-anterior optokinetic stimulation of the left eye. The most lateral region of this sheet (vertical lines) includes the area in which CFRs evoked by vestibular stimulation in the plane of the left anterior-right posterior semicircular canals ($L_{AC}.R_{PC}$) are found. This region starts 1 mm from the midline and extends 2 mm laterally. The width of the rabbit uvula-nodulus is approximately 3 mm. Only the Purkinje cells in the center receive innervation from vestibularly-modulated climbing fibers. **(C_{1-3})** Illustrations of the possible postural responses evoked by: (1) vestibular stimulation in the plane of the $L_{AC}.R_{PC}$, (2) optokinetic stimulation of the left eye in the posterior®anterior direction, and (3) vestibular stimulation in the plane of the $L_{PC}.R_{AC}$.

What would be the consequence of "inappropriate" sensory stimulation of these zones? Such "inappropriate" stimulation ("sensory conflict") occurs during externally imposed changes in linear acceleration or optokinetic stimulation and could lead to mixed activation patterns of sagittal zones within the topographic cerebellar map. The consequences of these zonal output patterns could be spatial disorientation and motion sickness. How could muscarinic receptor blockers reduce the effects of motion sickness-inducing visual and vestibular stimulation? Blocking the m_2 muscarinic receptors on granule cells would decrease the modulation of these cells by the cholinergic secondary afferents. This decreased modulation might also reduce the level of simple spike modulation of uvula-nodulus Purkinje cells evoked by stimulus-driven climbing fibers, thereby reducing inappropriate postural and autonomic responses.

These speculations are intended to focus further interest on a problem that, with the advent of improved molecular and biochemical methods, might eventually be solved. Additional knowledge of transmitter-specific pathways, particularly the cholinergic pathway to the cerebellum, should lead to a better understanding of how these pathways interact to regulate cerebellar function. This information will improve our understanding of how the cerebellum works and may clarify its role in spatial adaptation and motion sickness.

REFERENCES

1. Alley, K, Baker, R, and Simpson, JI, Afferents to the vestibulo-cerebellum and the origin of the visual climbing fibers in the rabbit, *Brain Res.,* 98:582,1975.
2. Alonso, R, Didier, M, and Soubrie, P, [^{3}H]N-methylscopolamine binding studies reveal M_2 and M_3 muscarinic receptor subtypes on cerebellar granule cells in primary culture, *J. Neurochem.,* 55:334,1990.
3. Appleyard, M, and Jahnsen, H, Actions of acetylcholinesterase in the guinea-pig cerebellar cortex *in vitro, Neuroscience,* 47:291,1992.
4. Appleyard, ME, Non-cholinergic functions of acetylcholinesterase, *Biochem.Soc.Trans.,* 22:749,1994.
5. Barmack, NH, GABAergic pathways convey vestibular information to the beta nucleus and dorsomedial cell column of the inferior olive, *Ann. NY Acad. Sci.,* 781:541,1996.
6. Barmack, NH, Baughman, RW, and Eckenstein, FP, Cholinergic innervation of the cerebellum of rat, rabbit, cat and monkey as revealed by choline acetyltransferase activity and immunohistochemistry, *J. Comp. Neurol.,* 317:233,1992a.
7. Barmack, NH, Baughman, RW, Eckenstein, FP, and Shojaku, H, Secondary vestibular cholinergic projection to the cerebellum of rabbit and rat as revealed by choline acetyltransferase immunohistochemistry, retrograde and orthograde tracers, *J. Comp. Neurol.,* 317:250,1992b.
8. Barmack, NH, Baughman, RW, Errico, P, and Shojaku, H, Vestibular primary afferent projection to the cerebellum of the rabbit, *J. Comp. Neurol.,* 327:521,1993a.
9. Barmack, NH, Fagerson, M, and Errico, P, Cholinergic projection to the dorsal cap of the inferior olive of the rat, rabbit and monkey, *J. Comp. Neurol.,* 328:263,1993b.
10. Barmack, NH, Fagerson, M, Fredette, BJ, Mugnaini, E, and Shojaku, H, Activity of neurons in the beta nucleus of the inferior olive of the rabbit evoked by natural vestibular stimulation, *Exp. Brain Res.,* 94:203,1993c.
11. Barmack, NH, Fredette, BJ, and Mugnaini, E, The parasolitary nucleus: A source of GABAergic vestibular information to the inferior olive of rat and rabbit, *J. Comp. Neurol.,* 392:352,1998.
12. Barmack, NH, Shojaku, H, Topography and analysis of vestibular-visual climbing fiber signals in the rabbit cerebellar nodulus., *Soc. Neurosci. Abst.,* 15:180,1989.
13. Barmack, NH, and Shojaku, H, Representation of a postural coordinate system in the nodulus of the rabbit cerebellum by vestibular climbing fiber signals, in: *Vestibular and Brain Stem Control of Eye, Head and Body Movements* (Shimazu H, and Shinoda Y Eds.), pp. 331-338. Tokyo and Basel: Japan Scientific Societies Press and Karger,1992a.
14. Barmack, NH, and Shojaku, H, Vestibularly induced slow oscillations in climbing fiber responses of Purkinje cells in the cerebellar nodulus of the rabbit, *Neuroscience,* 50:1,1992b.

15. Barmack, NH, and Shojaku, H, Vestibular and visual signals evoked in the uvula-nodulus of the rabbit cerebellum by natural stimulation, *J. Neurophysiol.*, 74:2573,1995.

16. Barmack, NH, and Young, WSI, Optokinetic stimulation increases corticotropin-releasing factor mRNA in inferior olivary neurons of rabbits, *J. Neurosci.*, 10:631,1990.

17. Blanks, RHI, Volkind, R, Precht, W, and Baker, R, Responses of cat prepositus hypoglossi neurons to horizontal angular acceleration., *Neuroscience,* 2:391,1977.

18. Boegman, RJ, Parent, A, and Hawkes, R, Zonation in the rat cerebellar cortex: patches of high acetylcholinesterase activity in the granular layer are congruent with Purkinje cell compartments, *Brain Res.,* 448:237,1988.

19. Bonner, TI, Domains of muscarinic acetylcholine receptors that confer specificity of G protein coupling, *Trends Neurolog. Sci.,* 13:48,1992.

20. Bradley, DJ, Ghelarducci, B, La Noce, A, Paton, JFR, Spyer, KM, Withington-Wray, DJ, An electrophysiological and anatomical study of afferents reaching the cerebellar uvula in the rabbit, *Exp. Physiol.,* 75:163,1990a.

21. Bradley, DJ, Ghelarducci, B, La Noce, A, and Spyer, KM, Autonomic and somatic responses evoked by stimulation of the cerebellar uvula in the conscious rabbit, *Exp. Physiol.,* 75:179,1990b.

22. Bradley, DJ, Ghelarducci, B, Paton, JFR, and Spyer, KM, The cardiovascular responses elicited from the posterior cerebellar cortex in the anaesthetized and decerebrate rabbit, *J. Physiol., (London)* 383:537,1987.

23. Brown, WJ, and Palay, SL, Acetylcholinesterase activity in certain glomeruli and Golgi cells of the granular layer of the rat cerebellar cortex, *Z. Anat. Entwickl-Gesch.,* 137:317,1972.

24. Caulfield, MP, Muscarinic receptors--Characterization, coupling and function, *Pharmacol. Ther.,* 58:319,1993.

25. Cohen, H, Cohen, B, Raphan, T, and Waespe, W, Habituation and adaptation of the vestibuloocular reflex: A model of differential control by the vestibulocerebellum, *Exp.Brain Res.,* 90:526,1992.

26. Crépel, F, and Dhanjal, SS, Cholinergic mechanisms and neurotransmission in the cerebellum of the rat, *Brain Res.,* 244:59,1982.

27. Csillik, B, Joo, F, and Kasa, P, Cholinesterase activity of archicerebellar mossy fibre apparatuses, *J. Histochem. Cytochem.,* 11:113,1963.

28. Danbolt, NC, Storm-Mathisen, J, Kanner, BI, An [$Na^+ + K^+$]coupled L-glutamate transporter purified from rat brain is located in glial cell processes, *Neuroscience,* 51:295,1992.

29. De la Garza, R, Freedman, R, and Hoffer, BJ, Kappa-bungarotoxin blockade of nicotine electrophysiological actions in cerebellar Purkinje neurons, *Neurosci. Lett.,* 99:95,1989a.

30. De la Garza, R, Freedman, R, and Hoffer, J, Nicotine-induced inhibition of cerebellar Purkinje neurons: Specific actions of nicotine and selective blockade by mecamylamine, *Neuropharmacology,* 28:495,1989b.

31. De Lacalle, S, Hersh, LB, and Saper, CB, Cholinergic innervation of the human cerebellum, *J. Comp. Neurol.,* 328:364,1993.

32. De Waele, C, Vibert, N, Baudrimont, M, and Vidal, PP, NMDA receptors contribute to the resting discharge of vestibular neurons in the normal and hemilabyrinthectomized guinea pig, *Exp. Brain Res.* 81:125,1990.

33. De Zeeuw, CI, Wentzel, P, and Mugnaini, E, Fine structure of the dorsal cap of the inferior olive and its GABAergic and non-GABAergic input from the nucleus prepositus hypoglossi in rat and rabbit, *J. Comp. Neurol.,* 327:63,1993.

34. del Toro, ED, Juiz, JM, Smillie, FI, Lindstrom, J, and Criado, M, Expression of a7 neuronal nicotinic receptors during postnatal development of the rat cerebellum, *Dev. Brain Res.,* 98:125,1997.

35. Dememes, D, Raymond, J, and Sans, A, Selective retrograde labeling of neurons of the cat vestibular ganglion with [^{3}H]D-aspartate, *Brain Res.,* 304:188,1984.

36. DeSouza, EB, Corticotropin-releasing factor receptors in the rat central nervous system: Characterization and regional distribution, *J. Neurosci.,* 7:88,1987.

37. DeSouza, EB, Insel, TR, Perrin, MH, Rivier, J, Vale, WW, and Kuhar, MJ, Corticotropin-releasing factor receptors are widely distributed within the rat central nervous system: An autoradiographic study, *J.Neurosci.,* 5:3189,1985.

38. Doi, K, Tsumoto, T, and Matsunaga, T, Actions of excitatory amino acid antagonists on synaptic inputs to the rat medial vestibular nucleus: An electrophysiological study in vitro, *Exp. Brain Res.*, 82:254,1990.

39. Douglas, WW, Histamine and 5-hydroxytryptamine (serotonin) and their antagonists. in Goodman and Gilman's the Pharmacological Basis of Therapeutics (Gilman AG, Goodman ALS, Rall TW, and Murad F Eds.), pp. 605-638. New York: Macmillan,1985.

40. Ehlert, FJ, and Tran, LLP, Regional distribution of M1, M2 and non-M1, non-M2 subtype of muscarinic binding sites in rat brain, *J. Pharmacol. Exp. Ther.*, 255:1148,1990.

41. Fagerson, MH, and Barmack, NH, Responses to vertical vestibular stimulation of neurons in the nucleus reticularis gigantocellularis in rabbits, *J. Neurophysiol.*, 73:2378,1995.

42. Fredette, BJ, and Mugnaini, E, The GABAergic cerebello-olivary projection in the rat, *Anat. Embryol. (Berlin)*, 184:225,1991.

43. Fukamauchi, F, Hough, C, and Chuang, D-M, Expression and agonist-induced down-regulation of mRNAs of m2- and m3-muscarinic acetylcholine receptors in cultured cerebellar granule cells, *J. Neurochem.*, 56:716,1991.

44. Fukamauchi, F, Saunders, PA, Hough, C, and Chuang, D-M, Agonist-induced down-regulation and antagonist-induced up-regulation of m_2- and m_3-muscarinic acetylcholine receptor mRNA and protein in cultured cerebellar granule cells, *Mol. Pharmacol.*, 44:940,1995.

45. Fushiki, H, and Barmack, NH, Topography and reciprocal activitiy of cerebellar Purkinje cells in the uvula-nodulus modulated by vestibular stimulation, *J. Neurophysiol.*, 1997, 78:3083,1997.

46. Gallo, V, Upson, LM, Hayes, WP, Vyklicky, JrL, Winters, CA, and Buonanno, A, Molecular cloning and developmental analysis of a new glutamate receptor subunit isoform in cerebellum, *J. Neurosci.*, 12:1010,1992.

47. Gerrits, NM, Voogd, J, and Magras, IN, Vestibular afferents of the inferior olive and the vestibulo-olivo-cerebellar climbing fiber pathway to the flocculus in the cat, *Brain Res.*, 332:325,1985.

48. Gorenstein, C, Bundman, MC, Bruce, JL, and Rotter, A, Neuronal localization of pseudocholinesterase in the rat cerebellum: Sagittal bands of Purkinje cells in the nodulus and uvula, *Brain Res.*, 418:68,1987.

49. Graf, W, Simpson, JI, and Leonard, CS, Spatial organization of visual messages of the rabbit's cerebellar flocculus. II. Complex and simple spike responses of Purkinje cells, *J. Neurophysiol.*, 60:2091,1988.

50. Hasler, WL, Kim, MS, Chey, WD, Stevenson, V, Stein, B, and Owyang, C, Central cholinergic and a-adrenergic mediation of gastric slow wave dysrhythmias evoked during motion sickness, *Am. J. Physiol. Gastrointest. Liver Physiol.*, 268:G539,1995.

51. Henry, RT, Connor, JD, and Balaban, CD, Nodulus-uvula depressor response: central GABA-mediated inhibition of alpha-adrenergic outflow, *Amer. J. Physiol.*, 1989.

52. Holopainen, I, and Wojcik, WJ, A specific antisense oligodeoxynucleotide to mRNAs encoding receptors with seven transmembrane spanning regions decreases muscarinic m_2 and gamma-aminobutyric acid$_B$ receptors in rat cerebellar granule cells, *J. Pharmacol. Exp. Ther.*, 264:423,1993

53. Ikeda, M, Houtani, T, Ueyama, T, and Sugimoto, T, Choline acetyltransferase immunoreactivity in the cat cerebellum, *Neuroscience*, 45:671,1991.

54. Illing, R-B, A subtype of cerebellar Golgi cells may be cholinergic, *Brain Res.*, 522:267,1990.

55. Ito, H, Sugimoto, T, Kobayashi, I, Takahashi, K, Katada, T, Ui, M, and Kurachi, Y, On the mechanism of basal and agonist-induced activation of the G protein-gated muscarinic K channel in atrial myocytes of guinea pig heart, *J. Gen. Physiol.*, 98:517,1991.

56. Jaarsma, D, Diño, M, Cozzari, C, and Mugnaini, E, Cerebellar choline acetyltransferase positive mossy fibres and their granule and unipolar brush cell targets: a model for central cholinergic nicotinic neurotransmission, *J. Neurocytol.*, 25:829,1996.

57. Jaarsma, D, Levey, AI, Frostholm, A, Rotter, A, Voogd, J, Light-microscopic distribution and parasagittal organisation of muscarinic receptors in rabbit cerebellar cortex, *J. Chem. Neuroanat.*, 9:241,1995.

58. Kamei, T, Nagai, T, McGeer, PL, and McGeer, EG, Evidence of an intracerebellar acetylcholinesterase-rich but probably non-cholinergic flocculo-nodular projection, Brain Res., 258:115,1983.

59. Kan, KSK, Chao, LP, and Eng, LF, Immunohistochemical localization of choline acetyltransferase in rabbit spinal cord and cerebellum, *Brain Res.*, 146:221,1978.

60. Kasa, P, and Silver, A, The correlation between choline acetyltransferase and acetylcholinesterase activity in different areas of the cerebellum of rat and guinea pig, *J. Neurochem.*, 16:389,1969.

61. Kawamura, K, and Onodera, S, Olivary projections from the pretectal region in the cat studied with horseradish peroxidase and tritiated amino acids axonal transport, *Arch. Ital. Biol.*, 122:155,1984.

62. Kohl, RL, and Homick, JL, Motion sickness: A modulatory role for the central cholinergic nervous system, *Neurosci. Biobehav. Rev.*, 7:73,1983.

63. Korte, G, and Mugnaini, E, The cerebellar projection of the vestibular nerve in the cat, *J. Comp. Neurol.*, 184:265,1979.

64. Lai, J, Waite, SL, Bloom, JW, Yamamura, HI, and Roeske, WR, The m2 muscarinic acetylcholine receptors are coupled to multiple signaling pathways via pertussis toxin-sensitive guanine nucleotide regulatory proteins, *J. Pharmacol. Exp. Ther.*, 258:938,1991.

65. Landis, DMD, and Reese, TS, Structure of the Purkinje cell membrane in staggerer and weaver mutant mice, *J. Comp. Neurol.*, 171:247,1977.

66. Lapchak, PA, Araujo, DM, Quirion, R, and Collier, B, Presynaptic cholinergic mechanisms in the rat cerebellum: Evidence for nicotinic, but not muscarinic autoreceptors, *J. Neurochem.*, 53:1843,1989.

67. Leonard, CS, Simpson, JI, and Graf, W, Spatial organization of visual messages of the rabbit's cerebellar flocculus. I. Typology of inferior olive neurons of the dorsal cap of Kooy. *J. Neurophysiol.*, 1988.

68. Li, M, Yasuda, RP, Wall, SJ, Wellstein, A, and Wolfe, BB, Distribution of m2 muscarinic receptors in rat brain using antisera selective for m2 receptors, *Mol. Pharmacol.*, 40:28,1991.

69. Loewy, AD, and Burton, H, Nuclei of the solitary tract: Efferent projections to the lower brain stem and spinal cord of the cat, *J. Comp. Neurol.*, 181:421,1978.

70. Lopez-Barneo, J, Darlot, C, Berthoz, A, and Baker, R, Neuronal activity in prepositus nucleus correlated with eye movement in the alert cat, *J. Neurophysiol.*, 47:329,1982.

71. Maekawa, K, and Simpson, JI, Climbing fiber responses evoked in vestibulocerebellum of rabbit from visual system, *J.Neurophysiol.*, 36:649,1973.

72. Maekawa, K, and Takeda, T, Origin of descending afferents to the rostral part of dorsal cap of inferior olive which transfers contralateral optic activities to the flocculus. A horseradish peroxidase study, *Brain Res.*, 172:393,1979.

73. Marani, E, Enzyme histochemistry. in: *Methods in Neurobiology*, Vol.1 (Lahue, R, Ed.), pp. 481-581, New York: Plenum, 1981.

74. Marani, E, and Voogd, J, An acetylcholinesterase band-pattern in the molecular layer of the cat cerebellum, *J. Anat.*, 124:335,1977.

75. Martin, LJ, Blackstone, CD, Levey, AI, Huganir, RL, and Price, DL, AMPA glutamate receptor subunits are differentially distributed in rat brain, *Neuroscience*, 53:327,1993.

76. Masumitsu, Y, and Sekitani, T, Effect of electric stimulation on vestibular compensation in guinea pigs, *Acta Otolaryngol. (Stockholm)*, 111:807,1991.

77. Matute, C, Wiklund, L, Streit, P, Cuenod, M, Selective retrograde labeling with **D**-[³H]-aspartate in the monkey olivocerebellar projection, *Exp. Brain Res.*, 66:445,1987.

78. McCrea, RA, and Baker, R, Anatomical connections of the nucleus prepositus of the cat, *J. Comp. Neurol*, 237:377,1985.

79. McFarland, JL, and Fuchs, AF, Discharge patterns in nucleus prepositus hypoglossi and adjacent medial vestibular nucleus during horizontal eye movement in behaving Macaques, *J. Neurophysiol.*, 68:319,1992.

80. Money, KE, Motion sickness, *Physiol. Rev.*, 50:1,1970.

81. Mount, HTJ, Dreyfus, CF, and Black, IB, Muscarinic stimulation promotes cultured Purkinje cell survival: A role for acetylcholine in cerebellar development, *J. Neurochem.*, 63:2065,1994.

82. Mugnaini, E, and Nelson, BJ, Corticotropin-releasing factor (CRF) in the olivo-cerebellar system and feline olivary hypertrophy. in: *The Olivocerebellar System in Motor Control* (Strata P, Ed.), pp 187-197. Berlin: Springer-Verlag.,1989.

83. Nelson, B, Barmack, NH, and Mugnaini, E, A GABAergic vestibular projection to rat inferior olive, *Soc. Neurosci. Abst.*, 12:255,1986. (Abstract)

84. Nelson, BJ, Adams, JC, Barmack, NH, and Mugnaini, E, A comparative study of glutamate decarboxylase immunoreactive boutons in the mammalian inferior olive, *J. Comp. Neurol.*, 286:514,1989.

85. Neustadt, A, Frostholm, A, and Rotter, A, Topographical distribution of muscarinic cholinergic receptors in the cerebellar cortex of the mouse, rat, guinea pig, and rabbit: A species comparison, *J. Comp. Neurol.*, 272:317,1988.

86. O'Malley, DM, Sandell, JH, and Masland, RH, Co-release of acetylcholine and GABA by the starburst amacrine cells, *J. Neurosci.*, 12:1394,1992.

87. Ojima, H, Kawajiri, S, and Yamasaki, T, Cholinergic innervation of the rat cerebellum: qualitative and quantitative analyses of elements immunoreactive to a monoclonal antibody against choline acetyl-transferase, *J. Comp. Neurol.*, 290:41,1989.

88. Pan, ZZ, and Williams, JT, Muscarine hyperpolarizes a subpopulation of neurons by activating an m2 muscarinic receptor in rat nucleus raphe magnus *in vitro. J. Neurosci.*, 14:1332,1995.

89. Paton, JFR, La, Noce, A, Sykes, RM, Sebastiani, L, Bagnoli, P, Ghelarducci, B, and Bradley, DJ, Efferent connections of lobule IX of the posterior cerebellar cortex in the rabbit - Some functional consider-ations, *J. Auton. Nerv. Syst.*, 36:209,1991.

90. Petralia, RS, and Wenthold, RJ, Light and electron immunocytochemical localization of AMPA- selective glutamate receptors in the rat brain, *J. Comp. Neurol.*, 318:329,1992.

91. Rasheed, BMA, Manchanda, SK, and Anand, BK, Effects of the stimulation of paleocerebellum on certain vegetative functions in the cat, *Brain Res.*, 20:293,1970.

92. Raymond, J, Nieoullon, A, Dememes, D, and Sans, A, Evidence for glutamate as a neurotransmitter in the cat vestibular nerve: radioautographic and biochemical studies, *Exp. Brain Res.*, 56:523,1984.

93. Richards, MH, Pharmacology and second messenger interactions of cloned muscarinic receptors, *Biochem. Pharmacol.*, 42:1645,1991.

94. Robertson, RT, Yu, BP, Liu, HH, Liu, NH, and Kageyama, GH, Development of cholinesterase histochem-ical staining in cerebellar cortex: Transient expression of "nonspecific" cholinesterase in Purkinje cells of the nodulus and uvula, *Exp. Neurol.*, 114:330,1991.

95. Rossiter, CD, Hayden, NL, Stocker, SD, and Yates, BJ, Changes in outflow to respiratory pump muscles produced by natural vestibular stimulation, *J. Neurophysiol.*, 76:3274,1996.

96. Rotter, A, Birdsall, NJM, Field, PM, and Raisman, G, Muscarinic receptors in the central nervous system of the rat. II. Distribution of binding of [^{3}H]propylbenzilylcholine mustard in the in the midbrain and hindbrain, *Brain Res. Rev.*, 1:167,1979.

97. Saint-Cyr, JA, and Courville, J, Projection from the vestibular nuclei to the inferior olive in the cat: An autoradiographic and horseradish peroxidase study, *Brain Res.*, 165:189,1979.

98. Sato, Y, and Barmack, NH, Zonal organization of the olivocerebellar projection to the uvula in rabbits, *Brain Res.*, 359:281,1985.

99. Shute, CCD, and Lewis, PR, Cholinesterase-containing pathways of the hindbrain: Afferent cerebellar and centrifugal cochlear fibres, *Nature*, 205:242,1965.

100. Smith, PF, and Darlington, CL, The NMDA antagonists MK801 and CPP disrupt compensation for unilateral labyrinthectomy in the guinea pig, *Neurosci. Lett.*, 94:309,1988.

101. Smith, PF, and Darlington, CL, Neurochemical mechanisms of recovery from peripheral vestibular lesions (vestibular compensation), *Brain Res. Rev.*, 16:117,1991.

102. Steinmetz, JE, Sears, LL, Gabriel, M, Kubota, Y, and Poremba, A, Cerebellar interpositus nucleus lesions disrupt classical nictitating membrane conditioning but not discriminative avoidance learning in rabbits, *Behav. Brain Res.*, 45:71,1991.

103. Takeda, T, and Maekawa, K, Olivary branching projections to the flocculus, nodulus and uvula in the rabbit. II. Retrograde double labeling study with fluorescent dyes, *Exp. Brain Res.*, 76:323,1989.

104. Terasawa, K, Otani, K, and Yamada, J, Descending pathways of the nucleus of the optic tract in the rat, *Brain Res.*, 173:405,1979.

105. Thompson, SJ, Schatteman, GC, Gown, AM, and Bothwell, M, A monoclonal antibody against nerve growth factor receptor: Immunohistochemical analysis of normal and neoplastic human tissue, *Am. J. Clin. Pathol.*, 92:415,1989.

106. Vilaró, MT, Wiederhold, K-H, Palacios, JM, and Mengod, G, Muscarinic M$_2$ receptor mRNA expression and receptor binding in cholinergic and non-cholinergic cells in the rat brain: A correlative study using *in situ* hybridization histochemistry and receptor autoradiography, *Neuroscience*, 47:367,1992.

107. Walberg, F, Descending connections from the mesencephalon to the inferior olive: An experimental study in the cat, *Exp. Brain Res.*, 21:145,1974.

108. Weiner, N, Atropine, scopolamine, and related antimuscarinic drugs. in: Goodman and Gilman's the Pharmacological Basis of Therapeutics (Gilman AG, Goodman ALS, Rall TW, and Murad, F, Eds.), pp. 130-144. New York: Macmillan,1985.

109. Whitham, EM, Challiss, RAJ, and Nahorski, SR, M_3 muscarinic cholinoceptors are linked to phosphoinositide metabolism in rat cerebellar granule cells, *Eur. J. Pharmacol. Mol. Pharmacol.,* 206:181,1991.
110. Wiklund, L, Toggenburger, G, and Cuenod, M, Aspartate: Possible neurotransmitter in cerebellar climbing fibers, *Science,* 216:78,1982.
111. Wylie, DR, De Zeeuw, CI, DiGiorgi, PL, and Simpson, JI, Projections of individual Purkinje cells of identified zones in the ventral nodulus to the vestibular and cerebellar nuclei in the rabbit, *J. Comp. Neurol.,* 349:448,1994.
112. Wylie, DR, De Zeeuw, CI, and Simpson, JI, Temporal relations of the complex spike activity of Purkinje cell pairs in the vestibulocerebellum of rabbits, *J. Neurosci.,* 15:2875,1995.
113. Wynn, PC, Hauger, RL, Holmes, MC, Millan, MA, Catt, KJ, and Aguilera, G, Brain and pituitary receptors for corticotropin releasing factor: Localization and differential regulation after adrenalectomy, *Peptides,* 5:1077,1984.
114. Zhang, N, and Ottersen, OP, In search of the identity of the cerebellar climbing fiber transmitter: Immunocytochemical studies in rats, *Can. J. Neurol. Sci.,* 20 Suppl., 3:S36,1993.

13

Physiology and Ultrastructure of Unipolar Brush Cells in the Vestibulo-Cerebellum

N.T. Slater, D.J. Rossi, M.R. Diño, D. Jaarsma, and E. Mugnaini

CONTENTS

13.1 SYNOPSIS

The unipolar brush cell (UBC) is closely tied to the vestibular system, albeit not exclusively. The term "unipolar brush" is descriptive of its single, large dendrite and the bunch of dendrioles on which it usually receives the synaptic input. The UBC has a characteristic set of morphological and neuronal phenotypes that distinguish it from other types of neurons in the cerebellar granular layer. UBCs are generated from the ventricular epithelium before differentiation of the granule cells; are intermediate in size between granule and Golgi cells; contain an assembly of ringlet subunits, numerous neurofilaments, and large, dense core vesicles; are secretogranin-positive and strongly calretinin-immunopositive, but are GABA- and glycine-immunonegative. They emit numerous, unusual, nonsynaptic appendages; have a single, thick dendrite; form a thicket of dendrioles at the dendritic tip or the soma; and receive an extraordinarily extensive synapse from a single MF. Furthermore, they form dendrodendritic synapses with granule cells and receive relatively few inhibitory Golgi boutons. It has been shown that UBCs express cytoplasmic and postsynaptic GluR2/3 at high levels, as well as postsynaptic GluR5 and NR1 (and presumably NR2) subunits. Moreover, they express mGluR1, mGluR1a, and mGluR2/3 at extrasynaptic sites.

Many of the biophysical and pharmacological properties of the giant mossy fiber-UBC synapse have been elucidated with patch-clamp pipettes in the slice preparation. The synaptic response is comprised of both a fast AMPA receptor-mediated EPSC, and a slower synaptic current which is mediated in part both by NMDA and AMPA receptors. The slow component of the EPSC in UBCs is very prolonged by comparison with the time course of glutamate receptor-mediated EPSCs in other CNS synapses. The prolonged late component of the synaptic current is consistent with the hypothesis that the MF-UBC synapse represents an ultrastructural specialization that serves to entrap released glutamate within the tortuous volume of the synaptic cleft. With repetitive activation, the EPSPs in UBCs fuse to form a plateau potential. This prolonged firing may be distributed by branches of the UBC axon to a large ensemble of granule cells and result in temporal summation and prolonged firing of these cells.

13.2 INTRODUCTION

The vestibulo-cerebellum is unique among the other cerebellar lobules in that it is directly innervated from sensory fibers derived from peripheral ganglia via the dorsal branch of the vestibular nerve[7,24,38] in addition to receiving sensory input already processed in one or more nuclei of the brainstem or spinal cord. Furthermore, while the cellular organization of the cortex in the vestibulo-cerebellum has traditionally been considered to be identical to that of other folia in the cerebellar vermis and hemispheres, it has recently been established that the nodulus, ventral uvula, flocculus, and ventral paraflocculus, some of which receive the bulk of primary vestibular nerve fibers, are particularly enriched with a very peculiar new type of cerebellar neuron, the unipolar brush cell (UBC).[45] UBCs are also enriched in the dorsal cochlear nucleus, a center that is innervated by primary auditory fibers and receives many other nonauditory inputs.[22]

13.3 MORPHOLOGY OF UNIPOLAR BRUSH CELLS.

The UBC is a small neuron that occupies the granular layer, and sometimes also the folial white matter, of the cerebellar cortex, as well as the granule cell domain of the cochlear nuclear complex. Recently, UBCs, which might be identical to a class of cerebellar interneurons previously described as "pale cells"[2, 21] and to the "mitt cells" in the cochlear nerve root of the chinchilla,[28] have been extensively studied with several light and electron microscopic methods.[5,9,10,15,18,22,26,27,30-32,45,47,53,55,64,68,69] UBCs show well-conserved morphological features among species, although their mode of distribution in the cerebellar lobules expands with mammalian evolution[18] (Diño, Willard, and Mugnaini, in preparation). In all mammals, UBCs are intermediate in size between granule and Golgi cells; emit numerous nonsynaptic appendages; possess numerous large, dense core vesicles; are rich in neurofilaments; and contain a peculiar organelle—termed the ringlet-subunit assembly[46] (Figure 13.1, main panel).

UBCs generally have a single, relatively thick dendritic stem that terminates in a brush-like formation of branchlets (or dendrioles) provided with nonsynaptic appendages, and shows little interspecies variation (Figure 13.1, insets). The brush usually establishes an elaborate asymmetric synapse with a single mossy fiber rosette, referred to herein as the MF-UBC synapse (Figure 13.2). Numerous actin filaments are present in the subsynaptic web and link the postsynaptic densities to the cytoskeleton of the UBC dendrioles.[19] This synapse is usually situated in a glomerulus, which may contain the brush of one or more UBCs, exclusively, or the brush of a UBC together with the dendrites of several granule cells. In some of the UBCs, the dendrite is very short or altogether absent (crossed arrow in inset to Figure 13.1), and the brush synapse may form directly on the cell body. Before the existence of the UBCs was recognized, this crenated apposition was attributed to Golgi cells, and was originally described as an *en marron* synapse.[11,25,43,44,50] The UBC branchlets contain presynaptic clusters of round vesicles and form asymmetric dendrodendritic synapses with

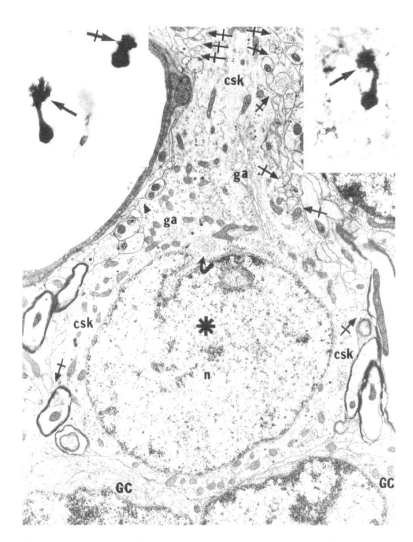

FIGURE 13.1 Electron micrographs of the soma and emerging dendrite of a UBC (asterisk) in the rat cerebellar nodulus. The nucleus (n) contains much less heterochromatin than that of granule cells (GC). The cytoplasm contains a special organelle consisting of nonmembranous riglet-subunits (curved arrow), an extensive Golgi apparatus, few cisterns of granular endoplasmic reticulum, and numerous neurofilaments interspersed among microtubules (csk). The somatodendritic surface emits numerous nonsynaptic appendages (crossed arrows). Insets: light micrographs of a rat UBC impregnated with the Golgi method (left), and a human UBC immunostained with calretinin antiserum (right); the cells emit single dendritic trunks terminating in a spray of branchlets or dendrioles (arrows). In the UBC labeled by crossed arrow (left inset), branchlets arise directly from the cell body. Main panel: X 15,000; insets: X 800.

granule cell dendrites, while the UBC axon forms large endings in the granular layer (Figure 13.4A) and may enter the white matter.[9,22,57,58] UBCs, which are immunonegative for both GABA and glycine, are presumed to be excitatory interneurons.[4,22,46,58] UBCs, or at least a proportion of them, also appear immunopositive for molecules of the intermediate filament family,[26,27] the secretogranin (chromogranin) family,[15,47] and of calretinin.[21,22,55,56] In the rat, the UBC population is generated prenatally for the most part.[1,61] UBCs have recently been grown in dissociated cell cultures, which should aid the characterization of their cell biological and molecular properties.[3,42]

13.4 ULTRASTRUCTURE OF THE MOSSY FIBER-UNIPOLAR BRUSH CELL SYNAPSE

The MF-UBC synapse is always of the asymmetric category (Figure 13.2) and is unusual in that the region of synaptic apposition is extremely extensive.[4]

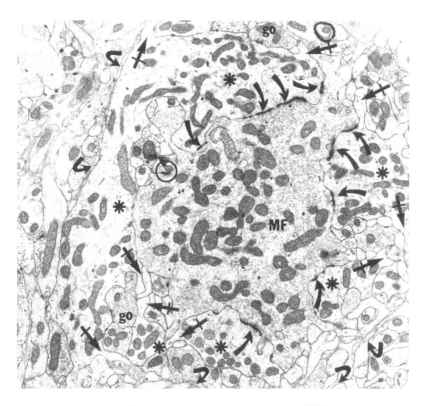

Figure 13.2 Electron micrographs of the soma and emerging dendrite of a UBC (asterisk) in the rat cerebellar nodulus. The nucleus (n) contains much less heterochromatin than that of granule cells (GC). The cytoplasm contains a special organelle consisting of nonmembranous riglet-subunits (curved arrow), an extensive Golgi apparatus, few cisterns of granual endoplasmic reticulum, and numerous neurofilaments interspersed among microtubules (csk). The somatodendritic surface emits numerous nonsynaptic appendages (crossed arrows). Insets: light micrographs of a rat UBC impregnated with the Golgi method (left), and a human UBC immunostained with calretinin antiserum (right); the cells emit single dendritic trunks terminating in a spray of branchlets or dendrioles (arrows). In the UBC labeled by crossed arrow (left inset), branchlets arise directly from the cell body. Main panel: X15,000; insets: X800.

This suggests that the physiology of transmission is different from that of other synapses previously subjected to careful biophysical and pharmacological examination (see below). The features of the MF-UBC synapse vary substantially in their detailed configuration. Some are characterized by long individual or serial synaptic junctions, each of which measures up to 10 mm in length and 1 to 3 µm in width, while others consist of numerous small synaptic junctions in the form of patches 0.2 to 0.6 mm in diameter and separated by seemingly nonspecialized plasma membrane appositions. This variation may reflect a degree of synaptic plasticity.

The majority of the mossy fibers terminating in the vestibulocerebellum, including the primary vestibular fibers38,52 and those synapsing on the UBCs, are likely to be glutamatergic.23,30,58,63 Some of the mossy fibers arising from the vestibular nuclei are immunopositive

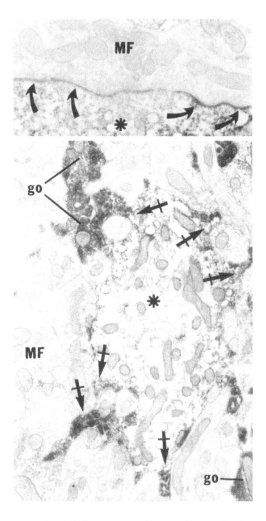

Figure 13.3 **(A)** AMPA receptor (GluR2/3) immunoreaction product marks the postsynaptic density (arrows) at an MF–UBC synapse in rat nodulus. Immunoreaction product is also present in the cytoplasm of the UBC branchlet (asterisk). X32,000. (Courtesy of Jaarsma, Wenthold, and Mugnaini.) **(B)** mGluR2/3 antibody densely labels nonsynaptic appendages (crossed arrows) of a dendriole of a rat UBC (asterisk) in cerebellar nodulus, as well as boutons of the Golgi axonal plexus (go). MF, mossy fiber rosette. X21,500. (Courtesy of Jaarsma, Diño, Ohishi, Shigemoto, and Mugnaini.)

for choline acetyltransferase (ChAT) and may have a component of cholinergic neurotransmission yet to be characterized.[7,8,31,48] Small boutons, which are presumed to belong to Golgi-cell axons because they contain pleomorphic synaptic vesicles, form both asymmetric and symmetric synapses on brush branchlets, where they may exert hyperpolarizing effects or shunt the excitatory mossy fiber EPSPs. Similar boutons form sparse, exclusively symmetric synapses elsewhere on the UBC, and they occur with progressively less frequency on the plasmalemma of the dendritic stem and the cell body. In general, the excitatory and inhibitory inputs tend to be distributed at the periphery of the short somato-dendritic compartment of the UBC.

13.5 ULTRASTRUCTURAL LOCALIZATION OF RECEPTOR SUBTYPES IN UNIPOLAR BRUSH CELLS

Recent immunocytochemical studies with the pre-embedding method utilizing subunit-specific antisera to glutamate receptors have shown that ionotropic glutamate receptors (Figure 13.3A) are primarily localized to postsynaptic densities of the giant MF -UBC synapses,[30] while metabotropic glutamate receptors (Figure 13.3B) are found on the nonsynaptic plasma membrane of the filopodial appendages, the dendritic trunk, the cell body, and the axolemma.[32,53,69]

UBCs express primarily the AMPA type GluR2/3 and the kainate subtype GluR 5. NMDA-R1 is present in the UBCs, but the forms of NMDA-R2 subtypes they express has yet to be established. UBCs express mGluR1, mGluR1a, and mGluR2/3.[32,53,69] The types of the presumed cholinergic receptors mediating the ChAT-positive mossy fiber input and the presumed GABAergic and gly-cinergic receptors mediating the Golgi-cell inputs remain to be established.

Since large, dense core vesicles, which may contain secretogranin molecules and unknown other peptides, may be released outside conventional presynaptic densities,[17] and the stems and nonsynaptic appendages of the brush branchlets in the glomeruli are often closely bunched together, it was hypothesized that UBCs may display autocrine secretion.[46] However, the receptors involved in this hypothetical function are unknown.

The main morphological and chemical phenotypes of the UBCs are summarized in Table 13.1.

TABLE 13.1

CHARACTERISTIC FEATURES OF UNIPOLAR BRUSH CELLS

1. Are generated perinatally from the ventricular epithelium
2. Are intermediate in size between granule and Golgi cells
3. Contain an assembly of ringlet subunits
4. Are rich in neurofilaments
5. Are secretogranin-positive and contain large dense core vesicles
6. Are strongly calretinin-immunopositive
7. Are GABA and glycine-immunonegative
8. Emit numerous, unusual, nonsynaptic appendages
9. Have a single, thick dendrite
10. Form a thicket of dendrioles at the dendritic tip or the soma
11. Receive an extraordinarily extensive synapse from a single MF
12. Express cytoplasmic and postsynaptic GluR2/3 at high levels
13. Express postsynaptic GluR5
14. Express postsynaptic NR1 (and presumably NR2) subunits
15. Express mGluR1, mGluR1a, and mGluR2/3 at extrasynaptic sites
16. Possibly express acetylcholine receptors
17. Form dendrodendritic synapses with granule cells
18. Receive relatively few inhibitory Golgi boutons

13.6 PROPERTIES OF SYNAPTIC TRANSMISSION

The unusual ultrastructural features of the MF-UBC synapse suggest that transmission at this synapse may also have some unusual features. This has been studied by Rossi et al.[57,58] and Kinney et al.[36] using patch-clamp recordings of UBCs in thin cerebellar slices of rat. When viewed in thin slices, UBCs can be distinguished from adjacent granule cells of the internal granular layer of the nodulus and uvula by their larger soma diameter and whole-cell capacitance, and a prolonged train of extracellular action potentials recorded in response to single white matter stimuli in cell-attached recordings. The inclusion of Lucifer Yellow in the recording pipette was also used to verify their characteristic morpholog0 (Figure 13.4A).

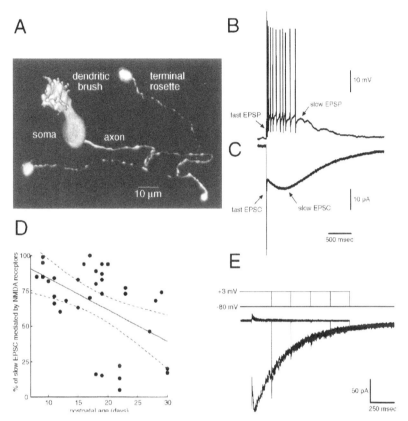

Figure 13.4 Morphology and synaptic responses of unipolar brush cells. **(A)** Confocal image of a Lucifer Yellow-filled UBC in which the axon and terminal rosettes were filled. **(B,C)** Recordings of the synaptic response to MF stimulation in another UBC obtained using current-clamp (B) and voltage-clamp (C) at a holding potential of -80 mV in the presence of external magnesium (1 nM). **(D)** Age dependence of the contribution of NMDA receptors to the slow EPSC. The relative contribution of NMDA receptors to the total charge underlying the slow EPSC was expressed as the percent block by NMDA receptor antagonists [D-AP5 (50 mM) or 7-chlorokynurenate (200 mM) in a Mg^{2+}-free medium], and is plotted against the postnatal age of the animal from which the recording was obtained. The continuous line is a regression fit to the data; dashed lines are 95% confidence limits. **(E)** Voltage-jump analysis of the time course of the conductance change during the slow EPSC. Traces are overlays of the current response for voltage jumps from the reversal potential of the EPSC (+3 mV) to -80 mV at varying time points after MF stimulation. Ohmic and capacitive currents have been digitally subtracted. The recording was made in a MG^{2+}-free medium, and the synaptic current is largely carried by NMDA receptors. (Adapted from Ref. 58.)

Electrical stimulation of MF afferents to UBCs evokes a large, long-lasting depolarization (Figure 13.4B). In all UBCs studied to date, the synaptic response was mediated by glutamate receptors. Voltage-clamp recordings demonstrate that the underlying synaptic current is comprised of both a fast AMPA receptor-mediated EPSC and a slower synaptic current mediated in part both by NMDA and AMPA receptors (Figures 13.4C, 5B).

Both the fast and slow AMPA receptor-mediated components of the EPSC (Figure 13.5B) were blocked by AMPA antagonists such as CNQX, were unaffected by NMDA receptor antagonists (D-AP5, 7-chlorokynurenate), and displayed linear *I-V* relations in the presence of external magnesium (Figure 13.5D). Conversely, the NMDA component displayed a slower initial rate of rise (Figure 13.5B, inset), was blocked by NMDA but not AMPA receptor antagonists, and displayed a prominent rectification at hyperpolarized membrane potentials in the presence of external magnesium (Figure 13.5C). Both the fast and slow components of the MF-evoked EPSC in UBCs were evoked in an all-or-none fashion at the same stimulus intensity, a result that is consistent with the ultrastructural observation that most UBCs receive innervation from a single MF.[46]

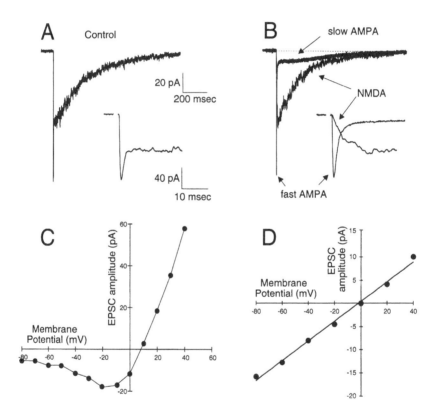

Figure 13.5 The synaptic current in UBCs is mediated both by AMPA and NMDA receptors. **(A)** The biphasic EPSC of a patch-clamped UBC at low and high (inset) time resolution in Mg^{2+}-free saline. **(B)** Overlay of the pharmacologically separated AMPA and NMDA components. The AMPA receptor-mediated EPSC was recorded in D-AP5 (50 mM) and 1 mM Mg^{2+}; and the NMDA component was recorded in a Mg^{2+}-free saline containing CNQX (10 mM). **(C)** I–V relations of the peak NMDA current (recorded in 1 mM Mg^{2+} and 10 mM CNQX) to show the voltage-dependent rectification of the NMDA component. **(D)** Lack of rectification of the slow AMPA receptor-mediated slow EPSC recorded in the presence of external Mg^{2+} (1 mM) in another UBC. (Adapted from Ref. 58.)

The slow component of the EPSC in UBCs was very prolonged in comparison with the time course of glutamate receptor-mediated EPSCs in other CNS synapses examined.[34] By pharmacologically isolating the two components of the EPSC (AMPA and NMDA receptor-mediated), the time course of each could be independently studied. The AMPA component was biphasic in the majority of neurons, being comprised of a rapidly activating and deactivating component similar in most respects to that seen at other glutamatergic synapses. This fast component was followed by a very slow component, which was observed either as a prolonged tail current or a late peak in the synaptic current occurring several hundreds of milliseconds after the fast component (e.g., Figure 13.4C) and decaying gradually over several seconds. By contrast, the NMDA receptor-mediated component did not display two distinct phases, but displayed a more slowly rising

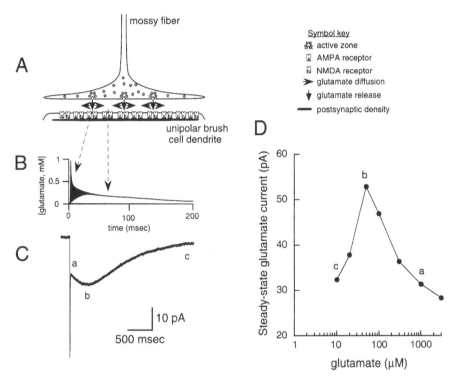

Figure 13.6 Hypothetical cellular mechanism underlying the slow AMPA receptor-mediated EPSC in unipolar brush cells. **(A)** Schematic illustration of the ultrastructural features of the MF–UBC synapse, showing multiple sites of glutamate release. **(B)** The hypothetical time course of changes in glutamate concentration within the cleft following release. Changes given by dark shading represent the more rapid rise and decay within regions of the cleft apposed to release sites, and light shading represents the slower diffusional exchange of glutamate into cleft regions between release sites. **(C)** The time course of an AMPA receptor-mediated EPSC in a UBC. **(D)** The steady-state dose-response curve for glutamate acting at AMPA receptors in excised patches of avian nucleus magnocellularis neurons. Immediately following release, a maximal concentration of glutamate is achieved, which produces a rapidly activating and desensitizing response, followed by a steady-state current reflective of the peak of the dose-response curve (a in C and D). As the concentration of glutamate in the cleft gradually declines due to diffusion and re-uptake, a larger steady-state current is seen (D, b), which is reflected by a late peak in the EPSC (C, b). Following this peak, a further decline in glutamate concentration results in a greater decline of occupancy, and consequently, a lower steady-state current and the EPSC decays (c in C and D). (A and C are adapted from Ref. 58. Data in D are redrawn from Ref. 54.)

activation to a peak from which the current slowly decayed over a period of several hundred milliseconds (Figure 13.5B).

The relative contributions of AMPA and NMDA receptors to the late component of the EPSC varied somewhat with developmental age. When studied in the absence of external magnesium, the contribution of NMDA receptors to the late component showed a modest decline with postnatal age over the range 8 to 30 days, although at all ages examined this current contributed >50% of the total synaptic charge transfer (Figure 13.4D). It is important to note, however, that in physiological concentrations of external magnesium (1 mM), the NMDA component will be largely suppressed at the resting membrane potential. The increase in the late AMPA receptor-mediated component with synaptic maturation will thus result in a gradually prolonged EPSP, with NMDA receptors contributing largely during periods of high afferent activity. At present, *in vivo* recordings of UBCs during behaviors such as head rotation have not been performed, so the normal rates of afferent fiber activation cannot be assessed.

A number of explanations can be advanced to explain the origin of the slow component of the EPSC, such as charge build-up in the proximal dendrites during the initial portion of the synaptic current,[6] or the activation of a slow inward current subsequent to dendritic calcium entry. However, in voltage-jump experiments in which UBCs were initially voltage-clamped at the reversal potential of the EPSC, the time course of the synaptic current following a jump to a hyperpolarized potential produced a family of overlapping current traces (Figure 13.4E). These results would indicate that the time course of the synaptic current thus reflects the time course of activation of the underlying glutamate receptor population rather than some other long-lasting synaptic phenomenon.[58]

The prolonged late component of the synaptic current is, therefore, consistent with the hypothesis that the MF-UBC synapse represents an ultrastructural specialization which serves to entrap glutamate within the tortuous volume of the synaptic cleft following release. At other glutamatergic synapses in the brain, the lifetime of glutamate in the synaptic cleft is believed to be brief, decaying with a time constant of approximately 1.2 ms by rapid diffusion into extracellular space where the concentration of ionotropic glutamate receptors is very low.[13,14,33,34] This brief pulse of glutamate will largely saturate both AMPA and NMDA receptors tethered to the postsynaptic density. A rapidly activating and deactivating AMPA component will be observed, as AMPA receptor-channels open quickly upon binding of glutamate; and following the clearance of glutamate from the cleft, the current will rapidly subside due to both dissociation from these low-affinity receptors and fast desensitization. The NMDA component will rise more slowly, peaking tens of milliseconds after glutamate has already fallen to levels too low to support a significant probability of rebinding, and will decay slowly as a consequence of prolonged binding of glutamate to these high-affinity receptors and some degree of desensitization. The rise-time of the NMDA receptor-mediated EPSC originates from the slow rates of transition of the agonist-bound, closed state of the receptor-channel complex to the open state.[39] Thus, the brief pulse of glutamate that occurs at typical small-diameter glutamatergic synapses will evoke a multi-component EPSC, as in UBCs, but the time course of these components will be dramatically faster than was observed at the MF-UBC synapse.

The very prolonged nature of the glutamatergic synaptic current in UBCs is without precedent in the CNS. A prolongation of the synaptic current due to the impaired diffusional escape of glutamate has been reported under conditions of high release probability, where individual terminals of a single afferent fiber are closely packed together. Examples of this include the parallel fiber-Purkinje cells synapse[6] or the large, calyceal synapses formed between auditory nerve terminals and second-order neurons of the avian nucleus magnocellularis.[49,66] However, the time course of the synaptic current in UBCs is more than several orders of magnitude greater in duration than at other glutamatergic synapses. Of particular note is the greatly prolonged time-to-peak of the slow AMPA component of the EPSC, which may be up to 500 milliseconds.

While the very slow overall time course of the slow component can be generally accounted for by a model in which glutamate becomes entrapped within the cleft following release, the cellular mechanism by which such a slow, monosynaptic AMPA component is produced is intriguing, since AMPA receptors desensitize very rapidly.[65,66] Two hypotheses have been advanced[37,58] to account for this observation. Initially, it was proposed that the late component may arise from the slow diffusion of glutamate from zones of the cleft immediately adjacent to presynaptic release sites to regions of the cleft between release sites where ionotropic receptors are also densely concentrated.

The fast component would thus be produced by the rapid activation and desensitization of AMPA receptors apposed to release sites, and the slow component would arise from the slower build-up of glutamate at distant sites. A test of this hypothesis was performed by the application of cyclothiazide, which blocks desensitization of AMPA receptors.[51,70] If slow diffusion alone accounted for the appearance of a second component, then both phases would be potentiated when desensitization was blocked. However, in the presence of cyclothiazide, the synaptic current, while greatly potentiated, decayed smoothly without evidence of a second component. This indicated that the appearance of a second component was itself dependent on desensitization. This somewhat paradoxical observation was explained in the following way. In some preparations the steady-state dose-response curve to glutamate acting at AMPA receptors is bell-shaped, with a decline in the response amplitude occurring when the concentration of glutamate is increased from 70 mM to 2 mM (Figure 13.6D).[54]

If a similar situation exists for AMPA receptors in UBCs, and the concentration of glutamate equilibrates relatively rapidly at all regions of the cleft following release relative to the slow time course of the EPSC, then a late, slow component would be observed as the cleft concentration gradually declines from an initial peak in the millimolar range down to levels where the steady-state current will first increase, and then decline.

A prediction of this second model is that, since the late component arises as a result of the interplay between desensitization kinetics and receptor occupancy, blockade of desensitization would abolish the late component, which was observed.[36,58] Another prediction is that the delivery of a second MF stimulus during the peak of the late component would transiently restore the cleft glutamate concentration to maximal levels, resulting in a lower steady-state current after an initial fast component, and this has also been demonstrated.[36,37] To further validate this hypothesis, excised patch experiments were conducted that demonstrated that the steady-state dose-response curve of glutamate on UBC membranes is indeed bell-shaped.[36] Further experiments in which the whole-cell steady-state inward current evoked by bath-applied L-glutamate was examined in UBCs also confirmed that this is the case (G.A. Kinney and N.T. Slater, unpublished observations). Neither of these hypotheses are mutually exclusive, and both require a very prolonged presence of glutamate in the synaptic cleft at concentrations high enough to allow significant rebinding of glutamate to AMPA and NMDA receptors. Surprisingly, in the presence of cyclothiazide, a single MF stimulus may evoke an AMPA receptor-mediated current lasting up to 15 seconds. Such a very prolonged activation of this low-affinity receptor suggests that the ultrastructural specialization of the synapse acts as a very effective trap of glutamate, and glutamate transporters may play a significant role in the removal of transmitter, a situation which does not occur at small diameter glutamatergic synapses.[29,59] Indeed, it was estimated that glutamate concentrations in the MF-UBC synaptic cleft decay with a time constant (t) of 800 ms, and extinguish at 5.4 seconds,[36] which is several orders of magnitude slower than the estimated decay rate at conventional glutamatergic synapses ($t = 1.2$ msec).[13] While glutamate transporters are present in the velate, glial periglomerular processes,[12] neuronal transporters have not been identified at the ultrastructural level within the MF-UBC synapse.[16] Inhibition of glutamate transport will prolong the late phase of the decay of the EPSC in UBCs, but this effect is not associated with an inward (reuptake) current in UBCs.[36] These results would indicate that, while glutamate transporters will contribute to the clearance of transmitter from the labyrinthine extracellular space within the glomerulus, this takes place at the astrocytic membrane which envelops the glomerulus, rather than by more rapid binding and eventual re-uptake at the postsynaptic membrane.

Likely sources of glutamatergic inputs are primary vestibular afferents, which have been demonstrated to innervate UBCs in the gerbil nodulus.[20] Furthermore, a recent study demonstrates that a subset of MFs that terminate on UBCs forming synapses with similar ultrastructural features to those described above contain high levels of ChAT immunoreactivity.[30,31] Furthermore, UBCs respond to bath-applied acetylcholine.[58] These data would suggest that a subset of MF-UBC synapses may utilize acetylcholine — alone, or in conjunction with glutamate — as a neurotransmitter.

13.7 UNIPOLAR BRUSH CELLS ARE INTERNEURONS OF THE INTERNAL GRANULAR LAYER

The soma of UBCs gives rise to a branching axon of variable diameter that wanders within the granular layer for distances of up to 200 mm to give rise to 2 to 4 large (4 to 8 mm diameter) rosette-like terminals of presumably presynaptic function. These basic features of the UBC axon and terminal rosettes have been visualized both by rapid Golgi staining methods[9] and by three-dimensional reconstruction of Lucifer Yellow-filled cells using a laser scanning confocal microscope.[57,58] UBCs are likely to be excitatory rather than inhibitory interneurons, as these cells are immunonegative for GABA and glycine and form dendrodendritic synapses with adjacent granule cell dendrites that are Type I asymmetric synapses with attached round vesicles.[22] The neurons that are postsynaptic to the UBC axon in the granular layer may be primarily granule cells, as these are the most numerous elements of this region, and rosette-like terminals make contact with granule cells.[44,50] Contacts between UBC terminals and granule cells have recently been demonstrated by electron microscopic examination of biocytin-filled UBCs in cerebellar slices, confirming this notion.[60] Furthermore, unusually long-lasting EPSCs have been observed in morphologically identified granule cells of the cerebellar nodulus and uvula, the time course of which would be consistent with that of a disynaptic response resulting from the activation of an MF-UBC-granule cell loop (D.J. Rossi, G.A. Kinney, and N.T. Slater, unpublished observations). UBCs are also postsynaptic to Golgi cells,[46,58] and so will receive feedback inhibition from these neurons. UBCs thus represent a powerful source of feedforward excitation within the granular layer of the vestibulo-cerebellum. Their localization within this region of the cerebellum would suggest that they may play an important role in certain aspects of vestibular and postural control. Moreover, the dense concentration of mGluRs and the calcium-binding protein, calretinin, in nonsynaptic appendages of the dendritic brush[22, 32] would indicate that these receptors may play a role in gating short- or long-term plasticity of transmission, as at other cerebellar synapses.[35,40] However, many outstanding questions remain with regard to the functional contribution of UBCs to motor control. The observation that UBCs are innervated by primary vestibular afferents is important in this regard. These fibers will fire at high frequencies during head rotation, and with repetitive activation the EPSPs in UBCs fuse to form a plateau potential.[62] This prolonged firing will be distributed to a large ensemble of granule cells and result in temporal summation and prolonged firing of these cells. Granule cells that are postsynaptic to UBCs will display a skewed input-output pattern, as they will continue to fire after the stimulus is withdrawn, producing a phase lag.

As yet, the sources of afferent MFs to UBCs are only partly established, and the properties of transmission to their postsynaptic targets have not been explored. Thus, despite the wealth of information that has been gathered regarding the ultrastructure and physiology of the MF-UBC synapse, the contribution made by UBCs to cerebellar control of motor coordination remains to be clarified. The unique synaptic ultrastructure and physiology make the UBCs an intriguing enigma in the brain, which represent an important addition to the mechanisms of synaptic integration within the cerebellum.

REFERENCES

1. Abbott, LC, and Jacobowitz, DM, Development of calretinin-immunoreactive unipolar brush-like cells and an afferent pathway to the embryonic and early postnatal mouse cerebellum, *Anat. Embryol.,* 191: 541,1995.
2. Altman, J, and Bayer, SA, Time of origin and distribution of a new cell type in the rat cerebellar cortex, *Exp. Brain Res.,* 29: 265,1977.
3. Anelli, R, Kettner, RE, and Mugnaini, E, Unipolar brush cells are present in cerebellar granule cell cultures, *Soc. Neurosci. Abstr.,* 22: 640,1996.
4. Aoki, E, Semba, R, and Kashiwamata, S, New candidates for GABAergic neurons in the rat cerebellum: An immunocytochemical study with anti-GABA antibody, *Neurosci. Letters.,* 68: 267,1986.

5. Arai, R, Winsky, L, Arai, M, and Jacobowitz, DM, Immunohistochemical localization of calretinin in the rat hindbrain, *J. Comp. Neurol.*, 310: 21,1991.

6. Barbour, B, Keller, BU, Llano, I, and Marty, A, Prolonged presence of glutamate during excitatory synaptic transmission to cerebellar Purkinje cells, *Neuron.*, 12: 1331,1994.

7. Barmack, NH, Baughman, RW, and Eckenstein, FP, Cholinergic innervation of the cerebellum of rat, rabbit, cat, and monkey as revealed by choline acetyltransferase activity and immunohistochemistry, *J. Comp. Neurol.*, 317: 233,1992a.

8. Barmack, NH, Baughman, RW, Eckenstein, FP, and Shojaku, H, Secondary vestibular cholinergic projection to the cerebellum of rabbit and rat as revealed by choline acetyltransferase immunohistochemistry, retrograde and orthograde tracers, *J. Comp. Neurol.*, 317: 250,1992b.

9. Berthié, B, and Axelrad, H Granular layer collaterals of the unipolar brush cell axon display rosette-like excrescences. A Golgi study in the rat cerebellar cortex, *Neurosci. Lett.*, 167: 161,1994.

10. Braak, E, and Braak, H, The new monodendritic neuronal type within the adult human cerebellar granule cell layer shows calretinin-immunoreacitivity, *Neurosci. Lett.*, 154: 199,1993.

11. Chan-Palay, V, and Palay, SL, The synapse en marron between Golgi type II neurons and mossy fibers in the rat's cerebellar cortex, *Z. Anat. Entwicklungsgeschichte*, 133: 274,1971.

12. Chaudhry, FA, Lehre, KP, Van Lookeren, Campagne, M, Ottersen, OP, Danbolt, NC, AND Storm-Mathisen, J, Glutamate transporters in glial plasma membranes: highly differentiated localizations revealed by quantitative ultrastructural immunocytochemistry, *Neuron.*, 15: 711,1995.

13. Clements, JD, Transmitter timecourse in the synaptic cleft: its role in central synaptic function, *Trends Neurosci.*, 19: 163,1996.

14. Clements, JD, Lester, RAJ, Tong, G, Jahr, CE, and Westbrook, GL, The time course of glutamate in the synaptic cleft, *Science*, 258: 1498,1992.

15. Cozzi, MG, Rosa, P, Greco, A, Hille, A, Huttner, WB, Zanini, A, De and Camilli, P, Immunohistochemical localization of secretogranin II in the rat cerebellum, *Neuroscience*, 28: 423,1989.

16. Danbolt, NC, The high affinity uptake system for excitatory amino acids in the brain, *Progress in Neurobiology*, 44: 377,1994.

17. De Camilli, P, and Jahn, R, Pathways to regulated exocytosis in neurons, *Ann. Rev. Physiol.*, 52: 625,1990.

18. Diño, MR, and and Mugnaini, E, Calretinin immunoreactive unipolar brush cells form parasagittal bands in the rabbit cerebellum, *Soc. Neurosci. Abstr.*, 25: 467,1995.

19. Diño, MR, Sekerkova, G, Cunha, SR, Binder, L, and Mugnaini, E, The cytoskeleton of unipolar brush cells of the mammalian cerebellum, *Soc. Neurosci. Abstr.*, 22:640.13,1996.

20. Diño, MR, Sekerkova, G, Perachio, AA, and Mugnaini, E, Unipolar brush cells are targets of primary vestibular fibers, *Soc. Neurosci. Abstr.*, 23:712.14,1997.

21. Floris, A, Dunn, ME, Berrebi, AS, Jacobowitz, DM, and Mugnaini, E, Pale cells of the flocculonodular lobe are calretinin positive, *Soc. Neurosci. Abstr.*, 18: 853,1992.

22. Floris, A, Diño, M, Jacobowitz, DM, and Mugnaini, E, The unipolar brush cells of the rat cerebellar cortex and cochlear nucleus are calretinin positive: a study by light and electron immunocytochemistry, *Anat. Embryol.*, 189: 495,1994.

23. Garthwaite, J, and Brodtbelt, AR, Glutamate as the principal mossy fibre transmitter in rat cerebellum: pharmacological evidence, *Eur. J Neurosci.*, 2: 177,1989.

24. Gerrits, NM, Epema, AH, van Linge, A, and Dalm, E, The primary vestibulocerebellar projection in the rabbit: absence of primary afferents in the flocculus, *Neurosci. Lett.*, 105: 27,1989.

25. Hámori, J, and Szentágothai, J, Participation of Golgi neuron processes in the cerebellar glomeruli: An electron microscopic study, *Exp. Brain Res.*, 2: 65,1966.

26. Harris, J, Moreno, S, Shaw, G, and Mugnaini, E, Unusual neurofilament composition in cerebellar unipolar brush neurons, *J. Neurocytol.*, 22: 663,1993.

27. Hockfield, S, A Mab to a unique cerebellar neuron generated by immunosuppression and rapid immunization, *Science*, 237: 67,1987.

28. Hutson, KA, and Morest, DK Fine structure of the cell clusters in the cochlear nerve root: stellate, granule and mitt cells offer insights into the synaptic organization of local circuit neurons, *J. Comp. Neurol.*, 371: 397,1996.

29. Isaacson, JS, and Nicoll, RA, The uptake inhibitor L-*trans*-PDC enhances responses to glutamate but fails to alter the kinetics of excitatory synaptic currents in the hippocampus, *J. Neurophysiol.*, 70: 2187,1993.

30. Jaarsma, D, Wenthold, RJ, and Mugnaini, E Glutamate receptor subunits at mossy fiber-unipolar brush cell synapses: light and electron microscopic immunocytochemical study in cerebellar cortex of rat and cat, *J. Comp. Neurol.*, 357: 145,1995.

31. Jaarsma, D, Diño, MR, Cozzari, C, and Mugnaini, E, Cholinergic mossy fibers and their granule cells and unipolar brush cells in rat cerebellar nodulus: a model for central nicotinic neurotransmission, *J. Neurocytol.*, 25: 829,1996.

32. Jaarsma, D, Diño, MR, Ohishi, H, Shigemoto, R, and Mugnaini, E, Metabotropic glutamate receptors of unipolar brush cells are primarily associated with non-synaptic appendages in rat cerebellar cortex and cochlear nuclear complex, *J. Neurocytol.*, 27:303,1998.

33. Jonas, P, Major, G, and Sakmann, B, Quantal components of unitary EPSCs at the mossy fibre synapse on CA3 pyramidal cells of rat hippocampus, *J. Physiol.*, 472: 615,1993.

34. Jonas, P, and Spruston, N, Mechanisms shaping glutamate-mediated excitatory postsynaptic currents in the CNS, *Curr. Opin. Neurobiol.*, 4: 366,1994.

35. Kinney, GA, and Slater, NT, Potentiation of NMDA receptor-mediated transmission in turtle cerebellar granule cells by activation of metabotropic glutamate receptors, *J. Neurophysiol.*, 69: 585,1993.

36. Kinney, GA, Overstreet, LS, and Slater, NT, Prolonged physiological entrapment of glutamate in the synaptic cleft of cerebellar unipolar brush cells, *J. Neurophysiol.*, 1997,78:1320,1997.

37. Kinney, GA, Rossi, DJ, and Slater, NT, Desensitization sculpts the time course of biphasic AMPA receptor-mediated synaptic currents in rat cerebellar unipolar brush cells, *Soc. Neurosci. Abstr.*, 21: 586,1995.

38. Korte, G, and Mugnaini, E, The cerebellar projection of the vestibular nerve in the cat, *J. Comp. Neurol.*, 184: 265,1979.

39. Lester, RAJ, Clements, JD, Westbrook, GL, and Jahr, CE, Channel kinetics determine the time course of NMDA receptor-mediated synaptic currents, *Nature*, 346: 565,1990.

40. Linden, DJ, Long-term synaptic depression in the mammalian brain, *Neuron*, 12: 547,1994

42. Marini, AM, Strauss, KI, and Jacobowitz, DM, Calretinin-containing neurons in rat cerebellar granule cell cultures, *Brain Res, Bull.*, 42: 279,1997.

43. Monteiro, RAF, Critical analysis on the nature of synapses en marron of the cerebellar cortex, *J. Hirnforschung*, 27: 567,1986.

44. Mugnaini, E, The histology and cytology of the cerebellar cortex. In Larsell O, Jansen J, Eds., *The Comparative Anatomy and Histology of the Cerebellum. The Human Cerebellum, Cerebellar Connections, and Cerebellar Cortex*. Minneapolis: The University of Minnesota Press, p. 201,1972.

45. Mugnaini, E, and Floris, A, The unipolar brush cell: a neglected neuron of the mammalian cerebellar cortex, *J. Comp. Neurol.*, 339: 174,1994.

46. Mugnaini, E, Floris, A, and Wright-Gross, M, The extraordinary synapses of the unipolar brush cell: An electron microscopic study in the rat cerebellum, *Synapse*, 16: 284,1994.

47. Munoz, DG, Monodendritic neurons: A cell type in the human cerebellar cortex identified by chromogranin A-like immunoreactivity, *Brain Res.*, 528: 335,1990.

48. Ojima, H, Kawajiri, S, and Yamasaki, T, Cholinergic innervation of the rat cerebellum: qualitative and quantitative analyses of elements immunoreactive to a monoclonal antibody against choline acetyltransferase, *J. Comp. Neurol.*, 290: 41,1989.

49. Otis, TS, Wu, Y-C, and Trussell, LO, Delayed clearance of transmitter and the role of glutamate transporters at synapses with multiple release sites, *J. Neurosci.*, 16: 1634,1996.

50. Palay, SL, Chan-Palay, V, *Cerebellar Cortex: Cytology and Organization*. New York: Springer-Verlag.,1974.

51. Partin, KM, Patneau, DK, Winters, CA, Mayer, ML, and Buonanno, A, Selective modulation of desensitization at AMPA versus kainate receptors by cyclothiazide and concanavalin, A. *Neuron*,11:1069,1993.

52. Petralia, RS, and Wenthold, RJ, Light and electron immunocytochemical localization of AMPA-selective glutamate receptors in the rat brain, *J. Comp. Neurol.*, 318: 329,1995.

53. Petralia, RS, and Wang, Y-X, Zhao H-M and Wenthold RJ Ionotropic and metabotropic glutamate receptors show unique postsynaptic, presynaptic, and glial localizations in the dorsal cochlear nucleus, *J. Comp. Neurol.*, 372:356,1996.

54. Raman, IM, and Trussell, LO, The kinetics of the response to glutamate and kainate in neurons of the avian cochlear nucleus, *Neuron*, 9:173,1993.

55. Résibois, A, and Rogers, JH, Calretinin in rat brain: an immunohistochemical study, *Neuroscience*, 46: 101,1992.

56. Rogers, JH, Immunoreactivity for calretinin and other calcium-binding proteins in cerebellum, *Neuroscience*, 31: 711,1989.

57. Rossi, DJ, Mugnaini, E, and Slater, NT, Novel time course of transmission at a giant glutamate synapse in rat vestibular cerebellum, *Soc Neurosci. Abstr.*, 20: 1508,1994.

58. Rossi DJ, Alford S, Mugnaini E, Slater NT Properties of transmission at a giant glutamatergic synapse in cerebellum. The mossy fiber-unipolar brush cell synapse, *J. Neurophysiol.*, 74: 24,1995.

59. Sarantis, M, Ballerini, L, Miller, B, Silver, RA, Edwards, M, and Attwell, D, Glutamate uptake from the synaptic cleft does not shape the decay of the non-NMDA component of the synaptic current, *Neuron*, 11: 541,1993.

60. Schuerger, RJ, Diño, MR, Liu, Y-B, Slater, NT, and Mugnaini, E,. Light and electron micrscopic identification of the axon terminals and postsynaptic targets of cerebellar unipolar brush cells, *Soc. Neurosci. Abstr.*, 23: in press,1977.

61. Sekerkova, G, and Mugnaini, E, Prenatal neurogenesis of cerebellar unipolar brush cells studied by bromodeoxyuridine and cell class specific markers, *Soc. Neurosci. Abstr.*, in press, 1977.

62. Slater, NT, Rossi, DJ, and Kinney, GA, Physiology of transmission at a giant glutamatergic synapse in cerebellum, *Prog. Brain Res.*, 114: 151,1997.

63. Somogyi, P, Halasy, K, Somogyi, J, Storm-Mathisen, J, and Ottersen, OP, Quantitation of immunogold labelling reveals enrichment of glutamate in mossy and parallel fibre terminals in cat cerebellum, *Neuroscience*, 19: 1045,1986.

64. Sturrock, RR, A quantitative histological study of Golgi II neurons and pale cells in different cerebellar regions of the adult and aging mouse brain, *Z. Mikrosk. Anat. Forsch. Leipzig.*, *104*: 705,1990.

65. Tang, C-M, Dichter, M, and Morad, M, Quisqualate activates a rapidly inactivating high conductance ionic channel in hippocampal neurons, *Science*, 243: 1474,1989.

66. Trussell, LO, and Fischbach, GD, Glutamate receptor desensitization and its role in synaptic transmission, *Neuron*, 3: 209,1989.

67. Trussell, LO, Zhang, S, and Raman, IM, Desensitization of AMPA receptors upon multiquantal neurotransmitter release, *Neuron*, 10:1185,1993.

68. Weedman, DL, Pongstaporn, T, and Ryugo, DK, Ultrastructural study of the granule cell domain of the cochlear nucleus in rats: mossy fiber endings and their targets, *J. Comp. Neurol.*, 369: 345,1996.

69. Wright, DD, Blackstone, CD, Huganir, RL, and Ryugo, DK, Immunocytochemical localization of the mGluRa metabotropic glutamate receptor in the dorsal cochlear nucleus, *J. Comp. Neurol.*, 364: 729,1996.

70. Yamada, K, and Tang, C-M, Benzothiadiazines inhibit rapid glutamate receptor desensitization and enhance glutamatergic synaptic currents, *J. Neurosci.*, 13: 3904,1993.

Section V

Neurochemical Organization of Sensory Inputs to the Vestibular System

14

Neuronal Circuitry and Neurotransmitters in the Pretectal and Accessory Optic Systems

Robert H.I. Blanks, Roland A. Giolli, and Johannes J.L. van der Want

CONTENTS

14.1. INTRODUCTION

This chapter is concerned with the neuronal circuitry and neurotransmitters of two vestibular/eye movement-related parts of the primary visual pathway: the nucleus of the optic tract (NOT), a part of the pretectal nuclear complex, and the terminal nuclei of the accessory optic system (AOS). Each of these nuclei receive retinal input predominantly or exclusively from the contralateral eye,[59,73,151] and nonretinal afferents from visual and oculomotor-related brainstem nuclei,[62] striate

0-8493-7679-3/00/$0.00+$.50
© 2000 by CRC Press LLC

cortex,[60,83,107,132] and regions of superior temporal, parieto-occipital, and frontal cortices.[83,108] In turn, the NOT and AOS neurons project heavily upon preoculomotor, precerebellar, and brainstem reticular formation nuclei[17,55,62,63,155] functioning in gaze control and visual-vestibular interaction.[10,15,36,64,65,81,106,112,128,131,139,159-161,172,186,192,193]

This chapter focuses on data obtained primarily from the NOT and AOS nuclei of mammals. Readers interested in the complete reviews of the pretectal complex or the AOS systems of nonmammalian species should consult other, more specific reviews.[24,26,49,65,115,159,186]

14.2 NEURONAL CIRCUITRY

14.2.1 NUCLEAR COMPONENTS

The pretectal nuclear complex of mammals consists of several different nuclei situated between the superior colliculus and the dorsal thalamus. For the purpose of gaze stabilization and visual-vestibular interaction, the most important of these is the NOT. Other pretectal nuclei include the anterior pretectal nucleus (APN), posterior pretectal nucleus (PPN), olivary pretectal nucleus (OPN), and suprageniculate pretectal nucleus.[151] The AOS of mammals consists of the classic dorsal, lateral, and medial terminal nuclei (DTN, LTN, MTN)[73] together with two related cell populations: one intercalated among the fibers of the superior fasciculus, posterior fibers[73] designated the nucleus of the superior fasciculus, posterior fibers (inSFp),[55] and another within the ventral midbrain tegmentum termed the "visual tegmental relay zone."[57] In nonprimate mammalian species (e.g., rat, rabbit, cat), the MTN is by far the largest of the AOS nuclei; whereas in nonhuman primates, the LTN is the predominate nucleus and the MTN is attenuated or absent.[17,189]

14.2.2 Efferent Projections of the Pretectal Nuclear Complex

The efferent projections of the mammalian NOT are well-established thanks, in large measure, to the developments of reliable methods of tract-tracing utilizing anterograde and retrograde axonal transport of molecules such as ^{3}H-amino acids or nucleosides, horseradish peroxidase, and *Phaseolus vulgaris-leucoagglutinin*. These projections have to date been investigated in a variety of species (*e.g., rat,*[27,98,120,171-173] *rabbit,*[86,111,120,170] *cat,*[2,38,39,93,100,114,130,144,184,189] *nonhuman primates*[11,12,48,72,129]). These studies demonstrate that the NOT sends major projections to the contralateral NOT and the AOS nuclei, as described in detail below. There are also projections to visual/oculomotor areas (inferior olive, prepositus hypoglossi, medial vestibular nucleus), NOT and AOS nuclei (LTN, MTN), nucleus reticularis tegmenti pontis, pontine gray (dorsolateral nucleus of the pons), and other retino-recipient zones (e.g., dorsal lateral geniculate nucleus (LGNd), superior colliculus, the pregeniculate nucleus, and zona incerta).

There is now overwhelming evidence implicating the NOT as the first critical relay of visual information from the retina to the vestibular and other brainstem nuclei related to horizontal optokinetic nystagmus (OKN). NOT neurons are speed- and direction-selective to whole-field visual motion in the horizontal direction.[32,37,82,84,153,161] Lesions of the NOT abolish horizontal OKN directed to the side of the lesion, while electrical stimulation of the NOT elicits horizontal OKN.[92,153,154]

The NOT and other parts of the horizontal optokinetic pathways in nonhuman primates are summarized in Figure 14.1.

Fibers from the contralateral retina target the NOT and AOS nuclei. From the NOT, visual signals pass to (1) the dorsal cap of Kooy of the inferior olive, (2) the nucleus prepositus hypoglossi and the medial vestibular nucleus, and (3) the dorsolateral nucleus of the pons.

The insert in Figure 14.1 shows the two components of horizontal OKN, that is, the slow build-up and decay of slow-phase eye velocity, and a rapid rise and fall in slow-phase velocity at the onset and termination of stimulus. It has been suggested that the first of these ("indirect path") may be related to activity in the brainstem velocity storage network; the second is suggested to be related

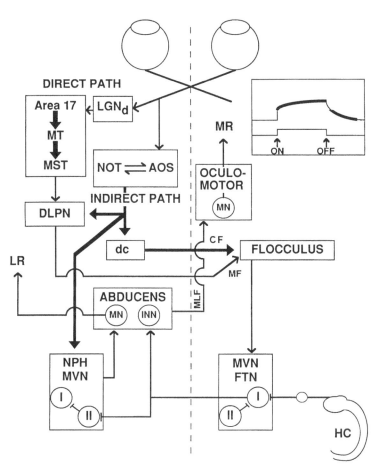

FIGURE 14.1 Summary of proposed pathways for horizontal optokinetic nystagmus (HOKN) in the nonhuman primate based on experimental data from nonhuman primates (redrawn from Mustari et al. [129]). The insert gives a typical HOKN slow-phase eye velocity profile with several important components. The initial fast rise and fall of HOKN velocity is governed by the so-called "direct path;" the subsequent slow build-up in slow-phase velocity associated with velocity storage and optokinetic afternystagmus, which outlasts the stimulus (thick lines in insert), is driven by the "indirect path." The nuclei associated with the direct path pass from the LGNd to the visual cortical areas and pontocerebellar pathways to the flocculus; the indirect path arises from the NOT and passes to the dorsal cap of the inferior olive (dc) from whence climbing fibers pass to the flocculus, and directly to the VOR-related neurons in the medial vestibular nucleus and nucleus prepositus hypoglossi.

to a pathway ("direct path") for smooth pursuit.[145] Although this distinction has been helpful in understanding the general connections in the horizontal optokinetic pathways, the reader should be aware of the differences and inconsistencies between data from the nonhuman primate (illustrated in Figure 14.1) and rat, where the NRTP is thought to be a relay of the visual signals from the NOT to the vestibular nuclei.[16,17,32,129] Even in the nonhuman primate, lesions of the NOT affect both the slow build-up and rapid rise in eye velocity during OKN,[153] indicating that the two pathways are not entirely separate and distinct.

The most widely studied of the pretectal ("indirect path") projections is the NOT/DTN pathway to the dorsal cap of the inferior olive from whence neurons project to the contralateral cerebellar flocculus (Figure 14.1).[87,118,160] This pathway in the rabbit has been implicated as carrying a visual retinal slip signal to the flocculus.[160] The second part of the NOT ("indirect path") pathway targets the ipsilateral prepositus hypoglossi and medial vestibular nuclei[129] via fibers descending, in part, within the medial longitudinal fasciculus.[98] These connections through the prepositus hypoglossi and vestibular nuclei may constitute the most important NOT link to eye movements because it is the most direct connection between the visual pathways and vestibulo-ocular neurons. The visual activation of vestibular nuclear neurons[183] and optokinetic afternystagmus (OKAN) are thought to be related to activity in this pathway.[52] Finally, there are important NOT connections with the dorsolateral nucleus of the pons targeting pontocerebellar neurons which also receive input from visual cortical Areas 17, medial-temporal and medial-superior temporal gyrus (for references, see above), and projecting to the contralateral flocculus. This is the so-called "direct pathway" related to smooth pursuit eye movements,[94,167] and controlling the rapid rise and fall of slow-phase eye movement velocity at the onset and termination of stimuli.[129,145]

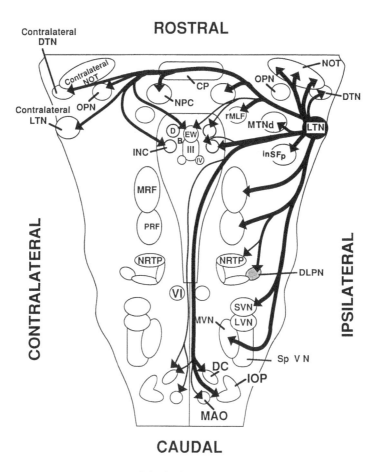

FIGURE 14.2 Schematic representation of the brainstem, summarizing the projections of the LTN in the marmoset monkey as demonstrated by [³H]-leucine light autoradiography.[17] The plane of the section depicted is approximately horizontal, but the nuclei are drawn in the transverse plane. Arrows depict the LTN projections; arrow size is in proportion to relative strength of projection.

14.2.3 Efferent Projections of the AOS nuclei

Since the first demonstrated efferent connection of the AOS to the inferior olive in rabbit,[112] there have been several anatomical surveys on the overall outflow of these nuclei (pigeon,[21,24] rat and rabbit,[18,55,62,63] nonhuman primate[7,17,113,128]). Across species, the AOS projections are remarkably similar to those of the NOT in that they target many of the same preoculomotor, precerebellar, and brainstem reticular formation nuclei. Figure 14.2 summarizes the overall projections of the largest nuclei of the AOS in the nonhuman primate, the LTN.[17]

The relative strength of each projection is given by the thickness of each line. As with the other species studied (pigeon, rat, rabbit), the largest projection of the AOS of the marmoset monkey is to the pretectal and AOS nuclei on the side of the injection; smaller contralateral projections are to the olivary pretectal nucleus and two of the AOS nuclei (DTN, LTN). Additional, moderate-to-heavy projections are to the ipsilateral accessory oculomotor nuclei (nucleus of posterior commissure, interstitial nucleus of Cajal, nucleus of Darkschewitsch) and nucleus of Bechterew, ipsilateral pontine and mesencephalic reticular formations, nucleus reticularis tegmenti pontis and basilar pontine complex (dorsolateral nucleus only), and a part of the ventral thalamus corresponding to the rostral interstitial nucleus of the medial longitudinal fasciculus. Last, there are two long, descending bundles targeting the dorsal cap and medial accessory olive of the inferior olivary complex, and the ipsilateral and contralateral and contralateral medial and superior vestibular nucleus.

Neurons of the AOS nuclei are speed- and direction-selective to slow, whole-field motion of the visual surround but respond preferentially to vertical motion of the visual surround.[10,64,106,159,160,185] In contrast to very compelling data on the NOT/dorsal terminal accessory optic nucleus (DTN) in serving as the first relay station in horizontal optokinetic nystagmus, the effects of lesions and/or electrical stimulation of the MTN[35,131] are conflicting and inconclusive, leaving the pathways related to vertical eye movements unsettled.[17,129] However, the AOS-inferior olivary pathway is likely responsible for mediating retinal slip signals to neurons in the rostral parts of the dorsal cap of the inferior olive, whereas the NOT projects to the caudal parts.[160] The AOS-prepositus and vestibular nuclear projection may serve, along with the NOT-vestibular projections, to mediate visual signals to vestibular nuclear neurons.[17]

14.3 NEURONAL MORPHOLOGY, SYNAPTOLOGY, AND TRANSMITTERS

14.3.1 Projection Neurons of NOT and AOS Nuclei

14.3.1.1 Neuronal Types and Synaptic Organization

Several types of neurons have been described in the pretectal nuclei of mammals.[91,136,152] One type is large and multipolar; these are generally located superficially in the nucleus with dendrites extending well into the overlying optic tract.[66] A more common, second type of small-to-medium neuron is distributed throughout the nuclei. Rostrally located NOT neurons project to the inferior olive; these show considerable variability in their dendritic pattern.[135]

GABAergic and non-GABAergic neurons are scattered throughout the NOT with no specific topography.[136,178] Typically, a GABAergic neuron has three to seven dendrites arising from soma (10 to 25 um diameter) and branching profusely throughout the NOT and adjacent nuclei. These extensively branching dendrites form a strongly interconnecting system linking the pretectal nuclei and, importantly, certain AOS nuclei.[135] The neuropil of the NOT and DTN is so interconnected that Gregory[66] considers these two as one functional entity designated as NOT/DTN. Finally, passing through the dendritic arborizations of the NOT/DTN are the perpendicularly arranged optic tract fibers. These arise from ganglion cells in the contralateral retina and terminate on the dendrites and somata of NOT and AOS neurons.

The synaptic organization of the NOT shares many of the same characteristics seen in the neuropil of the superior colliculus and LGNd, that is, four types of presynaptic terminals: two non-GABAergic (R-type, RLD-type) and two GABAergic (F-type, P-type). The R-type terminals are of retinal origin and form asymmetric contacts with NOT neurons that are either large and scalloped, or small and regular in outline; they contain spherical vesicles and electron-lucent mitochondria. Retinal terminals (R-type) in the pretectal and AOS nuclei in the rabbit appear to be glutamate-positive and GABA-negative.[133] There is evidence that small numbers may also contain substance P.[23] RLD-type terminals, the origin of which remains uncertain (but do not arise from the retina or cerebral cortex), have spherical vesicles, electron-dense mitochondria, and asymmetric synaptic contacts.

Within the AOS, it is estimated that the retinofugal terminals account for about 40% of the total axonal endings in the mouse MTN.[103] Unfortunately, it is not known whether the retinal terminals target GABAergic and/or non-GABAergic neurons of the AOS. This important information must be obtained in future experiments.

The two types of GABAergic terminals, F- and P- terminals, have contrasting morphology. F-type terminals possess flattened vesicles, opaque mitochondria, no ribosomes, symmetric contact zones, and show structural features that resemble axon terminals.[125] Several sources of afferents to the NOT/DTN could provide these GABAergic terminals, including the contralateral NOT or the ipsilateral MTN, LGNd, superior colliculus, or other pretectal/AOS nuclei (for references, see above). Proof of such connections requires combined retrograde axonal transport and GABA-immunocytochemistry. To date, there is only firm immunocytochemical support for the presence of GABAergic neuronal projection from the MTN to the NOT[58,175] and, reciprocally, from the NOT/DTN to the MTN.[177] Each of these projections terminates as F-type terminals on both GABAergic and non-GABAergic neurons of the NOT/DTN and MTN neurons.[175,177] Finally, F-type terminals can also receive retinal contacts, that is, F-type terminals can be postsynaptic to R-type terminals.

The second type of GABA-positive terminal is referred to as the P-type terminal. These are of dendritic origin and have pleomorphic vesicles, electron-lucent or opaque mitochondria, and asymmetric synaptic thickenings.[135] There are two types of neurons from which the P-type terminals may arise — GABAergic projection neurons and GABAergic local circuit neurons — but their distinct origin is undetermined.[135]

The main difference between the NOT and other visual relay nuclei, that is, LGNd, is that the NOT contains a relatively small number of F-type terminals, a paucity of triadic arrangements, and relatively small-sized R-type terminals. Nunes, Cardozo, and van der Want[135] argue that these differences may be related to retinal W-type ganglion cells which form the main retinal input to the NOT neurons of the rabbit, and which could be related to the identified direction-selective responses of units within the NOT. The direction-selective ganglion cells in the rabbit retina provide the error signal for stabilization of retinal images.[138] Similarly, in the cat, Stone and Fukuda[166] and Hoffmann and Stone[85] suggest that as few as 1140 to 2000 on-center, direction-selective retinal ganglion cells provide the major retinal input to the NOT.

14.3.1.2 NOT-Inferior Olive Projections

In the rat, rabbit, cat, and monkey, WGA-HRP injected into the IO-labeled neurons of the pretectal nuclei that are exclusively non-GABAergic.[87,111,118,181] The projection neurons to the inferior olive (IO) are segregated so that the NOT projects to the caudal part of the dorsal cap, whereas the AOS nuclei (MTN, LTN, ventral inSFp) project to the rostral of the dorsal cap.[17,160] Consistent with this topography, Barmack and Young[8] have shown that levels of corticotropin-releasing factor mRNA are increased in the caudal part of the dorsal cap, resulting from optokinetic stimulation in the horizontal plane which differentially activates the NOT.

14.3.1.3 Cerebral Cortical Afferents

Some authors suggest that the cerebral cortex projects to the pretectal complex and AOS nuclei only in more advanced mammalian species (e.g., carnivores and primates), whereas corticofugal is absent in other, "less-advanced" classes like the rodents and lagomorphs.[49,115,159,160,188] Whereas this may still be true of the AOS nuclei in the rat and rabbit,[60] recent evidence suggests that it is not true for the rodent NOT. Thus, portions of Areas 17 and 18 of the rat visual cortex have been recently shown to project upon the ipsilateral NOT/DTN,[157] and anterior pretectal nucleus[28] and Area 17 of the guinea pig has been found to project, in a loose topographical manner, onto the ipsilateral NOT and portions of adjacent anterior and posterior pretectal nuclei.[108] A recent, detailed study of projections of certain visual cortical areas to both pretectal and AOS nuclei in the macaque monkey shows complex patterns of neural connection.[108] The NOT receives input from medial prestriate cortex, Area 19, and superior temporal sulcus [upper bank (OAa) and lower bank (PGa)]; in contrast, the AOS nuclei receive projections exclusively from OAa and PGa.[108]

There are currently a number of important physiologic studies that deal with cortical control of the pretectal nuclei. After removal of the cortex, OKN remains, but it is weak and asymmetrical when elicited monocularly;[65,81,83,195] the relatively unchanged temporal-nasal component of monocular OKN, but loss of the temporal-nasal component, is taken as evidence that the visual cortex contributes to the nasal-temporal component of OKN, whereas the temporal-nasal component is derived from the retina. Molotchnikoff et al.[123] have shown that a disruption of corticofugal control of pretectal nuclei through cryoblockade of the visual cortex results in a profound decline in pretectal nerve discharge, implying that cortical-pretectal fibers are excitatory. Biochemical evidence for glutamate in these terminals is consistent with this interpretation.[51] Moreover, in an interesting study on the cat, Kitao and coworkers[96] demonstrated a cortico-pretectal-olivary pathway that extended from the motor cortex through the anterior pretectal nucleus to the inferior olivary nucleus. The presence of this latter pathway needs to be confirmed. Finally, Natal and Britto[131] reported that the rat cerebral cortex has a significant role in modulating the directional selectivity of MTN units in studies on the responses of near-complete decortication on MTN neurons to visual motion.

14.3.1.4 Other Pretectal Projections

Recent tract-tracing studies in the cat have demonstrated GABA-immunoreactive neurons projecting from the APN, NOT, and PPN to the superior colliculus.[2] Electron microscopic studies, combined with retrograde transport of WGA-HRP, now show conclusively that the NOT-superior colliculus projections are distributed to the superficial and deeper layers of the superior colliculus in rat.[134]

There are conflicting data on GABAergic projection from the pretectal nuclei to the LGNd. In this regard, Cucchiaro et al.[38] conclude that as many as 45% of the retrogradely labeled pretectal neurons from the cat LGNd are GABAergic. In contrast, Nabors and Mize[130] found that the cat pretecto-LGNd neurons are strictly non-GABAergic. Additional studies are required to resolve this discrepancy.

These conflicting results point to the usual difficulties in interpreting immunocytochemical data across studies. Differences between studies can arise because of differences in (1) sensitivities for detecting retrograde tracer, (2) the size of the injection site and the potential for uptake by passing fibers, (3) the sensitivity and specificity of the antibody used to identify GABA, and/or (4) the definition of boundaries between the nuclei under investigation. The latter is very important, given the nondistinct borders of the pretectal nuclei, including the NOT/DTN.[66,108,160]

14.3.2 Interneurons of the NOT and AOS Nuclei

Local circuit neurons (interneurons) are found among the projection neurons of the NOT and AOS nuclei. Local circuit neurons are assumed to have specific morphology[20,29,70,104] and show distinctive immunocytochemistry.[50,63,87,125,126,127,125,137] In the NOT/DTN, they are generally considered inhibitory and entirely GABAergic.[179] Using a light microscopy-GAD immunocytochemical technique, Giolli et al.[63] observed comparatively high numbers of GABAergic neurons in the NOT/DTN and AOS of the rat and gerbil; depending on the nuclei, the number of these neurons ranged from 21% to 72%. The GAD-positive neurons are statistically smaller than the mean size of the total population, and they are more spherical, but otherwise indistinguishable from the general population. Similarly, Horn and Hoffmann[87] showed that the GABAergic neurons of the NOT/DTN in the cat, rat, and monkey are significantly smaller than non-GABAergic projection neurons retrogradely labeled by injection of tracer into the inferior olive. However, local circuit neurons cannot be identified by size and/or GABA-positive nature because there are many medium and large GABAergic neurons in these nuclei (e.g., neurons of the MTN-NOT, NOT-MTN, and part of the NOT-superior colliculus pathway).[2,38,50,58,63,133,175]

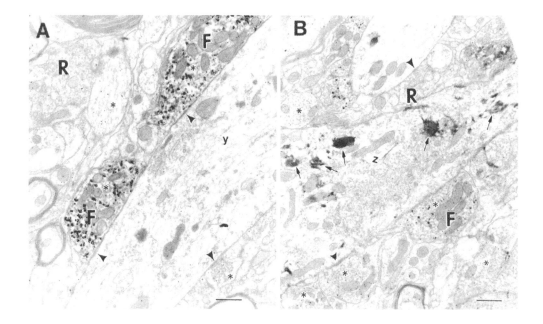

FIGURE 14.3 (A,B) Electron micrographs (EM) showing two adjacent sections through the rat MTN following triple-label experiments: *first,* the anterograde label phaseolus vulgaris leucoaggltuinin (PHAL) was injected into the NOT to demonstrate the NOT-MTN projection shown in Figure (A) as PHAL-labeled terminals; *second,* the retrograde label cholera toxin B horseradish peroxidase (CT$_b$ -HRP) was injected into the dorsal cap of the inferior olive to retrogradely label the MTN-inferior olive projecting neurons (B) , and *third,* the sections were processed immunocytochemically using a postembedding, immunogold protocol to demonstrate GABAergic neurons at the EM level (A and B). Retrograde and anterograde labels have been visualized with a preembedding protocol, which results in the appearance of electron-dense irregular precipitate. The following experimental features should be noted: (1) the dendritic process, passing from upper right to lower left in both A and B, contains CH$_b$-HRP (B; see large, irregularly shaped profiles marked with arrows) indicating that it belongs to a MTN neuron projecting to the ipsilateral inferior olive. (2) the PHAL-labeled axonal terminals contacting this dendrite (A; marked F and containing dark-staining, clumped material) indicate that the same MTN neurons receive projections from the NOT; (3) immunogold labeling is seen as small 15 nm particles distributed within the NOT-MTN terminals, indicating that the NOT-MTN projection is GABAergic (forming F-type terminals marked F). The MTN output neurons are also contacted by other GABAergic F-type and P-type terminals (indicated by asterisks), probably from local circuit neurons. Note that R-type terminals can be identified by their electron-lucent mitochondria and clear spherical vesicles (calibration bar in A,B is 0.5 mm).

14.3.3 Reciprocal Connections between NOT and AOS Nuclei

The pretectal and AOS nuclei are interconnected extensively in all mammalian species studied (*Lagomorpha* and *Rodentia*,[18,34,55,57,58,62,63,86,171] *Carnivora*,[14] *Primates*[7,17,31]). The major projection from the MTN to the NOT/DTN[18] arises from neurons of both ventral and dorsal subdivisions of the rat MTN and VTRZ.[58] Electron microscopic studies in the rat that combined injections of *Phaseolus vulgaris leucoagglutinin* into the MTN with GABA immuno-staining show that the MTN-NOT projection arises exclusively from GABAergic MTN neurons[175] and forms F-type terminals on both non-GABAergic and GABAergic neurons in the NOT/DTN (see Figures 3 and 4).[135,175,178]

In contrast, neither the LTN nor the inSFp send GABAergic projections to the NOT/DTN.[58] Similarly, there are reciprocal projections arising from GABAergic neurons in the NOT/DTN that produce F-type terminals on MTN neurons.[177]

14.4 OTHER NEUROTRANSMITTERS/NEUROMODULATORS IN NOT AND AOS NUCLEI

Retinofugal afferent terminals (R-terminals): the anatomical organization of the retinofugal projections to the NOT and the terminal accessory optic nuclei has been comprehensively studied in the rabbit and rat,[59,97,151,152] and the ultrastructural features of retinal terminals in visual nuclei, specifically the dorsal lateral geniculate nucleus[67,69,71,104,140,146,163] and superior colliculus,[9,80,109,119,158,180,187] have received substantial attention. Discussions by Nunes, Cardozo, and van der Want[135,136] and van der Want, Nunes, and Cardozo[178] have centered around comparisons of the fine structure of R-terminals within the dorsal lateral geniculate nucleus and superior colliculus contrasted with those in the NOT. Uniformly, the R-terminals in visual nuclei are described as having large profiles containing spherical synaptic vesicles and electron-lucent mitochondria, and synapsing on somata, dendrites, dendritic spines, and F-type axon terminals.

Studies designed to show the possible neurotransmitters in the retinofugal fibers are, at most, limited. The putative neurotransmitters that thus far have been identified are glutamate, aspartate, and substance P.[23,74,75,133,137] In this regard, these studies revealed that glutamate-like immunoreactivity is present in the goldfish optic tectum (superior colliculus of mammals)[90] and the macaque monkey dorsal lateral geniculate nucleus.[124,125] In these investigations, several types of retinal terminals of the goldfish optic tectum were noted to be glutamate-positive, while in the macaque LGNd, glutamate-immunoreactivity has been localized in three classes of axon terminals, one consisting of R-terminals. In the rabbit NOT, Nunes, Cardozo, and van der Want[136] have localized glutamate-immunoreactivity to R-terminals that form two types of glomerulus-like arrangements: (1) glutamate-positive R-terminals surrounded by small dendritic and axonal profiles, and (2) multiple glutamate-positive R-terminals surrounding a single dendrite or group of dendrites, presumably of interneuronal cells.

More recent evidence now shows that most rat retinal ganglion cells are labeled with antibodies against glutaminase, the synthesizing enzyme for glutamate,[118] supporting the tentative conclusion that glutamate is a major neurotransmitter in the primary optic projection system. Recently, De Vries et al.[41] and De Vries and Lakke[42] have demonstrated glutamate and aspartate to be major transmitters in the retinal ganglion cell projections to the suprachiasmatic nucleus and adjacent medial hypothalamic nuclei, both known visual visceral centers concerned with circadian rhymes.

As already indicated, N-acetylaspartylglutamate (NAAG) has also been reported to be a candidate as an excitatory neurotransmitter in retinal projection neurons.[121] In this regard, Molinar-Bode and Pasik[122] and Jones and Sillito[88] present evidence that NAAG is a neurotransmitter in the retinogeniculate pathway. The colocalization of NAAG with glutamate in retinohypothalamic axon terminals is discussed by De Vries et al.[41]

A. NOT-INFERIOR OLIVARY PATHWAY   **B. AOS-INFERIOR OLIVARY PATHWAY**

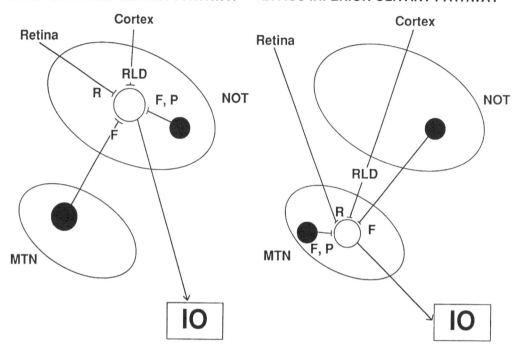

FIGURE 14.4 (A,B) The connectivity of the NOT and AOS is complex. However, one general pattern of synaptic connectivity is illustrated by the NOT-inferior olivary (A) and AOS-inferior olivary (B) paths (for references, see text). In this figure, excitatory neurons are illustrated with open circles, inhibitory neurons with filled circles. Note that the inferior olive projecting neurons of the NOT and MTN are excitatory, non-GABAergic; these output neurons receive excitatory synaptic terminals from the retina (R-type) and visual cortical areas (RLD-type), and inhibitory GABAergic terminals from local circuit neurons (mainly F-type but some P-type) and the system of inhibitory neurons interconnecting the NOT and AOS (F-type). (IO = the dorsal cap of Kooy, medial accessory olive and parts of the principal nucleus of the inferior olive).

The rat AOS nuclei also contain high levels of opioid[3,5,47,52,79,116] which, in radioligand binding studies, are identified as being almost exclusively of the m-opioid receptor type.[52] Monocular enucleation in young rats produces nearly a complete loss (>97%) of m-opioid receptor binding in the contralateral AOS nuclei. Similar results were obtained by monocular eye-patching (Figures 14.5 and 14.6).

Microdensitometric analyses of the MTN, NOT, and superior colliculus in the enucleation and eye-patching studies demonstrate a similar time course in receptor loss in all three visual areas, and some differences in the magnitude of receptor loss. Both experiments demonstrate that the m-opioid receptor is dependent on the presence of visual input. The effects are entirely contralateral, owing to the almost entirely crossed (about 97%) nature of the retinofugal projection.

Other putative neurotransmitter/neuromodulators are localized in varying abundance to neurons and/or retinofugal terminals on NOT and AOS nuclei. These include N-acetylaspartyl-glutamate (NAAG) (*rat*[121]), substance P plus NAAG (*rat,*[121] *rabbit*[23]), somatostatin (*rat*[102]), dopamine (*rat*[168]), and neurotensin (*pigeon*[22,25]). Substantial numbers of neurons of the AOS nuclei have been shown to receive retinofugal glutamate-positive terminals (*rabbit*[133]), NAAG (*rat*[121]), substance P (*rabbit*[23]), and catecholamines (*pigeon*[6,25,26,53,95]). Furthermore, Britto and colleagues[25] have demonstrated beautifully the presence of a large spectrum of neurotransmitters/neuromodulators which they find within neurons and/or axon terminals of the pigeon AOS nuclei. Finally, we have recently identified the presence of serotoninergic nerve fibers in the

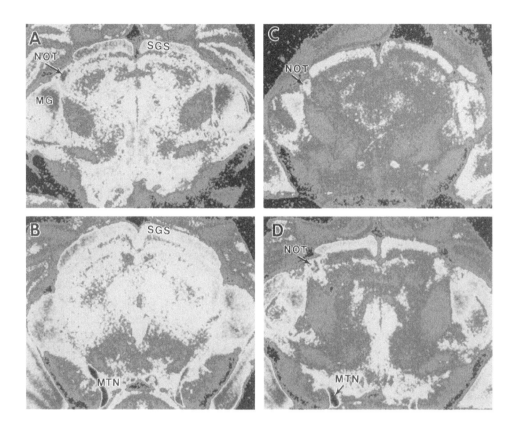

FIGURE 14.5 Computer-assisted, quantitative autoradiograms (A to D) of film exposed to radioactive ligand ([³H] DAGO) for m-opioid receptor. These transverse sections through the midbrain are from rats subjected to monocular eye-patching. Animals had the left eyelids sutured on postnatal day 11 (P11) before the lids normally opened (eyes open in control animals p12-p14), and the animals were killed after 5 days (D), 7 days (A,B), and 36 days (C). Areas containing high levels of m-opioid receptor are shown in red, followed by yellow, intermediate levels in green, and low levels in light and dark blue. A, C, and D show the decrease in radioligand receptor density in the NOT and superficial layers (stratus gresium superficiale (SGS)) of the superior colliculus contralateral to the patched eye (i.e., right side of each section). This contralateral preponderance is consistent with the almost entirely crossed (97%) retinofugal projection to the NOT and AOS in the rat (see text for references). Note the reduced density of m-opioid receptor in the MTN (B, D), the largest of the AOS terminal nuclei in the rodent. Similarly, the effects on the MTN are contralateral, owing to the crossed nature of the retinal input.

rat MTN and LTN (unpublished data, Giolli et al., 1995; but see Ref. 47) which agrees with findings on the nucleus of the basal optic root (nBOR) of the lizard[162,191] and chick.[150] Unlike the condition in the pigeon,[25] no neuropeptide Y reactive fibers have been reported within the terminal of the rat AOS nuclei.[40]

Swanson et al.[169] demonstrate nicotinic receptors (AChR) in rat medial terminal accessory optic nucleus, and in a more detailed study, Wada et al.[181] find the MTN of the rat to contain a sparse population of Alpha-4-1 and Alpha-4-2 AChR, as well as Beta-2 AChR. In addition, Dietl and colleagues[43,45] find the nBOR to contain one of the highest densities of muscarinic cholinergic receptors present in the pigeon brain, while in the chick, Krause et al.[99] demonstrate high concentrations of melatonin (ML1) receptor in retinal terminals of the nBOR and nucleus lentiformis mesencephali, the homologues, respectively, of the terminal AOS nuclei and NOT of mammals. Finally, both Dietl et al.[44] and Britto et al.[26] have shown that $GABA_A$ receptors are present in the pigeon nBOR.

14.5 SOME FUNCTIONAL CONSIDERATIONS

14.5.1 GABA

GABA is the major inhibitory neurotransmitter in the brain and plays a prominent role in the NOT and AOS pathways mediating visual-vestibular interaction. As reviewed in detail above, double-label retrograde and GABA immunocytochemistry in the rat and rabbit demonstrate that these nuclei are extensively interconnected by GABAergic projection neurons involving the NOT-MTN and NOT-MTN pathways. Similar pathways have been demonstrated in the nonhuman primate, but the main interconnections are between the NOT and LTN.[17,129] It has yet to be determined whether these interconnections in the nonhuman primate are also mediated via GABAergic projection neurons.

The first evidence for the inhibitory nature of the NOT-AOS interconnections came from microelectrode studies in the rabbit. Maekawa and Simpson[112] showed that electrical stimulation of the MTN and adjacent VTRZ in the rabbit inhibit the transmission of visual impulses from the optic chiasm through the NOT to the inferior olive and cerebellar flocculus. Similarly, neurons in the NOT of the rat are strongly inhibited by electrical stimulation of the MTN.[176] Iontophoretic application of bicuculline, a $GABA_A$ receptor antagonist, increases the spontaneous discharge of NOT/DTN neurons but has no effect on the inhibition invoked by electrical stimulation of the MTN. One explanation proposed to explain these observations is that the MTN inhibition of NOT neurons may be $GABA_B$-mediated.[176] Given the possible functional importance of $GABA_A$ and $GABA_B$ receptors in the MTN-NOT pathway, it is important to establish that the NOT in the rat contains an abundance of both $GABA_A$ and $GABA_B$ receptors.[19,33] The role of $GABA_B$-mediated inhibition will be extremely important to determine in future receptor-blocking studies, given the virtual absence of $GABA_A$ receptors in the rat MTN,[47] and only sparse-to-moderate concentrations of the $GABA_A$ receptors in the pigeon nBOR.[26]

Single unit studies in the pigeon[68] and rat[131] show that lesions of the pretectal complex modify the directional selectivity of neurons in the AOS nuclei, presumably by disruption of the inhibitory, GABAergic, pretectal-AOS interconnections. More recent experiments by Schmidt and colleagues[156] demonstrate that not only was spontaneous rate increased with application of bicuculline, but visual-evoked activity was increased as well. Direction-selectivity of NOT/DTN neurons to whole-field, moving stimuli was reduced for most neurons but not abolished. However, the difference between firing rates during stimulation in the preferred and nonpreferred direction did not change systematically with drug application. This was interpreted as GABAergic inputs being responsible for shaping the response properties of direction-selective NOT/DTN neurons instead of generating them.[156] The likely source of direction-selectivity of NOT/DTN neurons remains their retinal input from excitatory, on-center, direction-selective retinal ganglion cells.[139,85,161] Overall, these microelectrode studies in the rat indicate that direction-selectivity of NOT/DTN neurons receive GABAergic inhibition, which is most likely tonic and independent of the stimulus direction.

The GABA-mediated effects in the NOT assume greater importance on the basis of recent eye movement recording studies in the nonhuman primate.[36] Cohen and colleagues injected muscimol, a $GABA_A$ agonist, into the NOT at a site where electrical stimulation induced typical nystagmus with ipsilateral slow phase. After muscimol injection, a spontaneous nystagmus developed with a contralateral slow phase, and there was a reduction in slow-phase velocity and OKAN in the ipsilateral direction. These effects are entirely consistent with electrical stimulation and lesion studies[92,153,154] and are interpreted as resulting from GABAergic inhibition of NOT neurons mediating visual information to the velocity storage network.[36] Of additional interest, muscimol injections into the NOT produced a loss of habituation of the contralateral vestibuloocular reflex (VOR), and an inability to use vision to rapidly suppress the VOR to the contralateral side. These latter two effects were interpreted as muscimol causing a depression of NOT neurons projecting to the inferior olive and nucleus reticularis tegmenti pontis (NRTP), thereby reducing the climbing fiber and mossy fiber input to the vestibulo-cerebellum. A disruption of this activity to the nodulus/uvulae probably accounted for the muscimol effects on habituation and OKAN, given that both disappear after

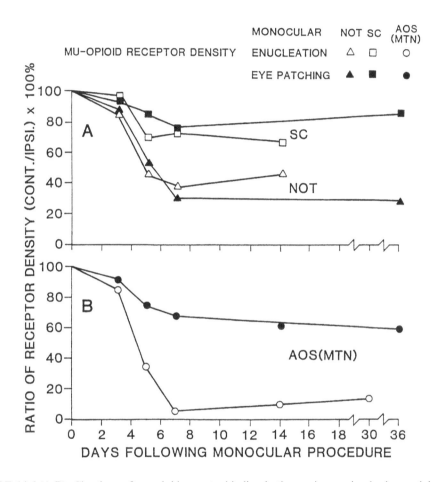

FIGURE 14.6 (A,B) Chartings of m-opioid receptor binding in three primary visual relay nuclei of the rat following two forms on monocular deprivation, eye enucleation (open symbols), and eye patching (filled symbols). Eye enucleation was performed on young adult animals and they were sacrificed on postoperative days 3, 5, 7, 14, and 30 (n = one animal/day). Eye-patching was accomplished by suturing the eyelids; animals were then sacrificed on 3, 5, 7, and 36 days following lid closure (n = one animal/day). Receptor density was determined from computer-assisted analysis of autoradiograms on comparable right-left portions of the nuclei.[56] Density was compared against known standards to express radioligand density in fmole/g tissue. Each data point gives the ratio of the contralateral:ipsilateral side, and was chosen given the almost totally crossed nature of the retinal projection to the NOT and AOS. These findings demonstrate activity-dependent regulation of m-opioid receptors in the primary visual relay nuclei. Note that enucleation is most effective in reducing m-opioid receptors in the AOS, whereas eye-patching is slightly more effective in the NOT. Additionally, the two protocols produce similar effects in SC and NOT, whereas the m-opioid receptor level in the AOS nuclei is considerably more sensitive to enucleation.

nodulouvulectomy,[182] whereas the muscimol effects on visual suppression of the VOR can be viewed as disruptions of the NOT-inferior olivary and NOT-NRTP pathways to the flocculus and paraflocculus.[36] The far-reaching consequences of the muscimol studies in the NOT are summarized by Cohen and colleagues,[36] who state that "...the profound effect of [GABAergic] inhibition of a visual structure [NOT], lying far in the central nervous system from the vestibular system on vestibular nystagmus, is of considerable interest clinically. It implies that vertigo, dizziness, a loss of gain adaptation, and a loss of habituation can occur on a transient or permanent basis from cerebral or brainstem lesions that do not directly involve vestibular structures."

14.5.2 OPIOID PEPTIDES

Opioid peptides have profound sensory effects[4,47] and are known to severely disrupt visual and oculomotor behavior in man, such as reduced visual sensitivity,[147] diminished gain of smooth pursuit, and hypometric saccades with reduced velocities.[148,149] These effects may result, in part, from diminished sensory capacity given the high density of opioid receptors in virtually all visual relay nuclei in the rat.[3,78,79]

The question of *mismatch* of neurotransmitter/neurotransmitter receptor appears to be quite relevant to the terminal AOS nuclei as studied in the rat. Thus, while these four visuomotor-related nuclei (MTN, LTN, DTN, inSFp) exhibit high concentrations of the mu opioid receptor (rat,[52] pigeon[54,52,142]), a sparse to moderate concentration of delta opioid receptors (47, but see Ref. 52), and a sparse concentration of k-opioid receptors[47,52] (but see Speciale et al.[164] who found high radioligand binding), these nuclei show only low concentrations of opiates.[46,47,79,105,116,117] Thus, an inverse, paradoxical relationship exists with regard to the AOS nuclei, and this could result from one or more factors (see[77,101]). *First,* the "mismatch" could result from the synthesis of large numbers of mu receptors in rat accessory optic neurons in response to opioid peptides arising from one or more sources of nonretinal afferents (the periaqueductal gray, ventral lateral geniculate nucleus, and NOT that contain moderate numbers of enkephalin-positive neurons).[46] As the rat AOS nuclei contain, at most, only minuscule levels of enkephalins and dynorphins,[46,117] this would require a slow, steady transport of opioid peptides to accessory optic neurons that is undetectable using available immunocytochemical methods. However, the finding that about 89% of all m-receptors disappear after seven days of monocular enucleation[56] suggests that these receptors are located presynaptically on retinofugal axon terminals and are not postsynaptic on AOS neurons. In agreement, Loughlin and Fallon[105] report that while the rat MTN shows high binding for m-opioid receptors, very little mRNA is expressed for these receptors, suggesting that the receptors are produced outside the MTN and reside on afferent terminals. *Second,* the "mismatch" could result from the distant release of opiate peptide, that is, a release from opiate-rich regions in the pretectum and/or midbrain tegmentum[46,117] to reach the AOS nuclei through parasynaptic diffusion in the manner proposed by Herkenham.[76]

The functional relationship between opioid receptors and putative neurotransmitters in the AOS has yet to be determined. However, it is possible that opioids interact at GABAergic neurons and axon terminals, and that such interaction produces neuronal inhibition comparable, perhaps, to that reported in the striatonigral and nigrostriatal systems of the rat.[1,89,165,174] The extraordinarily high levels of m-opioid receptors in the rat AOS nuclei, and the related opioid effects on visual sensitivity,[147] diminished gain of smooth pursuit, and saccadic velocities,[148,149] suggest that endogenous opiates may play a significant role in regulating visual transmission through the AOS nuclei; and through connections with the vestibular nuclei and precerebellar pathways, may account, in part, for such phenomena as the fluctuating gain of optokinetic nystagmus with arousal, and the adaptive plasticity of visuomotor reflexes.

As for the functional importance of opiates and opioid receptors, it is interesting that selective electrical stimulation of the anterior pretectal nuclei in the rat has been found by Roberts and Rees[141,143] to have an antinociceptive effect on lower portions of the body surface (tail region) — a perhaps expected finding considering that anatomical studies have shown that the anterior pretectal nucleus is interconnected with regions of spinal cord gray concerned with sensorimotor function.[13,110,190,1107]

14.6　SUMMARY AND CONCLUSIONS

A major breakthrough in our understanding of the functional role of the NOT and AOS nuclei has come from the systematic analysis of the circuitry and neurotransmitters of these nuclei. As summarized in this chapter, the neurons of the NOT and AOS nuclei receive excitatory input from retinal ganglion cells and visual cortical areas; in both cases, the candidate excitatory transmitter is glutamate. The NOT and AOS nuclei are extensively interconnected by largely GABAergic, but also non-GABAergic neurons.[54,61,175] These interconnecting GABAergic circuits play an exceedingly important and increasingly well-understood role in determining the receptive field properties[193] and, in the case of the NOT, in mediating oculomotor control of horizontal OKN, OKAN, and visual suppression and habituation of the slow phase of vestibular nystagmus.[36] Among the efferent projections, the NOT and AOS neurons projecting to the inferior olive are non-GABAergic, and it is assumed that the NOT neurons carrying visual information, ultimately, to the vestibular nuclei in support of visual-vestibular interaction are also non-GABAergic.

Studies employing $GABA_A$ agonists and antagonists have been very important in determining some of the cellular mechanisms involving the NOT and AOS pathways. The direction selectivity and resting discharge of NOT neurons, along with subsequent effects on vestibular habituation and visual suppression of VOR, are apparently mediated by $GABA_A$ receptors on NOT neurons. However, the local application of muscimol is ineffective in blocking the GABAergic inhibition of NOT neurons following electrical stimulation of the MTN, suggesting that these interconnecting pathways (e.g., MTN-NOT; NOT-MTN) are mediated by $GABA_B$ receptors. Receptor-blocking studies will be required to determine this important point. Finally, there is need to examine the pharmacological effects of the many other transmitters/neuromodulators localized within the NOT and AOS nuclei (e.g., opioids, NAAG, somatostatin, dopamine, neurotensin, and substance P). The short- and long-term effects of these agents on the excitability of NOT neurons may prove more important than formerly realized, given the evidence that tonic activity in the NOT (mediated through the inferior olivary and NRTP pathway to the vestibulo-cerebellum) is required to sustain vestibular habituation and normal functioning of velocity storage in support of OKN, OKAN, and the VOR.[36]

REFERENCES

1. Abou-Khalil, B, Young, AB, and Penney, JB, Evidence for the presynaptic localization of opiate binding sites on striatal efferent fibers, *Brain Res.,* 323: 21,1984.
2. Appell, PP, and Behan, M, Sources of subcortical GABAergic projections to the superior colliculus in the cat, *J. Comp. Neurol.,* 302:143,1990.
3. Atweh, SF, and Kuhar, MJ, Autoradiographic localization of opiate receptors in rat brain. II. The brainstem, *Brain Res.,* 129:1,1977.
4. Atweh, SF, and Kuhar, MJ, Distribution and physiological significance of opioid receptors in the brain, *Brit. Med. Bull.,* 9:47,1983.
5. Atweh, SF, Murrin, LC, and Kuhar, MJ, Presynaptic localization of opiate receptors in the vagal and accessory optic systems: An autoradiographic study, *Neuropharmacology,* 17:65,1978.
6. Bagnoli, P, Casini, G, Regional distribution of catecholaminergic terminals in the pigeon visual system, *Brain Res.,* 337,277,1985.
7. Baleydier, CM, Magnin, M, and Cooper, HM, Macaque accessory optic system: II. Connections with the pretectum, *J. Comp. Neurol.,* 302,405,1990.

8.   Barmack, NH, and Young, S, Optokinetic stimulation increases corticotropin-releasing factor mRNA in inferior olivary neurons of rabbits, *J. Neurosci.*, 10,631,1990.

9.   Behan, M, Identification and distribution of retinocollicular terminals in the cat: An electron microscopic and autoradiographic analysis, *J. Comp. Neurol.*, 199,1,1981.

10. Benassi, C, Biral, GP, Lui, F, Porro, CA, and Corazza, R, The interstitial nucleus of the superior fasciculus, posterior bundle (inSFp) in the guinea pig: Another nucleus of the accessory optic system processing the vertical retinal slip signal, *Vis. Neurosci.*, 2,377,1989.

11. Benevento, LA, Rezak, M, and Santos-Anderson, RM, An autoradiographic study of the projections of the pretectum in the rhesus monkey (*Macaca mulatta), Brain Res.*, 127,197,1977.

12. Benevento, LA, and Standage, GP, The organization of projections of the retinorecipient and nonretinorecipient nuclei of the pretectal complex and layers of the superior colliculus to the lateral pulvinar and medial pulvinar in the macaque monkey, *J. Comp. Neurol.*, 217,307,1983.

13. Berkley, KJ, and Mash, DC, Somatic sensory projections to the pretectum in the cat, *Brain Res.*, 158,445,1978.

14. Berson DM, and Graybiel, AM, Some cortical and subcortical fiber projections to the accessory optic nuclei in the cat, *Neuroscience,* 5,2203,1980.

15. Biral, GP, Porro, CA, Cavazzuti, M, Benassi, C, and Corazza, R, Vertical and horizontal visual whole-field motion differently affect the metabolic activity of the rat medial terminal nucleus, *Brain Res.,* 412,43,1987.

16. Blanks, RHI, Cerebellum. in: *Reviews of Oculomotor Research*, Vol. 2, (DA Robinson and H Collewijn, series eds; J Buttner-Ennever, volume Ed.). Elsevier Biomedical Press, Amsterdam, pp. 225,1988.

17. Blanks, RHI, Clarke, RJ, Lui, F, Giolli, RA, Pham, SV, and Torigoe, Y, Projections of the lateral terminal accessory optic nucleus of the common marmoset (*Callithrix jacchus), J. Comp. Neurol.*, 354,511,1995.

18. Blanks, RHI, Giolli, RA, and Pham, SV, projections of the medial terminal nucleus of the accessory optic system upon pretectal nuclei in the pigmented rat, *Exp. Brain Res.*, 48,228,1982.

19. Bowery, NG, Hudson, AL., and Price, GW, GABA$_A$ and GABA$_B$ receptors site distribution in the rat central nervous system, *Neuroscience,* 20,365,1987.

20. Braak, H, and Bachmann, A, The percentage of projection neurons and interneurons in the human lateral geniculate nucleus, *Human Neurobiol.,* 4,91,1985.

21. Brauth, SE, and Karten, HJ, Direct accessory optic projections to the vestibulo-cerebellum: A possible channel for oculomotor control systems, *Exp. Brain Res.*, 28,73,1977.

22. Brauth, SE, Kitt, CA, Reiner, A, Quirion, R, Neurotensin binding sites in the forebrain and midbrain of the pigeon, *J. Comp. Neurol.*, 253,358,1986.

23. Brecha, N, Johnson, D, Bolz, J, Sharma, S, Parnavelas, JG, and Lieberman, AR, Substance P-immunoreactive retinal ganglion cells and their central axon terminals in the rabbit, *Nature.*, 327,155,1987.

24. Brecha, N, Karten, HJ, and Hunt, SP, Projections of the nucleus of the basal optic root in the pigeon: An autoradiographic and horseradish peroxidase study, *J. Comp. Neurol.*, 189,615,1980.

25. Britto, LR, Hamassaki, DE, Keyser, KT, and Karten, HJ, Neurotransmitters, receptors, and neuropeptides in the accessory optic system: an immunohistochemical survey in the pigeon (*Columba livia), Vis. Neurosci.*, 3,463,1989.

26. Britto, LR, Keyser, KT, Hamassaki, DE, and Karten, HJ, Catecholaminergic subpopulation of retinal displaced ganglion cells projects to the accessory optic nucleus in the pigeon (*Columba livia), J. Comp. Neurol.*, 269,109,1988.

27. Brown, JT, Chan-Palay, V, and Palay, SL, A study of afferent input to the inferior olivary complex in the rat by retrograde axonal transport of horseradish peroxidase, *J. Comp. Neurol.*, 176,1,1977.

28. Cadusseau, J, and Roger, M, Cortical and subcortical connections of the pars compacta of the anterior pretectal nucleus in the rat, *Neurosci. Res.*, 12,83,1991.

29. Campbell, G, and Lieberman, AB, The olivary pretectal nucleus: experimental anatomical studies in the rat, *Phil. Trans. R. Soc. London.*, B 301,573,1985.

30. Cardozo, BN, Buijs, R, and van der Want, J, Glutamate-like immunoreactivity in retinal terminals in the nucleus of the optic tract in rabbits, *J. Comp. Neurol.*, 309,261,1991.

31. Carpenter, MB, and Pierson, RJ, Pretectal region and the pupillary light reflex. An anatomical analysis in the monkey, *J. Comp. Neurol.*, 149,271,1973.

32. Cazin, L, Precht, W, and Lannou, J, Firing characteristics of neurons mediating optokinetic responses to rat's vestibular neurons, *Pflugers. Arch. Physiol.*, 386,221,1980.

33. Chu, DCM, Albin, RL, Young, JB, and Penny, JB, Distribution and kinetics of $GABA_B$ binding sites in rat central nervous system: a quantitative autoradiographic study, *Neurosci.,* 34,431,1990.

34. Clarke, RJ, Giolli, RA, Blanks, RH, Torigoe, Y, and Fallon, JH, Neurons of the medial terminal accessory optic nucleus of the rat are poorly collateralized, *Vis. Neurosci.,* 2,269,1989.

35. Clement, G, and Magnin, M, Effects of accessory optic system lesions on vestibulo-ocular and optokinetic reflexes in the cat, *Exp. Brain Res.,* 55,49,1984.

36. Cohen, B, Reisine, H, Yokota, J-I, and Raphan, T, The nucleus of the optic tract. Its function in gaze stabilization and control of visual-vestibular interaction, *Ann. NY Acad. Sci.,* 656,277,1992.

37. Collewijn, H, Direction-selective units in the rabbit's nucleus of the optic tract, *Brain Res.,* 100,489,1975.

38. Cucchiaro, JB, Bickford, ME, and Sherman, SM, A GABAergic projection from the pretectum to the dorsal lateral geniculate nucleus in the cat. Neurosci 41,213,1991.

39. Cucchiaro JB, Uhlrich DJ, Sherman SM Ultrastructure of synapses from the pretectum in the A-laminae of the cat's lateral geniculate nucleus, *J. Comp. Neurol.,* 334,618,1993.

40. de Quidt, ME, and Emson, PC, Distribution of neuropeptide Y-like immunoreactivity in the rat central nervous system- II. Immunohistochemical analysis, *Neurosci.,* 18,545,1986.

41. De Vries, MJ, Nunes, Cardozo, B, van der Want, J, De Wolf, A, and Meijer, JH, Glutamate reactivity in terminals of the retinohypothalamic tract of the brown Norwegian rat: An electrophysiological and morphological study, *Brain Res.,* 612,231,1993.

42. De Vries, MJ, and Lakke, EA, Retrograde labeling of retinal ganglion cells and brain neuronal subsets by [^{3}H]-D-aspartate injection in the Syrian hamster hypothalamus, *Brain Res. Bull.,* 38,349,1995.

43. Dietl, MM., Cortes R, and Palacios, JM, Neurotransmitter receptors in the avian brain. II. Muscarinic cholinergic receptors, *Brain Res.,* 439,360,1988b.

44. Dietl, MM., Cortes, R, and Palacios, JM, Neurotransmitter receptors in the avian brain. III. GABA-benzodiazepine receptors, *Brain Res.,* 439,366,1988c.

45. Dietl, MM, and Palacios, JM, Neurotransmitter receptors in the avian brain. I. Dopamine receptors, *Brain Res.,* 439,354,1988a.

46. Fallon, JH, and Leslie, FM, Distribution of dynorphin and enkephalin peptides in the rat brain, *J. Comp. Neurol.,* 249,293,1986.

47. Fallon, JH, and Loughlin, SE, Substantia Nigra, in: *The Rat Nervous System*, 2nd edition (Paxinos G, Ed.), pp. 215,1995, San Diego: Academic Press,1995.

48. Feig, S, and Harting, JK, Ultrastructural studies of the primate lateral geniculate nucleus: morphology and spatial relationships of axon terminals arising from the retina, visual cortex (area 17), superior colliculus, parabigeminal nucleus, and pretectum of *Galago crassicaudatus, J. Comp. Neurol.,* 343,17,1994.

49. Fite, KV, Pretectal and accessory-optic visual nuclei of fish, amphibia and reptiles, *Brain Behav. Evol.,* 26,71,1985.

50. Fitzpatrick, D, Penny, GR, and Schmechel, DE, GAD-immunoreactive neurons and terminals in the lateral geniculate nucleus of the cat, *J. Neurosci.,* 4,1809,1984.

51. Fonnum, F, Storm-Mathisen, J, and Divac, I, Biochemical evidence for glutamate as neurotransmitter in corticostriatal and corticothalamic fibers in rat brain, Neurosci., 6,863,1981.

52. Fukushima, K, Kaneko, CRS, and Fuchs, AF, The neuronal substrate of integration in the oculomotor system, *Prog. Neurobiol.,* 39,609,1992.

53. Fuxe, K, and Ljunggren, K, Cellular localization of monoamines in the upper brain stem of the pigeon, *J. Comp. Neurol.,* 125,355,1965.

54. German, DC, Speciale, SG, Manaye, KF, and Sadeq, M, Opioid receptors in midbrain dopaminergic regions of the rat. I. Mu receptor autoradiography, *J. Neural. Transm.,* [Gen Sect] 91,39,1993.

55. Giolli, RA, Blanks, RHI, and Torigoe, Y, Pretectal and brainstem projections of the medial terminal nucleus of the accessory optic system of the rabbit and rat as studied by anterograde and retrograde neuronal tracing methods, *J. Comp. Neurol.,* 227,228,1984.

56. Giolli, RA, Blanks, RHI, Torigoe, Y, Clarke, RJ, Fallon, JH, and Leslie, FM, Opioid receptors in the accessory optic system of the rat: effects of monocular enucleation, *Vis. Neurosci.,* 5,497,1990a.

57. Giolli, RA, Blanks, RHI, Torigoe, Y, and Williams, DD, Projections of medial terminal accessory optic nucleus, ventral tegmental nuclei, and substantia nigra of rabbit and rat as studied by retrograde axonal transport of horseradish peroxidase, *J. Comp. Neurol.,* 232,99,1985a.

58. Giolli, RA, Clarke, RJ, Blanks, RHI, Torigoe, Y, and Fallon, JH, GABAergic and non-GABAergic projections of accessory optic nuclei, including the visual tegmental relay zone, to the nucleus of the optic tract and dorsal terminal accessory optic nucleus in rat, *J. Comp. Neurol.,* 319,349,1992.

59. Giolli, RA, and Guthrie, MD, The primary optic projections in the rabbit. An experimental degeneration study, *J. Comp. Neurol.,* 136,99,1969.

60. Giolli, RA, and Guthrie, MD, Organization of subcortical projections of visual areas I and II in the rabbit. An experimental degeneration study, *J. Comp. Neurol.,* 142,351,1971.

61. Giolli, RA, Peterson, GM, Ribak, CE, McDonald, HM, Blanks, RHI, and Fallon, JH, GABAergic neurons comprise a major cell type in rodent visual relay nuclei: an immunocytochemical study of pretectal and accessory optic nuclei, *Exp. Brain Res.,* 61, 194,1985b.

62. Giolli, RA, Torigoe, Y, and Blanks, RHI, Nonretinal projections to the medial terminal accessory optic nucleus in rabbit and rat: a retrograde and anterograde transport study, *J. Comp. Neurol.,* 269,73,1988a.

63. Giolli, RA, Torigoe, Y, Blanks, RHI, and McDonald, HM, Projections of the dorsal and lateral terminal accessory optic nuclei and of the interstitial nucleus of the superior fasciculus (posterior fibers), in the rabbit and rat, *J. Comp. Neurol.,* 277,608,1988b.

64. Grasse, KL, and Cynader, MS, Electrophysiology of medial terminal nucleus of the accessory optic system in the cat, *J. Neurophysiol.,* 48, 490,1982.

65. Grasse, KL, and Cynader, MS, The accessory optic system in frontal eyed animals. in: *The Neural Basis of Visual Function* (Leventhal AG, Ed.), pp. 111-139, Boca Raton: CRC Press,1991.

66. Gregory, KM, The dendritic architecture of the visual pretectal nuclei of the rat: A study with the Golgi-Cox method, *J. Comp. Neurol.,* 234,122,1985.

67. Guillery, RW, The organization of synaptic interconnections in the laminea of the dorsal lateral geniculate nucleus of the cat, *Z. Zellforsch*, 96,1,1969.

68. Hamassaki, DE, Gasparotto, OC, Nogueira, MI, and Britto, LRG, Telencephalic and pretectal modulation of the directional selectivity of accessory optic neurons in the pigeon, *Brazilian J. Med. Biol. Res.,* 21,649,1988.

69. Hamori, J, Pasik, T, Pasik, P, and Szentagothai, J, Triadic synaptic arrangements and their possible significance in the lateral geniculate nucleus of the monkey, *Brain Res.,* 80,379,1974.

70. Hamori, J, Pasik, P, and Pasik, T, Differential frequency of P-cells and I-cells in magnocellular and parvocellular laminae of monkey lateral geniculate nucleus: an ultrastructural study, *Exp. Brain Res.,* 52,57,1983.

71. Hamos, JE, Van Horn, SC, Raczkowski, D, and Sherman, SM, Synaptic circuits involving an individual retinogeniculate axon in the cat, *J. Comp. Neurol.,* 259,165,1987.

72. Harting, JK, Hashikawa, T, and van Lieshout, D, Laminar distribution of tectal, parabigeminal, and pretectal inputs to the primate dorsal lateral geniculate nucleus: connectional studies in *Galago crassicaudatus, Brain Res.,* 366,358,1986.

73. Hayhow, WR, Webb C, Jervie A The accessory optic fiber system in the rat, *J. Comp. Neurol.,* 115,185,1960.

74. Henke, H, Schenker, TM, and Cuenod, M, Uptake of neurotransmitter candidates by pigeon optic tectum, *J. Neurochem.,* 26,125,1976a.

75. Henke, H, Schenker, TM, and Cuenod, M, Effects of retinal ablation on uptake of glutamate, glycine, GABA, proline and choline in the pigeon tectum, *J. Neurochem.,* 26,131,1976b.

76. Herkenham, M, Mismatches between neurotransmitter and receptor localizations in brain, Observations and implications, *Neurosci.,* 23,1,1987.

77. Herkenham, M, and McLean, S Mismatches between receptor and transmitter localizations in the brain. in: *Quantitative Receptor Autoradiography* (Boast C, Snowhill EW, Altar CA, eds.), pp.137,1986, New York: Alan R. Liss,1986.

78. Herkenham, M, and Pert, CB, *in vitro* autoradiography of opiate receptors in rat brain suggests loci of "opiatergic" pathways, *Proc. Nat. Acad. Sci.,* 77,5532,1980.

79. Herkenham, M, and Pert, CB, Light microscopic localization of brain opiate receptors: A general autoradiographic method which preserves tissue quality, *J. Neurosci.,* 2,1129,1982,.

80. Hofbauer, H, Hollander, H, Synaptic connections of cortical and retinal terminals in the superior colliculus of the rabbit: An electron microscopic double labelling study, *Exp. Brain Res.,* 65,145,1986.

81. Hoffmann, K-P, Visual inputs relevant for the optokinetic nystagmus in mammals, *Brain Res.,* 64,75,1986.

82. Hoffmann, K-P, and Distler, C, Quantitative analysis of visual receptive fields of neurons in nucleus of the optic tract and dorsal terminal nucleus of the accessory optic system in macaque monkey, *J. Neurophysiol.,* 62,416,1989.

83. Hoffmann, K-P, Distler, C, and Erickon, Functional Projections from striate cortex and superior temporal sulcus to the nucleus of the optic tract (NOT) and dorsal terminal nucleus of the accessory optic tract (DTN) of macaque monkeys, *J. Comp. Neurol.,* 313,707,1991.

84. Hoffmann, K-P, and Schoppmann, A, A quantitative analysis of the direction-specific response of neurons in the cat's nucleus of the optic tract, *Exp. Brain Res.,* 42,146,1981.

85. Hoffmann, K-P, Stone, J, Retinal input to the nucleus of the optic tract of the cat assessed by antidromic activation of ganglion cells, *Exp. Brain Res.,* 59,395,1985.

86. Holstege, G, and Collewijn, H, The efferent connections of the nucleus of the optic tract and the superior colliculus in the rabbit, *J. Comp. Neurol.,* 209,139,1982.

87. Horn, AKE, and Hoffmann, KP, Combined GABA-immunocytochemistry and TMB-HRP histochemistry of pretectal nuclei projecting to the inferior olive in rats, cats and monkeys, *Brain Res.,* 409,133,1987.

88. Jones, HE, and Sillito, AM, The action of the putative neurotransmitters NAAG and homocysteate in cat dorsal lateral geniculate nucleus, *J. Neurophysiol.,* 68,663,1992.

89. Iwatsubo, K, Kondo, Y, Inhibitory effects of morphine on the release of preloaded 3H-GABA from rat substantia nigra in response to the stimulation of caudate nucleus and globus pallidus. in: *Characteristics and Functions of Opioids: Developments in Neuroscience.* (van Ree JM, Terenius L, eds.), pp. 357-358, Amsterdam: Elsevier/North Holland Biomedical Press,1978.

90. Kageyama, GH, and Meyer, RL, Glutamate-immunoreactivity in the retina and optic tectum of the goldfish, *Brain Res.,* 503,118,1989.

91. Kanaseki, T, and Sprague, JM, Anatomical organization of pretectal nuclei and tectal laminae in the cat, *J. Comp. Neurol.,* 158,319,1974.

92. Kato, I, Harada, K, Hasegawa, T, and Ikarashi, T, Role of the nucleus of the optic tract of monkeys in optokinetic nystagmus and optokinetic after-nystagmus, *Brain Res.,* 474,16,1988.

93. Kawamura, K, and Onodero, S, Olivary projections from the pretectal region in the cat studied with horseradish peroxidase and tritiated amino acids axonal transport, *Arch. Ital. Biol.,* 122,155,1984.

94. Keller, EL, and Heinen, SJ, Generation of smooth-pursuit eye movements: neuronal mechanisms and pathways, *Neurosci. Res.,* 11,79,1991.

95. Kiss, JZ, and Peczely, P, Distribution of tyrosine-hydroxylase (TH) - immunoreactive neurons in the diencephalon of the pigeon (*Columba livia domestica*), *J. Comp. Neurol.,* 257,333,1987.

96. Kitao, Y, Nakamura, Y, and Okoyama, S, An electron microscope study of the cortico-pretecto-olivary projection in the cat by a combined degeneration and horseradish peroxidase tracing technique, *Brain Res.,* 280,139,1983.

97. Klooster, J, van der Want, JJL, and Vrensen, G, Retinopretectal projections in albino and pigmented rabbits: An autoradiographic study, *Brain Res.,* 288,1,1983.

98. Korp, BJ, Blanks, RHI, and Torigoe, Y, Projections of the nucleus of the optic tract to the nucleus reticularis tegmenti pontis and praepositus hypoglossi nucleus in pigmented rat as demonstrated by anterograde and retrograde transport methods, *Vis. Neurosci.,* 2,275,1989.

99. Krause, DN, Siuciak, JA, and Dubocovich, ML, Unilateral optic nerve transection decreases 2-[125I]-iodomelatonin binding in retinorecipient areas and visual pathways of chick brain, *Brain Res.,* 654,63,1994.

100. Kubota, T, Morimoto, M, Kanaseki, T, and Inomata, H, Projection from the pretectal nuclei to the dorsal lateral geniculate nucleus in the cat: a wheat germ agglutinin-horseradish peroxidase study, *Brain Res.,* 421,30,1987.

101. Kuhar, MJ, The mismatch problem in receptor mapping studies, *Trends. Neurosci.,* 8,190,1985.

102. Laemle, LK, and Feldman, SC, Somatostatin -like immunoreactivity in subcortical and cortical visual centers of the rat, *J. Comp. Neurol.,* 233,452,1985.

103. Lenn, NJ, An electron microscopic study of accessory optic endings in the rat medial terminal nucleus, *Brain Res.,* 43,622,1972.

104. Lieberman, AR, Neurons with presynaptic perikarya and presynaptic dendrites in the rat lateral geniculate nucleus, *Brain Res.,* 59,35,1973.

105. Loughlin, SE, Leslie, FM, and Fallon, JH, Endogenous Opioid Systems, in: *The Rat Nervous System,* 2nd edition (Paxinos G, Ed.), pp. 975,1995. San Diego: Academic Press,1995.

106. Lui, F, Biral, GP, Benassi, C, Ferrari, R, and Corazza, R, Correlation between retinal afferent distribution, neuronal size, and functional activity in the guinea pig medial terminal accessory optic nucleus, *Exp, Brain Res.,* 81,77,1990.

107. Lui, F, Giolli, RA, Blanks, RH, and Tom, EM, Pattern of striate cortical projections to the pretectal complex in the guinea pig, *J. Comp. Neurol.,* 344,598,1994.

108. Lui, F, Gregory, KM, Blanks, RHI, and Giolli, RA, Projections of visual areas of the cerebral cortex to pretectal nuclei, terminal accessory optic nuclei, and superior colliculus in macaque monkey, *J. Comp. Neurol.,* 356,1,1995.

109. Lund, RD, Synaptic patterns of the superficial layers of the superior colliculus of the rat, *J. Comp. Neurol.,* 135,179,1969.

110. Mackel, R, and Noda, T, The pretectum as a site for relaying dorsal column input to thalamic VL neurons, *Brain Res.,* 476,135,1989.

111. Maekawa, K, and Kimura, M, Electrophysiological study of the nucleus of the optic tract that transfers optic signals to the nucleus reticularis tegmenti pontis- The visual mossy fiber pathway to the cerebellar flocculus, *Brain Res.,* 211,456,1981.

112. Maekawa, K, and Simpson, JI, Climbing fiber activation of Purkinje cells in the flocculus by impulses transferred through the visual pathway, *Brain Res.*, 39,245,1972.

113. Magnin, M, Courjon, JH, and Flandrin, JM, Possible visual pathways to the cat vestibular nuclei involving the nucleus prepositus hypoglossi, *Exp. Brain Res.,* 51,298,1983.

114. Magnin, M., Kennedy, H, and Hoffmann, K-P, A double-labeling investigation of the pretectal visuo-vestibular pathways, *Vis. Neurosci.,* 3,53,1989.

115. Mai, JK, The accessory optic system and the retino-hypothalamic system, A Review, *J. Hirnforschung,* 19,213,1978.

116. Mansour, A, Khachaturian, H, Lewis, ME, Akil, H, and Watson, SJ, Autoradiographic differentiation of mu, delta, and kappa opioid receptors in the rat forebrain and midbrain, *J. Neurosci.,* 7,2445,1987.

117. McLean, S, Skirboll, LR, and Pert, CB, Comparison of substance P and enkephalin distribution in the rat brain: An overview using radioimmunocytochemistry, *Neuroscience,* 14,837,1985.

118. Miguel-Hidalgo, JJ, Senba, E, Takatsuji, K, and Tohyama, M, Projections of tachykinin- and glutaminase-containing rat retinal ganglion cells. *Brain Res. Bull.,* 35,73,1994.

119. Mize, RR, Variations in the retinal synapses of the cat superior colliculus revealed using quantitative electron microscope autoradiography, *Brain Res.,* 269,211,1983.

120. Mizuno, NK, Mochizuki, K, Akimoto, C, and Matsushima, R, Pretectal projections to the inferior olive in the rabbit, *Exp. Neurol.,* 39,498,1973.

121. Moffett, JR, Williamson, LC, Neale, JH, Palkovits, M, and Namboodiri, MA, Effect of optic nerve transection on N-acetylaspartylglutamate immunoreactivity in the primary and accessory optic projection systems in the rat, *Brain Res.,* 538,86,1991.

122. Molinar-Bode, R, and Pasik, P, Amino acids and N-acetyl aspartyl glutamate as neurotransmitter candidates in the monkey retinogeniculate pathways, *Exp. Brain Res.,* 89,40,1991.

123. Molotchnikoff, S, Cerat, A, and Casanova, C, Influences of cortico-pretectal fibers on responses of rat pretectal neurons, *Brain Res.,* 446,67,1988.

124. Montero, VM, Quantitative immunogold analysis reveals high glutamate levels in synaptic terminals of retino-geniculate, cortico-geniculate, and geniculo-cortical axons in the cat, *Vis. Neurosci.,* 4,437,1990.

125. Montero, VM, and Singer, W, Ultrastructural identification of somata and neuronal processes immunoreactive to antibodies against glutamic acid decarboxylase (GAD) in the dorsal lateral geniculate nucleus of the cat, *Exp. Brain Res.,* 59,151,1985.

126. Montero, VM, and Wenthold, RJ, Quantitative immunogold analysis reveals high glutamate levels in retinal and cortical synaptic terminals in the lateral geniculate nucleus of the macaque, *Neurosci.,* 31,639,1989.

127. Montero, VM, and Zempel, J, Evidence for two types of GABA-containing interneurons in the A-laminae of the cat lateral geniculate nucleus: A double-label HRP and GABA-immunocytochemical study, *Exp. Brain Res.,* 60,603,1985.

128. Mustari, MJ, and Fuchs, AF, Response properties of single units in the lateral terminal nucleus of the accessory optic system in a behaving primate, *J. Neurophysiol.,* 61,1207,1989.

129. Mustari, MJ, Fuchs, AF, Kaneko, CR, and Robinson, FR, Anatomical connections of the primate pretectal nucleus of the optic tract, *J. Comp. Neurol.,* 349,111,1994.

130. Nabors, LB, and Mize, RR, A unique neuronal organization in the cat pretectum revealed by antibodies to the calcium-binding protein calbindin-D 28K, *J. Neurosci.,* 11,2460,1991.

131. Natal, CL, and Britto, LRG, The pretectal nucleus of the optic tract modulates the direction selectivity of accessory optic neurons in rats, *Brain Res.,* 419,320,1987.

132. Nauta, WJH, and Bucher, VM,   Efferent connections of the striate cortex in the albino rat, *J. Comp. Neurol.,* 100,257,1954.

133. Nunes, Cardozo, B, Buijs, R, and van der Want, J, Glutamate-like immunoreactivity in the retinal terminals in the nucleus of the optic tract in rabbits, *J. Comp. Neurol.,* 309,261,1991.

134. Nunes, Cardozo, JJ, Mize, RR, and van der Want, JJL, GABAergic and non-GABAergic neurons in the nuclei of the optic tract project to the superior colliculus: an ultrastructural retrograde tracer and immunocytochemical study in the rabbit, *J. Comp. Neurol.,* 350,646,1994.

135. Nunes, Cardozo, JJ, and van der Want, JJL, Synaptic organization of the nucleus of the optic tract in the rabbit: A combined Golgi-electron microscopic study, *J Neurocytol.,* 16,389,1987,.

136. Nunes, Cardozo, JJ, and van der Want, JJL, Ultrastructural organization of the retino-pretecto-olivary pathway in the rabbit. A combined WGA-HRP tracing and GABA immunocytochemical study, *J. Comp. Neurol.,* 291,313,1990.

137. Ottersen, OP, and Storm-Mathisen, J, GABA-containing neurons in the thalamus and pretectum of the rodent, *Anat. Embryol.,* 170,197,1984.

138. Oyster, CW, Barlow, HB, Direction-selective units in rabbit retina: Distribution of preferred directions, *Science,* 155,841,1967.

139. Oyster, CW, Takahashi, E, and Collewijn, H, Direction-selective retinal ganglion cells and control of optokinetic nystagmus in the rabbit, *Vision Res.,* 12,183,1972.

140. Rapisardi, SC, and Miles, TP, Synaptology of retinal terminals in the dorsal lateral geniculate nucleus in cat, *J. Comp. Neurol.,* 223,515,1984.

141. Rees, H, and Roberts, HM, The anterior pretectal nucleus: A proposed role in sensory processing, *Pain,* 53,121,1993.

142. Reiner, A, Brauth, SE, Kitt, CA, and Quirion, R, Distribution of mu, delta, and kappa opiate receptor types in the forebrain and midbrain of pigeons, *J. Comp. Neurol.,* 280, 359,1989.

143. Roberts, MH, and Rees, H, The antinociceptive effects of stimulating the pretectal nucleus of the rat, *Pain,* 25,83,1986.

144. Robertson, RT, Thompson, SM, and Kaitz, SS, Projections from the pretectal complex to the thalamic lateral dorsal nucleus of the cat, *Exp. Brain Res.,* 51,157,1983.

145. Robinson, DA, Visual-vestibular interaction in motion perception and the generation of nystagmus, *Neurosci. Res. Prog. Bull.,* 18,731,1980.

146. Robson, JA, Mason, CA, The synaptic organization of terminals traced from individual labeled retinogeniculate axons in the cat, *Neurosci.,* 4,99,1979.

147. Rothenberg, S, Peck, EA, Schottenfeld, S, Betley, GE, and Altman, JL, Methadone depression of visual signal detection performance, Pharmacol. Biochem. Behav., 11,521,1979.

148. Rothenberg, S, Schottenfeld, S, Gross, K, and Selkoe, D, Specific oculomotor deficit after acute methadone. I. Saccadic eye movements, *Psychopharmacology,* 67,221,1980a.

149. Rothenberg, S, Schottenfeld, S, Selkoe, D, and Gross, K, Specific oculomotor deficit after acute methadone. II. Smooth pursuit eye movements, *Psychopharmacol.,* 67,229,1980b.

150. Sako, H, Kojima, T, and Okado, N, Immunohistochemical study on the development of serotoninergic neurons in the chick: I. Distribution of cell bodies and fibers in the brain, *J. Comp. Neurol.,* 253,61,1986.

151. Scalia, F, The termination of retinal axons in the pretectal regions of mammals, *J. Comp. Neurol., 145*,223,1972.

152. Scalia, F, and V, Arango, Topographic organization of the projections of the retina to the pretectal region in the rat, *J. Comp. Neurol.,* 186,271,1979.

153. Schiff, D, Cohen, B, Buttner-Ennever, J, and Matsuo, V, Effects of lesions of the nucleus of the optic tract on optokinetic nystagmus and after-nystagmus in the monkey, *Exp. Brain Res.,* 79,225,1990.

154. Schiff, D, Cohen, B, and Raphan, T, Nystagmus induced by stimulation of the nucleus of the optic tract in the monkey, *Exp. Brain Res.,* 70,1,1988.

155. Schmidt, M, Mediation of visual responses in the nucleus of the optic tract in cat and rats by excitatory amino acid receptors, *Neurosci. Res.,* 12,111,1991.

156. Schmidt, M, Lewald, J, and van der Togt, C, Hoffmann K-P, The contribution of GABA-mediated inhibition to response properties of neurons in the nucleus of the optic tract in the rat, Eur J Neurosci 6,1656-1661.

157. Schmidt, M, Zhang, H-Y, and Hoffmann, K-P, OKN-related neurons in the rat nucleus of the optic tract and dorsal terminal nucleus of the accessory optic system receive direct cortical input, *J. Comp. Neurol.,* 330,147,1993.

158. Schonitzer, K, and Hollander, H, Retinotectal terminals in the superior colliculus of the rabbit: A light and electron microscopic analysis, *J. Comp. Neurol.,* 223,153,1984.

159. Simpson, JI, The accessory optic system, *Ann. Rev. Neurosci.,* 7,13,1984.

160. Simpson, JI, Giolli, RA, and Blanks, RHI, The pretectal nuclear complex and the accessory optic system. in: *Neuroanatomy of the Oculomotor System (Buttner-Ennever JA,* Ed.), pp. 333-362. Amsterdam: Elsevier/North Holland Biomedical Press,1988.

161. Simpson, JI, Soodak, RE, and Hess, R, The accessory optic system and its relation to the vestibulocerebellum. in: *Reflex Control of Posture and Movement. Prog. Brain Res.,* Vol. 50 (Granit R, Pompeiano O, Eds.), pp. 715-724, Amsterdam: Elsevier/North Holland Biomedical Press,1979.

162. Smeets, WJ, and Steinbusch, HW, Distribution of serotonin immunoreactivity in the forebrain and midbrain of the lizard Gekko gecko, *J. Comp. Neurol.,* 271,419,1988.

163. So, KF, Cambell, G, and Lieberman, AR, Synaptic organization of the dorsal lateral geniculate nucleus of the adult hamster, An electron microscope study using degeneration and horseradish peroxidase techniques, *Anat. Embryol. (Berl.),* 171,223,1985.

164. Speciale, SG, Manaye, KF, Sadeq, M, and German, DC, Opioid receptors in midbrain dopaminergic regions of the rat II. Kappa and delta receptor autoradiography, *J. Neural. Transm., [Gen Sect]* 91,53,1993.

165. Starr, MS, Multiple opiate receptors may be involved in suppressing gamma aminobutyratase in substantia nigra, *Life. Sci.,* 37,2249,1985.

166. Stone, J, and Fukuda, Y, Properties of cat retinal ganglion cells: a comparison of w-cells with x- and y-cells, *J. Neurophysiol.,* 37,722,1974.

167. Suzuki, DA, May, JG, Keller, EL, and Yee, RD, Visual motion response properties of neurons in dorsolateral pontine nucleus of alert monkey, *J. Neurophysiol.,* 63,37,1990.

168. Swanson, LW, and Hartman, BK, The central adrenergic system. An immunofluorescence study of the location of cell bodies and their efferent connections in the rat utilizing dopamine-beta-hydroxylase as a marker, *J. Comp. Neurol.,* 163,467,1975.

169. Swanson, LW, Simmons, DM, Whiting, PJ, and Lindstrom, J, Immunohistochemical localization of neuronal nicotinic receptors in the rodent central nervous system, *J. Neurosci.,* 7,3334,1987.

170. Takeda, T, and Maekawa, K, The origin of the pretecto-olivary tract. A study using the horseradish peroxidase method, *Brain Res.,* 117,319,1976.

171. Terasawa, K, Otani, K, and Yamada, J, Descending pathways of the nucleus of the optic tract in rat, *Brain Res.,* 173,405,1979.

172. Torigoe, Y, Blanks, RHI, and Precht, W, Anatomical studies on the nucleus reticularis tegmenti pontis in the pigmented rat. II. Subcortical afferents demonstrated by the retrograde transport of horseradish peroxidase, *J. Comp. Neurol.,* 234,88,1986.

173. Turlejski, K, Djavadian, RL, and Dreher, B, Parabigeminal, pretectal and hypothalamic afferents to rat's dorsal lateral geniculate nucleus. Comparison between albino and pigmented strains, *Neurosci. Lett.,* 160, 225,1993.

174. Turski, L, Havemann, U, Schwarz, M, and Kuschinsky, K, Disinhibition of nigral GABA output neurons mediates muscular rigidity elicited by striatal opioid receptor stimulation, *Life Sci.,* 31,2327,1982.

175. van der Togt, C, Nunes Cardozo, B, and van der Want, J, Medial terminal nucleus terminals in the nucleus of the optic tract contain GABA: An electron microscopical study with immunocytochemical double labeling of GABA and PHA-L, *J. Comp. Neurol.,* 312,231,1991.

176. van der Togt, C, and Schmidt, M, Inhibition of neuronal activity in the nucleus of the optic tract due to electrical stimulation of the medial terminal nucleus in the rat, *Eur. J. Neurosci.,* 6,558,1994.

177. van der Togt, C, van der Want, J, and Schmidt, M, Segregation of direction selective neurons and synaptic organization of inhibitory intranuclear connections in the medial terminal nucleus of the rat: an electro-physiological and immunoelectron microscopical study, *J. Comp. Neurol.*, 338,175,1993.

178. van der Want, JJL, and Nunes Cardozo, JJ, GABA immuno-electron microscopic study of the nucleus of the optic tract in the rabbit, *J. Comp. Neurol.*, 271,229,1988.

179. van der Want, JJL, Nunes Cardozo, JJ, and van der Togt, C, GABAergic neurons and circuits in the pretectal nuclei and the accessory optic system of mammals. in: *Progress in Brain Research*, Vol. 90 (Mize, RR, and Sillito, A, Eds.), pp. 283-305, Amsterdam: Elsevier Science Publishers B.V,1992.

180. Vrensen, G, and De Groot, D, Quantitative aspects of the synaptic organization of the superior colliculus in control and dark-reared rabbits, *Brain Res.*, 134,417,1977.

181. Wada, E, Wada, J., Boulter, J, Deneris, E, Heinemann, S, Patrick, J, and Swanson, LW, Distribution of Alpha2, Alpha3, Alpha4, and Beta2 neuronal nicotinic receptor subunit mRNAs in the central nervous system: A hybridization histochemical study in the rat, *J. Comp. Neurol.*, 284,314,1989.

182. Waespe, W, Cohen, B, and Raphan, T, Dynamic modification of the vestibulo-ocular reflex by the nodulus and uvula, *Science*, 228,199,1985.

183. Waespe, W, and Henn, V, Gaze stabilization in the primate. The interaction of the vestibulo-ocular reflex, optokinetic nystagmus, and smooth pursuit, *Rev. Physiol. Biochem. Pharmacol.*, 106,37,1987.

184. Walberg, FT, Nordby, T, Hoffmann, K-P, and Hollander, H, Olivary afferents from the pretectal nuclei in the cat, *Anat. Embryol.*, 161,291,1981.

185. Walley, RE, Receptive fields in the accessory optic system of the rabbit, *Exp. Neurol.*, 17,27,1967.

186. Wallman, J, McKenna, OC, Burns, S, Velez, J, and Weinstein, B, Relations of the accessory optic system and pretectum to optokinetic responses in chicken. in: *Progress in Oculomotor Research* (Fuchs AF, and Becker W, Eds.), pp. 435-442. Amsterdam: Elsevier/North Holland Biomedical Press,1981.

187. Weber, JT, and Harting, JK, The efferent projections of the pretectal complex: An autoradiographic and horseradish peroxidase analysis, *Brain Res.*, 194,1,1980.

188. Weber, JT, Pretectal complex and accessory optic system of primates, *Brain Behav. Evol.*, 26,117,1985.

189. Weber, JT, and Giolli, RA, The medial terminal nucleus of the monkey: Evidence for a "complete" accessory optic system, *Brain Res.*, 365,188,1986.

190. Willis, WD, Anatomy and physiology of descending control of nociceptive responses of dorsal horn neurons: comprehensive review, *Prog. Brain Res.*, 77,1,1988.

191. Wolters, JG, ten Donkelaar, HJ, Steinbusch, HW, and Verhofstad, AA, Distribution of serotonin in the brain stem and spinal cord of the lizard *Varanus exanthematicus:* an immunohistochemical study, *Neurosci.*, 14,169,1985.

192. Yucel, YH, Jardon, B, and Bonaventure, N, Involvement of ON and OFF retinal channels in the eye and head horizontal optokinetic nystagmus of the frog, *Vis. Neurosci.*, 2,357,1989.

193. Yucel, YH, Jardon, B, Kim, M-S, and Bonaventure, N, Directional asymmetry of the horizontal monocular head and eye optokinetic nystagmus: Effects of picrotoxin, *Vis. Res.*, 30,549,1990.

194. Zagon, A, Terenzi, MG, and Roberts, MHT, Direct projections from the anterior pretectal nucleus to the ventral medulla oblongata in rats, *Neuroscience*, 65,253,1995.

195. Zee, DS, Tusa, RJ, Butler, PH, Herman, SJ, and Gucer, C, Effects of occipital lobectomy upon eye movements in primates, *J. Neurophysiol.*, 58,883,1987.

ACKNOWLEDGMENTS

The authors are indebted to Carol Zizz for preparing the illustrations, and Thomas Pak and Marnie Dobson for editorial assistance. This work was funded, in part, by grants to R.A.G. and R.H.I.B. from the Hoover Foundation and National Science Foundation (IBN 9121376). R.H.I.B. is supported by a grant from the Rehabilitation Research and Development Service, Department of Veterans Affairs. RHIB and JLvdW are supported by a NATO travel grant (#CRG920203).

15

Somatosensory Influences on the Vestibular System

W.L. Neuhuber and S. Bankoul

CONTENTS

15.1 SYNOPSIS

Spinovestibular projections encompass primary and secondary afferents. Both classical and modern anatomical studies have demonstrated an almost exclusive upper cervical origin for the former, while the latter originate along the whole extent of the spinal cord. Primary afferents are exclusively ipsilateral, while second-order afferents provide bilateral input with a contralateral bias. Direct spinovestibular projections terminate in the medial and descending vestibular nuclei and in cell Groups X and F. Electrophysiological studies indicate prominent short latency influence also on the lateral vestibular nucleus, which is probably mediated by intrinsic and commissural vestibular neurons. The most likely transmitter of primary spinal afferents is an excitatory amino acid, while transmitters of second-order spinal afferents are unknown. Spinovestibular afferents are mainly related to vestibulo-spinal and, to a lesser extent, also vestibulo-oculomotor neurons. Spinovestibular connections, in particular from neck proprioceptors, play a significant role in head-to-trunk coordination and perception of head and body position and movement in space. They are also involved in eye-head coordination, although these mechanisms are less well understood. Neck proprioceptive afferents may also play a role in the pathogenesis of a variety of so-called vertebragenic syndromes after soft tissue lesions in the neck, in particular cervical vertigo.

15.2 INTRODUCTION

Control of posture and movements and orientation in space is a central task for all living organisms. During vertebrate phylogeny, several sense organs and neural networks devoted to these functions have evolved, the vestibular system being one of the oldest and most highly conserved. Land-dwelling vertebrates, especially those with flexible necks, rely particularly on the sophisticated cooperation of vestibular, visual, and proprioceptive systems.[16] The aim of this chapter is to briefly review available morphological and functional data on spinovestibular connections, and to focus on recent findings. Since reports on the neurochemistry of spinovestibular projections are almost lacking; emphasis will be made on hodological concepts.

15.3. HISTORICAL REVIEW

When Magnus published his seminal volume *Körperstellung* in 1924,[63] it became evident that posture control requires the intimate cooperation of vestibular, proprioceptive, and visual reflexes. Several reports had previously indicated an influence of neck afferents on equilibrium, posture control, and eye movements.[12,58,61] Neck reflexes acting on the limbs and on the oculomotor system had been defined which stabilize, in concert with vestibular reflexes, eye and body position during head movements.

Since the 1920s there has been much debate on the existence and reliable diagnosis of the so-called cervical nystagmus.[17] In analogy to vestibular vertigo, which is invariably accompanied by nystagmus, a correlation between subjective complaints of dizziness or vertigo and nystagmus was searched for in patients without vestibular lesions. Many of them had histories of soft-tissue neck injury. However, reports were conflicting and are still so today.[34,52] Nevertheless, three main theories have been proposed to explain the pathogenesis of cervical vertigo.

According to the vasculogenic concept, blood flow impairment in the vertebral arteries may result in transient or longer-lasting brainstem ischemia with ensuing vestibular and oculomotor symptoms. However, the time course of onset and cessation of symptoms is not consistent with a vascular origin. Even in patients with occlusion of one vertebral artery, nystagmus was inconsistently observed.[34]

A second theory proposes involvement of the sympathetic vertebral nerve. According to this model, irritation of sympathetic nerves leads to vasospasm and ischemia in brainstem, cerebellum, and inner ear.[34] However, as with the vasculogenic concept, this neurovasculogenic hypothesis suffers from similar shortcomings, and there is no convincing evidence for it.

The neurogenic hypothesis suggests improper stimulation of neck proprioceptors as the main pathogenetic factor of cervical nystagmus and vertigo. Although appealing and compatible with anatomical and experimental evidence,[12,17,58] a neck proprioceptive genesis of cervical nystagmus is not unanimously accepted, as is the existence of cervical nystagmus and vertigo.[24,34,52]

The proposed proprioceptive and vestibular interactions in posture and oculomotor control have been supported by early anatomical studies. Using degeneration methods, spinovestibular afferents, in particular from caudal levels of the cord, and primary afferents from upper cervical nerves to vestibular nuclei have been described.[21,30,41] Terminal areas were restricted to the caudal-most tips of the medial and descending vestibular nuclei and cell group x. A few degenerating fibers have also been found in the lateral vestibular nucleus.

15.4 SENSORS IN THE NECK: THE ROLE OF MUSCLE VERSUS JOINT AFFERENTS

Although early studies emphasized receptors in joint capsules and ligaments as key structures for proprioception,[65] more recent evidence has demonstrated the important role of muscle spindles, not only as afferents for neck reflexes, but also for the computation of position sense.[38,45,87,98] Neck muscles, in particular deep suboccipital muscles, contain spindles at an unusually high density.[82,92] Although these muscles are small, the high spindle density results in a high absolute number of muscle spindles.[87] Furthermore, spindles in neck muscles are arranged in a highly sophisticated manner. They typically occur in tandem (in series) of up to five spindles sharing a single long intrafusal fiber, in paired linkages or as compound spindles. Up to 50% of neck muscle spindles were found in tandem arrangements, whereas only some 10 to 20% of spindles in hindlimb muscles showed a similar complexity.[38]

In addition to low threshold muscle sensors, there are high threshold receptors in both neck muscles and joints. However, available data do not support any particular specialization which would distinguish them from high threshold sensors elsewhere in the musculoskeletal system.[70,86]

15.5 RECENT DATA ON SPINAL AFFERENTS TO THE VESTIBULAR NUCLEAR COMPLEX

15.5.1 SECOND-ORDER SPINAL AFFERENTS

In recent years, several studies have investigated spinovestibular projections utilizing both degeneration and tract tracing techniques in various species. The long-held view that the main spinal input to the vestibular nuclei arises from lumbar levels had to be changed after significant projections from the central cervical nucleus in the rostral cervical cord had been described.[21,27,67,69,95] In a comprehensive study, McKelvey-Briggs et al. reinvestigated the issue of both spinal origin and precise termination within the vestibular nuclear complex of second-order spinovestibular neurons using both retrograde and anterograde tracing methods in the cat.[67] In striking contrast to the classical view, they found the most significant projection arising from upper cervical levels, most notably from the contralateral central cervical nucleus. Ipsilateral pathways originated in Lamina VI of both rostral and caudal cervical segments. From all spinal levels investigated (C1-8, T9-12, lumbar enlargement), bilateral projections arose from neurons in Laminae IV, V, VII, and VIII. Thus, second-order spinovestibular neurons are located in laminae receiving proprioceptive (VI, VII, VIII), exteroceptive (IV, V), and also thin-calibre, probably nociceptive muscle and joint (IV, V) afferents. The cervical and lumbar enlargements contributed only modestly to the spinovestibular input, and thoracic projections were scarce. Similar findings have been reported for the rat where, in addition, a sparse ipsilateral projection from the central cervical nucleus could be found.[11]

Lateral and medial spinovestibular pathways have been described.[46,84] The lateral pathway originates at caudal levels of the spinal cord and courses together with the dorsal spinocerebellar tract. It is directed to Group X, the descending vestibular nucleus, and probably also the lateral vestibular nucleus. Most of the spinovestibular fibers apparently are not collaterals of spinocerebellar axons[21]. However, this issue has to be solved using double-tracing strategies. A medial spinovestibular pathway travels through the medial reticular formation and terminates in the medial vestibular nucleus.[46,84] It originates mainly in the central cervical nucleus.[95,96]

As has been shown for the cat, the termination sites of second-order spinal afferents are concentrated in cell group x, and in the medial and descending vestibular nuclei. A sparse projection is targeted at cell group f[67] (Figure 15.1).

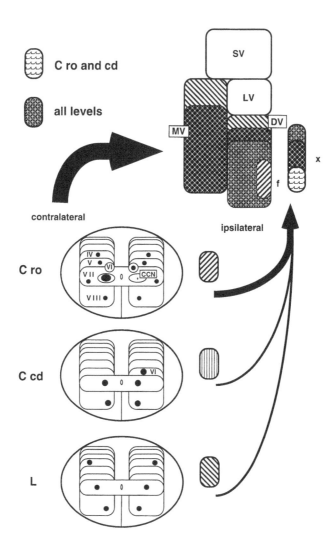

FIGURE 15.1 Origin and terminal distribution of second-order spinovestibular projections. Note the contralateral bias and the concentration of terminals in the caudal part of the vestibular nuclear complex (VNC). SV, LV, MV, DV: superior, lateral, medial, and descending (spinal) vestibular nucleus, respectively. x, f; cell groups x and f; CCN: central cervical nucleus; C ro, C cd: rostral and caudal cervical segments; L: lumbar enlargement. Roman numbers indicate Rexed's laminae. (According to data from References 11 and 67.)

It should be noted that although cell groups x and f are devoid of primary vestibular afferent terminations, they are classically included in studies on the vestibular nuclear complex.[25,46,69] Surprisingly, there is no convincing evidence for a direct spinal input to the lateral vestibular nucleus, as suggested by earlier reports. Likewise, spinovestibular terminals are absent from the superior vestibular nucleus and cell group y.

Within the various subnuclei, projections from different levels of the spinal cord are distributed in a largely overlapping manner. Nevertheless, some areas have a more restricted input (Figure

15.1). For example, the rostral medial and descending subnuclei receive only lumbar afferents, while cell group f is a specific target for fibers from upper cervical segments. The caudal four-fifths of the medial and descending vestibular nuclei, and the middle third of cell group x receive afferents from both upper cervical and lumbar levels, while the caudal two-thirds of the descending nucleus and the rostral third of cell group x are innervated by spinovestibular fibers from all levels.

In summary, second-order spinal afferents to the vestibular nuclear complex arise mainly from rostral cervical levels — in particular, the contralateral central cervical nucleus — and distribute to cell group x, the medial and descending vestibular nuclei, and cell group f.

15.5.2 PRIMARY SPINAL AFFERENTS

With the advent of sensitive anterograde and transganglionic tracing techniques, the issue of direct spinal ganglionic input to vestibular nuclei could be reassessed. Pfister and Zenker[78] were the first to describe a projection of afferents from the rat splenius capitis muscle to the descending vestibular nucleus using the HRP technique, thus confirming earlier studies.[30,41] Since then, several authors have reported spinal primary afferent terminations in rat,[74,77] cat,[6,9] and guinea pig[81] vestibular nuclei. Even in primates, a direct input from cervical spinal afferents to the medial and descending vestibular nuclei and cell group x has been described.[39,42]

Primary spinal afferents to vestibular nuclei originate almost exclusively from rostral cervical spinal ganglia (C2 and C3; C1 is frequently absent or contains only few afferent neurons), while fibers from caudal cervical (C4 to C8) levels are significantly sparser, and are even absent from thoracic and lumbosacral segments.[4,74,75] This could be evidenced by both anterograde and retrograde tracing. Transganglionic tracing demonstrated a muscular origin of cervical afferents to vestibular nuclei. Remarkably, a vestibular projection arises not only in epaxial,[74,81] but also hypaxial muscles, (e.g., the sternomastoid[74] and geniohyoid[73]). Cutaneous afferents obviously do not contribute to this projection. The issue of direct input from joint afferents could not be convincingly settled, since experimental isolation of, and tracer application to, nerve branches supplying cervical joints without contamination of muscular afferents proved impossible, at least in the rat.[74]

Since thin-calibre (Ad and C) muscle afferents are unlikely to ascend to supraspinal levels, terminals in vestibular nuclei most likely originate from thick-calibre afferents, probably from muscle spindles alone. Cell size distribution of rat and cat spinal ganglion neurons labeled retrogradely from the medial vestibular nucleus showing a maximum at 40 to 60 mm with a minority of small neurons is in accordance with this assumption.[9,10] However, this question awaits further clarification by, for example, intraaxonal staining of physiologically identified afferents.

In all species investigated, the termination sites of direct spinal primary afferent projections in the vestibular nuclear complex are in the caudal half of the medial and descending vestibular nuclei (Figures 15.2 and 15.3).

The fibers appear to represent collaterals of axons ascending in the most lateral portion of the cuneate fascicle to their main target in the external cuneate nucleus and traverse the ventral part of the descending vestibular nucleus. Patches of labeled terminals are found in cell groups x and f. In the medial vestibular nucleus, terminals are concentrated in a central "core" area as viewed in coronal sections (Figure 15.2). Single fibers continue their path to the nucleus prepositus hypoglossi. There are only a few aberrant fibers in the lateral, and none in the superior vestibular nuclei. Besides these vestibular targets, the main termination areas of thick-calibre neck muscle afferents are the central cervical (Figure 15.4) and the external (Figures 15.2 and 15.4) and medial cuneate nuclei.

In the spinal cord, further significant terminal labeling is found in Lamina VI (Figure 15.4) and in the ventral horn. No thick-calibre muscle afferents terminate within subnuclei of the spinal trigeminal complex,[1,74,81] which is, however, reached by cutaneous and probably also joint afferents.[1,74]

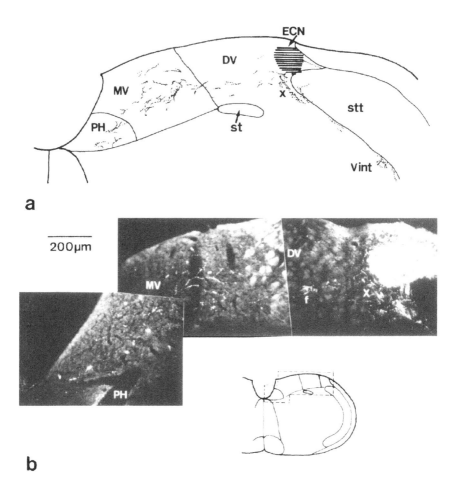

FIGURE 15.2 (A) Camera lucida drawing of a representative 40 mm coronal section adjacent to b. Note the distribution of labeled fibers to a "core" region in the MV. (B) Coronal section through the caudal vestibular nuclear complex after WGA-HRP injection into the C2 spinal ganglion in the rat. Note the patches of labeling in cell groups x and f, and labeled fibers in the descending (DV), and medial (MV) vestibular nuclei and in the nucleus prepositus hypoglossi (PH). The external cuneate nucleus is brightly labeled on the right.

Although terminal density in the vestibular nuclei appears to be quite sparse as compared to labeling in the external cuneate nucleus, the actual number of boutons might be higher. Transganglionic tracing is accompanied by a significant loss of tracer in spinal ganglion cell bodies due to degradation, probably resulting in underestimation of true central terminal density.[102] WGA-HRP, which had been used in anterograde tracing from dorsal root ganglia, preferentially labels thin calibre rather than thick-calibre afferents.[83] Thus, collaterals of muscle spindle afferents in vestibular nuclei might be underrepresented in these studies.

Besides spinal primary afferents, the vestibular nuclear complex receives a scant projection from trigeminal primary afferents arising in the mandibular area.[64] They are thought to contribute to head-neck coordination during mastication.

In summary, primary spinal afferents to vestibular nuclei enter the spinal cord almost exclusively through rostral cervical dorsal roots and project entirely ipsilaterally to the caudal half of the medial and descending vestibular nuclei, including groups x and f. Although they share the main termination area with second-order spinal afferents, the mainly contralateral origin of the latter has to be emphasized. Furthermore, second-order afferents originate throughout the length of the spinal

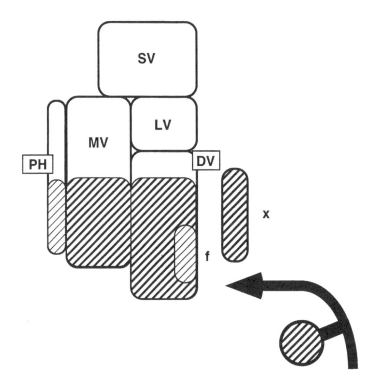

ipsilateral proprioceptive neck afferents C 2, 3, 4,...

FIGURE 15.3 Termination areas of primary spinal afferents in the vestibular nuclei. Note their concentration in cell group x and the caudal halves of the DV and MV. (According to data from References 4, 9, 42, 74, 77, 81.)

cord, although they are more numerous in upper cervical segments. A quantitative comparison in the rat showed that the number of dorsal root ganglion cells in C2 projecting to the medial vestibular nucleus equaled the number of central cervical nucleus neurons in segment C2 projecting to the same area.[75] Thus, it is reasonable to assume that primary spinal afferents to vestibular nuclei are functionally not less important than second-order spinovestibular afferents. Only thick caliber muscle afferents, most likely representing Ia neurons, contribute to the primary afferent projection.

15.6 NEUROCHEMICAL PHENOTYPES

To date, there is no information available on neurochemical characteristics of spinal ganglion cells and spinal cord neurons projecting to vestibular nuclei. However, it is reasonable to assume that glutamate and/or aspartate are the main transmitters, at least in primary afferents.[15,91,94] This assumption is supported by the demonstration of all the different types of glutamatergic receptors in vestibular nuclei[31] and by functional studies in the frog indicating glutamatergic synaptic transmission of proprioceptive signals in vestibular nuclei (see also Chapters 1 and 4, this volume).[79] Cell size distribution of dorsal root ganglion cells retrogradely labeled from vestibular nuclei argue against a major contribution from peptidergic afferents.[9,10] However, the issue of neurochemical phenotypes remains to be studied in more detail using retrograde labeling from vestibular nuclei combined with immunocytochemistry.

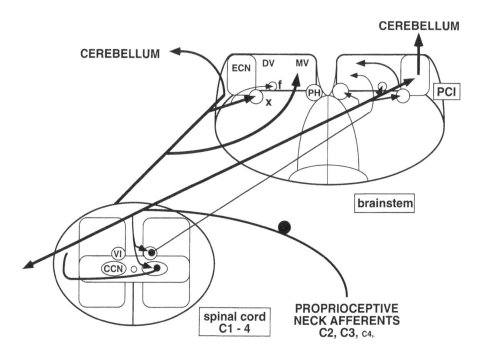

FIGURE 15.4 Summary diagram of spinovestibular projections from upper cervical segments emphasizing differences between primary (ipsilateral) and second-order (mainly contralateral) afferents. ECN: external cuneate nucleus; PCI: inferior cerebellar peduncle; PH: nucleus prepositus hypoglossi.

15.7 ELECTROPHYSIOLOGICAL STUDIES

Since the late 1950s, numerous studies using electrophysiological methods have provided evidence for somatosensory spinal input to vestibular nuclei. As a consistent feature, convergence of somatosenory and vestibular primary afferents was found on the majority of second-order vestibular neurons. While earlier descriptions focused on proprioceptive and exteroceptive signals from both fore and hindlimbs, emphasis was shifted to neck proprioceptors, in addition to limb afferents, by a seminal report of Fredrickson and colleagues.[44] The question of direct vs. indirect pathways for spinovestibular afferents was largely solved by the findings of Wilson et al., who showed that spinal input reached the vestibular nuclei even in cerebellectomized animals.[99] Subsequent studies have elaborated on details of spinovestibular projections, for example, ipsi- versus contralateral input origin in different parts of the body, or termination in various subnuclei of the vestibular complex.

There are both ispi- and contralatral inputs, mainly from proximal limb joints and neck, converging on the same or on separate vestibular units.[85] Both early (<6 m) and late responses in vestibular nuclei neurons have been described upon electrical stimulation of the contralateral C2 spinal ganglion.[20] In the target neurons, variable patterns of excitation and inhibition were elicited. The peripheral origin of somatosensory input to the vestibular nuclei appears to be rather complex. Although some authors emphasized proprioceptive over exteroceptive afferents,[44] muscle spindles were largely disfavored because adequate stimulation of hindlimb Ia afferents (sinusoidal muscle stretch) was ineffective in modulating units in cat lateral and medial vestibular nuclei.[79] Rather, group Ib, type II, and even III muscle and exteroceptive afferents were considered most effective in facilitating vestibular neurons.[79,101] Among proprioceptive sensors, joint afferents in limbs and neck were favored over spindle afferents in early studies. However, more recent investigations have demonstrated frequency-response dynamics of neck reflexes in forelimbs that match response characteristics of muscle spindles.[19,38,98] Likewise, neck muscle vibration that specifically stimulates muscle spindles was found to contribute, together with vestibular stimulation, to the subjective "straight ahead" orientation in man.[56] Thus, it is reasonable to assume that proprioceptive spinovestibular input, at least from the neck, derives mainly from muscle spindles.[98] The contribution of joint afferents is still to be determined.

Most investigations have centered on the lateral vestibular nucleus as a site of somatosensory-vestibular interaction. Both limb and neck units have been described, with a predominance of limb units.[85] However, the descending and medial vestibular nuclei also harbor significant numbers of neurons mediating somato-vestibular interaction.[3,85] Even in the superior vestibular nucleus, a few units with neck input have been described.[85] Convergence has been shown with both macular and canal input.[57,88]

Many vestibular neurons influenced by somatosensory afferents project through the vestibulospinal tracts to the spinal cord. Thus, somatosensory, in particular neck and vestibular afferent convergence, is thought to be of relevance in neck-to-limb reflexes. Vestibular and neck input typically cancel each other in vestibulo- and neck-to-forelimb reflexes.[57] On the other hand, the interaction between vestibulo- and cervico-collic reflexes is synergistic.[76] Since synergistic interaction of somatosensory and vestibular input is absent while antagonism between these two modalities is present in vestibular nuclei, they are supposed to play a role in vestibulo-somatic convergence in neck-to-forelimb but not cervico-collic reflexes.[57] The latter may be entirely integrated in the spinal cord, where neck proprioceptive and vestibular convergence occurs on spinal interneurons.[97]

Responses in vestibular nuclei upon somatosensory stimulation occur at relatively short latencies, suggesting rather direct pathways. Since stimulation of the contralateral C2 spinal ganglion was most effective in eliciting responses in the lateral vestibular nucleus,[20] a monosynaptic action of primary afferents is unlikely because of their ipsilateral course and the lack of significant primary afferent terminals in the lateral vestibular nucleus. More likely, primary afferents are relayed in the central cervical nucleus, which projects mainly to the contralateral vestibular nuclei. However, it is likely that yet another synapse may be associated with this pathway, since second-order spinovestibular afferents are sparse or even absent in the lateral vestibular nucleus.[67] The case may be different for neurons in the descending and medial vestibular nuclei and in cell group x. Both primary and second-order spinal afferent terminals have been demonstrated anatomically in these vestibular nuclei. At least in cell group x, ipsilateral electrical stimulation elicits responses that are relayed to the flocculus.[100]

There is yet another caveat associated with the correlation of the anatomy of spinal primary afferent terminal distribution in vestibular nuclei with data derived from electrophysiological or even more complex behavioral and psychophysical experiments. Investigations of impulse propagation into collaterals of primary afferents have shown propagation failure at many branch points.[93] It has been suggested that a $GABA_A$ mechanism is responsible for this kind of preterminal control. Since spinal primary afferent fibers in vestibular nuclei appear to represent collaterals of parent axons terminating heavily in the external cuneate nucleus, the question arises as to whether primary afferent volleys are transmitted efficiently enough to result in postsynaptic effects in vestibular neurons.

In summary, electrophysiological studies have demonstrated significant spinovestibular input, mainly from neck proprioceptors. These proprioceptive afferents converge with afferents from the labyrinth and are thought to play a significant role in head-to-trunk coordination. However, these data are not easily reconciled with anatomical evidence. In particular, pathways to the lateral vestibular nucleus appear to be less direct, although somatosensory signals arrive at short latencies.

15.8 CONNECTIVITY OF VESTIBULAR NEURONS CONTACTED BY SPINAL AFFERENTS AND FUNCTIONAL CONCEPTS

The main terminal areas of both primary and second-order spinal afferents reside in the descending and medial vestibular nuclei, in particular in their caudal parts and in cell group x. These subnuclei, in turn, harbor a variety of projection, commissural, and interneurons that open up a wide range of possible interactions for spinal afferents (Figure 15.5).

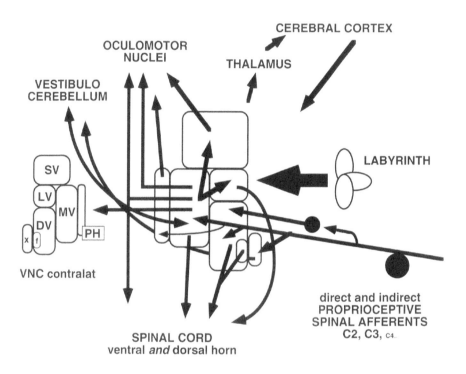

FIGURE 15.5 Synopsis of possible connectivity of primary and second-order spinovestibular afferent neurons in the vestibular nuclear complex (VNC). Best evidence exists for direct connections to vestibuo-spinal projection neurons. (Based on data presented in Reference 26, with inclusion of our own data.)

15.8.1 VESTIBULO-SPINAL NEURONS

In studies on vestibulo-somatosensory convergence, much attention has been focused on vestibulo-spinal neurons in the lateral vestibular nucleus.[19,20,88] As discussed above, spinovestibular pathways to this nucleus must be more indirect than what is suggested by the short poststimulus latencies, and may involve intrinsic vestibular connections (see below). However, vestibulo-spinal neurons located in the caudal halves of both the descending and medial vestibular nuclei and in cell group f, which give rise to the medial and caudal vestibulo-spinal tracts, may be directly contacted by spinal afferents. Recently, one of the authors has demonstrated anterogradely labeled primary afferent terminals in close association with vestibular neurons in the medial subnucleus that were retrogradely identified as vestibulo-spinal, following FluoroGold administration into the cervical spinal cord in the rat.[8] This correlates nicely with electrophysiological data obtained in the cat.[3] Of particular interest is a projection of vestibulo-spinal neurons in the caudal medial nucleus, not only to the cervical ventral horn, but also to the dorsal horn in the rat. It has been hypothesized that this cervico-vestibulo-cervical pathway may play a role in sensory processing in the dorsal horn, probably by inhibiting the transmission of nonproprioceptive afferent information via second-order spinovestibular neurons.[11]   Recent investigations suggest that similar cervico-vestibulo-cervical loops also exist in the cat.[9,35]

15.8.2 VESTIBULO-OCULOMOTOR NEURONS

Most second-order vestibulo-oculomotor neurons reside in the superior and rostral medial vestibular nuclei, and in a core region encompassing the border zones of the medial, lateral, and descending nuclei[26,84] (See also Chapter 9, this volume). These vestibular areas are reached by some second-order, but almost no primary spinal afferent fibers. However, a group of non-second-order, burst-tonic eye position-related neurons is located at more caudal levels at the border between the medial vestibular and prepositus hypoglossi nuclei.[26] These neurons might be within the termination area of primary spinal afferents. Furthermore, it should be noted that preoculomotor neurons are abundant in the nucleus prepositus hypoglossi,[66] an area that is also targeted by some primary neck afferents.[74,81] In addition, some of the vestibulo-spinal neurons that receive spinal input may send axonal collaterals to the oculomotor nuclei. This concept is compatible with the results of anterograde tracing experiments in which injections of PHA-L into the caudal medial vestibular nucleus labeled some terminal arborizations in the rat oculomotor nucleus (Bankoul, unpublished observations). There is also experimental evidence for neck and vestibular convergent input to motoneurons in the abducens nucleus, which is thought to be mediated via the vestibular nuclei.[48,90] However, direct connections of spinal afferents to vestibulo-oculomotor neurons may be of less functional importance.[3] There are conflicting reports on the cervico-ocular reflex,[13,18,22,33] on cervical nystagmus, and on cervical vertigo,[34,52] which may be explained by a less-stringent connectivity within the vestibular nuclei.

15.8.3. VESTIBULO-CEREBELLAR NEURONS

Cell group x, which receives dense projections from both primary and second-order spinal afferents, projects to the ipsilateral flocculus.[100] Thus, this projection complements the pathway that arises from the external cuneate and Clarke's nuclei, and which projects directly to the anterior and posterior vermis and intermediate hemispheres. Floccular and nodulus-uvular projection neurons are also concentrated in middle and caudal parts of the descending and medial vestibular nuclei, where primary spinal afferents terminate. This projection to the cerebellum is reciprocal to a certain extent.[26,84] Thus, spinal afferents probably also modulate vestibulo-cerebellar circuits at the level of the vestibular nuclei.

15.8.4. VESTIBULO-THALAMIC NEURONS

Vestibular neurons relaying labyrinthine input to the ventroposterior thalamus reside in caudal descending and medial vestibular nuclei.[84] Thus, vestibulo-somatosensory convergence, which is known to occur in thalamic[32] and cortical[71] neurons, may take place already in vestibular neurons which project through a thalamic relay to the cortex. Many vestibulo-thalamic neurons may send collaterals to the spinal cord.[54]

15.8.5 INTRINSIC AND COMMISSURAL VESTIBULAR NEURONS

Recent tracing studies have demonstrated that most of the intrinsic and commissural vestibular neurons reside in the caudal and parvocellular medial vestibular nucleus.[40] Spinal afferents may establish direct contacts with these neurons, although this has not been directly demonstrated. Nonetheless, a relay of spinovestibular afferents through intrinsic neurons projecting to the lateral vestibular nucleus could provide a reasonable explanation for the pronounced short latency effects in this nucleus in the absence of direct spinovestibular terminals. Likewise, relay of spinovestibular input through intrinsic and commissural connections could explain effects of somatosensory stimuli upon contralateral abducens motoneurons.[48,90] Transmission through these pathways may strongly depend on a variety of facilitatory and inhibitory influences within the vestibular nuclei. Thus, inconsistencies in the literature (e.g., the cervico-ocular reflex) may have their roots in the intrinsic complexity of the vestibular nuclear complex.

To summarize, available evidence favors a role for spinovestibular neurons, in particular neck afferents and their convergence with labyrinthine signals, in the coordination of vestibuo-spinal and neck reflexes. Although anatomical and functional connections to other systems (e.g., vestibulo-oculomotor, vestibulo-cerebellar, and vestibulo-thalamic) have been suggested, their significance is less certain.

15.9 CONCLUSION

15.9.1 Somatosensory Neck Afferents to Vestibular Nuclei: Part of a Peculiar "Neck Sensory Organ"?

Somatosensory afferents to vestibular nuclei originate mainly in upper cervical segments of the spinal cord. This is highlighted by the fact that primary afferents are relatively abundant from segments C2 and C3, but are virtually lacking from thoracic and lumbosacral levels of the spinal cord. Together with the high spindle density in suboccipital muscles, this connectivity pattern mirrors the important role played by neck sensors in perception and coordination of head-on-trunk position and movements and, in concert with vestibular sensors, of body motion in space.[72] Convergence of neck and vestibular afferents occurs in the cervical spinal cord and in a number of brain areas, the vestibular nuclei representing some of the important sites of this convergence.[72,75] The vestibular nuclei are not the only convergent structures that are directly concerned with motor control; neck-vestibular convergence on reticulospinal,[80] tectospinal,[2] and interneurons in the cervical spinal ventral horn[97] may be equally important. However, the vestibular nuclei are unique because they receive directly both primary neck and labyrinthine afferents. On the sensory side, neck proprioceptive-vestibular convergence has been described in the central cervical nucleus,[49] the external cuneate nucleus,[55] and in the thalamus.[32] Therefore, the cerebellum and cerebral cortex may receive multimodal input via the respective projections from the external cuneate and central cervical nuclei, and from the thalamus. In addition, extraocular proprioceptive and visual input is fed into these networks[5,7,28] (See also Chapter 13, this volume). Thus, interaction of all cues relevant for spatial orientation results in a reliable perception of body in space and appropriate steering of movements. Neck sensors and their projections to vestibular nuclei represent a significant component in this system. Since neck proprioceptors are very sensitive to head movements at low frequencies, thus "filling the gap" that is not covered by the labyrinth,[72] they may be considered, together with their projections, as the "cervical extension" of the vestibular apparatus.

15.9.2. Possible Clinical Implications

Patients suffering from post-traumatic soft tissue lesions at the cranio-cervical transition zone often present with a variety of symptoms which are difficult both to explain and treat successfully.[89] These include nonsystemic vertigo (cervical vertigo[24,47]), imbalance of gait, impaired hearing,[51] dysphonia,[50] and pain located in the occipital or retroorbital region.[47,60] Many of these patients profit from manipulations of upper cervical joints which are supposed to correct "functional deficits" in the musculoskeletal system.[43,68] Thus, involvement of proprioceptive and nociceptive afferents and their improper stimulation in lesioned tissues have long been postulated as a likely cause for the puzzling "vertebragenic" symptomatology.[24,34] A disordered neck proprioceptive input to vestibular nuclei and the ensuing computation with afferents from the labyrinth is supposed to result in transmission of error signals via vestibulothalamic neurons to the cortex. This may produce a subjective impression of vertigo. Other symptoms (e.g., disorders of gaze control[62] or vertebragenic dysphonia[50]) are more difficult to interpret, but may also find an explanation in disturbance of the intricate complexity of spino-vestibular interactions. Clearly, more research is needed.

15.9.3 UNSOLVED QUESTIONS AND FUTURE RESEARCH DIRECTIONS

Although recent studies have reshaped and extended our knowledge regarding spinovestibular systems, there are still a number of problems that await clarification. Furthermore, new questions have emerged. The assumption that primary cervical afferents represent collaterals of Ia fibers directed to the external cuneate nucleus should be verified by intraaxonal staining of physiologically identified afferents. Likewise, the contribution of Ib, II muscle, and II joint afferents to this projection must be elucidated. The question of whether second-order spinovestibular fibers are collaterals of spinoc-erebellar or spinothalamic tract axons must be solved using double-tracing strategies.

Since primary neck afferent projections have been shown mostly by WGA-HRP tracing, the density of collateral terminals in vestibular nuclei may have been underestimated because of a preferential demonstration of thin- over thick-calibre fibers with WGA-HRP.[83] Thus, primary neck afferent terminal density in vestibular nuclei should be reinvestigated using choleragenoid-HRP tracing which labels preferentially the central processes of thick-calibre afferents,[83] or using some of the newer dextran-labeled anterograde tracers.

The neurochemical coding of both primary and second-order spinovestibular afferents should be elucidated by combined retrograde tracing from vestibular nuclei and immunocytochemistry on upper cervical spinal ganglion neurons and the spinal cord. This approach could also contribute to solve the problem of functional types of these fibers. Since rapidly adapting mechanosensory neurons including muscle spindle afferents typically stain for calretinin and calbindin D-28k immunoreactivity,[36,37] their contribution to primary spinovestibular projections could be assessed. Likewise, the potential glutamatergic nature of primary spinovestibular afferents could be supported by immunohistochemical data.

Based on the distribution pattern of both primary and second-order spinovestibular terminals, and on results from electrophysiological studies, several inferences have been made as to their connectivity. Both morphological and electrophysiological data favor prominent direct connections to vestibulo-spinal neurons. In particular, the relationship between primary neck afferents and vestibulo-spinal neurons projecting to the cervical dorsal horn, together with the functional significance of this "new" descending pathway, should be elucidated. Preliminary results from double-tracing studies[8] should be extended to evaluate direct connections to, for example, vestibulo-oculomotor neurons. Since the exact pathway of spinal afferents to the lateral vestibular nucleus is still unclear, the connections of spinovestibular to intrinsic vestibular neurons should be clarified. Using transneuronal virus tracing[59] or stimulus-related immediate-early gene expression,[53] various levels of indirect connections could be studied.

In conclusion, further investigations on spinovestibular systems using both anatomical and electrophysiological methods should enable a better understanding of their functioning and their role in pathogenesis of so-called vertebragenic syndromes such as cervical vertigo and related disorders.

ACKNOWLEDGMENTS

Studies performed by the authors were supported by EMDO-Stiftung, Znrich and NIH grants NS 02619 and DC 02187 (to Dr. V. Wilson collaborating with S.B.). Thanks are due to Astrid Markus for critically reading the manuscript. Dedicated to Prof. Wolfgang Zenker on the occasion of his 70th birthday.

REFERENCES

1. Abrahams, VC, Richmond, FJ, and Keane, J, Projections from C2 and C3 nerves supplying muscles and skin of the cat neck: A study using transganglionic transport of horseradish peroxidase, *J. Comp. Neurol.,* 230:142,1984.

2. Abrahams, VC, and Rose, PK, Projections of extraocular, neck muscle and retinal afferents to the suprior colliculus in the cat: Their connections to cells of origin of tectospinal tract, *J. Neurophysiol.,* 38:10,1975.

3. Anastasopoulos, D, and Mergner, T, Canal-neck interaction in vestibular nuclear neurons of the cat, *Exp. Brain Res.,* 46:269,1982.

4. Arvidsson, J, and Pfaller, K, Central projections of C4-C8 dorsal root ganglia in the rat studied by anterograde transport of WGA-HRP, *J. Comp. Neurol.,* 292:349,1990.

5. Ashton, JA, Boddy, A, Dean, SR, Milleret, C, and Donaldson, IML, Afferent signals from cat extraocular muscles in the medial vestibular nucleus, the nucleus praepositus hypoglossi and adjacent brainstem structures, *Neuroscience.,* 26:131,1988.

6. Bakker, DA, Richmond, FJR, Abrahams, VC, and Courville, J, Patterns of primary afferent termination in the external cuneate nucleus from cervical axial muscles in the cat, *J. Comp. Neurol.,* 241:467,1985.

7. Balaban, CD, A projection from nucleus reticularis tegmenti pontis of Bechterew to the medial vestibular nucleus in rabbits, *Exp. Brain Res.,* 51:304,1983.

8. Bankoul, S, Cervical primary afferent input to vestibulo-spinal neurones projecting to the dorsal horn: A double labelling study in the rat, *Experientia.,* 50:A70,1994.

9. Bankoul, S, Goto, T, Yates, B, and Wilson, VJ, Cervical primary afferent input to vestibuo-spinal neurons projecting to the cervical dorsal horn: An anterograde and retrograde tracing study in the cat, *J. Comp. Neurol.,* 353:529,1995.

10. Bankoul, S, and Neuhuber, WL, A cervical primary afferent input to vestibular nuclei as demonstrated by retrograde transport of wheat germ agglutinin-horseradish peroxidase in the rat, *Exp. Brain Res.,* 79:405,1990.

11. Bankoul, S, and Neuhuber, WL, A direct projection from the medial vestibular nucleus to the cervical spinal dorsal horn of the rat, as demonstrated by anterograde and retrograde tracing, *Anat. Embryol.,* 185:77,1992.

12. Bárány, R, Augenbewegungen durch Thoraxbewegungen ausgelöst, *Zbl. Physiol.,* 20:298,1906.

13. Barmack, NH, Nastos, MA, and Pettorossi, VE, (1981) The horizontal and vertical cervico-ocular reflexes of the rabbit, *Brain Res.,* 24:261,1981.

14. Barré, JA, Le syndrome sympathique cervical postérieur, *Rev. Neurol.,* 45:248,1926.

15. Battaglia, G, and Rustioni, A, Coexistence of glutamate and substance P in dorsal root ganglion neurons of the rat and monkey, *J. Comp. Neurol.,* 277:302,1988.

16. Berthoz, A, Vidal, PP, and Graf, W, (Eds.) The Head-Neck Sensory Motor System, New York: Oxford UP,1992.

17. Biemond, A, deJong, JMB, V, On cervical nystagmus and related disorders, *Brain,* 92:437,1969.

18. Böhmer, A, and Henn, V, Horizontal and vertical vestibulo-ocular and cervico-ocular reflexes in the monkey during high frequency rotation, *Brain Res.,* 277:241,1983.

19. Boyle, R, and Pompeiano, O, Convergence and interaction of neck and macular vestibular inputs on vestibuo-spinal neurons, *J. Neurophysiol.,* 45:852,1981.

20. Brink, EE, Jinnai, K, Hirai, N, and Wilson, VJ Cervical input to vestibulocollic neurons, *Brain Res.,* 217:13,1981.

21. Brodal, A, Anatomy of the vestibular nuclei and their connections. in: *Handbook of Sensory Physiology VI/1,* (Kornhuber, HH, Ed.), pp. 239-351, Berlin: Springer,1974.

22. Bronstein, AM, and Hood, JD, The cervico-ocular reflex in normal subjects and patients with absent vestibular function, *Brain Res.,* 373:399,1986.

23. Brown, AG, Organization in the spinal cord, Berlin: Springer,1981.

24. Brown, JJ, (1992) Cervical contribution to balance: cervical vertigo. in: *The Head-Neck Sensory Motor System,* (Berthoz, A, Vidal, PP, Graf, W, Eds.), pp. 644,1992. New York: Oxford UP.

25. Burian, M, Gstoettner, W, and Mayr, R, Brainstem projection of the vestibular nerve in the guinea pig: An HRP (hoseradish peroxidase) and WGA-HRP (wheat germ agglutinin-HRP) study. J Comp Neurol 293:165-177.

26. Büttner-Ennever, JA, Patterns of connectivity in the vestibular nuclei. in: *Sensing and Controlling Motion. Vestibular and sensorimotor function, Ann. NY Acad. Sci.*, 656 (Cohen B, Tomko, DL, and Guedry F, Eds.), pp. 363-378. New York: NY Acad Sci.,1992.

27. Carleton, SC, and Carpenter, MB, Afferent and efferent connections of the medial, inferior and lateral vestibular nuclei in the cat and monkey, *Brain Res.*, 278:29,1983.

28. Cazin, L, Magnin, M, and Lannou, J, Non-cerebellar visual afferents to the vestibular nuclei involving the prepositus hypoglossal complex: An autoradiographic study in the rat, *Exp. Brain Res.*, 48:309,1982.

29. Cochran, SL, Kasic, P, and Precht, W Pharmacological aspects of excitatory synaptic transmission to second-order vestibular neurons in the frog, *Synapse*, 1:102,1987.

30. Corbin, KB, and Hinsey, JC, Intramedullary course of the dorsal root fibers of each of the first four cervical nerves, *J. Comp. Neurol.*, 63:119,1935.

31. De Waele, C, Mnhlethaler, M, and Vidal, PP, Neurochemistry of the central vestibular pathways, *Brain Res. Rev.*, 20:24,1995.

32. Deecke, L, Schwarz, DWF, and Fredrickson, JM, Vestibular responses in the rhesus monkey ventropos-terior thalamus. II. Vestibulo-proprioceptive convergence at thalamic neurons, *Exp. Brain Res.*, 30:219,1977.

33. Dichgans, J, Bizzi, E, Morasso, P, and Tagliasco, V, Mechanisms underlying recovery of eye-head coor-dination following bilateral labyrinthectomy in monkeys, *Exp. Brain Res.*, 18:548,1973.

34. Doerr, M, and Thoden, U, Gibt es einen zervikogenen Schwindel? In: *NeuroorthopΣdie, Vol. 5 (K^{n}gelgen B, Ed.)*, pp. 227-234, Berlin: Springer,1994.

35. Donevan, AH, Neuber-Hess, M, and Rose, PK, Multiplicity of vestibuo-spinal projections to the upper cervical spinal cord of the cat: A study with the anterograde tracer Phaseolus vulgaris leucoagglutinin, *J. Comp. Neurol.*, 302:1,1990.

36. Duc, C, Barakat-Walter, I, and Droz, B, Calbindin D-28k- and substance P-immunoreactive primary sensory neurons: Peripheral projections in chick hindlimbs, *J. Comp. Neurol.*, 334:151,1993.

37. Duc, C, Barakat-Walter, I, and Droz, B, peripheral projections of calretinin-immunoreactive primary sensory neurons in chick hindlimbs, *Brain Res.*, 622:321,1993.

38. Dutia, MB, The muscles and joints of the neck: their specialization and role in head movement, *Progr. Neurobiol.*, 37:165,1991.

39. Edney, DP, and Porter, JD, Neck muscle afferent projections to the brainstem of the monkey: implications for the neural control of gaze, *J. Comp. Neurol.*, 250:389,1986.

40. Epema, AH, Gerrits, NM, and Voogd, J, Commissural and intrinsic connections of the vestibular nuclei in the rabbit: A retrograde labeling study, *Exp. Brain Res.*, 71:129,1988.

41. Escolar, J, The afferent connections of the 1st, 2nd, and 3rd cervical nerves in the cat. An analysis by Marchi and Rasdolsky methods, *J. Comp. Neurol.*, 89:79,1948.

42. Fitz-Ritson, D, The direct connections of the C2 dorsal root ganglia in the macaca irus monkey: Relevance to the chiropractic profession, *J. Manip. Physiol. Ther.*, 8:147,1985.

43. Fitz-Ritson, D, The chiropractic management and rehabilitation of cervical trauma, *J. Manip. Physiol. Ther.*, 13:17,1990.

44. Fredrickson, JM, Schwarz, D, and Kornhuber, HH, Convergence and interaction of vestibular and deep somatic afferents upon neurons in the vestibular nuclei of the cat, *Acta. oto-laryng.*, 61:168,1965.

45. Gandevia, SC, McCloskey, DI, and Burke, D, Kinesthetic signals and muscle contraction, *Trends Neurosci.*, 15:62,1992.

46. Gerrits, NM, Vestibular nuclear complex. in: *The Human Nervous System* (Paxinos G, Ed.), pp. 863-888, San Diego: Academic Press,1990.

47. Gutmann G Klinik von posttraumatischen Funktionsst÷rungen der oberen HWS: Symptomenkombination und Symptomdauer, Frage der Latenz. In: Die Sonderstellung des Kopfgelenkbereichs (Wolff HD, Ed.), pp 129-148. Berlin: Springer,1988.

48. Hikosaka, O, and Maeda, M, Cervical effects on abducens motoneurons and their interaction with vestibular-ocular reflex, *Exp. Brain Res.*, 18:512,1973.

49. Hirai, N, Hongo, T, and Sasaki, S, Cerebellar projection and input organizations of the spinocerebellar tract arising from the central cervical nucleus in the cat, *Brain Res.*, 157:341,1978.

50. Hülse, M, Zervikale Dysphonie, *Folia Phoniat.*, 43:181,1991.

51. Hülse M Die zervikogene Hörstörung. HNO 42:604-613.

52. Hülse, M, Gibt es einen zervikogenen Schwindel?, in: *Neuroorthopädie*, Vol. 5 (Kügelgen, B, Ed.), pp. 207-225, Berlin: Springer,1994.

53. Hunt, SP, Pini, A, and Evan, G, Induction of c-*fos*-like protein in spinal cord neurons following sensory stimulation, *Nature*, 328:632,1987.

54. Isu, N, Sakuma, A, Kitahara, M, Ichikawa, T, Watanabe, S, and Uchino, Y, Extracellular recording of vestibulo-thalamic neuons projecting to the spinal cord in the rat, *Neurosci. Lett.*, 104:25,1989.

55. Jensen, DW, and Thompson, GC, Vestibular nerve input to neck and shoulder regions of lateral cuneate nucleus, *Brain Res.*, 280:335,1983.

56. Karnath, H-O, Sievering, D, and Fetter, M, The interactive contribution of neck muscle proprioception and vestibular stimulation to subjective "straight ahead" orientation in man, *Exp. Brain Res.*, 101:140,1994.

57. Kasper, J, Schor, RH, and Wilson, VJ, Response of vestibular neurons to head rotations in vertical planes. II. response to neck stimulation and vestibular-neck interaction, *J. Neurophysiol.*, 60:1765,1988.

58. de Kleijn, A, Tonische Labyrinth- und Halsreflexe auf die Augen, *Pflügers. Arch.*, 186:82,1921.

59. Kuypers, HGJM, and Ugolini, G Viruses as transneuronal tracers, *Trends Neurosci.*, 13:71,1990.

60. Lewit, K, Pathomechansimen des zervikalen Kopfschmerzes, *Psychiat. Neurol. Med. Psychol.*, (Leipzig), 29:261,1977.

61. Longet, FA, Sur les troubles qui surviennent dans l'ééquilibration, la station et la locomotion des animaux, après la section des parties molles de la nuque, *Gaz. Med. Paris.*, 13:565,1845.

62. Maeda, M, Clinical and experimental investigations of visually guided eye and head movement: Role of neck afferents. in: *The Head-Neck Sensory Motor System*, (Berthoz, A, and Vidal PP, and Graf, W, Eds.), pp. 648-653. New York: Oxford UP,1992.

63. Magnus, R, *Körperstellung*, Berlin: Springer,1924.

64. Marfurt, CF, and Rajchert, DM, Trigeminal primary afferent projections to "non-trigeminal" areas of the rat central nervous system, *J. Comp. Neurol.*, 303:489,1991.

65. McCouch, GP, Deering, ID, and Ling, TH, Location of receptors for tonic neck reflexes, *J. Neurophysiol.*, 14:191,1951.

66. McCrea, RA, and Baker, R, Anatomical connections of the nucleus prepositus of the cat, *J. Comp. Neurol.*, 237:377,1985.

67. McKelvey-Briggs, DK, Saint-Cyr, JA, Spence, SJ, and Partlow, GD, A reinvestigation of the spinovestibular projection in the cat using axonal transport techniques, *Anat. Embryol.*, 180:281,1989.

68. McKinney, MB, Treatment of cervical spine distortions in whiplash injuries, *Orthopäde* 23:287,1994.

69. Mehler, WR, and Rubertone, JA, Anatomy of the vestibular nucleus complex. in: *The Rat Nervous System*, vol. 2, (Paxinos, G, Ed.), pp.185-219, Sydney: Academic Press,1985.

70. Mense, S Nociception from skeletal muscle in relation to clinical muscle pain, *Pain*, 54:241,1993.

71. Mergner, T, Becker, W, and Deecke, L, Canal-neck interaction in vestibular neurons of the cat's cerebral cortex, *Exp. Brain Res.*, 61:94,1985.

72. Mergner, T, Siebold, C, Schweigart, G, and Becker, W, Perception of horizontal head and trunk rotation in space: role of vestibular and neck afferents. in: *The Head-Neck Sensory Motor System* (Berthoz, A, Vidal, PP, and Graf, W, Eds.), pp. 491-496. New York: Oxford UP,1992.

73. Neuhuber, WL, and Fryscak-Benes, A, Die zentralen Projektionen afferenter Neurone des Nervus hypoglossus bei der Albinoratte, *Verh. Anat. Ges.*, 81:981,1987.

74. Neuhuber, WL, and Zenker, W, The central distribution of cervical primary afferents in the rat, with emphasis on proprioceptive projections to vestibular, perihypoglossal and upper thoracic spinal nuclei, *J. Comp. Neurol.*, 280:231,1989.

75. Neuhuber, WL, Zenker, W, and Bankoul, S, Central projections of cervical primary afferents in the rat. Some general anatomical principles and their functional significance. in: *The Primary Afferent Neuron* (Zenker, W, and Neuhuber, WL, Eds.), pp. 173-188. New York: Plenum Press,1990.

76. Peterson, BW, Goldberg, J, Bilotto, G, and Fuller, JH, Cervicocollic reflex: Its dynamic properties and interaction with vestibular reflexes, *J. Neurophysiol.*, 54:90,1985.

77. Pfaller, K, and Arvidsson, J, Central distribution of trigeminal and upper cervical primary afferents in the rat studied by anterograde transport of horseradish peroxidase conjugated to wheat germ agglutinin, *J. Comp. Neurol.*, 268:91,1988.

78. Pfister, J, Zenker, W, The splenius capitis muscle of the rat, architecture and histochemistry, afferent and efferent innervation as compared with that of the quadriceps muscle, *Anat. Embryol.*, 169:79,1984.

79. Pompeiano, O, and Barnes, CD, Effect of sinusoidal muscle stretch on neurons in medial and descending vestibular nuclei, *J. Neurophysiol.*, 34:725,1971.

80. Pompeiano, O, Manzoni, D, Srivastava, UC, and Stampacchia, G, Convergence and interaction of neck and macular vestibular inputs on reticulospinal neurons, Neuroscience, 12:111,1984.

81. Prihoda, M, Hiller, M-S, and Mayr, R, Central projections of cervical primary afferent fibers in the guinea pig: An HRP and WGA/HRP tracer study, *J. Comp. Neurol.*, 308:418,1991.

82. Richmond, FJR, and Bakker, DA, Anatomical organization and sensory receptor content of soft tissues surrounding upper cervical vertebrae in the cat, *J. Neurophysiol.*, 48:49,1982.

83. Robertson, B, and Grant, G, A comparison between wheat germ agglutinin- and choleragenoid-horseradish peroxidase as anterogradely transported markers in central branches of primary sensory neurones in the rat with some observations in the cat, *Neuroscience*, 14:895,1985.

84. Rubertone, JA, Mehler, WR, and Voogd, J, The vestibular nuclear complex in: *The Rat Nervous System*, 2nd ed. (Paxinos G, Ed.), pp. 773-796. Sydney: Academic Press,1995.

85. Rubin, AM, Liedgren, SRC, Milne, AC, and Young, JA, Fredrickson JM Vestibular and somatosensory interaction in the cat vestibular nuclei, *Pfl°gers. Arch.*, 371:155,1977.

86. Schaible, HG, and Grubb, BD, Afferent and spinal mechanisms of joint pain, *Pain*, 55:5,1993.

87. Scott, SH, and Loeb, GE, The computation of position sense from spindles in mono- and multiarticular muscles, *J. Neurosci.*, 14:7529,1994.

88. Stampacchia, G, Manzoni, D, Marchand, AR, and Pompeiano, O, Convergence of neck and macular vestibular inputs on vestibuo-spinal neurons projecting to the lumbosacral segments of the spinal cord, *Arch. Ital. Biol.*, 125:201,1987.

89. Teasell, RW, and Shapiro, AP, Cervical whiplash injury. in: *Progress in Fibromyalgia and Myofascial Pain*, (V£r°y H, and Merskey H, Eds.), pp. 253-266. Amsterdam: Elsevier,1993.

90. Thoden, U, and Schmidt, P, Vestibular-neck interaction in abducens neurons. in: *Reflex Control of Posture and Movement* (Granit, R, and Pompeiano, O, Eds.), pp. 561-566. Amsterdam: Elsevier,1979.

91. Tracey, DJ, De Biasi, S, Phend, K, and Rustioni, A, Aspartate-like immunoreactivity in primary afferent neurons, *Neuroscience*, 40:673,1991.

92. Voss, H, Tabelle der absoluten und relativen Muskelspindelzahlen der menschlichen Skelettmuskulatur, *Anat. Anz.*, 129:562,1971.

93. Wall, PD, Do nerve impulses penetrate terminal arborizations? a presynaptic control mechanism, *Trends Neurosci.*, 18:99,1995.

94. Wanaka, A, Shiotani, Y, Kiyama, H, Matsuyama, T, Kamada, T, Shiosaka, S, and Tohyama, M, Glutamate-like immunoreactive structures in primary sensory neurons in the rat detected by a specific antiserum against glutamate, *Exp. Brain Res.*, 65:691,1987.

95. Wiksten, B, The central cervical nucleus in the cat. II. The cerebellar connections studied with retrograde transport of horseradish peroxidase, *Exp. Brain Res.*, 36:155,1979.

96. Wiksten, B, Further studies on the fiber connections of the central cervical nucleus in the cat, *Exp. Brain Res.*, 67:284,1987.

97. Wilson, VJ, Convergence of neck and vestibular signals on spinal interneurons, *Progr. Brain Res.*, 76:137,1988.

98. Wilson, VJ, Physiological properties and central actions of neck muscle spindles. in: *The Head-Neck Sensory Motor System* (Berthoz, A, Vidal, PP, Graf, W, Eds.), pp. 175-178. New York: Oxford UP,1992.

99. Wilson, VJ, Kato, M, and Thomas, RC, Excitation of lateral vestibular neurones, Nature, 206:96,1965.

100. Wilson, VJ, Maeda, M, Franck, JI, and Shimazu, H, Mossy fiber neck and second-order labyrinthine projections to rat flocculus, *J. Neurophysiol.*, 39:301,1976.

101. Wilson, VJ, Wylie, RM, and Marco, LA, Organization of the medial vestibular nucleus. Synaptic input to cells in the medial vestibular nucleus, *J. Neurophysiol.*, 31:166,1968.

102. Zenker, W, Mysicka, A, and Neuhuber, W, Dynamics of the transganglionic movement of horseradish peroxidase in primary sensory neurons, *Cell Tiss. Res.*, 207:479,1980.

16

Chemical Comparisons between Central Vestibular and Auditory Systems

Donald, A. Godfrey, Hongyan Li, C. David Ross, and Allan M. Rubin

CONTENTS

16.1. SYNOPSIS

The vestibular and auditory systems share similarly located peripheral mechanoreceptive sensory structures, but fundamentally different central pathways. The neurotransmitter chemistry of their primary and secondary brain regions has some similarities and some differences. For both, the ascending eighth nerve input may employ an excitatory amino acid as a transmitter. Also, both the vestibulo-cerebellum and dorsal cochlear nucleus contain granule cell parallel fiber systems that likely employ glutamate as a transmitter. Of the inhibitory amino acid transmitters, glycine tends to be more prominently represented in the auditory nuclei, and g-aminobutyrate (GABA) in the vestibular nuclei and vestibulo-cerebellum. Cholinergic feedback pathways appear to be a more significant part of information processing in the cochlear nucleus than in the vestibular nuclear complex.

16.2 INTRODUCTION

16.2.1 AUDITORY-VESTIBULAR INTERACTIONS

The obvious association of the ear with hearing dates from the earliest human experiences, but its association with balance was not realized until around 1870.[35] The close association of auditory and vestibular sensory organs in the inner ear, along with the transport of the coded sensory information about sounds and head position by adjacent portions of the eighth cranial nerve (Figure 16.1), provide the anatomical basis for the many clinical syndromes that include both auditory and vestibular symptoms.

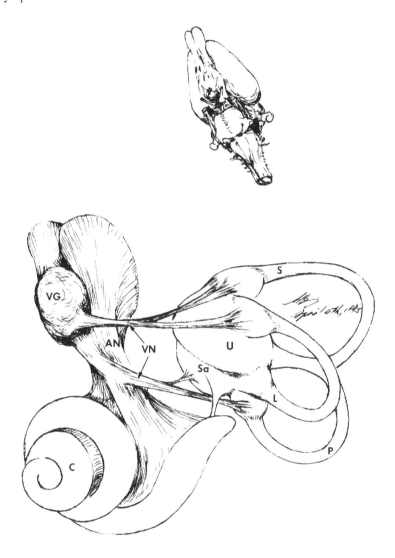

FIGURE 16.1 Drawing of the inner ear sensory organs of the rat, including the membranous labyrinth and the connecting auditory and vestibular nerves. The direction of view relative to the brain is shown by the small drawing of a rat brain above. Some details were checked against a published drawing.[34] Abbreviations are: AN, auditory nerve; C, cochlea; L, lateral semicircular canal; P, posterior semicircular canal; S, superior semicircular canal; Sa, saccule; U, utricle; VG, vestibular ganglion; VN, vestibular nerve, including superior (above) and inferior (below) divisions.

However, the auditory and vestibular components of the eighth nerve, once entering the brain, diverge: the auditory fibers enter the cochlear nucleus located lateral to the inferior cerebellar peduncle, while the vestibular fibers terminate in the vestibular nuclear complex located medial to the inferior cerebellar peduncle (Figure 16.2).

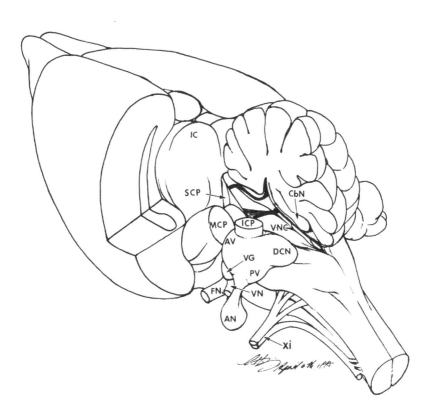

FIGURE 16.2 View of a rat brain from a lateral-dorsal-caudal position, with the left half of the cerebellum and the caudal part of the left cerebral hemisphere cut away to show the relative positions of the cochlear nucleus and vestibular nuclear complex. Abbreviations are: AN, auditory nerve; AV, anteroventral cochlear nucleus; CbN, cerebellar nodulus; DCN, dorsal cochlear nucleus; FN, facial nerve; IC, inferior colliculus; ICP, inferior cerebellar peduncle; MCP, middle cerebellar peduncle; PV, posteroventral cochlear nucleus; SCP, superior cerebellar peduncle; VG, vestibular ganglion; VN, vestibular nerve; VNC, vestibular nuclear complex; xi, eleventh cranial nerve (accessory).

From the primary eighth nerve nuclei, the major auditory and vestibular pathways project in fundamentally different directions. The auditory pathways generally ascend through the brainstem to the superior olivary complex, inferior colliculus, and medial geniculate, and then on to the auditory cortex. The vestibular pathways, on the other hand, go primarily to centers involved with motor activities: the spinal cord for influences on motorneurons involved in postural control, the motor nuclei controlling eye movements, and the vestibulo-cerebellum for coordination of such movements.

In view of the close peripheral association of auditory and vestibular systems, it has been of interest to search for more central interactions between them, and indeed some have been documented. When the olivo-cochlear bundle from the superior olivary complex to the cochlea was first studied, it was noted to follow the vestibular nerve, rather than the auditory, until reaching the inner ear.[70] Later, centrifugal fibers to the vestibular labyrinth were identified[18,40] deriving from near the vestibular nuclear complex.[76,93] These systems appeared to be entirely separate except for traveling the same course to the inner ear. More recently, however, evidence has been found for branches of the olivo-cochlear bundle to vestibular nuclei.[7]

Interaction between the ascending pathways has also been found; there is evidence for an auditory function of the saccule in fishes,[67] amphibians,[56] and mammals.[2,13,52,58,97] It was found that the auditory response activity of the vestibular nerve fibers in mammals was evoked only by relatively intense sounds with frequencies of 0.5 to 1 kHz.[54] Furthermore, firing rates could be increased, but thresholds were not decreased, by stimulation of the centrifugal fibers.[53] Such vestibular fiber responses to sound might constitute a physiological correlate of vestibular nerve fiber projections to the cochlear nucleus, particularly from the sacculus to granular regions.[8,38] A recent report additionally suggests projections to the superficial anteroventral cochlear nucleus from some small neurons with somata in the vestibular nerve root.[99] There is also evidence for the converse situation: some auditory nerve projections to the medial vestibular nucleus,[83] and responses to low-intensity sound of neurons in the lateral vestibular nucleus.[13]

Although these anatomical and physiological overlaps of the vestibular and auditory systems are intriguing, they appear to be minor relative to the largely independent functioning of the two eighth nerve sensory systems. A different, although related, question involves chemical similarities and differences between the systems. To what extent are the same neurotransmitters used at comparable parts of the systems? Are there similarities related to the similar peripheral connections of the systems, but differences related to their different requirements for stimulus coding? These chemical comparisons, specifically related to neurotransmitters, will be the focus of this chapter.

16.2.2 Introduction to Chemical Comparisons

Most chemical comparisons that can be made between the vestibular and auditory systems involve the more peripheral portions of the systems because most studies have dealt with these regions. Our main focus will be on the primary brain centers for the two systems, which are the transition points between the similarly located sensory organs and nerves and the very different central pathways (Figure 16.3).

We will also present some information for secondary brain centers.

In the auditory system, the primary brain center is the cochlear nucleus (CN), which is subdivided into anteroventral (AVCN), interstitial (auditory nerve root) (IN), posteroventral (PVCN), and dorsal (DCN) parts. The DCN is further divided into three layers: molecular (m), fusiform soma (f), and deep (d). In the vestibular system, the primary brain center is the vestibular nuclear complex (VNC), which is subdivided into medial (MVN), lateral (LVN), superior (SuVN), and spinal (SpVN) nuclei. Dorsal and ventral parts of MVN (MVNd and MVNv) and LVN (LVNd and LVNv) can be distinguished on histological and chemical bases. The secondary brain centers to be compared are the superior olivary complex (SOC) (auditory) and the vestibulo-cerebellum (VCb). Four major nuclei of the SOC will be included: the lateral superior olivary nucleus (LSO), superior paraolivary nucleus (SPN), medial nucleus of the trapezoid body (MNTB), and ventral nucleus of the trapezoid body (VNTB). Regions of the VCb included are the flocculus (CbF) and nodulus (CbN); data for their molecular (m), granular (g), and white matter (w) layers will be presented.

Our chemical comparisons will focus especially on major neurotransmitter candidates: the excitatory amino acids glutamate and aspartate, the inhibitory amino acids GABA and glycine, and acetylcholine. The amino acids probably represent the most prominent neurotransmitters of the brain, while acetylcholine is a well-established transmitter at both central and peripheral synapses. We will also comment on the monoamines norepinephrine and serotonin, two well-established transmitters which have some involvement in central auditory and vestibular centers.

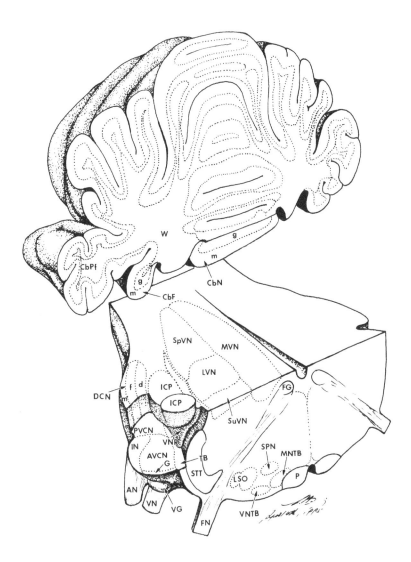

FIGURE 16.3 Semischematic view of a rat brain from a lateral-dorsal-rostral position, cut through in the coronal plane at the level of the facial nerve exit, with the cerebellum lifted away from the brain stem, which is cut in a horizontal plane just below the floor of the fourth ventricle. A second, more caudal, coronal cut at the ventral part of the cerebellar vermis was made to expose the nodulus. A deeper horizontal cut through part of the cochlear nucleus exposes its ventral parts, and a stump of the inferior cerebellar peduncle has been left. All regions for which data are presented in Figure 16.4 are thereby shown. Abbreviations are: AN, auditory nerve; AVCN, anteroventral cochlear nucleus; CbF, cerebellar flocculus, and CbN, cerebellar nodulus, with molecular (m), granular (g), and white matter (w) layers; CbPf, cerebellar paraflocculus; DCN, dorsal cochlear nucleus, with molecular (m), fusiform soma (f), and deep (d) layers; FG, facial genu; FN, facial nerve; G, granular region adjacent to AVCN; ICP, inferior cerebellar peduncle; IN, interstitial nucleus (auditory nerve root); LSO, lateral superior olivary nucleus; LVN, lateral vestibular nucleus; MNTB, medial nucleus of the trapezoid body; MVN, medial vestibular nucleus; P, pyramid; PVCN, posteroventral cochlear nucleus; SPN, superior paraolivary nucleus; SpVN, spinal vestibular nucleus; STT, spinal trigeminal nucleus; SuVN, superior vestibular nucleus; TB, trapezoid body; VG, vestibular ganglion; VN, vestibular nerve; VNR, vestibular nerve root; VNTB, ventral nucleus of the trapezoid body.

16.3 EXCITATORY AMINO ACIDS

As is common for sensory systems, there is evidence for glutamate, aspartate, or both as neurotransmitters of auditory and vestibular nerve fibers[32,69,79,92] (see Chapters 2 and 4, this volume). Glutamate concentrations are similar among AVCN, PVCN, DCNd, and the vestibular nuclei (Figure 16.4).

Somewhat higher glutamate concentrations are found in CN granular and superficial DCN layers. These appear related to evidence for glutamate as a transmitter of CN granule cells,[32,69,92] which have their somata in granular regions, including DCNf, and at least many of their axon terminals in DCNm.[57] All these regions of the CN have somewhat elevated glutamate concentrations (Figure 16. 4). The glutamate concentrations in the VCb are particularly high in granular and molecular layers, in correlation with the evidence that cerebellar granule cells use glutamate as a transmitter.[62,64,96] Relatively high glutamate concentrations also in the white matter may correlate with evidence for glutamatergic mossy fibers.[64] Slightly lower glutamate concentrations are found in the SOC. Thus, overall, the glutamate concentrations are rather similar among auditory and vestibular regions, except for higher values where there are granule cells.

The distribution of concentrations of glutamine, a major metabolic precursor of glutamate in a cycle involving both glia and neurons,[66,77] correlates rather well with that of glutamate (correlation coefficient 0.90) (Figure 16.4). Aspartate concentrations, like glutamate, do not show striking differences between vestibular and auditory regions but, unlike glutamate, are not elevated in regions containing granule cell axons and terminals (Figure 16.4). Distributions of glutamate, glutamine, and aspartate in guinea pig VNC are generally similar to those in the rat.[72]

Activities of the major enzymes of glutamate and aspartate metabolism, aspartate aminotransferase and glutaminase, have been measured in the guinea pig VNC (C.D. Ross, unpublished observations) and in the rat CN.[32] In both regions, the distributions are similar to that of glutamate. Compared to other fiber tracts, both the vestibular and auditory nerve roots contain relatively high aspartate aminotransferase activities, consistent with excitatory amino acid neurotransmission.[28] Unlike glutamate, however, glutaminase activity in the guinea pig is higher in the molecular layer of the DCN than in that of the cerebellum (C.D. Ross, unpublished observations).

Among excitatory amino acid receptor types, there appears to be a more prominent association of non-N-methyl-D-aspartate (non-NMDA) than of NMDA types with the vestibular and auditory nerve synapses[79,92] (see Chapters 2, 4, and 5). The binding density of 5,6-cyano-7-nitro-quinox-aline-2,3-dione (CNQX), representing locations of non-NMDA glutamate receptors, is particularly high in cerebellar and DCN molecular layers (Figure 16.4). Many of these non-NMDA receptors may be at the synapses formed by the parallel fiber axons of granule cells.[20,87]

An interesting difference between the VNC and CN concerns the effects of unilateral deafferentation on excitatory amino acid concentrations (Figure 16.5).

Results for guinea pig CN[89,91] show a decrease on the lesioned side, as compared to the unlesioned side, which is larger for aspartate than for glutamate and is maintained through 28 days of survival. Results for the rat VNC[48] show an initial decrease on the lesioned side relative to the unlesioned side, as in the CN, but slightly larger for glutamate than for aspartate. There is then a gradual decrease of the asymmetry between the two sides for both excitatory amino acids. These comparisons suggest that aspartate may be more prominently related to the auditory nerve than to the vestibular nerve, and that there may be more compensatory processes occurring in the VNC than in the CN following deafferentation. The latter difference may be related to the more extensive commissural connections between the vestibular than between the cochlear nuclei and/or to the more extensive connections with the cerebellum[11,25,36,55] (see Chapter 1, this volume). The finding that the recovery of symmetry is more prominent in LVN, which receives few commissural connections,[11] than in MVN suggests that the cerebellar influence may be more important.[48] A previous study found no glutamate or aspartate difference between lesioned and unlesioned sides of the VNC 10 months after unilateral labyrinthectomy in squirrel monkeys.[37] The gradual recovery over time of bilateral symmetry of glutamate and aspartate concentrations

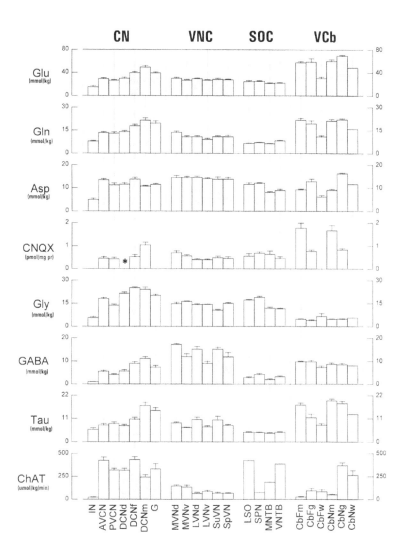

FIGURE 16.4 Bar graphs showing average data (mean + standard error) for concentrations of the amino acids glutamate (GLU), glutamine (GLN), aspartate (ASP), glycine (GLY), GABA, and taurine (TAU) (mmol/kg dry wt); CNQX binding densities (pmol/mg protein); and activities of choline acetyltransferase (ChAT) (mmol/kg dry wt/min) for rat auditory (CN and SOC) and vestibular (VNC and VCb) regions shown in Figure 16.3, with abbreviations as in that figure. All data except binding densities were obtained by chemical analysis of microdissected, freeze-dried tissue samples by the techniques of quantitative histochemistry.[26,50] Although various measures of sample size could be used, dry weight is the most accurate for this methodology and gives relative distributions not very different from what is seen using protein concentration.[26] Amino acid data for VNC and some of the CN values are from published studies;[32,46] those for the SOC have been reported;[49] the rest are not yet published. The data for CNQX binding are from H. Li, D.A. Godfrey, and A.M. Rubin, unpublished observations. The value in the DCNf column is actually for combined measurement of DCNf and DCNd (*); no data were obtained for CbFw or CbNw. The data for ChAT in CN and VCb are from previous publications;[27,30,73] those for SOC are averages from one rat, of which the values for LSO and VNTB are slightly lower than previously published values;[30] the data for VNC have been reported.[47]

in the VNC of rats after unilateral vestibular ganglionectomy (Figure 16.5) is consistent with this finding.

The greater postdeafferentation recovery of bilateral symmetry of excitatory amino acid concentrations in the VNC than in the CN may correlate with physiological findings after unilateral deafferentation. Cochlear destruction results in immediate loss of almost all spontaneous activity in the ipsilateral AVCN and PVCN neurons, which persists through 2 months after surgery.[41] By comparison, a partial loss of spontaneous activity in ipsilateral MVN shortly after labyrinthectomy shows some reversal even by 2 days after surgery.[78] Besides their probable association with the primary innervations of CN and VNC, glutamate and aspartate may also be associated with neurons projecting out of these nuclei.[17,31,86,92]

AMINO ACID CHANGES AFTER DEAFFERENTATION

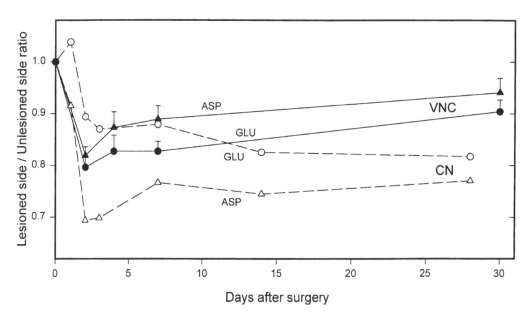

FIGURE 16.5 Comparison of deafferentation effects on aspartate (ASP, triangles) and glutamate (GLU, circles) concentrations in cochlear nucleus (CN, dashed lines with open symbols) and vestibular nuclear complex (VNC, solid lines with filled symbols). Data are presented as ratios of amino acid concentrations on the lesioned side compared to the contralateral (unlesioned) side. Data for CN are for guinea pig after cochlear destruction, from the study of Wenthold and Gulley.[91] Data for VNC are for rat after vestibular ganglionectomy, averaged with standard errors, from the data for individual nuclei[48]. Top of error bar for VNC GLU at 2 days is at bottom of ASP symbol.

16.4 INHIBITORY AMINO ACIDS AND TAURINE

Both GABA and glycine have been implicated as inhibitory transmitters in both the CN and VNC, in interneurons, descending innervation, and commissural connections[17,31,68,71,90] (see Chapters 6 and 17, this volume). However, the relative amounts of these two amino acids differ noticeably between auditory and vestibular regions (Figure 16.4). In auditory regions, ratios of glycine to GABA concentration range from about 2 in DCNm to over 6 in the LSO. By contrast, glycine-to-GABA ratios range from about 0.4 in CbFg to about 1.6 in LVNv. The differences in the relative distributions between VCb and SOC are particularly striking. The relatively high

GABA concentrations in the VNC, especially LVNd, MVNd, and SuVN, may be especially related to their major input from cerebellar Purkinje cells[11,17,23,55,61,71,85] (see Chapters 6 and 17), and the relatively high GABA concentrations in the cerebellar white matter may be particularly associated with the axons of the Purkinje cells. In the CN, there is a strong tendency for GABA and glycine concentrations to vary proportionately (correlation coefficient 0.92), in agreement with evidence for colocalization in many of its neurons.[1,42,90] This trend is not seen in the VNC, despite evidence for some colocalization[86] (see Chapter 7); there is actually some tendency toward an inverse relation.[46]

These distributional differences may be related to some fundamentally different roles for GABA and glycine between the vestibular and auditory systems. For example, there is evidence that some of the commissural neurons of the CN use glycine as a transmitter,[90] whereas there is more evidence for GABA than for glycine involvement in vestibular commissural inhibition[24] (see Chapter 6). Second, there is evidence that the Purkinje cells of the cerebellum use GABA as a transmitter,[17,23,61,71] whereas what are considered their counterparts in the DCN, the cartwheel cells,[95] appear to use glycine in addition to or in preference to GABA.[6,12,33,42,98] Thirdly, there is evidence for GABAergic projection neurons in the VNC, especially related to vestibulo-oculomotor pathways[17,71] (see Chapter 9), whereas there is so far no evidence for GABAergic neurons projecting from the CN to higher auditory nuclei.[31,90] Instead, many inhibitory ascending and descending auditory projections using GABA or glycine as a transmitter appear to originate in the SOC.[36,68]

Taurine concentrations were highest in the molecular layers of the VCb and DCN, and were often almost as high in granular regions. There has been some suggestion of correlations between taurine and GABA distributions in the VNC, possibly related to its innervation from the cerebellum.[46,51,63,86] Across all regions, however, the correlation between taurine and GABA concentrations was not high.

16.5 ACETYLCHOLINE

There is evidence for involvement of acetylcholine as a neurotransmitter in the CN, VNC, SOC, and VCb[17,31,71] (see Chapter 10 by Pompeiano). The enzyme of synthesis for acetylcholine, choline acetyltransferase (ChAT), has been used extensively as the most definitive marker for cholinergic neurons.[29] The activities of ChAT are much higher in the CN than in the VNC (Figure 16.4). The higher activities for vestibular nuclei reported by Burke and Fahn[9] may result from the lower-resolution sampling method used, coupled with the proximity of the cholinergic facial nerve fibers. Since the ChAT activities of facial nerve fibers[29] are about 50 times those of the VNC, inclusion of even a small amount of facial nerve tissue in VNC samples can make a large difference in measured activities. A similar problem can occur with sampling of the SOC from thick slabs of tissue[15] since the facial nucleus lies immediately caudal to it. The only vestibular region with ChAT activities approaching those of some of the auditory regions was the CbN, particularly its granular layer. The moderately high activity here, slightly less than that for whole rat brain (400 to 500 mmol/kg dry wt/min), has been associated with a cholinergic mossy fiber projection from the VNC, particularly MVN[3,4,59] (see Chapter 11 by Barmack). The ChAT activities for individual folia of rat measured by Barmack et al.[3] included highest values in the CbN, as in our results. However, the absolute values reported in that study are about 4 orders of magnitude higher than those shown here (Figure 16.4) and presumably reflect a calculation error.

While the origins of the ChAT activity in the CN have been associated primarily with a centrifugal innervation,[29-31] those of the ChAT activity in the VNC could not be determined by Burke and Fahn[10] (although the likelihood of including facial nerve tissue in the samples assayed would have obscured the results). A small amount should derive from its innervation by a few olivocochlear branches,[7] and some should derive from the pedunculopontine and laterodorsal tegmental nuclei.[94] The MVN, like the CN, contains significant amounts of muscarinic receptor binding[88] and neuropil staining for activity of acetylcholinesterase,[65] the enzyme which hydrolyzes acetylcholine and is especially concentrated at cholinergic synapses. Thus, the ChAT activities in MVN are unlikely to be totally contained within the neurons projecting to VCb.

The highest ChAT activities in the SOC are associated with the major nuclei of origin of the olivocochlear bundle in the rat: the LSO and VNTB.[84,93] The olivo-cochlear bundle tract and cochlear hair cell region of rat also contain high ChAT activities,[29] in line with considerable evidence that this pathway employs acetylcholine as a transmitter.[21] There is similar, although less complete, evidence for acetylcholine as a transmitter of the centrifugal fibers to the vestibular labyrinth,[40,44,76] which appear to travel with the olivo-cochlear fibers to the inner ear. Retrograde tracing studies indicate that the somata giving rise to the centrifugal vestibular fibers are located in several places, but most concentrated in a cluster just dorsolateral to the genu of the facial nerve root (Figure 16.3).[76,93] These neurons would therefore not be included in our samples of VNC, but measurements we have made of ChAT for this neuron cluster indicate activity about twice that of the LSO or VNTB. Although there is evidence for noncholinergic neurons grouped with cholinergic neurons in the cluster of vestibular centrifugal neurons,[76] our results so far suggest that the cholinergic neurons are in higher density here than in the nuclei containing the olivo-cochlear neurons.

16.6 MONOAMINES

There is some evidence for catecholamine and serotonin innervations of the VNC and CN, but little quantitative data that can be compared. Both the CN and VNC are reported to receive noradrenergic innervation from within or near the locus coeruleus[17,45,74,75] (see Chapter 10 by Pompeiano), in amounts that appear to be low in VNC, except for dorsal LVN, to moderate in CN compared to other brain regions.[16,81] A physiological study suggested that locus coeruleus projections are more influential on neuronal activities in the DCN than in the LVN.[14]

There is evidence for serotonergic innervation of both CN and VNC,[17,39,82] but the immunoreactivity for serotonin in both CN and VNC was considered weak relative to other brain regions.[80] Modest effects of serotonin on CN neuronal activities have been reported.[19]

16.7 DORSAL COCHLEAR NUCLEUS–VESTIBULOCEREBELLUM
COMPARISONS

A significant amount of evidence suggests similarities between superficial layers of the DCN and the cerebellar cortex.[6,22] Chemically, such similarities have been noted in the past.[27] Since the VCb, particularly the nodulus, receives some primary innervation from the vestibular nerve,[4] and since the DCN receives relatively less primary auditory innervation than the ventral CN,[60] an argument could be made for considering the superficial DCN as the auditory counterpart of the VCb cortex. It is therefore of interest to compare the chemistries of the molecular and granular layers of these two regions, keeping in mind that DCNf represents the granular layer of DCN since granule cells are by far the most abundant neurons contained within it.[95] Similarities between VCb and DCN molecular layers include relatively high CNQX binding (which increases after deafferentation. H. Li, D.A. Godfrey, and A.M. Rubin, unpublished observations), high glutamate, glutamine, GABA, and taurine concentrations, and moderate aspartate concentrations (Figure 16.4). Notable differences are much higher glycine concentration and ChAT activity in DCN. Comparison of DCN and VCb granular layers also revealed many similarities, including similar ChAT activity between DCN and CbN, but again, glycine concentration is much higher in DCN than in VCb cortex. Thus, the chemical data support the concept of many similarities between superficial DCN and cerebellar cortex, but also indicate some striking differences. The differences in glycine concentrations, with values about 5 times as high in DCN, might have important functional implications yet to be discovered. Also, although the GABA concentrations in the two regions are similar, those in superficial DCN probably include a larger contribution from centrifugal input.[5,31,43,68]

16.8 CONCLUSION

Although the primary inputs to both the vestibular and cochlear nuclei appear to contain excitatory amino acids, there is a tendency for aspartate to be more prominent in the auditory and glutamate in the vestibular input. The inhibitory pathways tend to be more glycine-dominated in the auditory and GABA-dominated in the vestibular brain centers. Possibly, this difference is important for timing or plasticity in the modulation of information processing in the two systems. Processing in the central auditory system shows more emphasis on timing, whereas the central vestibular system shows more evidence of plasticity. The more prominent cholinergic centrifugal innervation of the cochlear than of the vestibular nuclei may be part of a more extensive feedback refinement of auditory, as compared to vestibular, information processing. On the other hand, the vestibular nuclei have more prominent commissural interactions, consistent with earlier significant integration of information from both inner ears. The chemical results lend partial support to the concept that the superficial part of the dorsal cochlear nucleus might be the auditory equivalent of the vestibulocerebellar cortex.

ACKNOWLEDGMENTS

We are grateful to Tim Godfrey, Andrea Fulcomer, and Jun Liu for assistance with experiments. Supported by NIH grants DC00172 and DC02550.

REFERENCES

1. Altschuler, RA, Juiz, JM, Shore, SE, Bledsoe, SC, Helfert, RH, and Wenthold, RJ, Inhibitory amino acid synapses and pathways in the ventral cochlear nucleus. in: *The Mammalian Cochlear Nuclei: Organization and Function* (Merchyn MA, Juiz JM, Godfrey DA, and Mugnaini E, Eds.), pp. 211-224. New York: Plenum,1993.
2. Aran, J-M, Cazals, Y, Charlet, de Sauvage, R, Electrophysiological monitoring of the cochlea during and after total destruction of the organ of Corti, *Acta. Otolaryngol.,* (Stockholm) 89:376,1980.
3. Barmack, NH, Baughman, RW, and Eckenstein, FP, Cholinergic innervation of the cerebellum of rat, rabbit, cat, and monkey as revealed by choline acetyltransferase activity and immunohistochemistry, *J. Comp. Neurol.,* 317:233,1992.
4. Barmack, NH, Baughman, RW, Eckenstein, FP, and Shojaku, H, Secondary vestibular cholinergic projection to the cerebellum of the rabbit and rat as revealed by choline acetyltransferase immunohistochemistry, retrograde and orthograde tracers, *J. Comp. Neurol.,* 317:250,1992.
5. Batini, C, Compoint, C, Buisseret-Delmas, C, Daniel, H, and Guegan, M, Cerebellar nuclei and the nucleocortical projections in the rat: retrograde tracing coupled to GABA and glutamate immunohistochemistry, *J. Comp. Neurol.,* 315:74,1992.
6. Berrebi, AS, and Mugnaini, E, Alterations in the dorsal cochlear nucleus of cerebellar mutant mice. in: *The Mammalian Cochlear Nuclei: Organization and Function* (Merchán MA, Juiz JM, Godfrey DA, and Mugnaini E, Eds.), pp. 107-119. New York: Plenum,1993.
7. Brown, MC, Fiber pathways and branching patterns of biocytin-labeled olivocochlear neurons in the mouse brainstem, *J. Comp. Neurol.,* 337:600,1993.
8. Burian, M, and Gstoettner, W, Projection of primary vestibular afferent fibres to the cochlear nucleus in the guinea pig, *Neurosci. Lett.,* 84:13,1988.
9. Burke, RE, and Fahn, S, Choline acetyltransferase activity of the principal vestibular nuclei of rat, studied by micropunch technique, *Brain Res.,* 328:196,1985.

10. Burke, RE, Fahn, S, The effect of selective lesions on vestibular nuclear complex choline acetyltransferase activity in the rat, *Brain Res.,* 360:172,1985.
11. Büttner-Ennever, JA, Patterns of connectivity in the vestibular nuclei, *Ann. NY Acad. Sci.,* 656:363,1992.
12. Caspary, DM, Pazara, KE, Kᵃssl, M, and Faingold, CL, Strychnine alters the fusiform cell output from the dorsal cochlear nucleus, *Brain Res.,* 417:273,1987.
13. Cazals, Y, Erre, J-P, and Aurousseau, C, Eighth nerve auditory evoked responses recorded at the base of the vestibular nucleus in the guinea pig, *Hearing Res.,* 31:93,1987.
14. Chikamori, Y, Sasa, M, Fujimoto, S, Takaori, S, Matsuoka, I, Locus coeruleus-induced inhibition of dorsal cochlear nucleus neurons in comparison with lateral vestibular nucleus neurons, *Brain Res.,* 194:53,1980.
15. Contreras, NEIR, and Bachelard, HS, Some neurochemical studies on auditory regions of mouse brain, *Exp. Brain Res.,* 36:573,1979.
16. Cransac, H, Cottet-Emard, J-M, Pequignot, J-M, and Peyrin, L, Monoamines (noradrenaline, dopamine, serotonin) in the rat cochlear nuclei: endogenous levels and turnover, *Hearing Res.,* 90:65,1995.
17. DeWaele, C, Mnhlethaler, M, and Vidal, PP, Neurochemistry of the central vestibular pathways, *Brain Res. Rev.,* 20:24,1995.
18. Dohlman, G, Farkashidy, J, and Salonna, F, Centrifugal nerve-fibres to the sensory epithelium of the vestibular labyrinth, *J. Laryngol. Otol.,* 72:984,1958.
19. Ebert, U, and Ostwald, J, Serotonin modulates auditory information processing in the cochlear nucleus of the rat, *Neurosci. Lett.,* 145:51,1992.
20. Elias, SA, Yae, H, and Ebner, TJ, Optical imaging of parallel fiber activation in the rat cerebellar cortex: spatial effects of excitatory amino acids, *Neuroscience,* 52:771,1993.
21. Eybalin, M, Neurotransmitters and neuromodulators of the mammalian cochlea, *Physiol. Rev.,* 73:309,1993.
22. Floris, A, Diño, M, Jacobowitz, DM, and Mugnaini, E, The unipolar brush cells of the rat cerebellar cortex and cochlear nucleus are calretinin-positive: A study by light and electron microscopic immunocytochemistry, *Anat. Embryol.,* 189:495,1994.
23. Fonnum, F, Storm-Mathisen, J, and Walberg, F, Glutamate decarboxylase in inhibitory neurons. A study of the enzyme in Purkinje cell axons and boutons in the cat, *Brain Res.,* 20:259,1970.
24. Furuya, N, Yabe, T, and Koizumi, T, Neurotransmitters regulating vestibular commissural inhibition in the cat, *Acta. Otolaryngol., (Stockholm),* 481:205,1991.
25. Gacek, RR, A cerebellocochlear nucleus pathway in the cat, *Exp. Neurol.,* 41:101,1973.
26. Godfrey, DA, and Matschinsky, FM, Approach to three-dimensional mapping of quantitative histochemical measurements applied to studies of the cochlear nucleus, *J. Histochem. Cytochem.,* 24:697,1976.
27. Godfrey, DA, and Matschinsky, FM, Quantitative distribution of choline acetyltransferase and acetylcholinesterase activities in the rat cochlear nucleus, *J. Histochem. Cytochem.,* 29:720,1981.
28. Godfrey, DA, Bowers, M, Johnson, BA, and Ross, CD, Aspartate aminotransferase activity in fiber tracts of the rat brain, *J. Neurochem.,* 42:1450,1984.
29. Godfrey, DA, Park, JL, Dunn, JD, and Ross, CD, Cholinergic neurotransmission in the cochlear nucleus. in: *Auditory Biochemistry* (Drescher DG, Ed.), pp.163-183. Springfield, IL: Thomas,1985.
30. Godfrey, DA, Park JL, Rabe, JR, Dunn, JD, and Ross, CD, Effects of large brain stem lesions on the cholinergic system in the rat cochlear nucleus, *Hearing Res.,* 11:133,1983.
31. Godfrey, DA, Parli, JA, Dunn, JD, and Ross, CD, Neurotransmitter microchemistry of the cochlear nucleus and superior olivary complex. in: *Auditory Pathway* (Syka J, and Masterton RB, Eds.), pp. 107-121. New York: Plenum,1988.
32. Godfrey, DA, Ross, CD, Parli, JA, and Carlson, L, Aspartate aminotransferase and glutaminase activities in rat olfactory bulb and cochlear nucleus; comparisons with retina and with concentrations of substrate and product amino acids, *Neurochem. Res.,* 19:693,1994.
33. Golding, NL, Oertel, D, Glycinergic and GABAergic inputs excite cartwheel cells and inhibit fusiform cells in the dorsal cochlear nucleus, *Abstr. Assoc. Res. Otolaryngol.,* 194,1995.
34. Hardy, M, Observations on the innervation of the macula sacculi in man, *Anat. Rec.,* 59:403,1935.
35. Hawkins, JE, Jr., Auditory physiological history: a surface view. in: *Physiology of the Ear,* (Jahn AF, and Santos-Sacchi J, Eds.), pp. 1-28. New York: Raven,1988.

36. Helfert, RH, Snead, CR, and Altschuler, RA, The ascending auditory pathways. in: *Neurobiology of Hearing: the Central Auditory System*, (Altschuler, RA, Bobbin, RP, Clopton, BM, and Hoffman, DW, Eds.), pp. 1-25. New York: Raven,1991.

37. Henley, CM, and Igarashi, M, Amino acid assay of vestibular nuclei 10 months after unilateral labyrinthectomy in squirrel monkeys, *Acta. Otolaryngol. (Stockholm)*, 481:407,1991.

38. Kevetter, GA, and Perachio, AA, Projections from the sacculus to te cochlear nuclei in the mongolian gerbil, *Brain Behav. Evol.*, 34:193,1989.

39. Klepper, A, and Herbert, H, Distribution and origin of noradrenergic and serotonergic fibers in the cochlear nucleus and inferior colliculus of the rat, *Brain Res.*, 557:190,1991.

40. Klinke, R, and Galley, N, Efferent innervation of vestibular and auditory receptors, *Physiol. Rev.*, 54:316,1974.

41. Koerber, KC, Pfeiffer, RR, Warr, WB, and Kiang, NYS, Spontaneous spike discharges from single units in the cochlear nucleus after destruction of the cochlea, *Exp. Neurol.*, 16:119,1966.

42. Kolston, J, Osen, KK, Hackney, CM, Ottersen, OP, and Storm-Mathisen, J, An atlas of glycine- and GABA-like immunoreactivity and colocalization in the cochlear nuclear complex of the guinea pig, *Anat Embryol.*, 186:443,1992.

43. Kolston, J, Apps, R, and Trott, JR, A combined retrograde tracer and GABA-immunocytochemical study of the projection from nucleus interpositus posterior to the posterior lobe C2 zone of the cat cerebellum, *Eur. J. Neurosci.*, 7:926,1995.

44. Kong, W-J, Egg, G, Hussl, B, Spoendlin, H, and Schrott-Fischer, A, Localization of chat-like immunoreactivity in the vestibular endorgans of the rat, *Hearing Res.*, 75:191,1994

45. Kromer, LF, and Moore, RY, Norepinephrine innervation of the cochlear nuclei by locus coeruleus neurons in the rat, *Anat. Embryol.*, 158:227,1980.

46. Li, H, Godfrey, DA, and Rubin, AM, Quantitative distribution of amino acids in the rat vestibular nuclei, *J. Vest. Res.*, 4:437,1994.

47. Li, H, Squire, AB, Godfrey, DA, and Rubin, AM, Quantitative distribution of choline acetyltransferase activity in rat vestibular nuclear complex, *Soc. Neurosci. Abstr.*, 21:918,1995.

48. Li, H, Godfrey, TG, Godfrey, DA, and Rubin, AM, Quantitative changes of amino acid distributions in the rat vestibular nuclear complex after unilateral vestibular ganglionectomy, *J. Neurochem.*, (in press),1996.

49. Liu, J, Godfrey, TG, and Godfrey, DA, Distribution of amino acid neurotransmitters in rat superior olivary complex, *Soc. Neurosci. Abstr.*, 21:403,1995.

50. Lowry, OH, and Passonneau, JV, *A Flexible System of Enzymatic Analysis*, New York: Academic,1972.

51. Magnusson, KR, Madl, JE, Clements, JR, Wu, J-Y, Larson, AA, and Beitz, AJ, Colocalization of taurine- and cysteine sulfinic acid decarboxylase-like immunoreactivity in the cerebellum of the rat with monoclonal antibodies against taurine, *J. Neurosci.*, 8:4551,1988.

52. McCue, MP, Guinan, JJ Jr, Acoustically responsive fibers in the vestibular nerve of the cat, *J. Neurosci.*, 14:6058,1994.

53. McCue, MP, Guinan, JJ Jr, Influence of efferent stimulation on acoustically responsive vestibular afferents in the cat, *J. Neurosci.*, 14:6071,1994.

54. McCue, MP, Guinan, JJ Jr, Spontaneous activity and frequency selectivity of acoustically responsive vestibular afferents in the cat, *J. Neurophysiol.*, 74:1563,1995.

55. Mehler, WR, and Rubertone, JA, Anatomy of the vestibular nucleus complex. in: *The Rat Nervous System*, Vol. 2, Hindbrain and Spinal Cord (Paxinos, G, Ed.), pp. 185-219. Orlando: Academic,1985.

56. Moffat, AJM, and Capranica, RR, Auditory sensitivity of the saccule in the American toad *(Bufo americanus)*, *J. Comp. Physiol.*, 105:1,1976.

57. Mugnaini, E, Warr, WB, and Osen, KK, Distribution and light microscopic features of granule cells in the cochlear nuclei of cat, rat, and mouse, *J. Comp. Neurol.*, 191:581,1980.

58. Murofushi, T, Curthoys, IS, Topple, AN, Colebatch, JG, and Halmagyi, GM, Responses of guinea pig primary vestibular neurons to clicks, *Exp. Brain Res.*, 103:174,1995.

59. Ojima, H, Kawajiri, S-I, and Yamasaki, T, Cholinergic innervation of the rat cerebellum: qualitative and quantitative analyses of elements immunoreactive to a monoclonal antibody against choline acetyltransferase, *J. Comp. Neurol.*, 290:41,1989.

60. Osen, KK, Course and termination of the primary afferents in the cochlear nuclei of the cat: an experimental anatomical study, *Arch. Ital. Biol.*, 108:21,1970.

61. Otsuka, M, Obata, K, Miyata, Y, and Tanaka, Y, Measurement of g-aminobutyric acid in isolated nerve cells of cat central nervous system, *J. Neurochem.*, 18:287,1971.

62. Ottersen, OP, and Storm-Mathisen, J, Neurons containing or accumulating transmitter amino acids. in: *Handbook of Chemical Neuroanatomy*, Vol. 3, Classical Transmitters and Transmitter Receptors in the CNS, Part II (Bj¨rklund A, H¨kfelt T, Kuhar MJ, eds), pp. 141-246. Amsterdam: Elsevier,1984.

63. Ottersen, OP, Madsen, S, Storm-Mathisen, J, Somogyi, P, Scopsi, L, and Larsson, L-I, Immunocytochemical evidence suggests that taurine is colocalized with GABA in the Purkinje cell terminals, but that the stellate cell terminals predominantly contain GABA: a light- and electronmicroscopic study of the rat cerebellum, *Exp. Brain Res.*, 72:407,1988.

64. Ottersen, OP, Laake, JH, and Storm-Mathisen, J, Demonstration of a releasable pool of glutamate in cerebellar mossy and parallel fibre terminals by means of light and electron microscopic immunocytochemistry, *Arch. Ital. Biol.*, 128:111,1990.

65. Paxinos, G, and Watson, C, The rat brain in stereotaxic coordinates, New York: Academic,1986.

66. Peng, L, Hertz, L, Huang, R, Sonnewald, U, Petersen, SB, Westergaard, N, Larsson, O, and Schousboe, A, Utilization of glutamine and of TCA cycle constituents as precursors for transmitter glutamate and GABA, *Dev. Neurosci.*, 15:367,1993.

67. Popper, AN, and Fay, RR, Sound detection and processing by teleost fishes: a critical review, *J. Acoust. Soc. Amer.*, 53:1515,1973.

68. Potashner, SJ, Benson, CG, Ostapoff, E-M, Lindberg, N, and Morest, DK, Glycine and GABA: transmitter candidates of projections descending to the cochlear nucleus. in: *The Mammalian Cochlear Nuclei: Organization and Function* (Merchán MA, Juiz JM, Godfrey DA, Mugnaini E, Eds.), pp. 195-210. New York: Plenum,1993.

69. Potashner, SJ, Morest, DK, Oliver, DL, and Jones, DR, Identification of glutamatergic and aspartatergic pathways in the auditory system. in: *Auditory Biochemistry* (Drescher DG, Ed.), pp. 141-162. Springfield, IL: Thomas,1985.

70. Rasmussen, GL, The olivary peduncle and other fiber projections of the superior olivary complex, *J. Comp. Neurol.*, 84:141,1946.

71. Raymond, J, Dememes, D, and Nieoullon, A, Neurotransmitters in vestibular pathways. in: *Progress in Brain Research*, Vol. 76 (Pompeiano O, and Allum JHJ, Eds.), pp. 29-43. Amsterdam: Elsevier,1988.

72. Ross, CD, and Thompson, GC, Quantitative distributions of six amino acids in vestibular nuclei of bush baby and guinea pig, *Abstr. Assoc. Res., Otolaryngol.*, 15,1995.

73. Ross, CD, Smith, JT, and Godfrey, DA, Regional distributions of choline acetyltransferase and acetylcholinesterase activities in layers of rat cerebellar vermis, *J. Histochem. Cytochem.*, 31:927,1983.

74. Schuerger, RJ, and Balaban, CD, Immunohistochemical demonstration of regionally selective projections from locus coeruleus to the vestibular nuclei in rats, *Exp. Brain Res.*, 92:351,1993.

75. Schuerger, RJ, and Balaban, CD, Origins of noradrenergic projections to the vestibular nuclei, *Abstr. Assoc. Res. Otolaryngol.*, 100,1995.

76. Schwarz, DWF, Satoh, K, Schwarz, KH, and Fibiger, HC, Cholinergic innervation of the rat's labyrinth, *Exp. Brain Res.*, 64:19,1986.

77. Shank, RP, and Aprison, MH, Present status and significance of the glutamine cycle in neural tissues, *Life Sci.*, 28:837,1981.

78. Smith, PF, and Curthoys, IS, Neuronal activity in the ipsilateral medial vestibular nucleus of the guinea pig following unilateral labyrinthectomy, *Brain Res.*, 444:308,1988.

79. Smith, PF, De Waele, C, Vidal, P-P, and Darlington, CL, Excitatory amino acid receptors in normal and abnormal vestibular function, *Molec. Neurobiol.*, 5:369,1991.

80. Steinbusch, HWM, Serotonin-immunoreactive neurons and their projections in the CNS. in: *Handbook of Chemical Neuroanatomy*. Vol. 3, Classical Transmitters and Transmitter Receptors in the CNS, Part II (Bj¨rklund A, H¨kfelt T, and Kuhar MJ, Eds.), pp. 68-125. Amsterdam: Elsevier,1984.

81. Swanson, LW, and Hartman, BK, The central adrenergic system. An immunofluorescence study of the location of cell bodies and their efferent connections in the rat utilizing dopamine-b-hydroxylase as a marker, *J. Comp. Neurol.*, 163:467,1975.

82. Thompson, AM, Moore, KR, and Thompson, GC, Distribution and origin of serotoninergic afferents to guinea pig cochlear nucleus, *J. Comp. Neurol.*, 351:104,1995.

83. Tickle, DR, and Schneider, GE, Projection of the auditory nerve to the medial vestibular nucleus, *Neurosci. Lett.*, 28:1,1982.
84. Vetter, DE, and Mugnaini, E, Distribution and dendritic features of three groups of rat olivocochlear neurons, *Anat. Embryol.*, 185:1,1992.
85. Walberg, F, and Dietrichs, E, The interconnection between the vestibular nuclei and the nodulus: A study of reciprocity, *Brain Res.*, 449:47,1988.
86. Walberg, F, Ottersen, OP, and Rinvik, E, GABA, glycine, aspartate, glutamate and taurine in the vestibular nuclei: An immunocytochemical investigation in the cat, *Exp. Brain Res.*, 79:547,1990.
87. Waller, HJ, Godfrey, DA, and Chen, K, Responses of dorsal cochlear nucleus neurons to electrical stimulation of parallel fibers in rat brain stem slices, *Abstr. Assoc. Res., Otolaryngol.*, 12,1994.
88. Wamsley, JK, Lewis, MS, Young, WS, III, and Kuhar, MJ, Autoradiographic localization of muscarinic cholinergic receptors in rat brainstem, *J. Neurosci.*, 1:176,1981.
89. Wenthold, RJ, Glutamic acid and aspartic acid in subdivisions of the cochlear nucleus after auditory nerve lesion, *Brain Res.*, 143:544,1978.
90. Wenthold, RJ, Neurotransmitters of brainstem auditory nuclei. in: *Neurobiology of Hearing: the Central Auditory System* (Altschuler RA, Bobbin RP, Clopton BM, and Hoffman DW, Eds.), pp. 121-139. New York: Raven,1991.
91. Wenthold, RJ, and Gulley, RL, Aspartic acid and glutamic acid levels in the cochlear nucleus after auditory nerve lesion, *Brain Res.*, 138:111,1977.
92. Wenthold, RJ, Hunter, C, and Petralia, RS, Excitatory amino acid receptors in the rat cochlear nucleus. in: *The Mammalian Cochlear Nuclei: Organization and Function* (Merchán, MA, Juiz, JM, Godfrey, DA, and Mugnaini, E, Eds), pp. 179-194. New York: Plenum,1993.
93. White, JS, and Warr, WB, The dual origins of olivocochlear neurons in the albino rat, *J. Comp. Neurol.*, 219:203,1983.
94. Woolf, NJ, and Butcher, LL, Cholinergic systems in the rat brain: IV. Descending projections of the pontomesencephalic tegmentum, *Brain Res. Bull.*, 23:519,1989.
95. Wouterlood, FG, and Mugnaini, E, Cartwheel neurons of the dorsal cochlear nucleus: a Golgi-electron microscopic study in rat, *J. Comp. Neurol.*, 227:136,1984.
96. Young, AB, Oster-Granite, ML, Herndon, RM, and Snyder, SH, Glutamic acid selective depletion by viral induced granule cell loss in hamster cerebellum, *Brain Res.*, 73:1,1974.
97. Young, ED, Fernández, C, and Goldberg, JM, Responses of squirrel monkey vestibular neurons to audio-frequency sound and head vibration, *Acta. Otolaryngol.*, (Stockholm) 84:352,1977.
98. Zhang, S, Oertel D, Neuronal circuits associated with the output of the dorsal cochlear nucleus through fusiform cells, *J. Neurophysiol.*, 71:914,1994.
99. Zhao, HB, Parham, K, Ghoshal, S, and Kim, DO, Small neurons in the vestibular nerve root project to the marginal shell of the anteroventral cochlear nucleus in the cat, Brain Res., 700:295,1995.

Section VI

Neurochemical Basis of Plasticity and Adaption

17

Immediate Early Gene Expression in the Vestibular System

Galen D. Kaufman and Adrian A. Perachio

CONTENTS

17.1 SYNOPSIS

The characterization of inducible nuclear transcriptional control proteins and the genes that encode them (*housekeeping* genes, proto-oncogenes or immediate early genes, IEGs) provides the opportunity to monitor novel or changing genomic activity in a cell by detecting the expression and regulation of these proteins. This chapter will address the following points. First, a description of the current understanding of these "third-messenger" control systems, with an emphasis on the Fos/Jun group, will be presented. Then, using immunohistochemical and *in situ* hybridization experiments as examples, the utility of these molecules as markers for neurons changing their genomic expression program, with specific application to the vestibular system, will be discussed. Finally, we will suggest strategies for investigating the role of IEGs as molecular mechanisms underlying development and aging, repair and recovery, and motor learning strategies with respect to the vestibular system.

0-8493-7679-3/00/$0.00+$.50
© 2000 by CRC Press LLC

17.2. INTRODUCTION

What are immediate early genes? A simple answer is that IEGs are DNA sequences transcribed (copied to mRNA for protein expression) within minutes of receiving extracellular cues in order to regulate intracellular homeostasis. There is a long history of biological investigation concerning the mechanisms by which extracellular signals are transmitted across a cell membrane to activate second-messenger systems like cAMP or calcium within the cytoplasm. What was less understood until recently was how these events connected to transcriptional control of the genome within the cell nucleus. Early studies began to define these systems while searching for cell cycle genes using an *in vitro* pheochromocytoma PC12 cell induction model. Substances such as nerve growth factor, calcium, and potassium were found to stimulate c-*fos* expression.[49] Another route that brought awareness of these systems in the normal cell came via research regarding viral genomes which mimicked or interfered with eukaryotic transcription. This is the explanation behind the prefix v- (for viral) or c- (for cellular) in some of the IEG literature. Using the power of molecular techniques, this rapidly growing field of biology has produced the knowledge to link extracellular signals to intranuclear control; the proteins in the middle have become known as "third messengers" (transcribed and translated from IEGs) because of their rapid expression following extracellular stimuli. The diversity and variability that IEG proteins represent is only beginning to be appreciated. For an excellent review of the contemporary understanding of IEG regulation, see Reference 25.

17.3 MINUTES TO MONTHS

There are three general ways that transcriptional factors are activated: *post-translationally* by phosphorylation (e.g., the CREB ITF); with intracellular steroid-type *ligand-receptor interactions* (e.g., glucocorticoid receptors); or *transcriptionally*, that is, by the rapid transcription and translation of the factor's gene (e.g., c-*fos*, [25]). The latter category includes those genes we call immediate early genes (IEGs). In addition, there are three *modes* of IEG regulation in a cell.[50] The most well-known is the rapid transient response, where a signal(s) causes the transcription and translation of the protein within a few minutes, and the protein affects certain genes in the nucleus and is removed or destroyed again within a few hours. The second is a delayed but persistent upregulation of the protein. An example is the upregulation of Jun 12-24 hours following sciatic nerve ligation[20]. The third mode is the continuous expression of the protein in certain cell types and situations. For example, in the rat, neurons in the ventrolateral periaqueductal grey always exhibit some Fos expression.[31]

Research to date has identified several *classes* of IEG ITFs, based on their structures and the mechanisms they use to control gene promoter sequences. These classes are somewhat overlapping. They include proteins, the amino acid sequences of which confer similar functional properties to the tertiary structure of the molecule by way of specific motifs in the primary sequence of the peptide. These motifs can be thought of as functional cassettes, and include the basic-zipper, zinc-finger, ligand/receptor-dependent, and helix-loop-helix (HLH) protein motifs, as well as a basic amino acid sequence, the specificity of which determines to which sequence of DNA the molecule will bind (for review, see Refs. 7, 34, 38). These molecules form complexes with other ITFs or proteins in the cell nucleus and together, depending on many variables, some of which will be mentioned later, bind to specific segments of DNA to affect the molecular machinery of transcription. Many of the well-studied IEG proteins, for example those of the Fos and Jun families, are of the basic (leucine) zipper structure. These molecules have in common a 30-amino acid a-helix loop with four to five leucine residues regularly spaced every seven amino acids in the helix so that they are situated on one side of the helix. The attractive hydrophobic forces between two such molecules allows them to be connected along this row like a zipper. Other factors that share this motif are CCAAT/enhancer-binding proteins (C/EBP), cyclic AMP response element binding protein (CREB), activating transcription factor(s) (ATF), and the chicken oncogene product Myc.

Another class consists of proteins with the zinc-finger motif, for example, zif268 (i.e., Erg1, Krox24, TIS8, NGF1-A, or ZENK). This motif consists of many proline residues, preventing a-helix formation, and a fixed two-cysteine, two-histidine sequence that traps a zinc ion. The resulting structure is capable of sequence-specific DNA binding by virtue of the finger-like loops of basic amino acids in the primary sequence.[71] The nomenclature of ITFs in the literature is often redundant due to the parallel identification of identical sequences in different model systems. Proteins in the nuclear receptor class have homology with steroid and thyroid hormone receptors, requiring a specific ligand for activation. NGF1-B shares structural homology with this class. The helix-loop-helix (HLH) class has a three-part linear structure but also contains a leucine-zipper region and is predicted to dimerize with like molecules; its members include Myc and mouse muscle determination factor (MyoD). However, different members of one class do not necessarily cross-dimerize with any other member, so that groups of proteins can be defined within a class, for example, the Fos/Jun group. It has been shown using mutational analyses that dimerization is necessary for DNA binding, but is independent of DNA binding. These basic motifs are conserved across species as diverse as human, rat, chicken, maize, and yeast.

The focus of this brief review will be on IEG proteins with the capability to bind to nucleic acid sequences in a base-specific manner, and even more specifically on Fos, Jun, and a few other IEGs like zif268. Many IEGs, however, can code for other types of proteins with roles outside the nucleus, including signal transduction and metabolic control enzymes like phosphatases, G-proteins, and cytokine-like molecules.[51] Regarding DNA binding proteins, it might be useful to think of these dimers as analogous to the immunoglobins,[38] the hypervariable regions of which can be created to recognize a multitude of precise protein epitopes. In contrast, these proteins recognize specific DNA palindromic epitopes based on the sequence information in the strand. The dynamic combination and control of DNA binding with other proteins, many of which are always present in the nucleus, allows for precise regulation of specific genes by controlling the initiation, rate, or absence of messenger RNA transcription from the DNA template in question.

17.4 IMMEDIATE EARLY GENES

17.4.1 Fos As a Model of IEG Expression

On these types, the most popular and well-studied IEG to date has been the c-*fos* gene and its protein product, Fos. The majority of vestibular-related IEG expression studies to date have utilized Fos. This is due to the knowledge and commercial availability of antibodies against Fos rather than any understanding of its precise effect on the transcription of certain genes. Rather, the popularity of Fos immunohistochemistry came about due to its relative ease of application to many sensory paradigms, and the basic assumption that this expression revealed a "genetic activity" in the cell, giving neuroscientists a preliminary indicator of altered cellular transcription. The single cell resolution and the temporal characteristics of expression — spanning immediate pharmacological or stimulus events to long duration changes associated with habituation or adaptation — were also very attractive to investigators seeking a fresh look at neural plasticity. Of course, drawbacks and limitations exist, and will be discussed in the following sections.

In order to put the function of Fos in the proper framework, we will first describe the genetic control elements that support Fos expression (see Figure 17.1).

Fos is a 55-kilodalton (k Da) protein,[13;15] the phosphorylation state of which varies inversely with its ability to bind DNA.[52] Its transcription is protein synthesis-independent: constitutive initiation factors are poised and ready to begin within minutes of a stimulus. At least two second messenger systems can activate Fos expression independently: those leading to activation of PKC, and those resulting in an increase in cytosolic calcium.[25] The location of Fos in the cell has been determined ultrastructurally, where it is suggested that Fos-positive nuclear domains correspond to

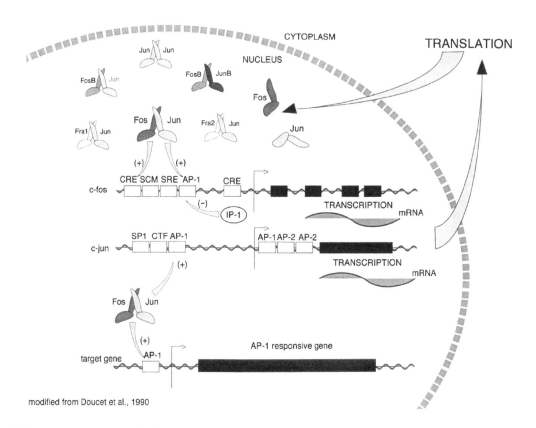

modified from Doucet et al., 1990

FIGURE 17.1 A schematic of Fos/Jun regulation and control (Modified from Ref. 15). The basic promoter sequence of the c-*fos* and c-*jun* immediate early genes are shown. Transcriptional initiation sites are designated by a vertical line across the gene sequence. IEG transcription proceeds by producing primary transcripts and mRNA, which move to the cytoplasm where the protein is translated. These proteins move back to the nucleus where they find dimer partners that influence the transcriptional machinery of other genes. As an example, a model AP-1 responsive target gene is shown. See text for more details.

the actively transcribing regions of chromatin, while the nucleolus does not contain Fos.[37] The Fos protein functions in the cell nucleus by its dimerization along the leucine zipper region with one of a series of proteins from the c-*jun* family. These dimers, also known as activator proteins and sometimes associated with yet other proteins as a complex, then attach to an activator-protein-1 site (AP-1; also TIS or TRE), the DNA sequence of which is: 5'-TGACTCA-3'. The "foot" of each transcription factor associates with only 3-4 nucleotide bases. This binding actually bends the DNA strand and allows transcription and editing enzymes (e.g., RNA polymerase II) to initiate reproduction of the strand into messenger RNA. Figure 17.1 shows the main regulatory domains currently associated with the c-*fos* and c-*jun* genes (based on Ref. 15). It should be apparent that many control points exist for this critical genomic regulatory system, including negative feedback (*trans*-repression) of the c-*fos* gene by the Fos protein on SRE and AP-1 sites.[44] In addition, truncational analyses reveal that besides the basic and leucine zipper motifs, Fos and Jun have additional domains likely associated with yet other factors that can contribute to transcriptional activation.[1,2] Subtle differences in types of AP-1 complexes consisting of different dimer combinations have variable effects on the transcriptional machinery. As a further complexity, the Fos/Jun complex can also bind to another regulatory element, the cyclic AMP regulatory element (CRE sequence: 5'-TGACGTCA-3'), and other ITFs can bind to AP-1 (e.g., CREB), with different results for the transcription of certain genes.

A wide range of dimerization options exist for the IEG proteins available to any one cell. Fos can only dimerize with a member of the c-*jun* family (heterodimers), while some Jun proteins can autodimerize (homodimers) and bind to DNA. Packing forces favor heterodimers in the leucine zipper structure,[60,61] so the Fos/Jun dimer is a stronger bond that a Jun homodimer. In general, the Fos/Jun dimers have a longer half-life than Jun/Jun dimers; the different binding affinities have been ranked.[65] The strength of the dimer association also affects subsequent DNA binding strength; the resulting transactivational activity is much higher with heterodimers containing Fos than with other dimer combinations. Some cells possess ITFs capable of binding other ITFs but which lack a DNA binding region, thereby creating another subtle regulatory capability by occupying otherwise functional dimer space.[38] In addition, the different combinations of dimers each have different effects on the initiation and maintenance of transcription in different genes, so that a complex range of regulatory capabilities is realized very quickly. This complexity is one of the problems in trying to understand the specific relationship of one IEG protein to a specific gene's expression. Not only can different neurotransmitter receptors activate different IEGs (e.g., muscarinic activation of JunB but not Jun,[25] but the *degree of activation* of the *same* receptor can recruit different IEGs.

Additional mechanisms of regulation include a host of phosphorylation and second-messenger cascades. The details of calcium entry and its spatial compartmentalization during Fos induction are an important factor in IEG regulation.[43] At least four different proteins have been shown to regulate the SRF element in the c-*fos* promoter region alone.[62] There is good evidence that these regulatory cis elements (i.e., DNA strand being transcribed) are poised, constitutively occupied with proteins.[14] The histone proteins, especially H1 which also recognizes specific DNA structures and acts as a eukaryotic suppressor, can also be involved in the regulation of DNA binding. Fos and other transcriptional activators like CTF/NF-I might also be a part of the cascade which utilizes nuclear receptor (ligand-dependent) action.[59] Ligand-receptor ITFs can also bind to Fos/Jun dimers and modulate their activity. The oxidative state of transcription factors affect their activity, introducing yet another level of control.[50] The oxidative state of a conserved cysteine residue in the basic region of both the Fos and c-jun protein has been shown to regulate DNA binding activity *in vitro*.[2] There also exists an inhibitory mechanism to block AP-1 activation. It has been shown in some cells that a 30 – 40 inhibitory protein, IP-1, blocks Fos/Jun heterodimer DNA binding, and that this inhibition takes place only with the un phosphorylated form of the protein, which might serve the role of a transcriptional anti-oncogene.[4] The CREB ITF can also block Fos/Jun binding at AP-1 sites due to its sequence homology with the CRE site.

It is mostly from an historical context that Fos deserves its reputation, but other IEGs are beginning to be recognized, understood, and appreciated for their unique roles. These include other Fos family transcription factors, as well as those of the c-*jun* family, zif268, and CREB. It is important to understand that each ITF has different expression kinetics and binding affinities for regulatory DNA sites and for other ITFs; the choreography of these different factors leads to the complex regulation of different genes. In order to appreciate the diversity, a brief introduction to these proteins follows.

17.4.2 OTHER IMMEDIATE EARLY GENES

Cellular homologs of the c-*fos* gene exist; Fos B, delta-Fos B, and at least two so-called "Fos-related antigens" (Fra1 and 2). Fos B has 70% sequence homology with Fos, and shares the leucine heptad/basic motif.[67] Two splice variations of this gene have been reported.[53] A truncated form of Fos B can inhibit Fos/Jun activity.[54] Fos B regulates c-*jun* and other target genes in a manner similar to Fos, but differs in several other ways. For example, Walther et al.[73] showed an increase in Fos B mRNA in the nucleus caudalis of the spinal tract of the trigeminal nerve ipsilateral to trigeminal ganglion electrical stimulation which they hypothesize to up-regulate proenkephalin transcription via its affinity with Jun B. Fra1 and Fra2 were not increased in their paradigm. It should be noted that some polyclonal antibodies raised against Fos recognize these related homologous proteins as well.

Although intimately linked with Fos regulation because of its functional requirement for heterodimerization, proteins in the c-*jun* family can have distinct function and expression patterns. Unlike Fos, Jun positively autoregulates the c-*jun* gene.[3] Jun expression is increased 12 to 24 hours after sciatic nerve transection or other axotomy[20,40,51] with an increased expression lasting weeks or months until regeneration is complete. Jun is also upregulated, although with a different time course, centrally in the substantia nigra following 6-hydroxydopamine lesions of catecholamine neurons.[26] Homologues of Jun called Jun B and Jun D have different expression kinetics and roles than Jun For example, Jun B is inducible like Jun, probably via NMDA glutamate receptors, but Jun D is usually expressed constitutively. Jun B is associated with cell differentiation, while Jun affects cell proliferation.[66] The Fos/Jun B dimer has been shown to act as a transcriptional repressor (see Ref. 15 for review). In general, Jun and Fra ITFs have longer induction and expression times than Fos.

Another IEG-related protein gaining prominence in the literature is zif268 (i.e. erg1, Krox24, TIS8, NFG1-A, ZENK, see analogs in Ref. 23. This IEG shares many of the same temporal and regulatory characteristics with the Fos protein, although the genes are located on different chromosomes.[71] Its DNA recognition site is 5-GCGTGGGGCG-3'.[9] It has been reported that zif268 can be regulated by methylation.[68] Fos can also down- regulate zif268 via the SRE site.[21] Zif268 expression exhibits dynamic patterns during development in the rat brain.[22] In the adult, expression of this protein appears to be driven constitutively in rat visual cortex by visual stimulation,[75] and decreases rapidly following a disruption of that input with intravitreal tetrodotoxin or eye-patching. The authors suggest that zif268 might be a better marker for cerebral cortical activity than c-*fos*, because the latter does not respond as rapidly to alterations of physiological activity. In the visual cortex of vervet monkeys, as little as 2 hours of monocular deprivation produced zif268 immuno-labeling (IL) patterns identical to ocular dominance columns observed with other techniques.[8] Zif268, (i.e., NGF1-A), should not be confused with NGF1-B, which is an ITF sharing steroid receptor homology. A DNA response element (5'-AGGTCA-3') has been defined for NGF1-B whose sequence is also contained in the regulatory domains of the genes for rat corticotropin releasing factor (CRF), oxytocin (OT), and arginine vasopressin (AVP).[7] It is interesting to note that the first two genes happen to lack AP-1 promoter sites.

CREB (cyclic AMP response element binding protein) is a 43-kD transcription factor activated post-translationally by protein kinase A or Ca++/calmodulin kinase. CREB binds to CRE sites as a monomer, or more strongly as a dimer, and can also bind AP-1 sites. Its detection with respect to its phosphorylation state could serve as another potential activity marker.[50]

17.4.3 LIMITATIONS AND DRAWBACKS OF USING IEG EXPRESSION

An often overheard sentiment regarding IEG studies is, "so what does it mean?" That is, since one is typically unsure of the targets or effects of a particular IEG change, how can the observation have any relevance to a specific problem? The functional heterogeneity and diversity of the IEG response described in this chapter begs the question. Despite the fact that many of the specific targets *are* being identified (see the following section), what it means is that the cell has recruited a powerful transcriptional signal and needs to do *something* which involves a long-term change in the phenotype. The basic drawbacks of protein IL, specifically Fos IL, as a neural activity marker have been outlined.[16] They consist of the following problems. There is an apparent mismatch with 2-deoxyglucose studies which measure glucose utilization as a function of metabolic activity.[18,59] One possible explanation is that the 2-deoxyglucose technique can detect axonal and dendritic activity, while Fos IL only reveals activity in the cell nucleus. Another concern is that some cells possess constitutive or basal expression of the Fos protein, and this would need to be well-defined in order to superimpose a change due to another factor. Alternatively, a decrease in this expression could also signify functional importance. Another problem is that the strength of activation necessary to elicit expression is not always clear. Also, the time course of expression must be defined and understood in relation to behavior or causal factors. Immunohistochemical techniques are subject to the non-specificity of polyclonal antibodies. One way this can be circumvented is by using transgenic ß-galactosidase expression inserted into the c-*fos* reading frame, so that a simple colorimetric reaction indicates c-*fos* expression only.[50]

The difficulty in identifying the specific genes that any one IEG protein regulates lies in the complex tangle of regulatory possibilities in which that protein can participate. It has been suggested that a better conceptual model of the IEG regulatory process consists of dynamic chemical-like shifts of equilibrium between different states of genetic control as the relative quantity of a specific IEG protein is increased or decreased. However, studies using induction systems and other techniques have approached the issue and determined some of the target genes, for example, prodynorphin expression as a result of Fos[17,70] and suggest others (e.g., oxytocin and vasopressin,[42]). Braselmann et al.[5] induced Fos target genes in 3T3 fibroblasts which included the related Fra1, a secreted protein (Fit-1), a biosynthetic enzyme (ODC), and two membrane-associated proteins (annexin II and V). In addition, their research suggested that Fos activity can be dominant over nerve growth factor function by blocking differentiation. Other possible targets of IEGs in the CNS include proenkephalin, TRH, CCK, GAD, GAP-43, somatostatin, the interleukins, TNFa, TrkB, neuropeptide Y, hsp70, nitric oxide synthase, galanin, and calcitonin gene-related peptide.[25] Some of these possibilities are based on a temporal relationship with other genes following AP-1 activation.[17,15] others on predicted sequence binding.

A very important consideration for any *in situ* experiment designed to detect IEGs is what pharmacological biases are being applied to the cells, since different anesthetics result in different patterns of expression, (see, for example, Ref. 39). For minimal anesthetic-induced Fos IL in rat brain, one group[72] recommends a fentanyl/midazolam cocktail over other anesthetics like a-chloralose, urethane, halothane, and pentobarbital sodium. Another Fos study chose urethane over pentobarbital based on the urethane's lesser effect on sympathetic and parasympathetic reflexes, despite its strong activation of autonomic regions.[35] Our own studies showed that both urethane (500 mg/ml, 1 g/kg I.P.), a non-specific agent often used for electrophysiological recordings,[45] and a Nembutal™ (sodium pentobarbital)/ketamine mixture (20 mg/kg I.P. and 30 mg/kg I.M.), strongly induced IEG expression in the rodent brainstem. In particular, urethane caused intense Fos IL of the solitary nucleus and the area postrema. Labeled cells were also found in the inferior vestibular nuclei and throughout the reticular formation. Ketamine, which blocks NMDA cation channels, can change the expression patterns of Fos from vestibular stimuli in specific inferior olive subnuclei.[33] Blocking NMDA receptors or the administration of sodium pentobarbital also abolishes the basal expression of zif268 in some brain areas.[75]

17.4.4 QUANTIFYING IEG EXPRESSION

There are problems in the quantification of IEG expression. For IL, the simplest method is to simply count the number of immunopositive nuclei in a defined region. It quickly becomes apparent that some neurons stain more darkly than others. In other words, different cells are expressing different relative amounts of the protein in question. How, then, do we address relative amounts, or the intensity of staining? The variability of immunostaining between different reactions, fixation conditions, and individual animals, as well as real differences in the copy number of Fos proteins expressed in different neurons, makes this a very difficult task for *in vivo* studies. One strategy utilized a normalized intensity measurement from an area which exhibits constitutive expression of the protein in question, for example, constitutive Fos expression in the lateroventral periaqueductal grey of rats.[31] This value can then be considered as an internal control using an image analysis system to generate cross-treatment expression intensity values. Another approach is quantitative polymerase chain reaction (PCR) to amplify an mRNA signal with reference to a non-regulated control gene like actin or tubulin, although this technique brings about another list of potential artifacts.[12] There are many potential control steps regarding the regulation of gene product synthesis:[63] (1) transcription of the primary transcript by RNA polymerase II, (2) intron editing to messenger RNA, (3) translation of the mRNA to protein, and (4) the degradation rate of the mRNA in question. It has been shown that a common degradation signal for mRNA is a string of uridine residues, which can be regulated with splicing.[6] In addition, the actual amount or number of mRNA strands being synthesized will affect the time course and relative importance of the product in question for a particular cell. For example, a cell might have approximately 20 copies of the Fos mRNA in a basal state and upregulate that amount to only a few hundred copies after a stimulus, while a cell directed to produce a neuropeptide for secretion might need mRNA in the 10^4 copy quantity.[63]

In addition to the above limitations, several characteristics of Fos expression require a careful interpretation of using its detection as a specific marker for neuronal activity. One study found no correlation between hippocampal neuronal burst firing and Fos expression, or the magnitude of that expression.[36] It has also been shown that the temporal pattern of a stimulus input to a cell can differentially express Fos in mouse dorsal root ganglion cells, despite the total input (number of stimuli) remaining the same.[69] Whether this is a temporal correlate of the absolute calcium concentration allowed to build up intracellularly, or a function of some other refractory period is not known. Regardless, Sheng et al. present the interesting hypothesis that IEG detection not only represents a general change in cell activity, but more explicitly a specific response to the specific pattern or form of a stimulus. One last issue is the problem of correlational measures, or how one establishes causality between IEG expression and other cellular indices or behavior. The only definitive correlate of Fos expression to date is prodynorphin transcription.[70] The application of antisense oligonucleotides against specific ITFs can begin to address these correlations (see the following section). Without careful *in vitro* analyses with a model gene induction system, *in vivo* studies must rely on temporal correlates and the logic of blocking or ablation experiments to establish only a possible connection between the expression of one ITF and a behavioral or other biochemical measure. This should not preclude such studies from being done, but their interpretation must be based on contemporary information of the molecular biology of the system in question.

In spite of these drawbacks, the utility of IEG detection has rapidly produced a wide variety of studies using these techniques. IEG expression has been used to examine the sequelae of various stimuli on the CNS, including visual, auditory, somatosensory, motor, olfactory, proprioceptive, and vestibular stimuli. Neurologic studies ranging from electroacupuncture[41] to circadian rhythms[11] have yielded new information using these protocols. In addition, generated activity like seizures or cognitive processes like memory,[64] and developmental and regeneration research[19] have been explored using IEG detection protocols.

17.5 IEG EXPRESSION IN THE VESTIBULAR SYSTEM

The experimental design of any study that seeks to relate IEGs to vestibular function faces several challenges. First, it is imperative to demonstrate that the stimulus that produces IEG expression acts via the vestibular neuroepithelium and that the response is dependent on the integrity of the labyrinth. Second, the exact nature of the effective stimulus should be determined. That is, one must determine the optimal vector(s) of angular or linear acceleration. Third, the temporal relationship between manipulation of vestibular input and measurement of IEG-related effects should be consistent with the known time course of the onset and duration of activation of the gene(s) of interest. *Also*, the cascade of molecular events must be considered. How can we link these measures to the IEG expression? Finally, the identity of the reactive neurons should be developed in terms of their inputs and outputs and their functional relationship to vestibular-related behavioral events.

Several studies have now demonstrated the usefulness of IEG expression experiments in delineating some aspects of vestibular function, and have revealed several novel findings. The expression of Fos in the rat brainstem was determined after a hypergravity centripetal acceleration stimulus[28,30] and during vestibular compensation following the chemical destruction of one labyrinth.[29] The combination of the hemilabyrinthectomy and hypergravity treatments demonstrated vector-specific brainstem patterns of Fos IL relative to the orientation of the restrained animal's intact labyrinth in the centripetal field.[31] Finally, galvanic (DC) electrical stimulation of one labyrinth was shown to mimic the patterns of Fos IL observed from the previous treatments.[33] Table 17.1 provides a synopsis of these results for several important vestibular-related brainstem nuclei, and the major findings are discussed in the following sections.

TABLE 7.1

Nucleus	Acute UL:	Chronic UL	UL toward/2 2 G	G	UL away/2 G	Cathodal	Anodal
IO beta	C **	-	*	I **	-	I **	C **
IO dmcc	+/-	-	***		-	+/-	+/-
IO drosal cap	C+/-	-	*	I *	-	I **	C **
Medial vestibular	I *	-	**	C **	-	C **	I **
Spinal vetibular	I **	-	**	C **	-	C **	I **
Prepositus hypoglossus	C **	C *	*	C *	C *	I **	C **

17.5.1 HYPERGRAVITY MODEL

The hypergravity experiments revealed, among other findings, that the dorsomedial cell column (dmcc), a small, midline subnucleus of the inferior olive that sends climbing fibers to the nodulus and uvula, is strongly recruited with Fos expression independent of the other olivary subnuclei, implicating this group of cells in the adaptation process to a change in the gravitoinertial force. Additional regions of increased Fos expression that were independent of other factors like restraint stress or angular acceleration were the y, medial, and descending vestibular nuclei, the dorsolateral periaqueductal grey, and the Darkschewitsch nucleus. This IL was not observed in animals with a bilateral labyrinthectomy subjected to the same stimuli. The destruction of the vestibular neuroepithelium in these animals was confirmed histologically. Control experiments looked at animals that were rotated on the axis of rotation (these animals received only brief angular and a 0.15-G centripetal component of acceleration), only restrained, or only subjected to the skull-cap surgery, to rule out IL not associated with a labyrinth input. Another variable that accounted for differences in IL was whether the animal's head was restrained during the hypergravity stimulus. Additional head movements during centripetal acceleration would cause intense cross-coupling of canal and otolith inputs and create a highly complex vestibular input stimulus.

When an animal was rotated so that the long axis of the body was perpendicular to the centripetal vector in the horizontal plane, there were asymmetries in some of the measured nuclei, presumably from the asymmetric vector across the distributed polarity of otolith hair cells in each labyrinth. When the skull was restrained during the stimulus, both asymmetry and total IL cell counts were slightly increased in several vestibular-related nuclei. However, other stress- or autonomic-related nuclei (e.g., the lateroventral periaqueductal grey) showed *decreases* in Fos IL in head-restrained animals compared to head-free animals. Blood pressure changes associated with the additional restraint might also have been a factor in these areas, as they are known to be involved with cardiovascular reflexes (see, for example, Ref. 74).

To precisely define the laterality of the system, further experiments were conducted to determine that the hypergravity Fos expression depended not only on the integrity of the labyrinth, but on the vector of acceleration across the intact labyrinth of a chronically hemilabyrinthectomized (unilateral lesion) animal. Rats hemilabyrinthectomized 14 days earlier (and determined to have little to no residual Fos expression as a result of the labyrinthectomy) were rotated off-axis in different orientations. The results clearly implicate the otolith organs as the source of the activation stimulus. The orientation of the animal with respect to the resultant gravito-inertial force dramatically changed the pattern of Fos expression in certain brainstem nuclei like the dmcc, prepositus and vestibular nuclei (Figure 17.2; see also Ref. 31).

However, in rats and gerbils rotated with an interaural centripetal vector and with intact labyrinths, the dmcc also has IL bilaterally, but some asymmetries can still be observed in the medial and inferior vestibular nuclei and the prepositus hypoglossus. These findings suggest that

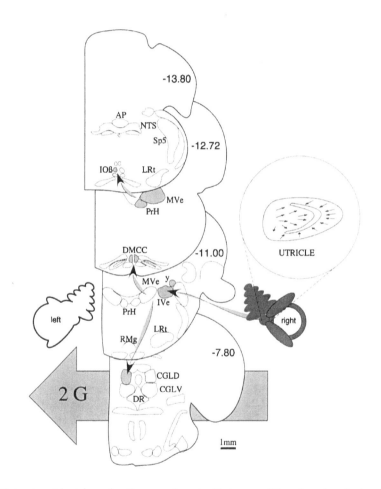

FIGURE 17.2 A rat centripetal acceleration experiment with two conditions based on body orientation. Both animals received a left unilateral labyrinthectomy 14 days earlier, and are in a compensated stage with respect to this lesion behaviorally. One is head and body restrained with its right ear toward the axis of rotation; the other with its left ear toward the axis. They are gravitoinertial-adapted by rotating 50 cm from the axis at ~360°/s for 90 min with a centripetal vector of 2 G in the dark. Below each positioning diagram is the corresponding Fos IL in a transverse brainstem section at the level of the inferior olivary dorsomedial cell column (D), and the medial (M) and inferior (I, descending) vestibular nuclei. Besides the contralateral prepositus hypoglossi (P), only the animal with its lesioned ear toward the axis of rotation has significant IL in these structures. These findings imply differential activation based on the polarity of a subset of hair cells in the otolith organs (Ref. 31).

one pathway providing the bilateral input to the dmcc originates in the portion of the utricle lateral to the striola. More recently, we have found that acceleration applied along the dorsoventral head axis, perpendicular to the major plane of the utricle, results in a bilaterally symmetric pattern of Fos IL in the medial and y vestibular nuclei, and within the dmcc (Marshburn, Kaufman, and Perachio, unpublished observations). These data suggest convergence from both otolith organs upon subsets of central neurons. The symmetrical labeling is consistent with symmetric bilateral labyrinth stimulation. Finally, to correlate this IEG expression with behavior, we began experiments using our protocols but also measuring the horizontal VOR with corneal search coils. Rats adapted for 90 minutes to a change in the gravitoinertial force and subsequently shown to express Fos in the dmcc and other vestibular-related areas also had increased torsional and vertical components in the vector of their horizontal VOR following the adaptation. (unpublished data, Ref. 27).

17.5.2 ACUTE HEMILABYRINTHECTOMY

The acute labyrinthectomy experiments (24 hours) revealed that the patterns of Fos expression in the vestibular nuclei (medial nucleus) were symmetrically opposite to that of spontaneous activity levels in second-order neurons in the same area and time in the gerbil.[55-58] Thus, at early time points, neurons *ipsilateral* to the lesion have recruited Fos presumably to initiate genetic program changes in those neurons supporting vestibular compensation. These findings support the general idea of central compensation, but give new insights into the locations and strategy for this activity. These findings have since been replicated.[10a] Cirrelli et al.[10b] used a surgical hemilabyrinthectomy to describe more time points (3, 6, and 24 hours) of Fos protein and mRNA expression. At only 3 hours, there was an increase in Fos expression in the ventral medial vestibular nucleus ipsilateral to the lesion, a region with projections to oculomotor areas. By 6 hours post-labyrinthectomy, this IL had virtually disappeared, and the distribution of IL returned to the pattern observed previously at 24 hours,[29] that is, dorsal ipsilateral medial vestibular and contralateral prepositus hypoglossi IL (see Table 17.1). These findings emphasize the dynamic and specific processes occurring in different populations of vestibular cells affected by the loss of input from one ear.

The time course of IEG activation has a large range. By judicious selection of candidate ITFs, a temporal relationship can be established that relates IEG expression to the development of a functional change such as the recovery of function following hemilabyrinthectomy (vestibular compensation). Although the majority of studies on IEG expression have concentrated on factors that are relatively rapidly and transiently expressed, long term molecular sequelae might also be examined (for example, the c-*jun* family proteins during hair cell regeneration).

More recent studies have shown that galvanic stimulation in precise labyrinth locations and of appropriate polarity and intensity can mimic the Fos expression patterns seen in the natural response to hypergravity or hemilabyrinthectomy.[33] Some of the same experiments have now been concluded looking for the zif268 ITF. In general, although the basal levels of expression were very different in location and amount, a comparison of the stimulus-induced expression between Fos and zif268 was remarkably similar. This might have to do with the similar expression time course and control mechanisms that these two IEGs share.[25] Taken together, these findings have strengthened our understanding of the cellular basis of vestibular plasticity, and provide several useful models for future studies in this area.

The results of the sequence of experiments described above begin to demonstrate the sorts of causal linkages that can be proffered to explain specific mechanisms in a particular cell group. For example, in the dmcc, hypergravity induces Fos and zif268 IL; the Fos IL can be blocked by ketamine (an NMDA channel blocker). However, in the beta subnucleus of the inferior olive, (1) acute hemilabyrinthectomy, but not hypergravity, causes contralateral Fos and zif268 expression, (2) ketamine potentiates that expression, and, (3) CRF mRNA is up-regulated in those neurons 4 days later.[32] Thus the dmcc and beta subnuclei respond to different vestibular "conditions" (gravitoinertial reference change versus compensation), and utilize different physiological strategies to effect the necessary changes, despite both subnuclei being homogenous cells of the same structure.

This is certainly a result of their different inputs and projections, and the functional roles (still undefined) that they play based on those connections.

Other studies have begun to extend our findings or explore other vestibular-related issues using IEGs. Recently, Beitz and Anderson (preliminary data) have shown that the sodium channel blocker tetrodotoxin (TTX) injected unilaterally into the middle ear bulla of rats can create a reversible blockage of labyrinthine function behaviorally in rats. This technique might represent a different way to probe the vestibular system during different stages of compensation by initiating a short-acting unilateral defect at different times. Preliminary results suggest that Fos IL in the vestibular nuclei resembles distributions observed after a chemical labyrinthectomy with sodium arsanilate,[29] although some differences are apparent.

Miller and Ruggiero[48] used Fos IL to identify brainstem neurons associated with the emetic reflex in cats given nauseogenic drugs.[48] Their results showed no differences in vestibular nuclei IL between control and treatment animals, and a broad pattern of activation in a band from the area postrema and solitary nucleus to the lateral tegmental field. These results likely reflect the effects of pharmacological agents used as the emetic stimulus. Experiments utilizing emesis-provoking motion might generate other patterns.

17.6 STRATEGIES FOR FUTURE INVESTIGATION

17.6.1 IMPLICATIONS FOR PLASTICITY: REGENERATION, HABITUATION, ADAPTATION, AND COMPENSATION

Within the field of vestibular research, the issues related to development and aging, repair and recovery, and motor learning suggests a number of lines of investigation that could examine the role of IEGs. The detection and quantification of transcriptional elements can link single-cell anatomy with changes in function in a time frame previously unavailable. Several types of studies in other areas have begun to use IEGs for questions relating to plasticity in the nervous system. The combination of IEG detection with other markers can associate function of the IEG to correlated activity of the cell. For example, Chaudhuri et al.[8] combined zif268 IL with cytochrome oxidase histochemistry in area 17 visual cortex of vervet monkey, showing the coincident expression of these two markers in neurons whose metabolic activity was high compared to adjacent areas experiencing visual suppression. Double-labeling with IL for calcium-binding proteins (calbindin and parvalbumin) revealed a negative correlation, suggesting that zif268 is expressed only in excitatory neurons. Other anatomical markers helped to define the cell type demonstrating zif268 IL. Their multi-probe strategy allowed them to make the observation of activity-dependent zif268 expression within a subset of excitatory neurons. Similar work might help to assign more specific function to vestibular-related cortices. Another study used IEGs to identify anatomical and temporal correlates of habituation to restraint stress in pituitary afferent areas in the rat brainstem.[47] They observed stressor-specific habituation-attenuated increases in Fos and Jun B mRNA, but not zif268, Jun, or Jun D mRNA, in rat cortex that was independent of adrenelectomy.

Much of contemporary neurobiology is channeled toward understanding mechanisms of adaptation. The studies described in the previous sections have begun to address vestibular compensation and adaptation to a new gravitoinertial reference. There is a wide range of vestibular-related questions which IEG expression can help to answer; for example, In what part of the brain and using which signaling pathway(s) are the neurons that govern VOR and VCR gain? What about those neurons involved with compensation? How does previous motion experience reduce motion sickness susceptibility? Can we recruit more function from senescent neurons? These sorts of efforts will eventually lead to an increased ability to initiate or control desirable forms of plasticity in the human patient under a wide variety of pathologies and conditions.

17.6.2 POTENTIAL STRATEGIES FOR IDENTIFYING FUNCTIONAL CORRELATES OF IEG EXPRESSION

Several techniques for blocking precise areas of activity in order to rule out function exist. For example, local muscimol blocks can temporarily inhibit $GABA_A$ receptive neuron activity and implicate specific nuclear groups. However, a more versatile technique that utilizes the genetic code is the developing field of antisense oligonucleotides. Hooper et al.[24] perfused sulfur-modified (phosphorothioate) antisense oligonucleotides (which hybridize with the promoter region of the c-*fos* gene) to attenuate D-amphetamine-induced Fos expression in rat striatum. They observed that unilateral striatal Fos attenuation resulted in circling behavior. Recent experiments in our lab using small 250-nl injections of anti-Fos phosphorothioate 20mers have shown sequence specific behavioral effects in gerbils 1 hour following hemilabyrinthectomy (submitted). The behavioral measures (ipsilateral or contralateral circling) could be correlated with histological verification of Fos suppression in regions (vestibular, prepositus, and IO) known to otherwise express Fos in response to a unilateral labyrinth lesion. This experimental paradigm will be a powerful way of addressing the behavioral importance of specific transcriptional factors or genes. The list of other candidate targets is enormous and efficient research will select targets based on the previous demonstration of causal relationships for the system in question.

Additional strategies will likely use transgenic technology to create conditional gene knockouts or dominant negative suppressers.[51] PCR amplification, differential display-RT-PCR, and subtractive hybridization strategies[46] will be used to identify tissue- and region-specific genes up- and down-regulated following a specific stimulus. The vestibular field should embrace these new technologies, as their specificity and power to dissociate the tangle of interactions within the cell nucleus will be necessary to understand the mechanisms of modifiability inherent in current research interests.

ABBREVIATIONS

IEG, immediate early gene; IL, immunolabeling; ITF, inducible transcription factor; CREB, cyclic-AMP response element binding protein; CRE, cyclic-AMP response element; TPA, tetradeconyl phorbol acetate; TRE, TPA response element, or, TIS, TPA-inducible sequence; cAMP, cyclic adenosine monophosphate; PKC, protein kinase C; DNA, deoxyribonucleic acid; mRNA, messenger ribonucleic acid; VOR, vestibulo-ocular reflex; VCR, vestibulo-collic reflex; dmcc, dorsomedial cell column of the inferior olivary (IO) nucleus.

REFERENCES

1. Abate, C, Luk, D, and Curran, T, Transcriptional regulation by Fos and Jun in vitro: Interaction among multiple activator and regulatory domains, *Mol. Cell, Biol.,* 11 (7) ,3624,1991.
2. Abate, C, Patel, L, III, FJR, and Curran, T Redox Regulation of Fos and Jun DNA-Binding Activity in Vitro, *Science,* 249 ,1157,1990.
3. Angel, P, Hattori, K, Smeal, T, and Karin, M The jun proto-oncogene is positively autoregulated by its product, Jun/AP-1, *Cell,* 55,875,1988.
4. Auwerx, J, and Sassone-Corsi, P, IP-1: A Dominant Inhibitor of Fos/Jun Whose Activity is Modulated by Phosphorylation, *Cell,* 64,983,1991.
5. Braselmann, S, Bergers, G, Wrighton, C, Graninger, P, Supertifurga, G and Busslinger, M, Identification of fos target genes by the use of selective induction systems, *J. Cell. Sci.,*1992.
6. Caput, D, Beutler, B and Hartog, K, Identification of a common nucleotide sequence in the 3'-untranslated region of mRNA molecules specifying inflammatory mediators, *Proceedings of the National Academy of Sciences,* 83,1670,1986.

7.   Chan, R, Brown, ER, Ericsson, A, Kovacs, KJ, and Sawchenko, PE, A comparison of two immediate-early genes, c-*fos* and NGFI- B, as markers for functional activation in stress-related neuroendocrine circuitry. *J. Neurosci.,* 13 (12),5126,1993.

8.   Chaudhuri, A, Matsubara, JA, and Cynader, MS, Neuronal activity in primate visual cortex assessed by immunostaining for the transcription factor zif268, *Visual Neurosci.,* 12 (1),35,1995.

9.   Christy, B, and Nathans, D, DNA binding site of the growth factor-inducible protein zif268, *Proceedings of the National Academy of Sciences,* 86,8737,1989.

10.a. Cirelli, C, Pompeiano, M, Dascanio, P, and Pompeiano, O, Early c-*fos* expression in the rat vestibular and olivocerebellar systems after unilateral labyrinthectomy, *Arch. Ital. Biol.,* 131 (1),71,1993. 10. b. Cirelli, C., et al., C-*fos* expression in the rat brain after unilateral labyrinthectomy and its relation to the uncompensated and compensated stages, *Neuroscience,* 1996. 70(2),p. 515,1996.

10.b. Colwell, CS, Kaufman, CM, Menaker, M, and Ralph, MR, Light-induced phase shifts and fos expression in the hamster circadian system—the effects of anesthetics, *J. Biol. Rhythm.,* 8 (3),179,1993.

10.   Crisan, D, Cadoff, EM, Mattson, JC, and Hartle, KA, Polymerase chain reaction: Amplification of DNA from fixed tissue, *Clin. Biochem.,* 23,489,1990.

11.   Curran, T, Sonnenberg, JL, Macgregor, P, and Morgan, JI, Transcription factors on the brain - fos, jun, and the Ap-1 binding site, *Neurotoxicity of Excitatory Amino Acids,* 4 (175),175,1990.

12.   Dey, A, Nebert, DW, and Ozato, K, The AP-1 Site and the cAMP- and Serum Response Elements of the c-*fos* gene are constitutively occupied *in vivo*, *DNA Cell Biol.,* 10 (7),537,1991.

13.   Doucet, JP, Squinto, SP, and Basan, NG, Fos-Jun and the primary genomic response in the nervous system: possible physiological role and pathophysiological significance. in: *Molecular Neurobiology,* Vol. (Bazan, NG) pp 27-55. *The Humana Press Inc.,*1990.

14.   Dragunow, M, and Faull, R, The use of c-*fos* as a metabolic marker in neuronal pathway tracing, *J. Neurosci. Methods.,* 29 (3),261,1989.

15.   Draisci, G, and Iadarola, MJ, Temporal analysis of increases in c-fos, preprodynorphin and preproenkephalin in mRNA's in rat spinal cord, *Mol. Brain Res.,* 6,31,1989.

16.   Duncan, GE, and Stumpf, WE, Brain activity patterns: Assessment by high resolution autoradiographic imaging of radiolabeled 2-deoxyglucose and glucose uptake, *Prog. Neurobiol.,* 37 (4),365,1991.

17.   Felipe, CD, Jenkins, R, O'Shea, R, Williams, TSC, and Hunt, SP, The role of immediate early genes in the regeneration of the nervous system, *Advances in Neurology,* 59,263,1993.

18.   Fiallosestrada, CE, Kummer, W, Mayer, B, Bravo, R, Zimmermann, M, and Herdegen, T, Long-lasting increase of nitric oxide synthase immunoreactivity, NADPH-diaphorase reaction and c-JUN co- expression in rat dorsal root ganglion neurons following sciatic nerve transection, *Neurosci. Lett.,* 150 (2),169,1993.

19.   Gius, D, Cao, X, Rausher, FJII, Cohen, DR, Curran, T, and Sukhatme, VP, Transcriptional activation and repression by Fos are independent functions: The C terminus represses immediate-early gene expression via CArG elements, *Molecular Cell Biology,* 10,4243,1990.

20.   Herms, J, Zurmohle, U, Schlingensiepen, R, Brysch, W, and Schlingensiepen, KH, Developmental expression of the transcription factor Zif268 in rat brain, *Neurosci, Lett,* 165 (1-2),171,1994.

21.   Herschman, HR, Expression of transiently induced genes in cells of the nervous system in response to growth factors. in: *Regulation of Gene Expression in the Nervous System*, Vol. pp. 93-107. Wiley-Liss, Inc,1990.

22.   Hooper, ML, Chiasson, BJ, and Robertson, HA, Infusion into the brain of an antisense oligonucleotide to the immediate-early gene c-*fos* suppresses production of fos and produces a behavioral effect, Neuroscience 63 (4),917,1994.

23.   Hughes, P, and Dragunow, M, Induction of immediate-early genes and the control of neurotransmitter-regulated gene expression within the nervous system, *Pharmacological Reviews*, 47 (1),133,1995.

24.   Jenkins, R, Oshea, R, Thomas, KL, and Hunt, SP, c-jun expression in substantia-nigra neurons following striatal 6-hydroxydopamine lesions in the rat, Neuroscience, 53 (2),447,1993.

25.   Kaufman, GD, and Anderson, JH, *The vestibulo-ocular reflex in the rat after exposure to centripetal acceleration, Society for Neuroscience Abstracts,* Miami, Florida, 23rd *Annual Meeting,*1993.

26.   Kaufman, GD, Anderson, JH, and Beitz, A, Activation of a specific vestibulo-olivary pathway by centripetal acceleration in rat, *Brain Res.,* 562 (2),311,1991.

27.   Kaufman, GD, Anderson, JH, and Beitz, AJ, Brainstem Fos Expression Following Unilateral Labyrinthectomy in the Rat, *NeuroReport,* 3 (10),829,1992a.

28. Kaufman, GD, Anderson, JH, and Beitz, AJ, FOS-defined activity in rat brainstem following centripetal acceleration, *J. Neuroscience,* 12 (11),4489,1992b.
29. Kaufman, GD, Anderson, JH, and Beitz, AJ Otolith-brain stem connectivity - evidence for differential neural activation by vestibular hair cells based on quantification of FOS expression in unilateral labyrinth-ectomized rats, *J. Neurophysiol.,* 70 (1),117,1993.
30. Kaufman, GD, Anderson, JH, and Beitz, AJ, Hemilabyrinthectomy causes both an increase and a decrease in corticotropin releasing factor mRNA in rat inferior olive, *Neurosci. Lett.,* 165 (1-2),144,1994.
31. Kaufman, GD, and Perachio, AA, Translabyrinth Electrical Stimulation for the Induction of Immediate Early Genes in the Gerbil Brainstem, *Brain Research,* 646,345,1994.
32. Kerkhoff, E, Bister, K, and Klempnauer, KH, Sequence-Specific DNA Binding by Myc Proteins, *Proc. Natl. Acad. Sci. USA,* 88 (10),4323,1991.
33. Krukoff, TL, Morton, TL, Harris, KH, and Jhamandas, JH, Expression of c-*fos* protein in rat brain elicited by electrical stimulation of the pontine parabrachial nucleus, *The Journal of Neuroscience,* 12 (9),3582,1992.
34. Labiner, DM, Butler, LS, Cao, Z, Hosford, DA, Shin, CS, and Mcnamara, JO, Induction of c-*fos* messenger RNA by kindled seizures - complex relationship with neuronal burst firing, J. Neurosci., 13 (2),744,1993.
35. Lafarga, M, Martinezguijarro, FJ, Berciano, MT, Blascoibanez, JM, Andres, MA, Mellstrom, B, Lopezgarcia, C, and Naranjo, JR, Nuclear fos domains in transcriptionally activated supraoptic nucleus neurons, *Neuroscience,* 57 (2),353,1993.
36. Lamb, P, and Mcknight, SL, Diversity and specificity in transcriptional regulation - the benefits of hetero-typic dimerization, *Trends Biochem. Sci.,* 16 (11),417,1991.
37. Lanteri-Minet, M, Isnardon, P, DePommery, J, and Menetrey, D, Spinal and hindbrain structures involved in visceroception and visceronociception as revealed by the expression of Fos, Jun and Krox-24 proteins, *Neuroscience,* 55 (3),737,1993.
38. Leah, J, Herdegen, T,, and Bravo, R Selective expression of Jun proteins following axotomy and axonal transport block in peripheral nerves in the rat; Evidence for a role in the regeneration process, *Brain Research,* 566,198,1991.
39. Lee, J-H, and Beitz, AJ, The distribution of brainstem and spinal cord nuclei associated with different frequencies of electroacupuncture analgesia, *Pain,* 52,11,1993.
40. Leng, G, Luckman, SM, Dyball, R, Hamamura, M, and Emson, PC, Induction of c-*fos* in magnocellular neurosecretory neurons - a link between electrical activity and peptide synthesis, *Neurohypophysis: A Window on Brain Function,* 689,133,1993.
41. Lerea, LS, and Mcnamara, JO, Ionotropic glutamate receptor subtypes activate c-*fos* transcription by distinct calcium-requiring intracellular signaling pathways, *Neuron,* 10 (1),31,1993.
42. Lucibello, FC, Lowag, C, Neuberg, M, and Muller, R, Trans-repression of the mouse c-*fos* promoter: a novel mechanism of Fos-mediated trans-regulation, Cell 59,999,1989.
43. Maggi, CA, and Meli, A, Suitability of urethane anesthesia for physiopharmacological investigations in various systems. Parts 1, 2, and 3, *Experientia.,* 42,109,1986;52,292,1986;42,531,1986.
44. Marechal, D, Forceille, C, Breyer, D, Delapierre, D, and Dresse, A, A subtractive hybridization method to isolate tissue- specific transcripts - application to the selection of new brain-specific products, *Anal. Biochem.,* 208 (2),330,1993.
45. Melia, KR, Ryabinin, AE, Schroeder, R, Bloom, FE, and Wilson, MC, Induction and habituation of immediate early gene expression in rat brain by acute and repeated restraint stress, *J. Neurosci.,* 14 (10),5929,1994.
46. Miller, AD, and Ruggiero, DA, Emetic reflexes are revealed by expression of the immediate- early gene c-*fos* in the cat, *J. Neurosci.,* 14 (2),871,1994.
47. Morgan, JI, and Curran, T, Stimulus-transcription coupling in neurons: role of cellular immediate-early genes, *TINS,* 12 (11),459,1989.
48. Morgan, JI, and Curran, T, The immediate-early gene response and neuronal death and regeneration, *The Neuroscientist,* 1 (2),68,1995a.
49. Morgan, JI, and Curran, T, Immediate-early genes: ten years on, *Trends Neurolog. Sci.,* 18 (2),66,1995b.
50. Muller, R, Bravo, R, Muller, D, Kurz, C, and Reiz, M, Different types of modification of c-*fos* and its associated protein p39: modulation of DNA binding by phosphorylation, *Oncogene Research,* 2,19,1987.

51. Mumberg, D, Lucibello, FC, Schuermann, M, and Muller, R, Alternative splicing of FosB transcripts results in differentially expressed messenger RNAs encoding functionally antagonistic proteins, *Gene. Develop.*, 5 (7),1212,1991.

52. Nakabeppu, Y, and Nathans, D, A naturally occurring truncated form of fosb that inhibits Fos/Jun transcriptional activity, *Cell 64*, (4),751,1991.

53. Newlands, SD, and Perachio, AA, Compensation of horizontal canal related activity in the medial vestibular nucleus following unilateral labyrinth ablation in the decerebrate gerbil. I. Type I neurons, *Exp. Brain Res.*, 82 (2),359,1990a.

54. Newlands, SD, and Perachio, AA, Compensation of horizontal canal related activity in the medial vestibular nucleus following unilateral labyrinth ablation in the decerebrate gerbil. II. Type II neurons, *Exp. Brain Res.*, 82 (2),373,1990b.

55. Newlands, SD, and Perachio, AA, Compensation of horizontal canal related activity in the medial vestibular nucleus following unilateral labyrinth ablation in the decerebrate gerbil: Type I & II neurones, *Exp. Brain Res.*, 82,359,1990c.

56. Newlands, SD, and Perachio, AA, Effect of T2 spinal transection on compensation of horizontal canal related activity in the medial vestibular nucleus following unilateral labyrinth ablation in the decerebrate gerbil, *Brain Res.*, 541 (1),129,1991.

57. Oikarinen, J, Histone H1 and the regulation of transcription by nuclear receptors, *FEBS*, 294 (1,2),6,1991.

58. Oshea, EK, Klemm, JD, Kim, PS and Alber, T X-ray structure of the GCN4 leucine zipper, a two-stranded, parallel coiled coil, *Science*, 254 (4),539,1991.

59. Oshea, EK, Rutkowski, R, and Kim, PS, Mechanism of specificity in the fos-jun oncoprotein heterodimer, *Cell*, 68 (4),699,1992.

60. Radtke, J, Dooley, S, Blin, N, and Unteregger, G, A set of four nuclear proteins binds to a dna sequence within the FOS promoter region, *Mol. Biol. Rep.*, 15 (2),87,1991.

61. Roberts, JL, Quantitative analysis of neuronal gene expression. in: *Progress in Brain Research*, Vol. 100 (Bloom, F, Ed.) pp. 33-37 Elsevier Science B.V.,1994.

62. Rose, S, How chicks make memories - the cellular cascade from c-*fos* to dendritic remodelling, *Trends Neurosci.*, 14 (9),390,1991.

63. Ryseck, RP, and Bravo, R, c-JUN, JUN-B, and JUN-D differ in their binding affinities to AP-1 and CRE consensus sequences - Effect of FOS proteins, *Oncogene*, 6 (4),533,1991.

64. Schlingensiepen, KH, Schlingensiepen, R, Kunst, M, Klinger, I, Gerdes, W, Seifert, W, and Brysch, W, Opposite functions of *jun*-b and c-*jun* in growth regulation and neuronal differentiation, *Dev. Genet.*, 14 (4),305,1993.

65. Schuermann, M, Jooss, K, and Muller, R, fosB is a transforming gene encoding a transcriptional activator, *Oncogene*, 6 (4),567,1991.

66. Seyfert, VL, McMahon, SB, Glenn, WD, Yellen, AJ, Sukhatme, VP, Cao, YM, and Monroe, JG, Methylation of an immediate-early gene as a mechanism for B cell tolerance induction, *Science*, 250,797,1990.

67. Sheng, HZ, Fields, RD, and Nelson, PG, Specific regulation of immediate early genes by patterned neuronal activity, *J. Neurosci. Res.*, 35 (5),459,1993.

68. Sonnenberg, JL, III, FJR, Morgan, JI, and Curran, T, Regulation of Proenkephalin by Fos and Jun. Science 246,1622,1989.

69. Sukhatme, V, Cao, X, Chang, L, Tsai-Morris, C, Stamenkovich, D, Ferreira, P, Cohen, D, Edwards, S, Shows, T, Curran, T, LeBeau, M and Adamson, E A zinc finger-encoding gene coregulated with c-*fos* during growth and differentiation, and after cellular depolarization, *Cell*, 53,37,1988.

70. Takayama, K, Suzuki, T and Miura, M The comparison of effects of various anesthetics on expression of fos protein in the rat brain, *Neurosci. Lett.*, 176 (1),59,1994.

71. Walther, D, Takemura, M, and Uhl, G, Fos family member changes in nucleus caudalis neurons after primary afferent stimulation: enhancement of *fos* B and c-fos, *Molecular Brain Research*, 17,155,1993.

72. Wilson, A, and Kapp, BS, Midbrain Periaqueductal Gray Projections to the Dorsomedial Medulla in the Rabbit, *Brain Res. Bull.*, 27 (5),625,1991.

73. Worley, PF, Christy, BA, Nakabeppu, Y, Bhat, RV, Cole, AJ, and Baraban, JM, Constitutive expression of zif268 in neocortex is regulated by synaptic activity, *Proc. Natl. Acad. Sci. USA*, 88 (12),5106,1991.

18

The Contributions of Excitatory and Inhibitory Amino Acid Neurotransmitters to Vestibular Plasticity

Paul F. Smith and Cynthia L. Darlington

CONTENTS

18.1 INTRODUCTION

In considering how the vestibular system functions on a neurochemical level, it is critical to recognize the importance of plasticity in response to changing environmental demands (both internal and external to the nervous system). Since plasticity is occurring continuously, it is impossible to distinguish it from "normal" function.[4] Although much has been learned about the neurophysiological basis of vestibulo-ocular reflex (VOR) plasticity under conditions of changing visual stimuli (e.g., Ref. [90,91]), at present there are few data relating to the possible neurochemical substrates of this form of vestibular plasticity. In contrast, research into the neurochemical substrates of recovery from vestibular lesions (vestibular compensation) has undergone a wave of intense interest over the past decade (for reviews see Refs. 20,142,146,153). For this reason, this chapter focuses on the contributions of excitatory and inhibitory amino acids to vestibular plasticity following peripheral vestibular lesions. However, it is hoped that some of the data and ideas to be discussed will have relevance for other forms of vestibular plasticity.

0-8493-7679-3/00/$0.00+$.50

18.2 PERSPECTIVES ON VESTIBULAR COMPENSATION

In order to discuss excitatory and inhibitory amino acid neurotransmitters in relation to vestibular compensation, it is important to elucidate the theoretical perspective from which the available data will be interpreted.

We assume first that vestibular compensation, the process by which behavioral recovery occurs following a peripheral vestibular lesion, is a stratified process in which different symptoms compensate at different rates, and different categories of symptoms (e.g., static versus dynamic) compensate to different extents. Although static ocular motor and postural symptoms (i.e., those that persist in the absence of head movement) compensate to a high degree, in general, static ocular motor symptoms such as spontaneous ocular nystagmus compensate more rapidly than static postural symptoms. In contrast, dynamic ocular motor and postural symptoms (i.e., those that occur as a result of head movement) compensate more slowly and to a lesser extent than their static counterparts; in general, dynamic postural symptoms may compensate more completely than dynamic ocular motor symptoms (for reviews see Refs. 16,17,82,128,139; see Figure 18.1).

STATIC SYMPTOMS OF UVD IN MAMMALS

spontaneous ocular nystagmus

yaw head tilt

roll heat tilt

* symptoms compensate quickly in most species

DYNAMIC SYMPTOMS OF UVD IN MAMMALS

dysfunction of vestibuloocular reflexes

dysfunction of vestibulospinal reflexes

* symptoms compensate slowly and less completely

FIGURE 18.1 The major categories of ocular-motor and postural symptoms following unilateral vestibular deafferentation (UVD) in mammals.

Because of this stratification of vestibular compensation, it is likely that different mechanisms are responsible for different aspects of compensation (see Ref. [122] for recent evidence). For example, many of the static symptoms may compensate relatively quickly as a result of largely "automatic" cellular processes,[27] whereas some aspects of dynamic compensation may occur as a result of sensory substitution and the generation of new motor programs which may, in some cases, require volitional cognitive strategies (for a review see Ref. 17).

We assume that the vestibular compensation process is triggered by the inactivation, rather than the degeneration, of the primary vestibular afferents following unilateral vestibular deafferentation (UVD). Although the vestibular nerve is known to degenerate much more rapidly following vestibular neurectomy than following surgical labyrinthectomy (e.g., Refs. 12,136), many studies have found that the severity of the behavioral symptoms, their rate of compensation and the changes that occur in the neuronal activity of the ipsilateral vestibular nerve and vestibular nucleus complex (VNC), are similar.[12,68,136,138] Kunkel and Dieringer[79] have recently reported that the electrophysiological changes that occur in the frog VNC are similar following pre- or post-ganglionic vestibular nerve transection. Furthermore, vestibular compensation can be accelerated by the application of cathodal electrical stimulation to the vestibular nerve ipsilateral to the labyrinthectomy.[98] Taken together, these data support the assumption that it is the functional inactivation of the vestibular nerve, rather than its degeneration, that is the critical stimulus for vestibular compensation.[79]

In this chapter we also assume that static vestibular compensation is associated with a return of resting activity to neurons in the VNC ipsilateral to the UVD (see Figures 18.2 and 18.3).

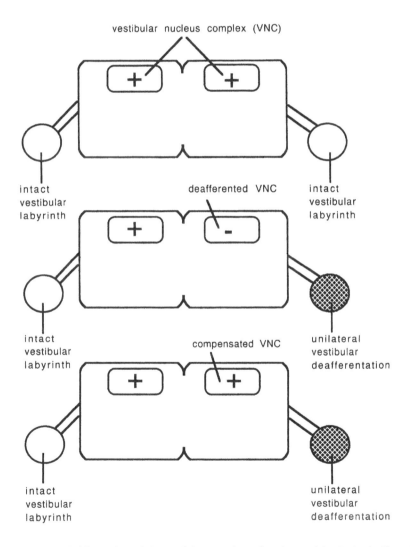

FIGURE 18.2 Schematic illustration of the partial restoration of resting activity in the ipsilateral vestibular nucleus complex (VNC) following UVD. In the labyrinthine-intact state, the level of resting activity in the two VNC is approximately symmetrical. Immediately following UVD, there is a large decrease in resting activity in the ipsilateral VNC. During the development of vestibular compensation, a partial restoration of resting activity occurs in the VNC on the ipsilateral side.

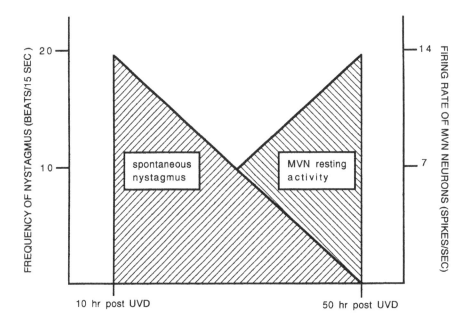

FIGURE 18.3 Schematic illustration of the correlation between the disappearance of spontaneous nystagmus and the recovery of ipsilateral type I MVN neuron resting activity in the guinea pig following UVD.

Although the degree of recovery of resting activity is controversial, the majority of studies have demonstrated a substantial recovery of type I neuron resting activity in the ipsilateral medial vestibular nucleus (MVN) following UVD[56,102,103,116,121-123,137,138] for a discussion, see Refs. 111,122,140); a slower and more limited recovery of resting activity has been documented in the ipsilateral lateral vestibular nucleus (LVN).[81,176,113,180] Consistent with these results are data from both 2-deoxyglucose[40,92,94] and cytochrome oxidase[71] studies. Finally, lesions of the ipsilateral VNC have been shown to prevent compensation or result in decompensation.[155,167] Waespe et al.[174] have reported a recovery of type I neuron resting activity in the bilateral MVN of the alert monkey following a bilateral vestibular neurectomy; this result demonstrates that the contralateral, "intact" labyrinth is not causally related to the return of resting activity in the ipsilateral VNC following UVD (cf. Refs. 18,47,116,138,149).

18.3 EXCITATORY AMINO ACIDS IN THE VNC

A large body of evidence now supports the hypothesis that at least one of the neurotransmitters used by the vestibular nerve is an excitatory amino acid, probably glutamate. In 1984, Raymond and colleagues[117] reported that cat presynaptic vestibular nerve fibers take up glutamate and that the extent of the glutamate uptake substantially decreases following vestibular nerve transection. Studies of glutamate immunoreactivity in frog and rat vestibular nerve are consistent with this finding.[119,158] High concentrations of glutamate and aspartate have been found in the VNC,[57,87,120,175] with glutamate and aspartate concentrations decreasing after removal of the ipsilateral Scarpa's ganglion[88] and glutamate concentrations increasing following vestibular nerve stimulation[178]. Excitatory amino acid receptor binding sites have been documented in VNC subnuclei,[88,89,99,114,166] as have mRNA for the N-methyl-D-aspartate (NMDA) receptor.[171]

Over the past decade, numerous *in vitro* electrophysiological studies have reported that the kainate/a-amino-3-hydroxy-5-methyl-4-isoxazole-propionic acid (AMPA) excitatory amino acid receptor subtypes, rather than the NMDA receptor, mediate synaptic transmission between the vestibular nerve and the VNC[10,11,14,35,49,76,86] (see Figure 18.4); however, most of the mammalian studies conducted so far have used the MVN subnucleus. In some studies of the frog VNC, a late, NMDA receptor-mediated component of the vestibular nerve-evoked excitatory postsynaptic potential (EPSP) has been observed in low Mg^{2+} artificial cerebrospinal fluid[14,76] (for a review see Ref. 153). Recently, the hypothesis that vestibular nerve input to the MVN is mediated solely by kainate/AMPA receptors has been questioned by data from patch-clamp studies, which suggest that a substantial component of the vestibular nerve-induced EPSP can be blocked by an NMDA receptor antagonist.[74,161] A calcium fluorescence study, in which vestibular nerve-induced increases in the intracellular calcium concentrations of second-order MVN neurons were reduced by an NMDA receptor antagonist, is also consistent with the patch clamp data.[160] However, both the patch clamp and calcium fluorescence studies were conducted using brainstem slices from very young rats and it is possible that the increased expression of NMDA receptors in the developing nervous system (for a review see Ref. 95) could account for the apparent discrepancy between the previous intracellular studies and the patch-clamp studies.[21] In some cases,[74] bicuculline was used to block $GABA_A$ receptor-mediated inhibition within the slice; this would have increased the depolarization of second order MVN neurons, thus reducing the Mg^{2+} blockade of the NMDA receptor-associated ion channel and increasing NMDA receptor function (for a review see Ref. 95). Since *in vivo*, type I MVN neurons are normally released from tonic GABAergic inhibition during ipsilateral angular head acceleration, bicuculline would simulate "normal" circumstances only if the degree of disinhibition produced corresponded to the GABAergic disinhibition incurred during ipsilateral head rotation. It should also be stressed that most of these *in vitro* studies have focused on the MVN, and it remains to be seen whether vestibular nerve input to the LVN and the other VNC subnuclei is also mediated by excitatory amino acid receptors (for a review, see Ref. 21). Recent studies by Dieringer and colleagues using frog suggest that some of the discrepancies among the previous studies may be due to differences in the activation of VNC NMDA receptors by different groups of afferents in the vestibular nerve,[157,158] with thick fibers co-activating both NMDA and non-NMDA receptors (see also Ref. [170]).

Aside from the question of whether VNC NMDA receptors contribute to the mediation of input from the vestibular nerve, it is still unclear what other functions they might serve. Numerous *in vitro* electrophysiological studies have shown that many second-order VNC neurons are depolarized by NMDA, an effect that can be blocked by NMDA receptor antagonists.[11,14,49,76,85,86,132,143,150] It is possible that NMDA receptors partially mediate input from the spinal cord[14] or commissural input from the contralateral VNC.[14,35,76]

At present, the least understood subtype of excitatory amino acid receptor in the VNC is the metabotropic subtype. Vibert et al.[168] have reported that application of the metabotropic agonist, trans-amino-cyclopentyl-1,3-dicarboxylate (ACPD), resulted in depolarization of MVN neurons in brainstem slices. Kinney et al.[73] reported that the metabotropic agonist, 1S,3R -ACPD, blocked EPSPs induced by electrical stimulation of the vestibular nerve or the medial longitudinal fasciculus. However, De Waele et al.[171] failed to locate mRNA for the metabotropic receptor subtype, mGluR1, in the MVN. Recently, Darlington and Smith[26] have reported that between 35 and 60% of MVN neurons in brainstem slices responded to the metabotropic agonist, 1S,3R-ACPD, at concentrations ranging from 10^{-10} to 10^{-6} M.

There have been few *in vivo* studies of excitatory amino acid receptors in the VNC. De Waele et al.[173] reported that unilateral injection of the NMDA receptor antagonist, D-2-amino 5-phosphonovaleric acid (D-APV), but not the kainate/AMPA receptor antagonist, 6-cyano-7-nitroquinoxaline-2,3-dione (CNQX), into the VNC caused ocular motor and postural symptoms similar to the effects of UVD. These results suggested that NMDA receptors, but not kainate/AMPA receptors, contribute to the resting discharge of VNC neurons (see also Ref. 9).

Although there have been few studies of the allosteric binding sites of NMDA receptor complexes in the VNC, a recent behavioral study has examined strychnine-insensitive glycine binding sites[5] (see Figures 18.4 and 18.5A).

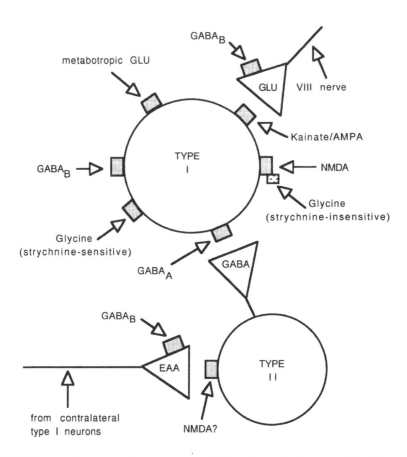

FIGURE 18.4 Schematic diagram indicating the probable location of excitatory and inhibitory amino acid receptors in the VNC on the basis of current evidence. Glu, glutamate. Other abbreviations as in text.

Unilateral cannula injection into the VNC of D-serine, a selective agonist for the strychnine-insensitive glycine binding site on the NMDA receptor, caused VOR asymmetries that were suggestive of hyperactivity in the perfused VNC; the selective antagonist, 7-chlorokynurenate, had an opposite effect[5]. Electrophysiological and immunohistochemical studies in frog also support the hypothesis that glycine is co-released by the vestibular nerve and acts on strychnine-insensitive glycine binding sites on NMDA receptor complexes on VNC neurons.[119,120,157,158,159]

18.4 EXCITATORY AMINO ACIDS IN THE VNC DURING VESTIBULAR COMPENSATION

To date, there have been relatively few biochemical studies of excitatory amino acids in the VNC during vestibular compensation. In the largest study of its kind conducted to date, Li et al.[88] found a complex range of changes in amino acid concentrations taking place in the bilateral VNC between 2 and 30 days post-UVD. Significant decreases in glutamate were observed in many regions of the VNC by 2 days post-UVD; many of these decreases were maintained up to 30 days post-UVD. More symmetrical glutamate levels between the bilateral VNC at 30 days post-UVD were largely a result of lower glutamate levels on the intact side rather than a recovery of glutamate levels on the lesioned side. Similar, but smaller, changes in aspartate concentrations were observed. Henley and Igarashi[57] found normal glutamate levels in the ipsilateral VNC at 10 months post-UVD;

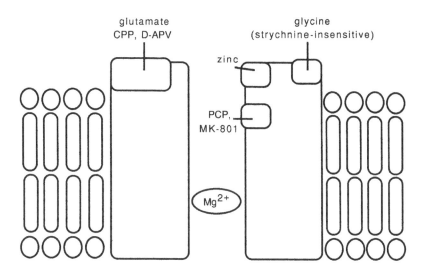

FIGURE 18.5 (A) Schematic diagram indicating the current understanding of the structure of the NMDA receptor complex. PCP, phencyclidine. Other abbreviations as in text. (B) Schematic diagram indicating the current understanding of the structure of the GABA$_A$ receptor complex. ETOH, ethanol; BDZ, benzodiazepine.

however, given the speed of static compensation (for a review see Ref. 17), it seems unlikely that this slow recovery of glutamate levels within the ipsilateral VNC could be causally related to static compensation. Consistent with this view, Raymond et al.[118] found that the number of glutamate binding sites in the ipsilateral VNC did not increase following UVD. Likewise, De Waele et al.[171] could find no evidence for an increase in NMDA receptor mRNA following UVD, although the authors acknowledged that they could detect only certain NMDA receptor subunits and therefore it was possible that other NMDA receptor subtypes, without those subunits, could have been up-regulated. Immunohistochemical studies[87] of AMPA receptors suggest that there are no regional asymmetries between the bilateral VNC between 4 and 30 days post-UVD. However, relative to the intact side, small decreases in the GluR2 and GluR3 AMPA subtypes were observed in the ipsilateral VNC by 7 days post-UVD; most of these asymmetries had disappeared between 14 and 30 days post-UVD[87]. Most recently, Li et al.[89] have reported that both AMPA and NMDA receptors show large decreases in the ipsilateral VNC, relative to the intact side, between 4 and 30 days post-UVD.

Electrophysiological studies also do not support large increases in excitatory amino acid receptors during vestibular compensation. Using frog brainstem explants, Knopfel and Dieringer[77] reported that the increase in the efficacy of the vestibular commissural input to the ipsilateral VNC, which is associated with vestibular compensation in frog,[34] was not due to an increase in the contribution of NMDA receptors. Similarly, Smith and Darlington[143,144] found that ipsilateral MVN neurons in brainstem slices from compensated guinea pigs responded to NMDA agonists and antagonists in a similar way to MVN neurons from labyrinthine intact animals (see Figure 18.6A). However, given the extent of the deafferentation involved in such *in vitro* studies, these results need to be further investigated *in vivo*.

Most of the data that exists on excitatory amino acid receptors during vestibular compensation is behavioral. Many of these studies were motivated by the revelation in the 1980s that long-term potentiation (LTP) is induced by NMDA receptor activation (for a review, see Ref. 7). It therefore seemed quite natural to speculate that NMDA receptors might be involved in inducing or maintaining the vestibular compensation process.

Smith and Darlington[141] reported that a single i.p injection of the non-competitive NMDA receptor/channel antagonist, (+)-5-methyl-10,11-dihydro-5H-dibenzo[a,d]cyclohepten-5,10-imine

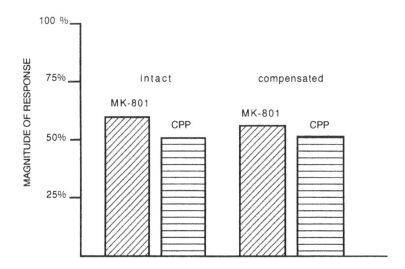

FIGURE 18.6 (A) Mean decrease in firing rate for single MVN neurons in brainstem slices exposed to the NMDA receptor antagonists MK-801 or CPP (10^{-8} M in both cases). "Intact" indicates slices from labyrinthine-intact guinea pigs. "Deafferented" indicates ipsilateral MVN in slices from guinea pigs that received a previous UVD and had compensated. (B) Schematic illustration of decompensation of spontaneous nystagmus following an i.p. injection of an NMDA receptor antagonist at 50 h to 10 days post-UVD, but not at 2 weeks to 2 months post-UVD. Saline or pentobarbital (Nembutal) injections had no effect. (From Ref. 141.)

maleate (MK-801, 0.5 mg/kg), or the competitive receptor antagonist, 3-(($\pm$)-2-carboxypiperazin-4-yl)-propyl-1-phosphonic acid (CPP, 1 mg/kg), resulted in decompensation of spontaneous nystagmus when the injection was delivered at 2 to 3 days post-UVD. This loss of compensation was not obviously related to the non-specific ataxic and sedative side effects of MK-801, since CPP produced very little ataxia or sedation but still produced decompensation. Injections of vehicle or pentobarbital did not produce decompensation (see Figure 18.6B). In further studies, Darlington and Smith[22] found that systemic injections of MK-801 or CPP during the first 24 hr post-UVD disrupted the development of vestibular compensation. Similar results have been obtained with systemic injections of MK-801 in rat,[72,75] frog[42] and goldfish.[42] The most direct evidence that NMDA receptor antagonists disrupt compensation by acting on the ipsilateral VNC has come from cannula studies. Sansom et al.[124] found that injection of CPP into the ipsilateral VNC, or IVth ventricle close to the ipsilateral VNC, at 2 to 3 days post-UVD, resulted in decompensation. De Waele et al.[173] obtained similar results with the competitive NMDA receptor antagonist, DL-APV. Interestingly, it has been found that NMDA receptor antagonists disrupt compensation only for a limited period following UVD; in guinea pig, rat, frog, and goldfish, injections of NMDA receptor antagonists had no effect beyond a species-specific critical period following the UVD.[22,42,75]

More recently, Sansom et al.[125] have reported that i.p. injection of MK-801 before UVD results in a decrease in the static symptoms of UVD in guinea pig. Pettorossi et al.[112] also found that when DL-APV administration began before the UVD and continued during vestibular compensation, static vestibular symptoms were reduced. Darlington et al.[32] have replicated Sansom et al.'s original results using MK-801 pretreatment; however, the effects of pre-UVD administration of other NMDA receptor antagonists such as CPP and *cis*-4-(phosphonomethyl)-piperidine-2-carboxylic acid (CGS 19755) are much more complex and difficult to interpret.[126] Recently, Aoki et al.[2] have also reported that pre-UVD administration of MK-801 in guinea pig results in a reduction in SN frequency. In contrast, Flohr and Luneburg[42] have reported that pre-UVD administration of MK-801 delayed compensation in the goldfish.

In the only gene knock-out study of glutamate receptors in relation to vestibular compensation which has been reported to date, Funabiki et al.[44] observed that compensation was retarded in mice deficient in the d-2-glutamate receptor subunit, which is expressed only in ionotropic glutamate receptors in the cerebellum. The functional signifcance of this finding is unclear at present.

Taken together, the results of the biochemical, electrophysiological, and behavioral studies of excitatory amino acid receptors during vestibular compensation, suggest that these receptors do not undergo any large increases in number, affinity, or efficacy following UVD which might explain the return of resting activity to the ipsilateral VNC. Rather, it seems more likely that NMDA receptors might play a key role in the induction of vestibular compensation processes, after which they become redundant. The experiments in which NMDA receptor/channel antagonists have been administered before UVD may provide a clue as to how NMDA receptors might be involved in the induction of vestibular compensation. One possibility is that, at the time of the UVD, injury discharges in the vestibular nerve result in a large release of glutamate into the ipsilateral VNC, causing excessive activation of kainate/AMPA receptors, thereby opening both NMDA receptor channels and L-type voltage-sensitive calcium channels through increased depolarization.[160] If this were the case, one reason why so many VNC neurons might become silent immediately following the UVD, despite still receiving excitatory input from many other areas of the CNS, may be because of an excessive calcium influx through NMDA receptor ion channels and L-type calcium channels. Such increased intracellular calcium concentrations might lead to diaschisis or "neural shock" (for a review see Ref. 37), thus exacerbating the effects of the UVD. If this hypothesis were correct, then NMDA receptors in the ipsilateral VNC would partially mediate the expression of the UVD symptoms, and in that sense could be regarded as having an important role in the induction of the vestibular compensation process. This hypothesis is consistent with evidence that NMDA receptor/calcium channel antagonists, voltage-sensitive calcium channel antagonists and calcium-dependent enzyme inhibitors, administered at or close to the time of the UVD, reduce UVD symptoms[23,32,84,112,125-127,165] (but see Ref. 51). Also consistent with this hypothesis is the evidence that anesthetizing the peripheral vestibular receptors before surgical UVD results in a reduction in the labyrinthine syndrome.[25] Finally, a number of drugs that are known to modulate calcium channels and/or intracellular calcium concentrations have been shown to facilitate vestibular compensation (e.g., short fragments of the adrenocorticotrophic hormone (ACTH);[41,52,54,63] Ginkgo biloba and ginkgolides;[81,96] synthetic steroids[69,177]). Recent studies suggest that short ACTH fragments may in fact interact with the NMDA receptor.[32,53,156]

18.5 INHIBITORY AMINO ACIDS IN THE VNC

The inhibitory amino acids, GABA and glycine, were among the earliest neurotransmitters to be identified within the VNC.[13,19,64-67,107-109,162,163]). There is immunohistochemical evidence for GABA and glycine within the VNC (e.g[106,120,164,171,175]); immunohistochemical and binding studies also support the existence of GABA$_A$ and GABA$_B$ binding sites on VNC neurons.[8,58,60,179] The most recent binding studies suggest that GABA$_B$ binding sites are also localized both on presynaptic primary vestibular afferents[59] and on presynaptic commissural fibers from the contralateral VNC[97]. Binding studies have confirmed the existence of the benzodiazepine recognition sites which are normally associated with the GABA$_A$ receptor[8,179] (see Figure 18.4).

GABA and glycine receptors have been studied extensively in the LVN. GABA$_A$ receptors have been shown to mediate monosynaptic inhibition of LVN neurons by Purkinje cells in the cerebellar flocculus.[13,19,43,61,64-67,107-109,162,163] There is some evidence that certain peptides (e.g motilin, somatostatin, and met-enkephalin) may enhance the action of GABA on LVN neurons, and it has been speculated that GABA may be released by Purkinje cell axons with a peptide as co-transmitter.[13]

Glycine receptors on LVN neurons may mediate inhibition by the midbrain or spinal cord, but the results are unclear.[19,109,163] Some other amino acids (b-alanine, imidazole acetic acid, and taurine) have also been shown to inhibit LVN neurons.[13,109]

GABA and glycine receptors have been studied less extensively in the MVN. It has been shown that antagonists for GABA$_A$ receptors reduce commissural inhibition of type I MVN neurons.[45,100,115] Although it has been speculated that glycine might also be involved in mediating commissural inhibition,[115] recent studies have disputed this.[45] These results are consistent with the hypothesis that type II MVN neurons inhibit type I neurons by releasing GABA, which acts on GABA$_A$ receptors.[45,154]

Ethanol has been shown to potentiate the action of GABA on GABA$_A$ receptors on type I MVN neurons.[100] It has recently been reported that MVN type I neurons respond with an increase in firing rate to the glucocorticoid, dexamethasone; the effect could be blocked by the glucocorticoid receptor antagonist, RU38486.[177] Since GABA$_A$ receptors are known to have an allosteric binding site for steroids, it is possible that dexamethasone exerts its effects via the GABA$_A$ receptor complex on type I neurons (see Figure 18.5B). Interestingly, both the steroids, dexamethasone[177] and methylprednisolone,[69] have been demonstrated to facilitate vestibular compensation.

It has been confirmed that NMDA receptors in the MVN have an allosteric binding site for glycine;[5,83] therefore, glycine may serve a dual role in the MVN - as an inhibitory transmitter at strychnine-sensitive glycine binding sites and as a modulator of NMDA receptor function at strychnine-insensitive glycine binding sites on NMDA receptor complexes (see Figure 18.5A).

In vitro electrophysiological studies have confirmed the existence of GABA$_A$, GABA$_B$, and glycine receptors in the MVN.[36,50,55,73,151,152,169] GABA$_B$ receptors appear to be distributed both pre- and postsynaptically.[36,55,60,169] One possible function of presynaptic GABA$_B$ receptors would be to regulate the amount of glutamate released by the vestibular nerve (see Figure 18.4).

The presence of GABA$_A$ receptors in the VNC probably accounts for the disruptive effects of benzodiazepines and ethanol on the vestibular reflexes[6,62,110,129,147]. The presence of GABA$_B$ receptors accounts for the disruptive action of the GABA$_B$ receptor agonist, baclofen, on the VOR.[15,105]

18.6 INHIBITORY AMINO ACIDS IN THE VNC DURING VESTIBULAR COMPENSATION

GABA levels have been reported to increase in the ipsilateral LVN and decrease in the contralateral LVN at 3-6 days following UVD.[164] More recently, Li et al.[88] have reported that GABA concentrations decrease in the ipsilateral VNC subnuclei shortly following UVD, gradually recovering toward 30 days post-UVD and in some cases increasing in the dorsal MVN and LVN. Calza et al.[8] found only a transient imbalance in benzodiazepine binding sites between the deafferented and contralateral VNC immediately following the UVD; between 1 and 90 days post-UVD, no significant differences in benzodiazepine binding were detected. Calza et al.[8] inferred that the lack of difference in benzodiazepine binding indicated no change in the number of GABA$_A$ receptors. However, the benzodiazepine and GABA binding sites on the GABA$_A$ receptor complex can be regulated independently of one another (see, for example, Ref. 48); therefore, it is conceivable that the number, affinity, and/or efficacy of GABA binding sites could have changed even though benzodiazepine binding sites did not.

Dieringer and Precht[34] have reported that the development of vestibular compensation in the frog is correlated with an increase in the efficacy of commissural inhibition of neurons in the deafferented VNC: whereas galvanic depolarization of the contralateral VIIIth nerve rarely evoked inhibitory postsynaptic potentials (IPSPs) in deafferented VNC neurons immediately following UVD, at 60 days post-UVD, IPSPs were readily evoked. This increase in the frequency of IPSPs was blocked by systemic administration of the non-competitive GABA$_A$ receptor antagonist, picrotoxin, suggesting that the IPSPs were a result of GABA acting on GABA$_A$ receptors on type I VNC neurons. It is unclear whether this strengthening of commissural inhibition in the frog is in any way related to the increase in GABA concentrations reported by Thompson et al.[164] in the deafferented LVN of the squirrel monkey.

In mammalian spec*ies*, there is limited evidence for an increase in the commissural inhibition of deafferented type I MVN neurons.102,121,116 Precht et al.116   reported that the threshold for inhibiting type I neurons in the ipsilateral MVN by depolarization of the contralateral vestibular nerve, was lower than in the labyrinthine-intact animal. However, Ried et al.[121] could not replicate this *re*sult. Smith and Curthoys[138] found that commissurectomy resulted in a large increase in the resting activity of ipsilateral MVN neurons at 8 to 12 months post-UVD. This increase in resting activity was greater than that observed in the MVN following acute contralateral labyrinthectomy[137] (see Figure 18.7)

Studies in frog have indicated that injection of GABA$_A$ receptor agonists causes decompensation, whereas GABA$_A$ receptor antagonists cause overcompensation.[39,40] However, these results are difficult to interpret because of the use of systemic injections..

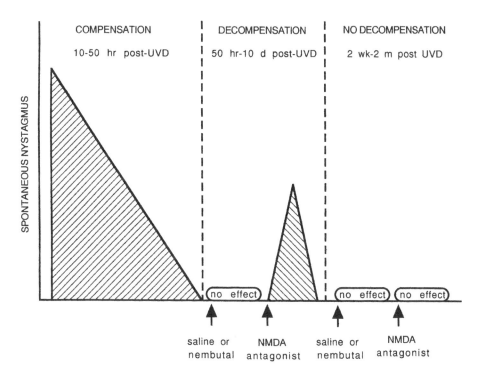

FIGURE 18.7 Schematic diagram illustrating the resting activity of ipsilateral MVN neurons in compensated guinea pigs at 52 to 60 hs post-UVD or 8 to 12 months post-UVD, before and after transection of the brainstem commissures. (From Ref. 138.)

18.7 CONCLUSIONS

18.7.1 EXCITATORY AMINO ACIDS

At present, there is no convincing evidence that changes in excitatory amino acid neurotransmitters or their receptors underlie the maintenance of static or dynamic vestibular compensation. Most studies have failed to find evidence in favor of increased release of excitatory amino acids,[87] or increased affinity, efficacy, or up-regulation of their receptors in the ipsilateral VNC.[14,77,118,173,143,144,171] In other studies, the changes that have been documented appear to develop too slowly to account for the rapid restoration of resting activity in the ipsilateral VNC and the development of static compensation (see Ref. [57]). Given the lack of evidence for changes in other neurotransmitter/neuromodulator systems that might account for static compensation (e.g., Refs.

8,31), and the general consensus that reactive synaptogenesis would be too slow (e.g., Refs. 46,78; for discussion see Refs. 17,139), it is reasonable to speculate that the intrinsic properties or voltage-operated channels of ipsilateral VNC neurons might be responsible for the resumption of resting activity.[25,27,30,36,70,133,134] In support of this "intrinsic mechanism" hypothesis,[27] the resting activity which recovers in the ipsilateral VNC has been shown to persist to a substantial extent following transection of brainstem and cerebellar commissural inputs[116,138], decerebration[103,113,116,176], spinal transection[3,104], and has even been shown to persist when the MVN on the compensated side is removed and maintained *in vitro*.[30,31,143,144] If this hypothesis is correct, the contribution of NMDA receptors in the ipsilateral VNC to static vestibular compensation may be only in the induction phase, if they respond to excessive glutamate release as a consequence of the UVD. In this way, NMDA receptors may be partially responsible for the severity of the hypoactivity which is observed in the ipsilateral VNC at the time of the UVD; since this hypoactivity can be considered the stimulus for vestibular compensation, NMDA receptors in the ipsilateral VNC may function as a biochemical "error" signal, inducing the compensation processes which follow. Why NMDA receptor antagonists disrupt compensation for a limited period following UVD remains unclear. However, it is possible that NMDA receptors in the ipsilateral VNC mediate a degree of synaptic excitation during the early stages of static compensation, and that exposure to an NMDA receptor antagonist during this fragile stage in the recovery of resting activity causes decompensation (see Ref. [148] for a detailed discussion). Particularly interesting in this respect is the recent report that decompensation caused by injection of an NMDA receptor/channel antagonist post-UVD in rat is associated with the re-expression of the Fos immediate early gene protein.[75]

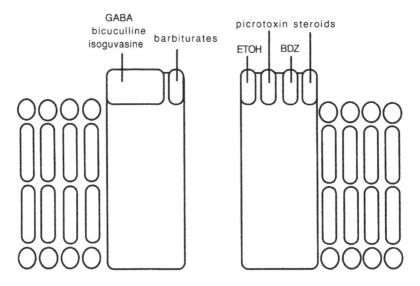

FIGURE 18.8 Schematic diagram illustrating a possible increase in type II neuron inhibition of ipsilateral type I neurons during vestibular compensation.

18.7.2 INHIBITORY AMINO ACIDS

There is little reason to believe that changes in inhibitory amino acid transmitters make a substantial contribution to the recovery of resting activity in ipsilateral VNC neurons that is associated with static compensation. Most electrophysiological studies in mammalian species indicate that the recovery of resting activity is independent of commissural inputs[102,103,116,121,137,138,149]; it has also been shown that a similar recovery of resting activity occurs after bilateral vestibular neurectomy in rhesus monkeys.[174]

On the other hand, it is clear that commissural inhibition, via the brainstem and cerebellar commissures, is critical for a limited recovery of the dynamic responses of VNC neurons, and therefore for the dynamic aspects of vestibular compensation. For example, modulation of deafferented MVN and LVN neurons via inhibition and disinhibition by commissural inputs from the intact VNC is probably the main mechanism by which dynamic compensation takes place.[102,103,116,121,137,138,176] Whether the efficacy of this commissural inhibition, via GABA acting on $GABA_A$ receptors, is greater than normal, is unclear (Figure 18.8). However, some electrophysiological evidence is consistent with this hypothesis[34,116,121,137,138]. Recent models of vestibular compensation would also favor this view.[1,38,102,103]

ACKNOWLEDGMENTS

This research was supported by Project Grants from the Health Research Council of New Zealand and the New Zealand Neurological Foundation (to PS and CD).

REFERENCES

1. Anastasio, TJ, Simulating vestibular compensation using recurrent back-propagation, *Biol. Cybern.*, 66: 389,1992.
2. Aoki, M, Miyata, H, Mizuta, K, and Ito, Y, Evidence for the involvement of NMDA receptors in vestibular compensation, *J. Vest. Res.*, 6: 315,1996.
3. Azzena, GB, Mameli, O, and Tolu, E, Vestibular units during decompensation, *Experientia*, 33: 234,1977.
4. Baker, R, Neuronal Mechanisms of Adaptation: A Viewpoint. In: *Adapative Processes in the Visual and Oculomotor System* (Keller, EL, and Zee, DS, Eds.), pp. 419-426. New York: Pergamon,1985.
5. Benazet, M, Serafin, M, de Waele, C, Chat, M, Lapeyre, P, and Vidal, PP, *Invivo* modulation of the NMDA receptor by its glycinergic site in the guinea pig vestibular nuclei, *Soc. Neurosci., Abstr,* 19: 136,1993.
6. Blair, SM, and Gavin, M, Brainstem vestibular commissures and control of time constant of vestibular nystagmus, *Acta. Otolaryngol.*, 91: 1,1981.
7. Bliss, TVP, and Collingridge, GL, A synaptic model of memory: long-term potentiation in the hippocampus, *Nature, Lond.*, 361: 31,1993.
8. Calza, L, Giardino, L, Zanni, M, Galetti, R, Parchi, P, and Galetti, G, Involvement of Cholinergic and GABAergic Systems in Vestibular Compensation. In: *Vestibular Compensation: Facts, Theories and Clinical Perspectives*, (Lacour M, Toupet M, Denise P, and Christen Y, Eds.), pp. 189-199, Paris: Elsevier,1989.
9. Caria, MA, Melis, F, Podda, MV, Solinas, A, and Deriu, F, Does long-term potentiation occur in guinea pig Deiter's nucleus?, *NeuroReport* 7: 2303,1996.
10. Capocchi G, Della, Torre, G, Grassi, S, Pettorossi, VE, and Zampolini, M, NMDA-receptor-mediated long-term modulation of electrically evoked field potentials in the rat medial vestibular nuclei, *Exp. Brain Res.*, 90: 546,1992.
11. Carpenter, DO, and Hori, N, Neurotransmitter and peptide receptors on medial vestibular nucleus neurons, *Ann. NY Acad. Sci.*, 656: 668,1992.
12. Cass, SP, and Goshgarian HG Vestibular compensation after labyrinthectomy and vestibular neurectomy in rats, *Otolaryngol. Head Neck Surg.*, 104: 14,1991.
13. Chan-Palay, V, Ito, M, Tongroach, P, Sakurai, M, and Palay, S, Inhibitory effect of motilin, somatostatin, [Leu]enkephalin, [Met]enkephalin and taurineonneuronsofthelateralvestibularnucleus:interactionswith gamma-aminobutyric acid, *Proc. Natl. Acad. Sci., USA*, 79: 3355,1982.
14. Cochran, SL, Kasik, P, and Precht, W, Pharmacological aspects of excitatory synaptic transmission to second-order vestibular neurons in the frog, *Synapse*, 1: 102,1987.
15. Cohen, B, Helwig, D, and Raphan, T, Baclofen and velocity storage: A model of the effects of the drug on the vestibulo-ocular reflex in the rhesus monkey, *J. Physiol.*, 393: 701,1987.

16. Curthoys, IS, and Halmagyi, GM, Behavioural and neural correlates of vestibular compensation. in: Bailliere+s Clinical Neurology, Vol. 1 (2), pp. 345-372, *London: Bailliere Tindall,*1992.

17. Curthoys, IS, and Halmagyi, GM, Vestibular compensation: A review of oculomotor, neural and clinical consequences of unilateral vestibular loss, *J. Vest. Res.,* 5: 67,1995.

18. Curthoys, IS, Harris, RA, and Smith, PF, The effect of unilateral labyrinthectomy on neural activity in the guinea pig vestibular nuclei. in: *The Vestibular System: Neurophysiologic and Clinical Research* (Graham MD, and Kemink JL ,Eds.), pp. 677-687, *New York: Raven Press,*1987.

19. Curtis, DR, Duggan, AW, and Felix, D, GABA and inhibition of Deiter's neurones, *Brain Res.,* 23: 117,1970.

20. Darlington, CL, Flohr H, and Smith, PF, Molecular mechanisms of brainstem plasticity: the vestibular compensation model, *Mol. Neurobiol.,* 5: 355,1991.

21. Darlington, CL, Gallagher, JP, and Smith, PF, *In vitro* electrophysiological studies of the vestibular nucleus complex, *Prog. Neurobiol.,* 45: 335,1995.

22. Darlington, CL, and Smith, PF, NMDA receptor antagonists disrupt the development of vestibular compensation in the guinea pig, *Eur. J. Pharmacol.,* 174: 273,1989.

23. Darlington, CL, and Smith, PF, Pre-treatment with a calcium channel antagonist facilitates vestibular compensation, *NeuroReport,* 3: 143,1992.

24. Darlington, CL, and Smith, PF, Middle ear procaine injection before surgical labyrinthectomy reduces nystagmus, *NeuroReport,* 4: 1353,1993.

25. Darlington, CL, and Smith, PF, The intrinsic properties of vestibular nucleus neurons and recovery of motor function following peripheral vestibular deafferentation: Is there a link? *Hum. Mov. Sci.,* 12: 195,1993.

26. Darlington, CL, and Smith, PF, Metabotropic glutamate receptors in the guinea pig medial vestibular nucleus *in vitro, NeuroReport,* 6: 1799,1995.

27. Darlington, CL, and Smith, PF, Recovery of static vestibular function following peripheral vestibular lesions (vestibular compensation) in mammals: the intrinsic mechanism hypothesis, *J. Vest. Res.,* 6: 185,1996.

28. Darlington, CL, Smith, PF, and Gilchrist, DPD, Comparison of the effects of ACTH-(4-10) on medial vestibular nucleus neurons in brainstem slices from labyrinthine-intact and compensated guinea pigs, *Neurosci. Lett.,* 145: 97,1992.

29. Darlington, CL, Smith, PF, and Gilchrist, DPD, The contributions of vestibular studies to understanding the beneficial effects of short ACTH fragments in the CNS: Reply to Pranzatelli., *Exp. Neurol.,* 128: 155,1994.

30. Darlington, CL, Smith, PF, and Hubbard, JI, Neuronal activity in the guinea pig medial vestibular nucleus *in vitro* following chronic unilateral labyrinthectomy, *Neurosci. Lett.,* 105: 143,1989.

31. Darlington, CL, Smith, PF, and Gilchrist, DPD, Comparison of the effects of ACTH-(4-10) on medial vestibular nucleus neurons in brainstem slices from labyrinthine-intact and compensated guinea pigs, *Neurosci. Lett.,* 145: 97,1992.

32. Darlington, CL, Smith, PF, and Gilchrist, DPD, The effects of combined pre-treatment with an NMDA receptor antagonist and post-treatment with a synthetic ACTH-(4-9) fragment on spontaneous nystagmus compensation following vestibular deafferentation, *Neurosci. Res. Commun.,* 14: 133,1994.

33. Dieringer, N, and Precht, W, Modification of synaptic input following unilateral labyrinthectomy, *Nature, London,* 269: 431,1977.

34. Dieringer, N, Precht, W, Mechanism of compensation for vestibular deficits in the frog. II. Modifications of the inhibitory pathway, *Exp. Brain Res.,* 36: 329,1979.

35. Doi, K, Tsumoto, T, and Matsunaga, T, Actions of excitatory amino acid anatgonists on synaptic inputs to the rat medial vestibular nucleus: an electrophysiological study *in vitro, Exp. Brain Res.,* 82: 254,1990.

36. Dutia, MB, Johnston, AR, and McQueen, DS, Tonic activity of rat medial vestibular nucleus neurones *in vitro* and its inhibition by GABA, *Exp. Brain Res.,* 88: 466,1992.

37. Feeney, DM, and Baron, JC, Diaschisis, *Stroke,* 17: 817,1986.

38. Fetter, M, and Zee, DS, Recovery from unilateral labyrinthectomy in rhesus monkey, *J. Neurophysiol.,* 59: 370,1988.

39. Flohr, H, Abeln, W, and Luneburg, U, Neurotransmitter and neuromodulator systems involved in vestibular compensation. in: Adapative Mechanisms In Gaze Control: Facts and Theories (Berthoz A, Melvill Jones G, Eds.), pp. 269-277, Amsterdam: Elsevier,1981.

40. Flohr, H, Bienhold, H, Abeln, W, and Macskovics, I, Concepts of vestibular compensation. in *Lesion-Induced Neuronal Plasticity in Sensorimotor Systems* (Flohr H, Precht W, Eds.), pp. 153-172, *Amsterdam: Springer*,1981.

41. Flohr, H, and Luneburg, U, Effects of ACTH-(4-10) on vestibular compensation, *Brain Res.*, 248: 169,1982.

42. Flohr, H, and Luneburg, U, Role of NMDA receptors in lesion-induced plasticity, *Arch. Ital. Biol.*, 131: 173,1993.

43. Fonnum, F, Storm-Mathisen, J, and Walberg, F, Glutamate decarboxylase in inhibitory neurons. A study of the enzyme in Purkinje cell axons and boutons in the cat, *Brain Res.*, 20: 259,1970.

44. Funabiki, K, Mishina, M, and Hirano, T, Retarded vestibular compensation in mutant mice deficient in d-2 glutamate receptor subunit, *NeuroReport*, 7: 189,1995.

45. Furuya, N, Yabe, T, and Koizumi, T, Neurotransmitters in the vestibular commissural system of the cat, *Ann. NY Acad. Sci.,* 656: 594,1992.

46. Gacek, RR, Lyon, MJ, and Schoonmaker, J, Ultrastructural changes in vestibulo-ocularneuronsfollowing vestibular neurectomy in the cat,*Ann. Otol. Rhinol. Otolaryngol.*, 97: 42,1988.

47. Galiana, HL, Flohr, H, and Melvill-Jones, G, A re-evaluation of intervestibularnuclear coupling: its role in vestibular compensation, *J. Neurophysiol.*, 51: 258,1984.

48. Gallager, DW, Lakoski, JM, Gonsalves, SF, and Rauch, SL, Chronic benzodiazepine treatment decreases postsynaptic GABA sensitivity, *Nature*, 308: 74,1984.

49. Gallagher, JP, Lewis, MR, and Shinnick-Gallagher, P, An electrophysiological investigation of the rat medial vestibular nucleus *in vitro*. in: Contemporary Sensory Neurobiology, (Correia MJ, Perachio AA, Eds.) pp. 293-304, *New York: A.R. Liss,*1985.

50. Gallagher, JP, Phelan, KD, and Shinnick-Gallagher, P, Modulation of excitatory transmission at the rat medial vestibular nucleus synapse, *Ann. NY Acad. Sci.,* 656: 630,1992.

51. Gilchrist, DPD, Darlington, CL, and Smith, PF, The effects of flunarizine on ocular motor and postural compensation following peripheral vestibular deafferentation in guinea pig, *Pharmacol. Biochem. Behav.,* 44: 99,1993.

52. Gilchrist, DPD, Darlington, CL, and Smith, PF, A dose-response analysis of the beneficial effects of the ACTH-(4-9) analogue Org 2766 on behavioural recovery following unilateral labyrinthectomy in guinea pig, *Brit. J. Pharmacol.,* 111: 358,1994.

53. Gilchrist, DPD, Darlington, CL, and Smith, PF, Org 2766 treatment prevents disruption of vestibular compensation by an NMDA receptor antagonist, *Eur. J. Pharmacol.,* 252: R1,1994.

54. Gilchrist, DPD, Smith, PF, and Darlington, CL, ACTH-(4-10) accelerates ocular motor recovery in the guinea pig following vestibular deafferentation, *Neurosci. Lett.,* 118: 14,1990.

55. Grassi, S, Zampolini, M, Della, Torre, G, Capocchi, G, and Pettorossi, VE, GABA-mediated vestibular long-term depression (LTD), *Eur. J. Neurosci. Suppl.,* 7: 219,1994.

56. Hamann, KF, and Lannou, J, Dynamic characteristics of vestibular nuclear neurons responses to vestibular and optokinetic stimulation during vestibular compensation in the rat, *Acta. Otolaryngol. (Stockholm) Suppl.,* 455: 1-19,1988.

57. Henley, CM, and Igarashi, M, Amino acid assay of vestibular nuclei 10 months after unilateral labyrinthectomy in squirrel monkeys, *Acta. Otolaryngol. (Stockholm) Suppl.,* 481: 407,1991.

58. Hironaka, T, Morita, Y, Hagihira, S, Tateno, E, Kita, H, and Tohyama, M, Localization of GABA$_A$- receptor a1 subunit mRNA-containing neurons in the lower brainstem of the rat, *Mol. Brain Res.,* 7: 335,1990.

59. Holstein, GR, Gannon, PJ, Martinelli, GP, and Cohen, B, L-Baclofen-sensitive GABA$_B$ binding sites are present on primary vestibular afferents in the medial vestibular nucleus, *Soc. Neurosci. Abstr.,* 20: 570,1994.

60. Holstein, GR, Martinelli, GP, and Cohen, B, L-baclofen-sensitive GABA$_B$ binding sites in the medial vestibular nucleus localized by immunocytochemistry, *Brain Res.,* 581: 175,1992.

61. Houser, CR, Barber, RP, and Vaughn, JE, Immunocytochemical localization of glutamic acid decarboxylase in the dorsal lateral vestibular nucleus: Evidence for an intrinsic and extrinsic GABA-ergic innervation, *Neurosci. Lett.,* 47: 213,1984.

62. Hutchinson, M, Smith, PF, and Darlington, CL, Tolerance to the ataxic effects of diazepam in guinea pig is not associated with a reduced sensitivity of $GABA_A$ receptors in the vestibular nucleus, *Eur. J. Pharmacol.*, 301: 83,1996.

63. Igarashi, M, Ishikawa, K, Ishii, M, and Schmidt, K, Effect of ACTH-(4-10) on equilibrium compensation after unilateral labyrinthectomy in the squirrel monkey, *Eur. J. Pharmacol.*, 119: 239,1985.

64. Ito, M, Highstein, SM, and Fukuda, J, Cerebellar inhibition of the vestibulo-ocular reflex in rabbit and cat and its blockage by picrotoxin, *Brain Res.*, 17: 524,1970.

65. Ito, M, Kawai, N, and Udo, M, The origin of cerebellar-induced inhibition of Deiter's neurones III. Localization of the inhibitory zone, *Exp. Brain Res.*, 4: 310,1968.

66. Ito, M, Kawai, N, Udo, M, and Sato, N, Cerebellar-evoked disinhibition in dorsal Deiter's neurones, *Exp. Brain Res.*, 6: 247,1968.

67. Ito, M, and Yoshida, M, The origin of cerebellar-induced inhibition of Deiter's neurones I. Monosynaptic initiation of the inhibitory postsynaptic potential, *Exp. Brain Res.*, 2: 330,1966.

68. Jensen, DW, Survival of function in the deafferented vestibular nerve, *Brain Res.*, 273: 175,1983.

69. Jerram, A, Darlington, CL, and Smith, PF, Methylprednisolone enhances vestibular compensation of spontaneous ocular nystagmus following unilateral labyrinthectomy in guinea pig, *Eur. J. Pharmacol.*, 275: 291,1995.

70. Johnston, AR, MacLeod, NK, and Dutia, MB, Ionic conductances contributing to spike repolarization and after-potentials in rat medial vestibular nucleus neurones, *J. Physiol. (London)*, 481: 61,1994.

71. Kevetter GA, and Perachio, AA, Cytochrome oxidase histochemistry in Scarpa+s ganglion after hemila-byrinthectomy, *Neurosci. Lett.*, 175: 141,1994.

72. Kim, MS, Jin, BK, Chun, SW, Lee, MY, Lee, SH, Kim, JH, and Park, BR, Role of vestibulocerebellar N-methyl-D-aspartate receptors for behavioral recovery following unilateral labyrinthectomy, *Neurosci. Lett.*, 222: 171,1997.

73. Kinney, GA, Peterson, BW, and Slater, NT, Long-term synaptic plasticity in the rat medial vestibular nucleus studied using patch-clamp recording in an *in vitro* brain slice preparation, *Soc. Neurosc. Abstr.*, 19: 1491,1993.

74. Kinney, GA, Peterson, BW, and Slater, NT, The synaptic activation of N-methyl-D-aspartate receptors in the rat medial vestibular nucleus, *J. Neurophysiol.*, 72: 1588,1994.

75. Kitihara, T, Takeda, N, Saika, T, Kubo, T, and Kiyama, H, Effects of MK-801 on Fos expression in the rat brainstem after unilateral labyrinthectomy, *Brain Res.*, 700: 182,1995.

76. Knopfel, T, Evidence for N-methyl-D-aspartic acid receptor-mediated modulation of the commissural input to central vestibular neurons of the frog, *Brain Res.*, 426: 212,1987.

77. Knopfel, T, and Dieringer, N, Lesion-induced vestibular plasticity in the frog: are N-methyl-D-aspartate receptors involved? *Exp. Brain Res.*, 72: 129,1988.

78. Korte, GE, and Friedrich, VL, The fine structure of the feline superior vestibular nucleus: identification and synaptology of the primary vestibular afferents, *Brain Res.*, 176: 3,1979.

79. Kunkel, AW, and Dieringer, N, Morphological and electrophysiological consequences of unilateral pre-versus post-ganglionic vestibular lesions in the frog, *J. Comp. Physiol.*, A 171: 621,1994.

80. Kumoi, K, Saito, N, and Tanaka, C, Immunohistochemical localization of gamma-aminobutyric acid and aspartate-containing neurons in the guinea pig vestibular nuclei, *Brain Res.*, 416: 22,1987.

81. Lacour, M, Ez-Zaher, L, and Raymond, J, Plasticity mechanisms in vestibular compensation in the cat are improved by an extract of Ginkgo Biloba (EGb 761), *Pharmacol. Biochem. Behav.*, 40: 367,1991.

82. Lacour, M, Toupet, M, Denise, P, and Christen, Y, (Eds.) *Vestibular Compensation: Facts, Theories and Clinical Perspectives, Paris: Elsevier*,1989.

83. Lapeyre, PNM, and de Waele, C, Dual glycinergic modulation of spontaneously active guinea pig medial vestibular nucleus neurons: An *in vitro* study, *Neurosci. Lett.*, 188: 155,1995.

84. Leinhos, P, and Flohr, H, Calcium antagonists and lesion-induced neural plasticity, *Eur. J. Neurosci. Suppl.*, 5: 74,1992.

85. Lewis, MR, Gallagher, JP, and Shinnick-Gallagher, P, An *in vitro* brain slice preparation to study the pharmacology of central vestibular neurons, *J. Pharmacol. Meth.*, 18: 267,1987.

86. Lewis, MR, Phelan, KD, Shinnick-Gallagher, P, and Gallagher, JP, Primary afferent excitatory transmission recorded intracellularly *in vitro* from rat medial vestibular neurons, *Synapse*, 3: 149,1989.

87. Li, H, Godfrey, TG, Godfrey, DA, and Rubin, AM, Quantitative changes of amino acid distributions in rat vestibular nuclear complex after unilateral vestibular ganglionectomy, *J. Neurochem.*, 66: 1550,1996.

88. Li, H, Godfrey, TG, Godfrey, DA, and Rubin, AM, Immunohistochemical study of the distributions of AMPA receptor subtypes in rat vestibular nuclear complex after unilateral deafferentation, *Ann. NY Acad. Sci.*, 653,1995.

89. Li, H, Godfrey, DA, and Rubin, AM, Quantitative autoradiography of 5-[^{3}H]6-cyano-7-nitro-quinoxaline-2,3,dione and (+)-3-[^{3}H]dizocilpine maleate binding in rat vestibular nuclear complex after unilateral deafferentation, with comparison to cochlear nucleus, *Neurosci.*, 77: 473,1997.

90. Lisberger, SG, Pavelko, TA, and Broussard, DM, Neural basis for motor learning in the vestibulo-ocular reflex of primates. I. Changes in the responses of brainstem neurons, *J. Neurophysiol.*, 72: 928,1994.

91. Lisberger, SG, and Sejnowski, TJ, Motor learning in a recurrent network model based on the vestibulo-ocular reflex, *Nature, Lond.*, 360: 159,1992.

92. Llinas, R, and Walton, K, Vestibular compensation: A distributed property of the central nervous system. in: *Integration In The Nervous System* (Asanuma, H and Wilson, VJ, Eds.), pp. 145-166, Tokyo: Igaku-Shon,1979.

93. Lopez, I, Baloh, RW, and Honrubia, V, Glycine receptor immunolocalization in the chinchilla vestibular nuclear complex, *Soc Neurosci. Abstr.*, 20: 569,1994.

94. Luyten, WHML, Sharp, FR, and Ryan, AF, Regional differences of brain glucose metabolic compensation after unilateral labyrinthectomy in rats: A [^{14}C]2-deoxyglucose study, *Brain Res.*, 373: 68,1986.

95. McBain, CJ, Mayer, and ML, N-methyl-D-aspartic acid receptor structure and function, *Physiol. Revs.*, 74: 723,1994.

96. Maclennan, K, Smith, PF, and Darlington, CL, Ginkgolide B accelerates vestibular compensation of spontaneous ocular nystagmus following unilateral labyrinthectomy in guinea pig, *Exp. Neurol.*, 131: 273,1995.

97. Martinelli, GP, Holstein, GR, and Cohen, B, L-Baclofen-sensitive GABA$_B$ binding sites are present on some commissural axon terminals in the medial vestibular nucleus (MVN), *Soc. Neurosci. Abstr.*, 20: 570,1994.

98. Matsumitsu, Y, and Sekitani, T, Effect of electric stimulation on vestibular compensation in guinea pigs, *Acta. Otolaryngol. (Stockholm)* 111: 807,1991.

99. Monaghan, DT, and Cotman, CW, Distribution of N-methyl-D-aspartate-sensitive L-[3H] glutamate binding sites in rat brain, *J. Neurosci.*, 5: 2909,1985.

100. Nakamura, J, Sasa, M, and Takaori, S, Ethanol potentiates the effect of gamma-aminobutyric acid on medial vestibular nucleus neurons responding to horizontal rotation, *Life Sci.*, 45: 971,1989.

101. Nakao, S, Sasaki, S, Schor, RH, and Shimazu, H, Functional organization of premotor neurons in the cat medial vestibular nucleus related to slow and fast phases of vestibular nystagmus, *Exp. Brain Res.*, 45: 371,1982.

102. Newlands, SD, and Perachio, AA, Compensation of horizontal canal-related activity in the medial vestibular nucleus following unilateral labyrinth ablation in the decerebrate gerbil. I. Type I neurons, *Exp. Brain Res.*, 82: 359,1990.

103. Newlands, SD, and Perachio, AA, Compensation of horizontal canal-related activity in the medial vestibular nucleus following unilateral labyrinth ablation in the decerebrate gerbil. II. Type II neurons, *Exp. Brain Res.*, 82: 373,1990.

104. Newlands, SD, and Perachio, AA, Effect of T2 spinal transection on compensation of horizontal canal-related activity in the medial vestibular nucleus following unilateral labyrinth ablation in the decerbrate gerbil, *Brain Res.*, 541: 129,1991

105. Niklasson, M, Tham, R, Larsby, B, and Eriksson, B, Effects of GABA$_B$ activation and inhibition on vestibulo-ocular and optokinetic responses in the pigmented rat, *Brain Res.*, 649: 151,1994.

106. Nomura, I, Senba, E, Kubo, T, Shiraishi, T, Matsunaga, T, Tohyama, M, Shiotani, Y, and Wu, JY, Neuropeptides and gamma-aminobutyric acid in the vestibular nuclei of the rat: an immuohistochemical analysis, I. Distribution. *Brain Res.*, 311: 109,1984.

107. Obata, K, Ito, M, Ochi, R, and Sato, N, Pharmacological properties of the postsynaptic inhibition by Purkinje cell axons and the action of GABA on Deiter's neurons, *Exp. Brain Res.*, 4: 43,1967.

108. Obata, K, and Takeda, K, Release of GABA into the fourth ventricle induced by stimulation of the cat's cerebellum, *J. Neurochem.*, 16: 1043,1969.

109. Obata,, K, Takeda, K, and Shinozaki, H, Further study on pharmacological properties of the cerebellar-induced inhibition of Deiter's neurons, *Exp. Brain Res.*, 11: 327,1970.

110. Padoan, S, Korttila, K, Magnusson, M, Pyyko, I, and Schalen, L, Reduction of gain and time constant of vestibulo-ocular reflex in man induced by diazepam and thiopental, *J. Vest. Res.*, 1: 97,1990.

111. Perachio, AA, Newlands, SD, Reply to: P.F. Smith and I.S. Curthoys, *Exp. Brain Res.*, 86: 681,1991.

112. Pettorossi, VE, Della, Torre, G, Grassi, S, Zampolini, M, Capocchi, G, and Errico, P, Role of NMDA receptors in the compensation of ocular nystagmus induced by hemilabyrinthectomy in the guinea pig, *Arch. Ital. Biol.*, 130: 303,1992.

113. Pompeiano, O, Xerri, C, Gianni, S, and Manzoni, D, Central compensation of vestibular deficits. II. Influences of roll tilt on different size lateral vestibular neurons after ipsilateral labyrinth deafferentation, *J. Neurophysiol.*, 52: 18,1984.

114. Popper, P, Rodrigo, JP, Alvarez, JC, Lopez, I, and Honrubia, V, Expression of the AMPA-selective receptor subunits in the vestibular nuclei of the chinchilla, *Mol. Brain Res.*, 44: 21,1997.

115. Precht, W, Schwindt, PC, and Baker, R, Removal of vestibular commissural inhibition by antagonists of GABA and glycine, *Brain Res.*, 62: 222,1973.

116. Precht, W, Shimazu, H, and Markham, CH, A mechanism of central compensation of vestibular function following hemilabyrinthectomy, *J. Neurophysiol.*, 29: 996,1966.

117. Raymond, J, Nieoullon, A, Dememes, D, and Sans, A, Evidence for glutamate as a neurotransmitter in the cat vestibular nerve, *Exp. Brain Res.*, 56: 523,1984.

118. Raymond, J, Touati, J, and Dememes, D, Changes in the glutamate binding sites in the rat vestibular nuclei following hemilabyrinthectomy, *Soc. Neurosci. Abstr.*, 15:,1989.

119. Reichenberger, I, and Dieringer, N, Size-related colocalization of glycine- and glutamate-immunoreactivity in frog and rat vestibular afferents, *J. Comp. Neurol.*, 349: 603,1994.

120. Reichenberger, I, Straka, H, Ottersen, OP, Streit, P, Gerrits, NM, and Dieringer, N, Distribution of GABA, glycine and glutamate immunoreactivities in the vestibular nuclear complex of the frog, *J. Comp. Neurol.*, 377:149,1997.

121. Ried, S, Maioli, C, and Precht, W, Vestibular nuclear neuron activity in chronically labyrinthectomized cats, *Acta. Otolaryngol. (Stockholm)*, 98:1,1984.

122. Ris, L, Capron, B, De Waele, C, Vidal, PP, Godaux, E, Dissociations between behavioural recovery and restoration of vestibular activity in the unilabyrinthectomized guinea pig, *J. Physiol.*, 500:509,1997.

123. Ris, L., De Waele, C, Serafin, M, Vidal, PP, Godaux, E, Neuronal activity in the ipsilateral vestibular nucleus following unilateral labyrinthectomy in the alert guinea pig, *J. Neurophysiol.*, 74:2887,1995.

124. Sansom, AJ, Darlington, CL, and Smith, PF, Intraventricular administration of an N-methyl-D-aspartate, receptor antagonist disrupts vestibular compensation, *Neuropharmacol.*, 29:83,1990.

125. Sansom, AJ, Darlington, CL, and Smith, PF, Pre-treatment with MK-801 reduces spontaneous nystagmus following unilateral labyrinthectomy, *Eur. J. Pharmacol.*, 220:123,1992.

126. Sansom, AJ, Darlington, CL, and Smith, PF, Comparison of the effects of pre-treatment with competitive or non-competitive NMDA receptor antagonists on vestibular compensation, *Pharmacol. Biochem. Behav.*, 46: 807,1993.

127. Sansom, AJ, Darlington, CL, Smith, PF, Gilchrist, DPD, Keenan, CJ, and Kenyon, R, Injection of calmidazolium chloride into the ipsilateral vestibular nucleus or IVth ventricle reduces spontaneous ocular nystagmus following unilateral labyrinthectomy in guinea pig, *Exp. Brain Res.*, 93:271,1993.

128. Schaefer, KP, and Meyer, DL, Compensation of vestibular lesions. in: *Handbook of Sensory Physiology* (Kornhuber HH, ed.), Vol. VI(2), pp. 463-490, *Berlin: Springer,*1974.

129. Scott, SJ, Smith, PF, and Darlington, CL, Quantification of the depressive effects of diazepam on the guinea pig righting reflex, *Pharmacol. Biochem. Behav.*, 47:739,1994.

130. Serafin, M, Khateb, A, de Waele, C, Vidal, PP, and Muhlethaler, M, Low threshold calcium spikes in medial vestibular nuclei neurones *in vitro*: A role in the generation of the vestibular nystagmus quick phase *in vivo*? *Exp. Brain Res.*, 82:187,1990.

131. Serafin, M, Khateb, A, de Waele, C, Vidal, PP, and Muhlethaler, M, Medial vestibular nucleus in the guinea pig: NMDA-induced oscillations, *Exp. Brain Res.*, 88:187,1992.

132. Serafin, M, Ris, L, Bernard, P, Muhlethaler, M, Godaux, E, and Vidal, PP, Neuronal correlates of vestibulo-ocular reflex adaptation in the alert guinea pig, *Eur. J. Neurosci.*, 11:1827,1999.

133. Serafin, M, De Waele, C, Khateb, A, Vidal, PP, Muhlethaler, M, Medial vestibular nucleus in the guinea pig. I. Intrinsic membrane properties in brainstem slices, *Exp. Brain Res.,* 84:417,1991.

134. Serafin, M, De Waele, C, Khateb, A, Vidal, PP, and Muhlethaler, M, Medial vestibular nucleus in the guinea pig II. Ionic basis of the intrinsic membrane properties in brainstem slices, *Exp. Brain Res.,* 84: 426,1991.

135. Shimazu, H, and Precht, W, Inhibition of central vestibular neurons from the contralateral labyrinth and its mediating pathway, *J. Neurophysiol.,* 29: 467,1966.

136. Sirkin, DW, Precht, W, Courjon, JH, Initial, rapid phase of recovery from unilateral vestibular lesion in rat is not dependent on survival of central portion of vestibular nerve, *Brain Res.,* 302: 245,1984.

137. Smith, PF, and Curthoys, IS, Neuronal activity in the contralateral medial vestibular nucleus of the guinea pig following unilateral labyrinthectomy, *Brain Res.,* 444: 295,1988.

138. Smith, PF, and Curthoys, IS, Neuronal activity in the ipsilateral medial vestibular nucleus of the guinea pig following unilateral labyrinthectomy, *Brain Res.,* 444: 308,1988.

139. Smith, PF, and Curthoys, IS, Mechanisms of recovery from unilateral labyrinthectomy, *Brain Res. Revs.,* 14: 155,1989.

140. Smith, PF, and Curthoys, IS, Comments to: S.D. Newlands and A.A. Perachio: Neuronal activity in the medial vestibular nuclei following unilateral labyrinthectomy, *Exp. Brain Res.,* 86: 679,1991.

141. Smith, PF, and Darlington, CL, The NMDA receptor antagonists MK-801 and CPP disrupt compensation for unilateral labyrinthectomy in the guinea pig, *Neurosci. Lett.,* 94: 309,1988.

142. Smith, PF, and Darlington, CL, Neurochemical mechanisms of recovery from peripheral vestibular lesions (vestibular compensation), *Brain Res. Revs.,* 16: 117,1991.

143. Smith, PF, and Darlington, CL, Comparison of the effects of NMDA receptor antagonists on medial vestibular nucleus neurons in brainstem slices from labyrinthine-intact and chronically labyrinthectomized guinea pigs, Brain Res., 590: 345,1992.

144. Smith, PF, and Darlington, CL, The effects of N-methyl-D-aspartate (NMDA receptor antagonists on vestibular compensation in the guinea pig: *in vivo* and *in vitro* studies. in: *The Head/Neck Sensorimotor System* (Berthoz A, Graf W, and Vidal PP, Eds.), pp. 631-635, New York: Oxford University Press,1992.

145. Smith, PF, and Darlington, CL, The pharmacology of the vestibular system. In: *Bailliere's Clinical Neurology: Neurotology* (Baloh R, Ed.), Vol. 3, No. 1, pp. 467-484, London: Bailliere Tindall,1994.

146. Smith, PF, and Darlington, CL, Can vestibular compensation be enhanced by drug treatment? *J. Vest. Res.,* 4: 169,1994.

147. Smith, PF, and Darlington, CL, Rapid tolerance to the depressive effects of diazepam on guinea pig motor control using divided doses, *Pharmacol. Biochem. Behav.,* 48: 535,1994.

148. Smith, PF, and Darlington, CL, The contribution of NMDA receptors to lesion-induced plasticity in the vestibular nucleus, *Prog. Neurobiol.,* 53:517,1997.

149. Smith, PF, Darlington, CL, and Curthoys, IS, Vestibular compensation without brainstem commissures in the guinea pig, *Neurosci. Lett.,* 65: 209,1986.

150. Smith, PF, Darlington, CL, and Hubbard, JI, Evidence that NMDA receptors contribute to synaptic function in the guinea pig medial vestibular nucleus, *Brain Res.,* 513: 149,1990.

151. Smith, PF, Darlington, CL, and Hubbard, JI, Evidence for inhibitory amino acid receptors on guinea pig medial vestibular nucleus neurons *in vitro, Neurosci. Lett.,* 121: 244,1991.

152. Smith, PF, Darlington, CL, and Hubbard, JI, Direct evidence for inhibitory amino acid receptors in the medial vestibular nucleus, *Int. J. Neurosci.,* 57: 276,1991.

153. Smith, PF, de Waele, C, Vidal, PP, and Darlington, CL, Excitatory amino acid receptors in normal and abnormal vestibular function, *Mol. Neurobiol.,* 5: 369,1991.

154. Spencer, RF, Wenthold, RJ, and Baker, R, Evidence for glycine as an inhibitory neurotransmitter of vestibular, reticular and prepositus hypoglossi neurons that project to the cat abducens nucleus, *J. Neurosci.,* 9: 2718,1989.

155. Spiegel, EA, Demetriades, TD, Die zentrale compensation des labyrinthverlustes, *Pflug. Arch. Ges. Physiol.,* 210: 215,1925.

156. Spruijt, BM, Josephy, M, Van Rijzingen, I, Maaswinkel, H, The ACTH-(4-9) analog Org 2766 modulates the behavioral changes induced by NMDA and the NMDA receptor antagonist AP5, *J. Neurosci.,* 14: 3225,1994.

157. Straka, H, and Dieringer, N, Uncrossed disynaptic inhibition of second-order vestibular neurons and its interaction with monosynaptic excitation from vestibular nerve afferent fibers in the frog, *J. Neurophysiol.,* 76: 3087,1996.

158. Straka, H, Debler, K, and Dieringer, N, Size-related properties of vestibular afferent fibers in the frog: differential synaptic activation of N-methyl-D-aspartate and non-N-methyl-D-aspartate receptors, *Neuroscience,* 70: 697,1996.

159. Straka, H, Reichenberger, I, and Dieringer, N, Size-related properties of vestibular afferent fibers in the frog: uptake of and immunoreactivity for glycine and aspartate/glutamate, *Neuroscience,* 70: 685,1996.

160. Takahashi, Y, Takahashi, MP, Tsumoto, T, Doi, K, and Matsunaga, T, Synaptic input-induced increase in intraneuronal calcium in the medial vestibular nucleus of young rats, *Neuroscience Res.,* 21: 59,1994.

161. Takahashi, Y, Tsumoto, T, and Kubo, T, NMDA receptors contribute to afferent synaptic transmission in the medial vestibular nucleus of young rats, *Brain Res.,* 659: 287,1994.

162. Ten, Bruggencate, G, and Enberg, I, Effects of GABA and related amino acids on neurones in Deiter+s nucleus. Brain Res 14: 533-536.

163. Ten Bruggencate G, Enberg I The effect of strychnine on inhibition in Deiter+s nucleus induced by GABA and glycine, *Brain Res.,* 14: 536,1969.

164. Thompson, GC, Igarashi, M, and Cortez, AM, GABA imbalance in squirrel monkey after unilateral vestibular end-organ ablation, *Brain Res.,* 370: 182,1986.

165. Tolu, E, Mameli, O, Caria, MA, and Melis, F, Improvement of vestibular plasticity in the guinea pig with a calcium entry blocker, *Acta. Otolaryngol. (Stockholm), Suppl.* 460: 72-79,1988.

166. Touati, J, Raymond, J, and Dememes, D, Quantitative autoradiographic characterization of L-[^{3}H] glutamate binding sites in rat vestibular nuclei, *Exp. Brain Res.,* 76: 646,1989.

167. Uemura, T, and Cohen, B, Vestibulo-ocular reflexes: effects of vestibular nuclear lesions. in: *Basic Aspects Of Central Vestibular Mechanisms* (Brodal AA, Pompeiano O, Eds.), *Prog. Brain Res.,* 37, pp. 515-528. Amsterdam: Elsevier,1972.

168. Vibert, N, Serafin, M, Khateb, A, Vidal, PP, and Muhlethaler, M, Effects of amino acids on medial vestibular neurones in guinea pig, *Soc. Neurosci. Abstr.,* 18,1992.

169. Vibert, N, Serafin, M, Vidal, PP, and Muhlethaler, M, Effects of baclofen on medial vestibular nucleus neurons in guinea pig brainstem slices, *Neurosci. Lett.,* 183: 193,1995.

170. Vidal, PP, Babalian, A, De Waele, C, Serafin, M, Vibert, N, and Muhlethaler, M, NMDA receptors of the vestibular nuclei neurones, *Brain Res. Bull.,* 40: 347,1996.

171. De Waele, C, Abitbol, M, Chat, M, Menini, C, Mallet, J, and Vidal, PP, Distribution of glutamatergic receptors and GAD mRNA-containing neurons in the vestibular nuclei of normal and hemilabyrinthectomized rats, *Eur. J. Neurosci.,* 6: 565,1994.

172. De Waele, C, Muhlethaler, M, and Vidal, PP, Neurochemistry of central vestibular pathways: A review, *Brain Res. Revs.,* 20: 24,1995.

173. De Waele, C, Vibert, N, Baudrimont, M, and Vidal, PP, NMDA receptors contribute to the resting discharge of vestibular neurons in the normal and hemilabyrinthectomized guinea pig, *Exp. Brain Res.,* 81: 125,1990.

174. Waespe, W, Schwarz, U, Wolfensberger, M, Firing characteristics of vestibular nuclei neurons in the alert monkey after bilateral vestibular neurectomy, *Exp. Brain Res.,* 89: 311,1992.

175. Walberg, F, Ottersen, OP, and Rinvik, E, GABA, glycine, aspartate, glutamate and taurine in the vestibular nuclei: an immunocytochemical investigation in the cat, *Exp. Brain Res.,* 79: 547,1990.

176. Xerri, C, Gianni, S, Manzoni, D, and Pompeiano, O, Compensation of central vestibular deficits. I. Response characteristics of lateral vestibular neurons to roll tilt after ipsilateral labyrinth deafferentation, *J. Neurophysiol.,* 50: 428,1983.

177. Yamanaka, T, Sasa, M, Amano, T, Miyahara, H, and Matsunaga, T, Role of glucocorticoid in vestibular compensation in relation to activation of vestibular nucleus neurons, *Acta. Otolaryngol. (Stockholm) Suppl.,* 519: 168,1995.

178. Yamanaka, T, Sasa, M, and Matsunaga, T, Release of glutamate from the vestibular nerve in the medial vestibular nucleus as a neurotrasnmitter: *in vivo* microdialysis, *Acta. Otolaryngol. (Stockholm) Suppl.,* 520: 92,1995.

179. Zanni, M, Giardino, L, Toschi L, Galetti, G, and Calza, L, Distribution of neurotransmitters, neuropeptides, and receptors in the vestibular nuclei complex of the rat: an immunocytochemical, in situ hybridization and quantitiative receptor autoradiographic study, *Brain Res. Bull.,* 36: 443,1995.

180. Zennou-Azogui, Y, Borel, L, Lacour, M, Ez-Zaher, L, and Ouakine, M, Recovery of head postural control following unilateral vestibular neurectomy in the cat, *Acta. Otolaryngol. (Stockholm) Suppl.,* 509: 1-19,1993.

Section VII

Summary

19

General Concepts and Principles

Alvin J. Beitz and John H. Anderson

CONTENTS

19.1 INTRODUCTION

The vestibular system is involved with both sensing and controlling movement and orientation in space. The vestibular hair cells detect the movement and position of the head, and the central nervous system uses the information to control posture, stabilize vision, and gives rise to a perception of the orientation of the body in space. The semicircular canals and the otolith organs detect the angular and linear accelerations of the head in three dimensions. This information is conveyed via the eighth nerve to the vestibular nuclei, where the information converges with visual and somatosensory information to provide an internal 3-D representation of head position and movement in space. The basic anatomical and physiological features of this system are summarized in Chapter 1. and 2. It is clear from the subsequent chapters in this volume that a large number of neurotransmitters and neuromodulators are involved in the control of movement and posture. The data presented indicate that excitatory and inhibitory amino acids play important transmitter roles in both peripheral and central vestibular structures. Other transmitters and neuropeptides appear to have more subtle neuromodulatory roles. The following sections summarize some of the major points made in the different chapters in this book and attempt to present some general concepts related to the neurochemical organization of the vestibular system.

19.2 EXCITATORY AMINO ACIDS AND GLUTAMATE RECEPTORS: FROM THE EAR TO THE BRAIN

The deflection of the hair cells' stereocilia opens ionic channels, causing cations to cross the cell membrane and depolarize the hair cell. This depolarization further opens additional voltage-dependent Ca^{+2} channels, resulting in an increased transmitter efflux. This transmitter, thought to be glutamate, appears to bind with kainate- and AMPA-type subsynaptic receptors, resulting in the opening of a channel permeable to both Na^+ and K^+, which acts to short-circuit the postsynaptic membrane and depolarize the afferent (see Chapter 3). Studies have demonstrated the existence of kainate, AMPA, NMDA, and metabotropic glutamate receptors on the membranes of type I hair cells, suggesting that glutamate might act not only as the neurotransmitter at the synapse between hair cells and the afferent terminals, but also might act on presynaptic autoreceptors to modulate the release of transmitter from the hair cell.

The vestibular afferents project to the vestibular nuclei and cerebellum. As summarized in Chapter 3, immunocytochemical methods have detected the presence of glutamate, aspartate, glycine, and substance P in vestibular nerve afferent fibers. While each of these substances appears to play a role in the transmission between afferent and central vestibular neurons, compelling evidence suggests that glutamate is the putative transmitter of the afferent fibers. This is supported by electrophysiological, biochemical, pharmacological, and molecular data, as discussed in Chapters 5, 6, 10, 16 and 18. There is further support from data showing that neurons in the vestibular nuclei express a number of glutamate receptor subtypes in their membranes. Thinner vestibular afferent fibers appear to activate predominantly AMPA/kainate receptors, whereas thicker afferents activate AMPA/kainate, as well as NMDA receptors. In addition, metabotropic glutamate receptors are activated. Recent work suggests that metabotropic glutamate receptors play a dual role in the medial vestibular nucleus: they inhibit glutamate release from first-order vestibular afferents under basal conditions and facilitate NMDA dependent plasticity[6].

Thus, both hair cells and vestibular afferents appear to utilize glutamate as their major transmitter. Glutamate also appears to be a major transmitter of second-order vestibular neurons that project to extraocular motor nuclei, many vestibular commissural neurons some spinovestibular and vestibulocerebellar neurons as well as granule cells within the vestibulo-cerebellum (see Chapters 5, 10, 13, 16 and 18). Such widespread distribution of glutamate and its receptors throughout the vestibular system is perhaps not surprising, considering the fact that glutamate is the most prevalent excitatory transmitter in the central nervous system.[4,17] The predominance of this amino acid transmitter in both first order afferents and vestibular nuclei neurons has several important implications: 1) glutamate receptors may be altered following inflammation or damage to the labyrinth or eighth nerve and therefore contribute to the compensation process; and 2) glutamate receptors represent a potential target for pharmacological treatment of various vestibular disorders.

With respect to the first implication, the available data indicate that following unilateral labyrinthectomy or deafferentation, glutamate receptors do not undergo any large increase in number, affinity, or efficacy. Rather, it seems more likely that glutamate receptors, and in particular the NMDA receptor subtype, might play a key role in the induction of the compensation processes following a labyrinthectomy (Chapter 18). Thus, NMDA receptor antagonists cause a decompensation in early phases of compensation, but have no such effect on the long-term maintenance of the compensated state. It is concluded that the NMDA receptor channel complex plays a role analogous to that assumed in associative long-term potentiation. The plastic changes underlying vestibular compensation seem to be induced by a Hebbian-like mechanism in the postsynaptic neuron, whereas the expression of a permanent change in synaptic efficacy is presynaptic and probably involves non-NMDA synapses. It is possible that the NMDA synapse performs a transient function by inducing a "sensory substitution" process in which the system learns to replace the missing labyrinth input with information from several other sensory sources, including the contralateral labyrinth, the visual, and somatosensory systems. Alternatively, the pattern of electrical activity in the vestibular nerve at the time of UL may induce a form of long term depresson (LTD) in neurons of the ipsilateral vestibular nuclei, which decreases their response to other, remaining, synaptic inputs.[12] This heterosynaptic LTD may be dependent on the activation of NMDA receptors at the time of UL, and this NMDA receptor-mediated LTD would be partly responsible for inducing the symptoms of UL.[12] The process of vestibular compensation would occur as a result of the gradual dissipation of LTD, allowing synaptic inputs from other areas of the central nervous system and the intrinsic properties of ipsilateral vestibular neurons to generate a partial recovery of resting activity. While NMDA receptors are clearly implicated in the process of vestibular compensation, the exact role played by individual NMDA receptor subtypes (NR1, NR2 and NR3) or other glutamate receptor subtypes in the compensation process requires further study. In particular, important information could be gained from the use of knockout and transgenic mice which target a particular receptor subtype, transporter, channel, or other relevant protein in combination with the use of appropriate electrophysiological, behavioral, molecular, or pharmacological analyses.

With respect to the possibility of using glutamate agonists or antagonists to treat vestibular disorders, some insight into their possible effect may be gleaned from data obtained in behavioral studies that have examined the effects of administration of glutamatergic agents on the vestibular compensation process. The results thus far suggest that administration of glutamate, particularly NMDA, antagonists can delay or alter vestibular compensation. In this regard, Darlington and Smith[1] reported that systemic injections of MK-801 or CPP during the first 24 hr post-unilateral vestibular deafferentation disrupted the development of vestibular compensation. While the majority of studies report that glutamate antagonists affect the compensation process, the use of different species, of different antagonists for the same receptor subtype, and of different doses of antagonists have made the data difficult to compare and in many cases controversial. The possible use of glutamate receptor antagonists or anti-sense to treat vestibular disorders in human or veterinary medicine awaits further experimentation.

19.3 INHIBITORY AMINO ACIDS: GABA AND GLYCINE

The central vestibular and oculomotor systems, and in particular the vestibular nuclear complex (VNC), are areas in which the inhibitory amino acid transmitters, GABA and glycine, coexist and may even colocalize. GABA clearly plays a major role in the commissural projections between the vestibular nuclei, in several structures that participate in the vestibulo-ocular reflex, and in the accessory optic system (see Chapters 7, 10, 14, 16 and 18). From a functional standpoint GABA is associated predominantly with ascending second-order vestibular axons that project to the oculomotor and trochlear nuclei, and that participate in the production of the vertical VOR. In contrast, glycine is localized primarily in descending axons of second-order vestibular neurons that project to the abducens nuclei and the spinal cord, and that participate in the horizontal VOR (see Chapters 7 and 10). This functional and neurochemical distinction may also correlate with the differential roles of GABA and glycine in vestibular commissural inhibition. The majority of studies to date support the concept that type II medial vestibular neurons, which mediate commissural inhibition, inhibit ipsilateral type I neurons via GABA acting on GABAA receptors rather than GABAB or glycine receptors (see Chapter 7 and the recent study by Furuya and Koizumi[5]).

There is little data to support the notion that changes in inhibitory amino acid transmitters make a substantial contribution to the partial recovery of resting activity in vestibular neurons associated with the static symptoms following unilateral vestibular deafferentation (see Smith and Darlington, Chapter 18). On the other hand, it is clear that inhibition, via the brainstem commissural fibers and cerebellovestibular projections, is critical for a limited recovery of the dynamic responses of VNC neurons and the dynamic behavior of vestibular reflexes. For example, modulation of deafferented MVN and LVN neurons via inhibition and disinhibition by commissural inputs from the intact VNC is probably the main mechanism by which dynamic compensation takes place (see Chapter 18). Several lines of evidence suggest that the efficacy of this commissural inhibition, via GABA acting on GABAA receptors, is greater than normal in the compensated state.

19.4 OTHER TRANSMITTERS AND NEUROMODULATORS

19.4.1 Acetylcholine and monoamines

As alluded to above, neuropeptides and transmitters, other than amino acids, appear to play primarily neuromodulatory roles in the vestibular system. Acetylcholine, for instance, appears to be used in the synapse between the brainstem efferent fibers and hair cells, producing rapid efferent actions, including excitation and inhibition with the specific effects being dependent on the species and the acetylcholine receptor subtypes present (see Chapter 4). In addition it is associated with ascending pathways which originate from the caudal medial and inferior vestibular nuclei and which project bilaterally as mossy fibers onto granule cells in the uvula-nodulus (see Chapter 11). Acetylcholine has been proposed to act as a neuromodulator in the cerebellar cortex rather than as a transmitter. In this regard, iontophoretic application of acetylcholine or acetylcholine agonists produce increases in the response of Purkinje cells to glutamate or GABA. Thus, acetylcholine input to the cerebellum may facilitate the action of conventional afferent systems by increasing the sensitivity of the Purkinje cells to the corresponding excitatory and inhibitory neurotransmitters, providing a mechanism for enhancing information processing within cerebellar circuits (see Chapter 11). Similarly, evidence is presented in Chapters 8 and 11 indicating that noradrenergic and other monoaminergic systems also act as neuromodulators in the vestibular system, in some cases by producing long-lasting increases in responsiveness of their target neurons to conventional excitatory (glutamate) and inhibitory (GABA) responses. Many of the monoaminergic neurotransmitters can activate metabotropic receptors which act through second messenger systems and have a slower timecourse for the resulting effects.

In Chapter 11 Pompeiano further summarizes experiments involving the local microinjection of noradrenergic and cholinergic agonists and antagonists, either into the locus coeruleus-complex and the related pontine tegmental region or into different regions of the cerebellar cortex. The results show that these structures exert a prominent role in the gain regulation of both the vestibulo-spinal (VSR) and the vestibulo-ocular reflexes (VOR). Thus, several of these neurotransmitters serve to modulate the activity caused by the excitatory and inhibitory amino acid transmitters, and appear to play an important role in vestibular function by increasing or decreasing the adaptive changes in gain of the VSR or the VOR elicited by various types of sensory stimulation (see Chapter 11). It is not surprising, therefore, that in the vestibular system various agonists and antagonists of the monoaminergic receptors are commonly used to treat vertigo and motion sickness, or to accelerate vestibular compensation following surgical sections of the vestibular nerve (Chapter 8).

19.4.2 Neuropeptides

In addition to traditional neurotransmitters, like acetylcholine, norepinephrine, and serotonin, there are a large number of neuropeptides that have been reported to be associated with the vestibular system (Chapters 2 and 9). As reviewed above, there is ample evidence to suggest that somatostatin, enkephalin, substance P, and perhaps CCK play important roles in vestibular function. While several other neuropeptides are localized within the vestibular complex or have been shown physiologically or behaviorally to affect vestibular function, more investigation is required to clarify their role as neuromodulators in the vestibular system.

19.4.3 "Non-Neural" Peptides

In addition to neuropeptides, there is a growing number of "non-neural" peptides that have been found in the brain and that must also be considered as potential candidates for exerting an influence on vestibular function. These include a large number of peptides from the chemokine and cytokine families,[7,16,18] immunophilins,[13] and heat shock proteins,[14] as well as, a growing list of representatives from the growth factor families, including nerve growth factor (NGF), brain derived neurotrophic factor (BDNF), glial-derived neurotrophic factor (GDNF) and neurotrophic factor 3 (NT-3), neurotrophic factor 4/5 (NT-4/5), neurturin and persephin.[8,10] The majority of these peptides have not been tested with regard to their influence on vestibular function, particularly following inflammation or damage to the inner ear or vestibular nerve. In this regard, data shows that pretreatment of vestibular ganglion neuronal cultures with the neurotrophins, NT-3, NT-4/5, and BDNF, but not NGF, prevented or reduced the neurotoxicity of the ototoxic drugs, cisplatin and gentamicin[21]. This suggests that these three neurotrophins are survival factors for vestibular ganglion neurons and might be important in the therapeutic prevention of vestibular ganglion neuronal loss caused by injury and ototoxins. On the other hand, a recent study by Sukhov et al.[15] found that neurons in the prepositus hypoglossi and inferior vestibular nucleus, that are involved in gaze, are responsive to NGF. Finally, recent work by Smith and co-workers[11] indicates that unilateral vestibular deafferentation induces BDNF protein expression in the guinea pig lateral, but not medial, vestibular nuclei. Collectively, these and other studies in the literature indicate that, while neurotrophic factors are clearly involved in development of the vestibular system, they may also play a role in normal vestibular function, as well as in vestibular plasticity and compensation. The details of these possible roles in vestibular function and compensation remain to be elucidated.

19.4.4 STEROIDS

Steroids and their receptors are another group of possible modulators of vestibular function that were not covered in this book. Clearly, vestibular neurons express steroid receptors and are influenced by steroid hormones, which are present in the extracellular milieu surrounding neurons. Recent work has shown, for instance, that the orphan steroid nuclear receptors, Nurr1 and Nur77 are expressed in neurons of the vestibular system.[19] Nur77 is expressed in the medial and lateral vestibular nuclei and in cerebellar Purkinje cells. Nurr1 is not expressed in the vestibular nuclei, but is found in cerebellar granule cells and Purkinje cells. The overlapping and different distubutions of these two steroid/thyroid hormone receptors in the vestibular system suggest that they might regulate different sets of responsive genes. Recent work has shown differential effects of neurosteroids on the vestibular GABAergic system. Yamamoto et al.[20] have shown that the neurosteroid, dehydroepiandrosterone sulfate, suppresses GABAergic inhibitory effects on MVN neurons through both $GABA_A$ and $GABA_B$ receptors, resulting in disturbance of control of neuronal activity in the MVN and consequent development of vertigo. In contrast, Okada et al.[9] have shown that 20-hydroxyecdysone potentiates the action of GABA, probably by acting directly on the $GABA_A$ receptor in MVN neurons. With respect to vestibular compensation, Alice et al.[1] have shown that dexamethasone does not accelerate compensation of spontaneous nystagmus following a surgical unilateral labyrinthectomy, and that dexamethasone and methylprednisolone have a direct action on a minority of MVN neurons. Clearly, the role of steroid receptors in vestibular function could be a fruitful area for further study. It will be especially important to distinquish the possible effects of adrenal steroids from sex steroids, and to focus on possible gender differences with respect to the effect of steroids on vestibular function (see for example, Darlington and Smith[3]).

19.5 FUTURE DIRECTIONS

Much of what we know regarding the neurochemical organization of the vestibular system has been learned over the past decade. It is now clear that excitatory and inhibitory amino acids and their receptors play a major role in vestibular function, while acetylcholine, monoamines, and neuropeptides play somewhat secondary, neuromodulatory roles. While the major transmitters and receptors have been identified, the molecular events that underlie normal function, and in particular the molecular events that underlie vestibular compensation and adaptation to altered gravity states, are only beginning to be understood. In this regard, Chapter 17 has reviewed several transcription factors that are up-regulated in response to vestibular compensation or hypergravity stimulation.

However, our knowledge of the cascade of second messengers and the details of the transcription processes that are associated with these events is rudimentary. Clearly, future studies must be aimed at uncovering the molecular events that are responsible for the changes in the physiological properties of vestibular neurons that occur during vestibular compensation, following exposure to altered gravity states, or in association with specific vestibular disorders. Under these conditions, cellular transcription might be altered to ultimately produce an increase or decrease in the expression of specific channel or receptor proteins. Alternatively, channel conductance or receptor properties might be modified by the expression of mRNA splice variants or by the expression of different combinations of receptor or channel subunits. Studies aimed at elucidating these events will increase our understanding of how the vestibular system adapts to new types of environments, disease states, and the aging process.

19.6 REFERENCES

1. Alice, C, Paul, AE, Sansom, AJ, Maclennan, K, Darlington, CL, and Smith, PF, The effects of steroids on vestibular compensation and vestibular nulceus neuronal activity in the guinea pig, *J. Vestib. Res.*, 8,201,1998.
2. Darlington, CL, Smith, PF, NMDA receptor antagonists disrupt the development of vestibular compensation in the guinea pig, *Eur. J. Pharmacol.*, 174,273,1989.
3. Darlington, CL, and Smith, PF, Further evidence for gender differences in circularvection, *J. Vestib, Res.*, 8,151,1998.
4. Fonnum, F, Glutamate: A neurotransmitter in mammalian brain, *J. Neurochem.*, 42,1,1984.
5. Furuya, N, and Koizumi. T, Neurotransmitters of vestibular commissural inhibition in the cat, *Acta. Otolaryngol. (Stockholm)*, 118,64,1988.
6. Grassi, S, Malfagia, C, and Pettorossi, VE, Effects of metabotropic glutamate receptor block on the synaptic transmission and plasticity in the rat medial vestibular nuclei, *Neuroscience*, 87,159,1998.
7. Hansen, MK, Taishi, P, Chen, Z, and Krueger, JM, Vagotomy blocks the induction of interleukin1b (IL-1b) mRNA in the brain of rats in response to systemic IL-1b, *J. Neurosci.*, 18,2247,1998.
8. Milbrandt, J, de Sauvage, FJ, Fahrner, TJ, Baloh,, RH, Leitner, ML, Tansey,, MG, Lampe, PA, Heuckeroth,, RO, Kotzbauer, PT, Simburger, KS, Golden, JP, Davies, JA, Vejsada, R, Kato,, AC, Hynes, M, Sherman, D, Nishimura, M, Wang, L-C, Vandlen, R, Moffat, B, Klein, RD, Poulsen, K, Gray, C, Grarces, A, Henderson, CE, Phillips, HS, and Johnson, EM, Persephin, a novel neurotrophic factor related to GDNF and Neurturin, *Neuron.*, 20,245,1998.
9. Okada, M, Ishihara, K, Sasa, M, Izumi, R, Yajin, K, and Harada, Y, Enhancement of GABA-mediated inhibition of rat medial vestibular nucleus neurons by the neurosteroid 20-hydroxyecdynsone, *Acta. Otolaryngol, (Stockholm)*, 118,11,1998.
10. Russell, DS, Neurotrophins: Mechanisms of Action, *The Neuroscientist,* 1,3,1995.
11. Smith, PF, Darlington, CL, Yan, Q, and Dragunow, M, Unilateral vestibular deafferentation induces brain-derived neurotrophic factor (BDNF) protein expression in the guinea pig lateral, but not medial, vestibular nuclei, *J. Vestib. Res.,* 8:443,1998.
12. Smith, PF, and Darlington, CL, The contribution of N-methyl-D-aspartate receptors to lesion-induced plasticity in the vestibular nucleus, *Prog. Neurobiol.,* 53,517,1997.
13. Snyder, SH, Lai, MM, and Burnett, PE, Immunophilins in the nervous system, *Neuron,* 21,283,1998.
14. Srivastava, PK, Menoret, A, Basu, S, Binder, RJ, and McQuade, KL, Heat shock proteins come of age, *Immunity,* 8,657,1998.
15. Sukov, RR, Cayouette, MH, Radeke, MJ, Feinstein, SC, Blumberg, D, Rosenthal, A, Price, DL, and Koliatsos, VE, Evidence that perihypoglossal neurons involved in vestibular-auditory and gaze control functions respond to nerve growth factor, *J. Comp. Neurol.,* 383,123,1997.
16. Ward, SG, Bacon, K, and Westwick, J, Chemokines and T-Lymphocytes: More than an attraction, Immunity, 9,1,1998.
17. Watkins, JC, Krogsgaard-Larsen, P, and HonorÈ, T, Structure-activity relationships in the development of excitatory amino acid receptor agonists and competitive antagonists, *Trends, Pharmacol. Sci.,* 11,25,1990.

18. Watkins, LR, Maier, SF, and Goehler, LE, Cytokine-to-brain communication: a review and analysis of alternative mechanisms, *Life Sci.,* 57,1001,1995.
19. Xiao, Q, Castillo, SO, and Nikodem, VM, Distribution of messenger RNAs for the orphan nuclear receptors Nurr1 and Nur77 (NGFI-B) in adult rat brain using in situ hybridization, *Neuroscience*, 75,221,1996.
20. Yamamoto, T, Yamanaka, T, and Matsunaga, T, Effects of the neurosteroid dehydroepiandrosterone sulfate on medial vestibular neurons, *Acta. Otolaryngol. (Stockholm),* 118,185,1998.
21. Zheng, JL, Steward, RR, and Gao, WQ, Neurotrophin 4/5, brain-derived neurotrophic factor, and neurotrophin-3 promote survival of cultured vestibular ganglion neurons and protect them against neurotoxicity of ototoxins, *J. Neurobiol.*, 28,330,1995.

Index

A

N-Acetylaspartylglutamate (NAAG), 313, 314
Acetylcholine (ACh), 36, 62, 76, 212
    agents, intraflocular microinjection of, 251
    enzyme responsible for synthesizing, 35
    involvement of as neurotransmitter, 355
    synthetic enzyme for, 271
Acetylcholinesterase (AChE), 62
ACh, see Acetylcholine
AChE, see Acetylcholinesterase
ACh systems, see Noradrenergic and ACh systems, control
    of vestibulospinal and vestibulo-ocular reflexes by
ACTH, see Adrenocorticotropic hormone
Adenosine 5¢-triphosphate (ATP), 80
Adrenergic system, central, 168
b-Adrenoceptors, 229
Adrenocorticotropic hormone (ACTH), 188
Afferents
    first-order vestibular, 269
    primary spinal, 333
    responses of to efferent activation, 68
    second-order spinal, 331
    spinovestibular, 330
Afterhyperpolarization (AHP), 164
AHP, see Afterhyperpolarization
Amines, biogenic, 212
Amino acid(s)
    excitatory, 352, 393, 407
    immunocytochemistry, 52
    inhibitory, 152, 155, 354, 391, 394, 409
    neurotransmitters, see Excitatory and inhibitory amino
        acid neurotransmitters, contribution of to
        vestibular plasticity
Amino acid transmitters, inhibitory, 143–157
    colocalization and coexistence of inhibitory amino acid
        neurotransmitters, 152
    connectivity, neurocytology, and ultrastructure of
        vestibular neurons, 144
    GABA and glycine receptors in vestibular nuclei,
        152–155
    GABA and glycine in vestibular nuclei, 145–152
        calcium binding proteins, 149–152
        GABA, 145–149
        glycine, 152
    transporters for inhibitory amino acid
        neurotransmitters, 155–157
a-Amino-3-hydroxy-5-methyl-4-isoxazole-propionic acid
    (AMPA), 34, 386, 388
    component, monosynaptic, 296
    current, 138
    receptor(s), 53, 98, 128, 208, 294

    localization, 106
    subtypes, 105
    subunit, 104
D-2-Amino-5-phosphonovaleric acid (D-APV), 33
AMPA, see a-Amino-3-hydroxy-5-methyl-4-isoxazole-
    propionic acid
Anatomy of vestibular system, 3–18
    optokinetic inputs to vestibular system, 17–18
    vestibular nerve, 4–6
        labyrinth, 4
        termination of first-order afferents in vestibular
        nuclei, 4–5
        vestibular efferent system, 5–6
    vestibular nuclei, 6–17
        commercial intrinsic vestibular connections, 13
        cytoarchitecture and nomenclature, 6–8
        projections to and from cerebral cortex, 16
        projections to and from interstitial nucleus of
        Cajal, 15–16
        vestibulo-cerebellar interconnections, 13–15
        vestibulo-oculomotor pathways, 8–12
        vestibulospinal pathways, 12–13
Animal tilt
    response of LC neurons to, 241
    responses of pontine reticular and related medullary
        inhibitory RS neurons to, 243
Anterior pretectal nucleus (APN), 306
Antibody(ies)
    labels, 275
    polyclonal, 371
Anti-motion sickness drug, 252
Antiporters, 156
AOS nuclei, efferent projections of, 309
APN, see Anterior pretectal nucleus
D-APV, see D-2-Amino-5-phosphonovaleric acid
Aspartate, 50, 102, 237
D-Aspartate, 52
ATP, see Adenosine 5¢-triphosphate
ATPase, 156
Atropine, 277
Auditory efferents, 63
Auditory fibers, 349
Auditory systems, chemical comparisons between central
    vestibular and, 347–357
    acetylcholine, 355–356
    auditory-vestibular interactions, 348–350
    chemical comparisons, 350–351
    dorsal cochlear nucleus, 356
    excitatory amino acids, 352–354
    inhibitory amino acids and taurine, 354–355